LAKE STECHLIN

MONOGRAPHIAE BIOLOGICAE

VOLUME 58

Series Editor
H.J. DUMONT

LAKE STECHLIN

A temperate oligotrophic lake

Edited by
S. JOST CASPER

1985 SPRINGER-SCIENCE+BUSINESS MEDIA, B.V.

Library of Congress Cataloging in Publication Data

Main entry under title:

Lake Stechlin : a temperate oligotrophic lake.

(Monographiae biologicae ; v. 58)
Includes index.
1. Limnology--Germany (East)--Stechlin Lake.
2. Nuclear power plants--Environmental aspects--Germany
(East)--Stechlin Lake. 3. Stechlin Lake (Germany)
I. Casper, S. Jost. II. Series.
QP1.P37 vol. 58 [QH149] 574 s 84-25050
[574.5'26322'0943151]

ISBN 978-94-010-8930-2 ISBN 978-94-009-5506-6 (eBook)
DOI 10.1007/978-94-009-5506-6

Preface

On February 1, 1959 the Academy of Sciences of the German Democratic Republic founded the 'Forschungsstelle für Limnologie' (now Abteilung Limnologie des Zentralinstitutes für Mikrobiologie und Experimentelle Therapie der Akademie der Wissenschaften der DDR) in Jena and a research laboratory at Lake Stechlin at Neuglobsow near Rheinsberg used as field experimental station. The foundation followed the building of the first nuclear power plant in the GDR ('Kernkraftwerk I der DDR') in the period 1956–66 between Rheinsberg and Fürstenberg (Havel).

The nuclear power plant is situated between two stratified lakes – Lake Stechlin and Lake Nehmitz – far from any industrialized area in the midst of extensive woodland. The cooling water is taken from Lake Nehmitz and led away into Lake Stechlin, having passed through the cooling system of the plant. The temperature is on average 10° C above the ordinary level. From Lake Stechlin the water is led back to Lake Nehmitz (Schräder 1962). It was expected that the deep, oligotrophic Lake Stechlin would be loaded thermally and materially by the cooling water circuit, meaning that it would be 'eutrophized', while the slightly eutrophic Lake Nehmitz would be 'oligotrophized'. Thus there would be a solid basis for a research programme on fundamental limnological problems, the solution of which could facilitate problems associated with improvement of conditions in water resources.

The German Democratic Republic is not rich in water; it is one of those countries which have a very critical water balance. Only 880 m^3 water per year are available per capita. The natural water resource runs up to 17.7 thousand million cubic metres, but only 9 thousand million cubic metres are available in dry years, while the average quantity of water required per year is 8 thousand million cubic metres (Klapper et al. 1976; Lösel 1983). Like the other industrialized countries in the northern hemisphere, the GDR is faced with increasing environmental problems. The Institute of Water Management Berlin, several other research establishments of the Academy of Sciences and the universities concentrate their efforts on a sufficient and stable water supply for domestic, industrial and agricultural purposes.

From the beginning the main mission of the Limnological Research Station has been seen as multidisciplinary, long-term, fundamental research on the biogenic determined flow and transformation of energy and material, as well as on the structure and functioning of aquatic ecosystems, in particular of lakes. Our work has been done in the sound traditional line, which started with August Thienemann's (1926, 1933, 1955) view of the life of a lake. No function would occur in the absence of structure, and we need full insight into both to understand the lake as an ecosystem (Rodhe 1979). A most thorough knowledge of physical materials occurring in lakes will be one basis needed for rational prognostication of the best way to make use of water resources and of the application of suitable sanitation methods if conditions have deteriorated. Our scientific programme is subject to the needs of society.

Thanks are due to Dr. W. Junk Publishers, Dordrecht, for providing us the opportunity to use a volume of Monographiae Biologicae. We are most grateful to late Prof. Joachim Illies, the former editor-in-chief, and Prof. Dietrich Uhlmann for their kind encouragement, and to Mr. Wil R. Peters for the most pleasant cooperation. We thank Mr. Colin de la Motte-Sherman for improving the English text. We are indebted to Mr. Manfred Rückert for the graphic design of the whole work, to Mr. Wolfram Scheffler who took most of the field photographs and some of the microphotographs, to Mrs. Brigitte Koschel and Mr. Hans Frech for some graphics, and last not least to Mrs. Bärbel Schramm and Mrs. Rosmarie Casper for typing work and Mrs. Anita Lange for proof-reading. We owe Mr. Dr. Kurt Augsten and Mr. Theodor Peschke some REM work. The Fontane-Archiv Potsdam kindly permitted the reproduction of page 1 of the original issue of Theodor Fontane's famous novel 'Stechlin', and Mrs. Prof. Charlotte Jolles kindly procured the translation of the two extracts of it and of Fontane's 'Rambles in Mark Brandenburg'. The Museum für Ur- und Frühgeschichte Potsdam kindly permitted the reproduction of Fig. 1.8.

An important yardstick to measure the progress of work and the efficiency of a scientific institute is the quality and quantity of its research output. Thus the most apposite way of celebrating our Silver Jubilee is to take stock of the situation, to pool our scientific and social efforts and to compile more than 70 former and recent works, hitherto mostly published in 'Limnologica' since 1962, in the form of the present volume.

Research is done by individuals, and a lot of people – not all mentioned here – have been involved in this work. We owe all of them very much. In the following pages we have attempted to bring together a first survey of the limnology of Lake Stechlin and its surroundings. Naturally, there are obvious gaps. For some of the substantial topics of an investigation of a lake it will be evident that our knowledge is still rudimentary.

Nevertheless, we would be very contented with our work if we could win only half of the glory and honour which was offered to the late Sir Samuel Pickwick, Esq. P.P.K.M., whose 'Speculations on the Source of the Hampstead Ponds, with some observations on the Theory of Tittlebats' once 'agitated the scientific world' as Charles Dickens affirms authentically.*

Jena, 10 March 1984 S. Jost Casper

References

Klapper, H. et al. 1976. Protection of storage reservoirs; water quality management; sanitation of lakes; active environmental protection in the GDR. In: Institut für Wasserwirtschaft Berlin (ed): EUSTROSYM-booklet. Berlin.

Krausch, H.-D. & Zühlke, D. 1974. Das Rheinsberg-Fürstenberger Seengebiet. Werte unserer Heimat Band 25. Berlin: Akademie-Verlag.

Lösel, P. 1983. Wasser: Probleme für Forschung und Technik. spectrum 4/83: 5–7.

Rodhe, W. 1979. The life of lakes. Arch. Hydrobiol. Beih. Ergebn. Limn. 13: 5–9.

Schräder, Th. 1962. Die Forschungsstelle für Limnologie der Deutschen Akademie der Wissenschaften zu Berlin. Limnologica 1: 1–19.

Thienemann, A. 1926. Der Nahrungskreislauf im Wasser. Verh. Deut. Zool. Ges. 1926: 29–79.

Thienemann, A. 1933. Vom Wesen der Limnologie und ihrer Bedeutung für die Kultur der Gegenwart (Eröffnungsansprache). Verh. Int. Ver. Limn. 6: 21–27.

Thienemann, A. 1955. Die Binnengewässer in Natur und Kultur. Verständliche Wissenschaft Band 55: Heidelberg.

* Charles Dickens, The Posthumous Papers of the Pickwick Club, Vol. I. London: Chapman and Hall, Limited [s.a.], p. 1 and 2. 'Tittlebat' (*Gasterosteus aculeatus* L.) means 'Stichling' in German. 'Stichling' sounds like 'Stechlin' in English. In the opinion of some scientists of dishonest character, Charles Dickens would have had in mind Lake Stechlin when he reported upon Samuel Pickwick's limnological attempts – the first limnological investigations at all, hitherto not noticed by science – to trace to the source of the 'mighty Ponds of Hampstead'. But it is now known: 'Stechlin' is derived from the slavic *steklo* 'glass'. Lake Stechlin was named after water like clear glass, easy to see through (Krausch & Zühlke 1974).

List of contributors

Christina Babenzien, Akademie der Wissenschaften der DDR, Zentralinstitut für Mikrobiologie und Experimentelle Therapie, Jena, Abteilung Limnologie, Außenstelle Stechlinsee.

Dr. Hans-Dietrich Babenzien, Akademie der Wissenschaften der DDR, Zentralinstitut für Mikrobiologie und Experimentelle Therapie, Jena, Abteilung Limnologie, Außenstelle Stechlinsee.

Doz. Dr. sc. Jürgen Benndorf, Technische Universität Dresden, Sektion Wasserwesen, Bereich Hydrobiologie.

Dr. habil. S. Jost Casper, Akademie der Wissenschaften der DDR, Zentralinstitut für Mikrobiologie und Experimentelle Therapie, Jena, Abteilung Limnologie.

Dr. Dietrich Flößner, Akademie der Wissenschaften der DDR, Zentralinstitut für Mikrobiologie und Experimentelle Therapie, Jena, Abteilung Limnologie.

Dr. Peter Kasprzak, Akademie der Wissenschaften der DDR, Zentralinstitut für Mikrobiologie und Experimentelle Therapie, Jena, Abteilung Limnologie, Außenstelle Stechlinsee.

Dr. sc. Helmut Klapper, Institut für Wasserwirtschaft Berlin, Abteilung Oberflächengewässerschutz, Außenstelle Magdeburg.

Dr. Rainer Koschel, Akademie der Wissenschaften der DDR, Zentralinstitut für Mikrobiologie und Experimentelle Therapie, Jena, Abteilung Limnologie, Außenstelle Stechlinsee.

Dr. habil. Heinz-Dieter Krausch, Akademie der Wissenschaften der DDR, Zentralinstitut für Mikrobiologie und Experimentelle Therapie, Jena, Abteilung Limnologie.

Dr. Ludwig Krey, Akademie der Wissenschaften der DDR, Zentralinstitut für Mikrobiologie und Experimentelle Therapie, Jena, Abteilung Limnologie.

Dr. habil. Georg Mothes, Akademie der Wissenschaften der DDR, Institut für Geographie und Geoökologie, Bereich Hydrologie.

Dr. Gottfried Proft, Akademie der Wissenschaften der DDR, Zentralinstitut für Mikrobiologie und Experimentelle Therapie, Jena, Abteilung Limnologie.

Dr.-Ing. Frieder Recknagel, Technische Universität Dresden, Sektion Wasserwesen, Bereich Hydrobiologie.

Dr. Dieter Richter, Meteorologischer Dienst der DDR, Forschungsinstitut für Hydrometeorologie.

Dr. Diethelm Ronneberger, Akademie der Wissenschaften der DDR, Zentralinstitut für Mikrobiologie und Experimentelle Therapie, Jena, Abteilung Limnologie, Außenstelle Stechlinsee.

Wolfram Scheffler, Akademie der Wissenschaften der DDR, Zentralinstitut für Mikrobiologie und Experimentelle Therapie, Jena, Abteilung Limnologie, Außenstelle Stechlinsee.

Dr. habil. Wilfried Schönborn, Akademie der Wissenschaften der DDR, Zentralinstitut für Mikrobiologie und Experimentelle Therapie, Jena, Abteilung Limnologie.

Contents

Preface by S.J. Casper IX

List of contributors XI

PART ONE: THE HISTORICAL BACKGROUND

1. The Lake Stechlin area, past and present, and the Lake Stechlin research project
   by S.J. Casper, H.-D. Krausch and L. Krey 3
   - Lake Stechlin, Theodor Fontane and the legend of the Red Cock (S.J. Casper) 3
   - Morphology and morphogenesis of the Lake Stechlin area (L. Krey) 7
   - The plant cover of the Lake Stechlin area (H.-D. Krausch) 11
   - Settlement and economy of the Lake Stechlin area (H.-D. Krausch) 15
   - The Limnological Laboratory Stechlin and the Lake Stechlin research project (S.J. Casper) 19

PART TWO: THE HYDROGRAPHICAL, HYDROMETEOROLOGICAL AND CHEMICAL BACKGROUND

2. The lakes of the Lake Stechlin area: aspects of their morphometry
   by L. Krey 29
3. The hydrometeorology of the Lake Stechlin area
   by D. Richter and R. Koschel 41
   - Climatic conditions (D. Richter) 41
   - Water temperature (D. Richter) 47
   - Heat balance components (D. Richter) 56
   - Amount of evaporation (D. Richter) 64
   - Water balance components (D. Richter) 67
   - Light penetration (R. Koschel) 72

4. The chemical environment by G. Mothes, R. Koschel and G. Proft 87
- The phosphorus budget (R. Koschel and G. Mothes) 87
- The carbon budget (G. Proft) 101
- The nitrogen budget (G. Mothes and R. Koschel) 111
- The silicon budget (G. Mothes and R. Koschel) 116
- The oxygen budget (G. Mothes and R. Koschel) 118

PART THREE: THE BIOLOGICAL BACKGROUND

5. The plant communities by S.J. Casper, H.-D. Krausch and W. Scheffler 129
- Aquatic macrophytes in the Lake Stechlin area (H.-D. Krausch) 129
- The microphytobenthos (W. Scheffler) 149
- The phytoplankton (S.J. Casper) 157
6. The bacteria by H.-D. Babenzien and Christina Babenzien 197
7. The invertebrate communities by D. Flößner, P. Kasprzak, G. Mothes, D. Ronneberger and W. Schönborn 213
- The microzoobenthos (W. Schönborn) 213
- The meiozoobenthos (D. Flößner) 219
- The macrozoobenthos (G. Mothes) 230
- The zooplanktonic population (P. Kasprzak and D. Ronneberger) 243
8. The vertebrate communities by D. Flößner, D. Ronneberger and W. Scheffler 261
- Fish (D. Ronneberger) 261
- Birds (D. Flößner and W. Scheffler) 269
9. Flora and fauna of the Lake Stechlin area. Short comments upon the lists of taxa (appendix) by S.J. Casper and W. Schönborn 277

PART FOUR: THE METABOLISM OF THE LAKE STECHLIN AREA

10. The primary production by R. Koschel and W. Scheffler 287
- The primary production of the phytoplankton (R. Koschel) 287
- The primary production of the microphytobenthos (W. Scheffler) 314
11. The secondary production by P. Kasprzak and D. Ronneberger 323
12. Microbial activities at Lake Stechlin by H.-D. Babenzien and Christina Babenzien 347
13. Sediments and sedimentation by G. Mothes and G. Proft 375

- Sediments (G. Proft) 375
- Sedimentation (G. Mothes) 386

14. Fallen leaves in Lake Stechlin
by S.J. Casper, L. Krey and G. Proft 401

15. The ecosystem Lake Stechlin: the life of the lake
by R. Koschel, G. Mothes and S.J. Casper 411
- The ecological status of Lake Stechlin (S.J. Casper) 411
- Calcite precipitation and its role in the life of the lakes (R. Koschel) 412
- The nuclear power plant and its role in the life of Lake Stechlin (R. Koschel, G. Mothes and S.J. Casper) 419

16. The pelagic zone of Lake Stechlin: an approach to a theoretical model
by J. Benndorf, R. Koschel and F. Recknagel 433

17. Lake Stechlin area and society
by H. Klapper and R. Koschel 455

Appendix
Taxa observed in the Lake Stechlin area 485

Taxonomic index 531
Subject index 541

PART ONE

The historical background

Über Land und Meer.

№ 1.

79. Band. Vierzigster Jahrgang. Oktober 1897–1898.

Erscheint jeden Sonntag.

Preis vierteljährlich 3.50 M. Mit Postaufschlag 3.75 M.

Redakteur: Ernst Schubert in Stuttgart.

Stechlin.

Roman von

Theodor Fontane.

I.

Im Norden der Grafschaft Ruppin, hart an der mecklenburgischen Grenze, zieht sich von dem Städtchen Gransee bis nach Rheinsberg hin (und noch darüber hinaus) eine mehrere Meilen lange Seenkette durch eine menschenarme, nur hie und da mit ein paar alten Dörfern, sonst aber ausschließlich mit Förstereien, Glas- und Teeröfen besetzte Waldgegend. Einer der Seen, die diese Seenkette bilden, heißt „der Stechlin". Zwischen flachen, nur an einer einzigen Stelle steil und quaiartig ansteigenden Ufern liegt er da, rundum von alten Buchen eingefaßt, deren Zweige, von ihrer eignen Schwere nach unten gezogen, den See mit ihrer Spitze berühren. Hie und da wächst ein weniges von Schilf und Binsen auf, aber kein Kahn zieht seine Furchen, kein Vogel singt, und nur selten, daß ein Habicht drüber hinfliegt und seinen Schatten auf die Spiegelfläche wirft. Alles still hier. Und doch, von Zeit zu Zeit wird es an eben dieser Stelle lebendig. Das ist, wenn es weit draußen in der Welt, sei's auf Island, sei's auf Java, zu rollen und zu grollen beginnt oder gar der Aschenregen der hawaiischen Vulkane bis weit auf die Südsee hinausgetrieben wird. Dann regt

Theodor Fontane in seinem Arbeitszimmer.

Figure 1.1 Copy of the first page of the original issue of Fontane's novel 'Stechlin' containing the portrait of the poet. Photo Helga Bagemühl, published with the permission of Fontane-Archiv, Potsdam.

1. The Lake Stechlin area, past and present, and the Lake Stechlin research project

S.J. Casper, H.-D. Krausch & L. Krey

Lake Stechlin, Theodor Fontane and the legend of the Red Cock (S.J. Casper)

In the north of the County of Ruppin, hard by the Mecklenburg border stretches a chain of lakes several miles long, from the little town of Gransee to Rheinsberg (and even beyond it), in the midst of a thinly-populated woodland area. There are but a few ancient villages dotted here and there, otherwise the region is solely devoted to forestry and furnaces for the manufacture of glass and tar.

One of the lakes which make up this chain is called 'Stechlin'. It lies between low banks, rising steeply like a quay in one place only, and is ringed round by old beechtrees whose branches, bowed by their own weight, touch the lake with their tips.

Here and there appear some rushes and reeds, but no boat ploughs the water, no bird sings, and only seldom a hawk flies over, casting its shadow on the mirror-like surface. All is still. And yet from time to time things become lively on just this very spot ... when rumblings and grumblings begin far away out in the world, whether in Iceland, in Java, or even when the ashes from Hawaiian volcanoes rain far out over the southern seas. Then a stir begins here too ... (Fontane 1898, p. 1)

These are the first, fairly emotional sentences, which Theodor Fontane (1819–1898), the great realist of German literature of the late nineteenth century, wrote at the beginning of his old-age novel, 'Stechlin' (Fig. 1.1). The stage on which the actors appear from time to time is the fictitious village of Stechlin, situated near the shore of Lake Stechlin, the background of the whole action. It was Fontane's novel which made people familiar with the beautiful, lonely lake located far from inhabited places or towns. Only a short time later, the Berliners had made it their favoured week-end place and holiday-camp (see p. 475).

Some years previously during September 1873, Theodor Fontane had visited Lake Stechlin, which really fascinated him (Fig. 1.3). In the first volume of his famous 'Rambles in Mark Brandenburg' he wrote down his observations, experiences, and feelings. The old story of the Red Cock which lived in Lake Stechlin, and at that time haunted the region, was told him by his fellow-traveller Alexander Gentz and was expressed in the following way:

How transquil lies the Stechlin there! ... but people about here know some tales to tell about it.

It is one of the most distinguished of lakes, keeping up important connections. At the time of the

Casper, S. Jost (ed.), Lake Stechlin. A temperate oligotrophic lake. ISBN 978-94-010-8930-2.

Lisbon earthquake, whirlpools and funnels of water appeared, and waterspouts flinging spray danced between its banks. It reaches a depth of 400 feet and in more than one place the plummet cannot find the bottom.

It has moods, too, and must be studied, like a woman. It can put up with this, but not that; sometimes there is not a hand's breadth between what flatters it and what puts it in a rage. It goes without saying that the fishermen know it best. Here they may cast their nets, and all remains clear and bright on the surface, but it won't tolerate this ten paces further on, out of pure wilfulness; its countenance grows wrinkled and dark and a rumble of anger can be heard. Then it is time to steer clear of it and seek the shore. But if some reckless fellow is in the boat and seeks a confrontation, then misfortune follows and the Cock appears, red and angry, the Cock that sits below on the bottom of the Stechlin, and beats the lake with his wings until it foams and surges (Fig. 1.2); and he seizes the boat and screams and crows so that it re-echoes throughout the whole of the forest of Menz from Dagow to Roofen, even as far as Alt-Globsow. (Fontane 1876, p. 374–375)

This passage is the oldest source of the so-called Minack-legend of today. Fontane's 'dare-devil in the boat' was transformed into the shape and appearance of a rude and savage fisherman named Minack, who is said to have lived in the fisherman's cottage Stechlin. The landowners, Litzmann of Neuglobsow and Schall of Neuroofen, both very interested in the history of their region, had heard of fisherman Minack's allegedly awful death by drowning in Lake Stechlin. The story was told them by an old inhabitant of the neighbouring village of Menz, who had sworn by all that is holy that every word of what he said would be the truth (Haase 1887). In this form the legend of Lake Stechlin and its Red Cock has been taken down by the most prominent collectors of Brandenburgian legends (Haase 1887; v. Schulenburg 1912; Schmidt 1918; Burkhardt 1960).

Again and again people were intensely impressed by the Lake Stechlin legend, and many tried to find out the real background of the tale. As early as the end of the eighteenth century, the allegedly observed odd water movements were coupled with the terrible earthquake at Lisbon in 1755 in an old topography of County Ruppin (Bratring 1799; v. Petersen 1928). Between 1733 and 1826 a fisher-family, Minack, lived in Menz as the parochial register shows beyond doubt. But there are no hints of death by drowning in the family. The legend of the Red Cock is by nature a legend of a nix: nixes are typical figures in Brandenburgian legends and regularly characterized by redcoloured clothes (Schmidt 1918). Also other more realistic explanations of the legend have been discussed (Krausch 1968).

In the thirties of our century the mining engineer C. Hänel, stimulated by the fantastic tales of the mysterious eruptive activity of Lake Stechlin and rumours of its subterranean connections with waters in far distant parts of the earth, suggested the erection of a research station by the then German Research Community. The reasons for the allegedly observed geyserlike eruptions ought to be investigated. Hänel expected much of exact studies in the prognostication of

←

Figure 1.3 Lake Stechlin. View from the chimney of the nuclear power plant over the west and central basin of the lake. The landscape, as far as the eye can reach, is formed by wood and water. Photo W. Scheffler.

Figure 1.2 The Red Cock attacking fisherman Minack. Copy taken from: Anonymus, 1926. Grafschaft Ruppin. Ferdinand Hirt's Heimat-Lesehefte, Gruppe A: 3. und 4. Schuljahr. Breslau: Ferdinand Hirt.

earthquakes. He intended to use Lake Stechlin as a kind of 'pre-seismograph' (Thienemann 1935). In the end, nothing came of his plans.

Yet, legend and reality knocked at the door of science, and you never can tell if there would ever have been a limnological research station at Lake Stechlin without the work of Theodor Fontane and his tale of the Red Cock.

Morphology and morphogenesis of the Lake Stechlin area (L. Krey)

The Lake Stechlin area is situated in the northern part of the GDR (Fig. 1.4) on the southern border of the Mecklenburg Lake District at latitude 53° 10′ N and longitude 13°02′ E, roughly 100 km north of Berlin. It belongs to the 'Ostmecklen-

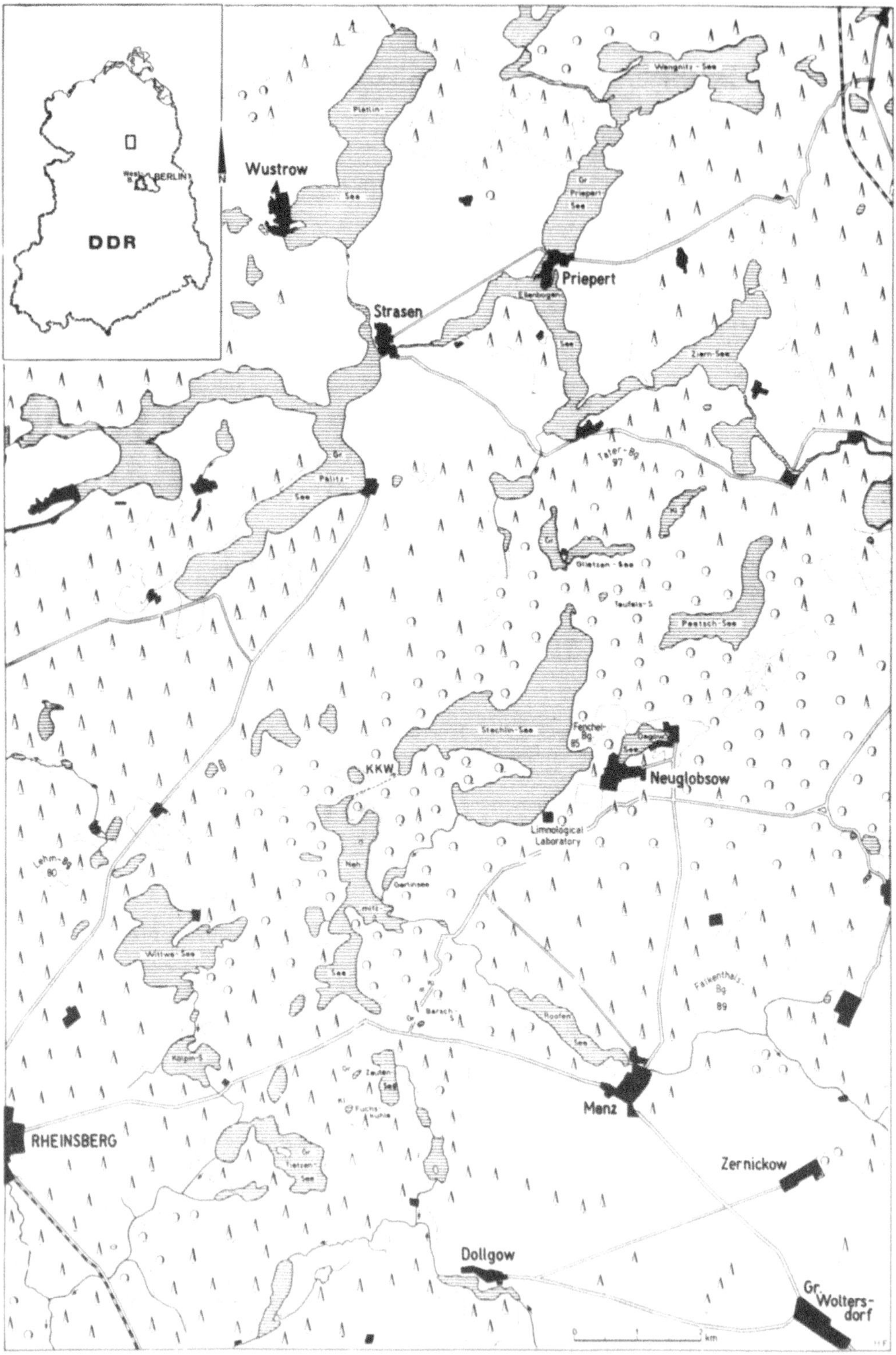

Figure 1.4 Map of the Lake Stechlin area.

burgische Kleinseeplatte' (Schultze 1955) as part of the Northern (Baltic) Land Ridge (Marcinek 1981).

The area studied covers a territory of nearly 50 square kilometres, extending to the north as far as the River Obere Havel and its chain of lakes (Lakes Menow, Ziern and Ellbogen), and to Lake Großer Pälitz, to the south to the country-town Rheinsberg and the village Menz (Fig. 1.4). Within this small area many lakes and pools of different trophic levels can be encountered. Its mean altitude is 72 m above sea-level (a. s. l.) (Heitmann, Richter & Schumann 1969). The highest elevations (up to 100 m) are in the east of the area (Stabschläger Heide). The Fenchelberg situated immediately on the east shore of Lake Stechlin reaches a height of 84.5 m a. s. l. The water level of the lakes is on average 60 m a. s. l. (Lake Stechlin 59.9 m a. s. l.).

If you approach the Lake Stechlin area from the south, wide stretches of fairly flat, sandy and loamy, open land intensively used for agriculture are traversed. Some big villages, a few little lakes and pools, and many small woods are dotted here and there. Near the village Menz vast forests dominated by pines will be met. The landscape becomes conspicuously hilly, and to the northeast of Lake Stechlin formations of gravel and sand deposits very similar to terminal moraines are found close to the shores of the lakes. The relief of the landscape varies between depressions and heights within short distances. This area is born in the ice-age.

The greatest part of North and Central Europe has been formed by the three Quaternary Scandinavian glaciations. During this era the ice margin changed its position repeatedly, resulting in glacial series of deposits. Great quantities of earth, gravel, rock, and sand, etc., were carried by the glaciers and deposited as their ground moraine and their terminal moraine respectively, or formed outwash plains and eroded melt water channels and valleys. Such glacial deposits, locally more than 200 m thick and mostly consisting of sand poor in nutrients or of sandy loam, resulted in the preglacial relief of the Lake Stechlin area not being visible on the surface (Krausch & Zühlke 1974).

The latest glacial period – the Weichselian – has strongly formed our study area by its forces and deposits mainly during the ice retreat. The terminal moraines of the 'Frankfurter Eisrandlage' set the south border. Within their hinterland the Rheinsberg Basin was eroded, forming a new floor of deposits (Marcinek 1981).

As a result of the relatively quick melting of the inland ice, dead ice was left in the grooves and depressions formed by the earlier ice action. On its moraines, which are still clearly visible today south of Fürstenberg, a new ice edge was formed (the 'Fürstenberger Staffel') about a thousand years after the Frankfurt ice-edge. The depressions filled with dead ice were covered with outwash in the Rheinsberg Basin (Marcinek 1981), and the melt-water created the 'Rheinsberg Gate' in the terminal moraine by breaking through it. Only the melting of the covered dead ice-blooks on the at first monotonous outwash surface led to the formation of the present day very varied lake and moor landscape around Lake Stechlin.

Within the 'Rheinsberg Basin', small ground moraines, large flat or slightly hilly outwash plains with shallow hollows, eroded melt-water channels often bordered with deep slopes and filled with lakes and pools, and dead ice-areas with many kettleholes and depressions occur (Heitmann, Richter & Schumann 1969). All lakes investigated were created by dead ice. Lake Stechlin and Lake Nehmitz are probably a 'combination' of the glacial-tunnel type (Heitmann & Schubert 1965) and the kettlehole type ('kombinierter Beckensee'; Marcinek & Nitz 1973). Through ice erosion, melt water channels or valleys filled with dead ice-blocks were built (Fig. 1.5). Subsequently they were covered by the outwash of the 'Fürstenberger Staffel' and temporarily prevented from melting. At the turning-point from late to post glacial era the dead ice melted and the outwash cover sank. Within the so formed cavities, ground-water lakes and bogs originated above substrata impervious to water.

Due to the glacially formed surface resulting in a high permeability of the mostly sandy soils, the Lake Stechlin area – like other glacial-born regions – possessed no visible natural surface discharge to the sea (Treichel 1957; Driescher 1983). It is an inland drainage area (Heitmann & Schubert 1965; Schumann 1968; Heitmann, Richter & Schumann 1969; Marcinek 1981; Fig. 2.2), the biggest one of the region.

Since the construction of the Polzow canal between 1745–1750 connecting the Lakes Stechlin, Nehmitz, Roofen, and the Wentow waters with the River Havel System, the Lake Stechlin area has drained to the North Sea. As a consequence, the ground-water level and the water level of Lake Stechlin and Lake Nehmitz sank considerably and the littoral belts were reduced. Today the former shore lines and littoral zones can still be dectected (Krausch 1963).

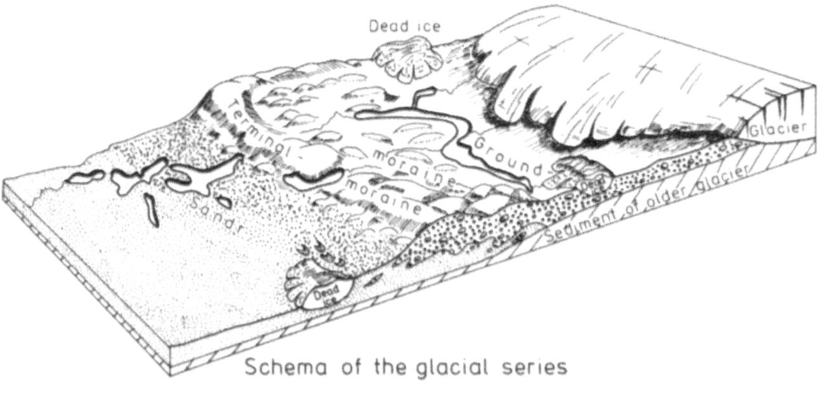

Figure 1.5 Diagrammatic section of the Lake Stechlin area during the latest glacial period, the Weichselian.

The plant cover of the Lake Stechlin area (H.-D. Krausch)

Present real vegetation

At the time when Theodor Fontane wrote his novel the Lake Stechlin area was an extensive woodland area intermingled only with a little open land in the vicinity of settlements, by some meadows, bogs and mires, and by many lakes and bog-pools.

The greatest part of the existing forest is dominated by Pine (*Pinus sylvestris*). Natural Pine-woods – mostly developed as *Myrtillo-Pinetum,* locally as *Cladonio-Pinetum* or *Peucedano-Pinetum* also – exist only in the south and the west of the area. Additionally manmade Pine-forests – *Oxalis*-Pine-forests and *Hypnum*-Pine-forests – take the place of former Beech-forests. *Picea-*, *Larix-*, and *Pseudotsuga*-forests are only to be found locally (Krausch 1970).

Of the native Beech- and Beech-Oak-forests, only a few small areas are preserved, mainly on the moraine-stands north of Neuglobsow, between Lake Stechlin and Lake Nehmitz, and in the Junkerbusch near Zernikow (Krausch 1970; Fig. 1.9).

Undrained hollows, in part sunk deeply in the surrounding land, have often developed oligotrophic *Sphagnum*-bogs. They have developed from formerly open water by silting and plant accumulation ('Verlandungs-Hochmoore'). Some of them now include residual lakes (for instance Lakes Großer and Kleiner Barsch, Große and Kleine Fuchskuhle, Lake Teufel, Lake Steutzen; see Figs. 2.12, 2.14). In these mostly acid, oligoproductive bog-pools *Drepanocladus fluitans, Nuphar lutea* and *Nymphaea alba* are the only macrophytic waterplants.

From open water to the shore the succession of plants begins partly with small belts of *Carex lasiocarpa,* partly with bog mosses of *Sphagnum* floating on the water surface. Only a few bog-pools richer in nutrients (for instance Lake Steutzen) show Reed-swamps with *Phragmites australis* and *Typha angustifolia.* As a rule a swing moor of *Sphagnum recurvum* is joining, mixed with *Carex limosa, Rhynchospora alba, Scheuchzeria palustris, Drosera rotundifolia, Eriophorum angustifolium* and other species. When the organic substratum has become more stabilized, *Eriophorum vaginatum, Pinus sylvestris, Betula pubescens,* and *Ledum palustre* appear, and a bog-Pine-wood *(Ledo-Pinetum)* develops (Fig. 1.6). Thus the bogs of the Lake Stechlin area belong to the type of subcontinental wooded raised bogs (Krausch 1968).

In the environs of mesotrophic lakes rich in lime (e.g. Lake Glawke) sedge-rich basic mires have developed. Although still existing with small residues in 1960, they have nearly vanished as a result of draining.

In valleys and on shores of the lakes with meso- to eutrophic peats, locally small stands of an Alder carr are preserved, for instance in the Ochsenkopf Bay of Lake Stechlin. Most of the Alder carrs of the Lake Stechlin area belong to the poor *Betula pubescens*-Alder carr (Fig. 1.7). Occasionally the richer subassociation of *Ranunculus repens* appears. However, mostly the Alder carr is cleared and

Figure 1.6 View into a wooded raised bog with Ledo-Pinetum at Lake Kleiner Barsch. Photo H.-D. Krausch.

Figure 1.7 Wet Alder-Birch-carr (*Carici elongatae-Alnetum glutinosae betuletosum pubescentis*) on the swamp eastern of Kamper Lake. Photo H.-D. Krausch.

changed into meadows (sedge rich poor varieties of the *Angelico-Cirsietum*, partly of the *Junco-Molinietum*).

In recent years remote glades have no longer been cultivated. These meadows developed into a *Carici-Agrostietum caninae* or into a *Caricetum ripario-acutiformis*. They will soon become wood again with Willows and Alders.

General phytogeographical situation

The area of Lake Stechlin lies in the transitional region between the oceanic and continental climatic zones of Europe, in which the influence of the oceanic zone predominates. Although species of typical oceanic distribution are rare (e.g. *Osmunda regalis, Lycopodiella inundata, Lonicera periclymenum*), oceanic species of large distribution belong to the essential components of landscape-forming plant-communities. These are, for instance, in Beech-forests *Fagus sylvatica, Viola riviniana, Hieracium* spp., and *Polypodium vulgare;* on heaths *Calluna vulgaris, Sarothamnus scoparius, Genista pilosa,* and *Sieglingia decumbens;* in dry grassland *Corynephorus canescens, Aira praecox, A. caryophyllacea, Teesdalia nudicaulis;* and in the weed communities *Arnoseris minima, Hypochoeris glabra,* and *Geranium pusillum.*

The high portion of boreal distributed species is characteristic for the Lake Stechlin area. Most of them grow in the mires and bogs, which are climatic cold stands, but also occur in the Pine-forests and in the oligotrophic waters. In addition to the general boreal species, mainly continental boreal species (e.g., *Pinus sylvestris, Ledum palustre, Ramischia secunda*) are represented, less frequently oceanic-boreal species (e.g., *Blechnum spicant, Empetrum nigrum*).

Thermophilic species of continental and submediterranean distribution do not occur often. They are only to be found locally in thermophilic woods and dry grassland on sunny and warm slopes (e.g., *Koeleria glauca, Festuca psammophila, Helichrysum arenarium, Pulsatilla pratensis, Salvia pratensis;* Krausch 1970).

Many water- and swamp-plants of the Lake Stechlin area are wide-spread (e.g., *Potamogeton lucens, P. natans, P. pectinatus, Myriophyllum verticillatum, M. spicatum, Nuphar luteum, Ceratophyllum demersum, Phragmites australis,* and *Typha angustifolia*). Additionally there also occur here a great number of boreal water plants (e.g., *Potamogeton rutilus, P. alpinus, P. nitens, P. praelongus, P. gramineus, P. filiformis, Myriophyllum alterniflorum, Sparganium minimum, Utricularia minor,* and *U. intermedia*). Thermophilic water plants are retreating. *Cladium mariscus,* a plant surviving from the Atlanticum stage (Table 1.1), is still numerous in the area. *Aldrovanda vesiculosa, Najas marina,* and *N. minor,* probably brought along by birds of passage from warm regions of Europe and Africa, subsisted only a short time and disappeared (Winter 1870; Retzdorff 1907; Müller-Stoll & Krausch 1959; Krausch 1964).

Table 1.1 Late Weichselian and postglacial development of vegetation in the Lake Stechlin area (according to Kloss 1980).

Pollen zones	Climatic stages	Time	Development of vegetation
X	Younger Subatlanticum	800 A.D.–present	See Table 1.2
IX	Older Subatlanticum	500 B.C.–800 A.D.	500–0: predominance of *Pinus;* later spread of *Fagus* and limitation of *Pinus* to the poorest sandy stands in S and W; *Fagus-Quercus*-forests on middle and rich stands; in the lowlands *Alnus*-carrs, in the oligotrophic flats Ledo-Pinetum and open bogs
VIII	Subboreal	2500–500 B.C.	Gradual spread of *Fagus* and *Carpinus;* increase of *Alnus;* temporary advance of *Picea.* Local influence of human settlement on vegetation during Bronze Age; increase of *Pinus,* decline of *Ulmus, Tilia, Corylus,* and *Fraxinus* at the end of the stage
VII, VI	Atlanticum	6000–2500 B.C.	On rich stands dense forests with *Quercus, Ulmus, Tilia,* and *Corylus;* on poor stands *Pinus-Quercus-Betula* forests; in the lowlands *Alnus* and *Fraxinus;* decline of *Ulmus* and *Fraxinus,* arrival of *Fagus* and *Carpinus* at the end of the stage
V	Boreal	7000–6000 B.C.	Pine-forests with *Corylus*-shrubs; spread of *Ulmus, Quercus* and *Tilia,* in the lowlands of *Alnus* and *Fraxinus,* locally of *Picea*
IV	Pre-Boreal	8200–7000 B. C.	Dense forests of *Betula, Populus, Salix* and *Pinus;* immigration of *Corylus* and other thermophilic shrubs and trees
III	Younger Dryas (Post-Alleröd)	9000–8200 B.C.	Birch-tundra; herbaceous plants predominating in the bottom-layer
II	Alleröd	10000–9000 B. C.	Dense Subarctic Birch- and Pine-forests; 3 cm sediment layer of volcanic ash (Eifel-eruption at 9500 B. C.)
I	Older Dryas (Pre-Alleröd)	15000–10000 B. C.	Arctic tundra; advance of *Pinus* and *Betula* in the Bölling-Interstadial

The Lake Stechlin area is far from traffic, resulting in the low portion of adventives and neophytes. About 1870 *Elodea canadensis* migrated from the rivers Havel and Rhin into the Lake Stechlin area (Winter 1870; Krausch 1964).

Settlement and economy of the Lake Stechlin area (H.-D. Krausch)

History of settlement

In primeval historical time the Lake Stechlin area was thinly populated, temporarily and only on its borders. During the Middle Ages more intensive settlement began.

During the tenth century northwestslavic settlers migrated from the regions on the rivers Rhin and Havel into the vast woodland called Lietze between the rivers Dosse and Havel. Here they founded, mostly in the vicinity of lakes, small villages, the economic basis of which was stock-farming, agriculture, and gathering.

After the conquest of this region by the Germans, especially as a consequence of the crusade against the Wends carried out in 1147 by German, Danish, and Polish feudal lords, a more or less close settlement begun by immigrants from the western regions such as Magdeburg, Altmark, and western Central Europe. This medieval agricultural expansion included, above all, the moraine-plateaus southeast of the Lake Stechlin area with soils favourable for corn-growing (Fig. 1.8).

After extensive clearing of the forests (Table 1.2), a dense network of well-planned big villages originated. At important strategic points, above all in protection of the then boundaries, castles were founded. Under their protection small country-towns such as Rheinsberg and Fürstenberg developed, whereas Menz (situated nearest to Lake Stechlin) remained a village. The centre of the Lake Stechlin area remained a woodland, with small villages dotted here and there near the lakes (Fig. 1.9) forming a natural boundary between the territories of Ruppin and Stargard.

In the Late Middle Ages a considerable retrogression of settlements occurred (Table 1.2). At the end of the fifteenth century 41 of the 46 medieval villages (see Fig. 1.9) were deserted. The Lake Stechlin area thus belongs to those regions of Central Europe with the greatest portion of deserted villages. The agrarian crisis of the Late Middle Ages together with epidemics (pestilence) and feudal wars caused this depopulation. The villages suffered from plunderings of Mecklenburgian robber knights. In 1422, for instance, they stole two bullocks in the village Stechlin, and in 1423 they totally plundered the villages Menz and Roofen. Then all villages on the terminal moraine zone and the outwash plains and many villages on the ground moraine plateaus were given up.

The resettlement of the deserted medieval communities began in the 16th century and increased in the middle of the 18th century in the course of the

Figure 1.8 Pot of the high medieval period, found in Lake Stechlin near the place of the deserted medieval village Stammstechlin. Photo D. Sommer, with the permission of the Museum für Ur- und Frühgeschichte Potsdam.

Table 1.2 The development of vegetation in the Lake Stechlin area since the Middle Ages.

Years	Period	Vegetation
1850–present day	Industrial age	Extensive afforestation of less productive sandy soils. Big increase of *Pinus*, reduction of *Fagus* and *Quercus*. Plantation of alien trees (*Picea, Larix, Pseudotsuga*). Drainage of swamps.
1500–1850	Early modern times	Local clearing of woods, intensification of agriculture, increasing use of forests for logging and pasture, in the second part of 18th century devastation of large forest lands, reduction of *Quercus* and *Fagus*, advance of *Pinus* and *Betula*.
1350–1500	Late Middle-Ages	Renewed advance of forests, especially of pioneer-trees (*Pinus, Betula* a. o.). Low-productive use of parts of the medieval deserted fields by remaining villages, local pastures with heaths and dry grassland.
1150–1350	High Middle-Ages	Extensive clearing, reduction of forests (Oak- and Beech-forests of the moraine-plateaus), clearing of the remaining forests for logging and pasture, extension of heaths and dry grassland.
800–1150	Early Middle-Ages	In the immediate neighbourhood of the villages small fields, pastures and cleared forests, otherwise dense primeval forests.

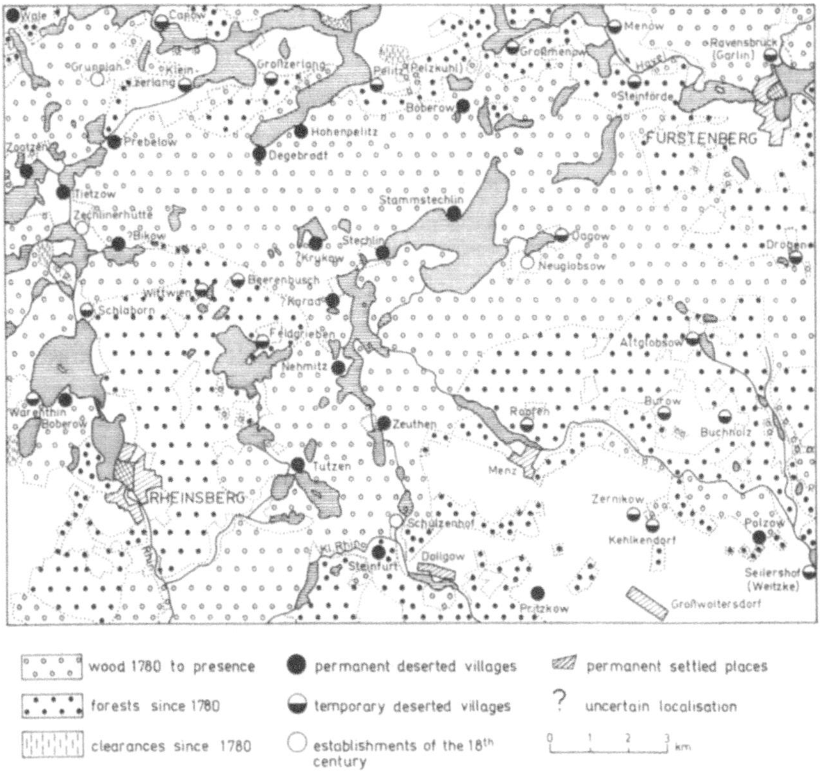

Figure 1.9 Development of settlements since Middle Ages and changes of forest areas since 1780 in the Lake Stechlin area (according to Krausch 1970).

Frederician colonization. Altogether, nearly half of the deserted medieval villages were resettled – for instance, Burow, Dagow and Altglobsow in 1751–53, Roofen in 1771 (see Fig. 1.9).

Economy – past and present

At all times fishing has played a leading part in the economy of the Lake Stechlin area.

In the eighteenth century the lakes were leased by their owners to professional fishermen. From 1708–1717 four fishermen of Lindow were the general leaseholders of the ten lakes of Amt Lindow, which also included Lakes Roofen and Globsow. Some years later fishermen were dwelling in special fishery houses on the Lakes Stechlin, Großer Tietzen, and Wittwe. The manner Groß Zerlang and subsequently the municipality of Menz, as owners of waters, also leased their lakes.

In 1898, when the Prussian state bought back the 'Lakes of Menz', near the old fishing cottage, a new solid fishing farm was established on the southeastern shore of Lake Stechlin. The Limnological Laboratory Stechlin has been situated there since 1959.

The lakes were not only used for fishing. Formerly the Charales were used as a substitute for dung. These macrophytic algae, vernacularly named 'Post', were pulled out of the water with large rakes into boats and taken to the shore, from where they were carried to the fields. Since animal fertilizers were not available the Charales, rich in lime, represented a valuable and cheap dung replacement. In Mecklenburg the utilization of Post as fertilizer occurred widely, and in 1800 the County Mecklenburg–Strelitz issued a special order to regulate the drawing of the Charales from the lakes. Post was used as fertilizer until the end of the nineteenth, locally until the middle of the twentieth century (Krausch 1962).

Since ancient times wood has been cut in the vast forests of the Lake Stechlin area (Table 1.2). The adjacent villages were allowed to taken wood for building or burning gratuitously or for a small sum.

Moreover, timber and firewood were needed outside the area and often transported by rafting.

The rivulet Kleiner Rhin was enlarged as a rafting canal which led to Lake Nehmitz. In the middle of the eighteenth century, when the population of the Prussian capital Berlin increased rapidly, more firewood was required. Therefore, from 1745–1751 a rafting canal called Polzow-canal was constructed, connecting the River Havel to the Wentow-Lakes, the Polzow brook, and the Lakes Nehmitz, Stechlin, and Dagow. In the following thirty years as a result of extensive woodcutting and exhaustion of the wood-stocks, rafting was ended. Since the beginning of the nineteenth century the forests have been re-cultivated (Table 1.2). In Rheinsberg, Fürstenberg, and other places sawmills were established to process the timber.

Until the middle of the nineteenth century some tar furnaces produced tar, pitch, pine oil, and other products by dry distillation of the resinous wood of the Pine.

During the eighteenth century some glass-works were established, at first in 1737 in Zechlinerhütte. Much wood was consumed by the furnaces and for production of potassium. The glass-work, founded in Altglobsow in 1752, was moved deeper into the forest to the shore of Lake Dagow in 1778, and subsequently the village Neuglobsow originated. The glassworks were in operation until 1880.

In earlier centuries the forests were also used as a pasture for animals. Through wood-pasture the broad-leafed tree population was reduced. In the second half of the nineteenth century all titles for wood-pasture were abolished.

At the beginning of the twentieth century the tourist industry developed in the Lake Stechlin area. After World War II it increased greatly, and now the wood- and lake-rich region between Rheinsberg and Fürstenberg is a much-frequented recreation area (see p. 475).

The Limnological Laboratory Stechlin and the Lake Stechlin research project (S.J. Casper)

The Limnological Laboratory Stechlin is situated on the south shore of Lake Stechlin (Fig. 1.10), within the area of the so-called 'Alte Fischerhütte', approximately 2 km west of the small village Neuglobsow. Its first director was Dr. Theodor Schräder. The Hydrometeorological Station of the Meterorological Service of the GDR is in the immediate neighbourhood.

The Laboratory consists of nine different types of building: the former dwelling-house of the fisherman and its wash-house, the former public-house with its living quarters and wash-house, an old shed, two wooden houses, and a boathouse (Fig. 1.10). Now the buildings are equipped with modern biological, microbiological, chemical, and radiochemical laboratories.

The 1959–1965 period

During the six years before the nuclear power plant was brought into operation, the research in the Lake Stechlin area, a nature reserve since 1938, was mainly of a basic character. For such a programme the region offers very suitable conditions, for within a small area lakes of different trophic levels, depending on location, can be found.

Lake Stechlin is highly transparent, deep enough and stratified, fairly large but easy to survey, and oligotrophic (see p. 29); Lake Nehmitz is fairly transparent, of medium size, moderately deep and stratified, rich-structured, and mesotrophic (see p. 31); Lake Dagow is not transparent, fairly shallow, small, and eutrophic (see p. 35). Many of the small lakes in the woodland are bog lakes or bog pools (see p. 37). As is to be expected, the physiographical situation which has been described is reflected in a corresponding flora and fauna (see p. 129).

More than 1900 taxa of bacteria, fungi, algae, macrophytes, protozoa, invertebrates, and vertebrates were observed and partially taxonomically investigated, their temporal and spatial distribution established, and their association analysed. Now Lake Stechlin and the neighbouring lakes belong to the best-known waters in Central Europe with respect to flora and fauna (Casper 1974). The qualitative and quantitative analysis of the stock of organisms in Lake Stechlin and its environ was a substantial and necessary prerequisite for the subsequent phase in which several different processes, mainly influenced by the cooling water circuit of the nuclear power plant, were described more thoroughly.

During this time five working groups – physiography, bacteriology, botany, zoology, and documentation – were formed (Schräder 1962) and a great many papers were published, mainly in 'Limnologica'.

Figure 1.10 View over the Limnological Laboratory Stechlin. In the background parts of the Hydrometeorological Research Institute and Lake Stechlin. Photo W. Scheffler.

Figure 1.11 Participants of a post-graduate training course of the UNESCO in front of the former 'Alte Fischerhütte' of the Limnological Laboratory Stechlin. Photo W. Scheffler.

The 1966–1975 period

In 1966 the nuclear power plant (Fig. 1.12) was brought into almost continous operation. 400 000 m^3 of heated water circulated daily through the Lake Nehmitz-Lake Stechlin-system, and as a result the heat and matter content of the surface waters of Lake Stechlin were increased.

Now the research programme was focused on the effect of the influx of the cooling water on the production and energy flow of the whole ecological system of the lake. Studies on standing crop biomass and primary production of phytoplankton, periphyton, and macrophytes, on oxygen and nutrient (phosphorus, nitrogen, silicon, carbon) budget, on utilization of nutrients by different biocoenotic links, and on sedimentation were carried out.

The studies were concentrated in the 'Primary and Secondary Production' work groups. In addition to many special papers, mainly published in 'Limnologica', a comprehensive account of the effect of the cooling water circuit on the lake system was published (Koschel & Mothes 1976).

In this period the research work on fundamental interrelations, mainly in the euphotic area of Lake Stechlin ecosystem, resulted in a fairly adequate insight into the material and energy balance and into the primary and secondary production contingent on nutrients and temperature. Lake Stechlin proved to be a most sensitive laboratory or test instrument of the biotic and abiotic changes in an aquatic ecosystem caused by various external influences. Thus Lake Stechlin was chosen as a standard water body for limnological research.

From 1966 onward the Laboratory participated in the International Biological Programme. In 1970 Dr. Theodor Schräder, having reached retirement age, was succeeded by Dr. S. Jost Casper.

The period after 1976

Owing to force of circumstances – some members of the staff had to concentrate their research on rivers (Saale, Ilm) in the southern parts of the GDR – the concept of long-term whole-lake research had to be partially postponed. Now there was a strong desire to fill in the gaps, and to find an integrated ecosystem approach to the limnological problems not only in the Lake Stechlin area but also in the whole pleistocene lake region in the northern parts of the GDR. From time to time the investigations were extended to Lake Müritz near Waren, to Lakes Breiter Luzin, Schmaler Luzin, and Haussee near Feldberg, and to Lake Tollense near Neubrandenburg. Nevertheless, the main mission of our scientific work was directed to the study of fundamental limnological problems in their broadest sense in the Lake Stechlin area.

Simultaneous studies of the processes controlling the production of organic matter by autotrophs and heterotrophs, its consumption by herbivores, and its

Figure 1.12 The nuclear power plant I of the GDR. Photo W. Scheffler.

decomposition by bacterial activity, were set up. Observations in the field and investigations under laboratory conditions were brought together. Efforts were strengthened in order to contribute to a technique of modelling to analyse the regularities in changes of equilibrium.

In the near future the role of benthal and sediments as a whole will be investigated thoroughly. The research of the 'Algology' work group will be focused on the causes of algal mass development, on the role of μ-algae, and on explaining the apparently accidental periodicity of abundant algal species in stratified lakes and storage reservoirs. The newly forme 'Microbiology' work group will study the degradation and utilization of organic substances by heterotrophic microorganisms, e.g. the in situ carbon turnover balance. The 'Foodchain and Production Studies' work group will estimate the population structure and density, fecundity, mortality, growth, standing crop biomass and production for the most important fishfood species, e. g. the planktonic copepods, cladocerans, and rotifers. The interrelations between the fish populations and their food organisms will be tested in nature as well as under experimental conditions to find out biologically sound sanitation methods ('biomanipulation') for loaded waters. The results of all these studies will be used in a simulation model, called SALMO and developed at the Technical University Dresden (Benndorf & Recknagel 1982), in order to describe and to obtain more insight into the dynamics of lake ecosystems and their subsystems.

The scientific programme of the Limnological Laboratory Stechlin is integrated into the national research programme on limnetic ecology performed by various institutes under the auspices of the Ministry of Higher Education in cooperation with the Ministry of Environment and Water Management and the Academy of Sciences of the GDR. The Limnological Laboratory Stechlin collaborates with a number of institutes in the GDR. The collaboration with the Laboratory of Hydrobiology of the Dept. of the Water Science of the Techn. University Dresden and with the Institute of Water Management Berlin requires a special mention. Training courses for undergraduate hydrobiology students from universities are organized at Neuglobsow.

International contacts are also numerous, especially with institutes in the USSR and the ČSSR. Direct bilateral cooperation exists with the Czechoslovakian Academy of Sciences, especially with the Biological Centre at Budéjovice, with the Botanical Institute at Průhonice near Praha and its Hydrobotanical Laboratory at Třeboň. The Limnological Laboratory Stechlin participates in training courses for postgraduate students organized by the UNESCO (Fig. 1.11).

In the scientific programme of the Limnological Laboratory Stechlin fundamental and applied limnology – by no means two 'academic' points of view but topical objectives of limnology (Thienemann 1933) – are brought together to do both fundamental research towards a theory of limnetic ecosystems and applied research towards a rational management of water quality. Whether we will succeed in our theoretical and practical activities – and we must achieve success in view of both the increasing water pollution and the decreasing water resources in many parts of the world – will depend mainly on the ability of limnologists to learn to look also with the 'eyes' of *Dinobryon* or *Synura,* to understand the complexity of aquatic ecosystems. Limnologists must find out the exact scale to weigh the different 'interests' of nature and society. Protection of waters does not mean prevention of usage but protection for rational humanist use by society.

References

Benndorf, J. & Recknagel, F. 1982. Problems of application of the ecological model SALMO to lakes and reservoirs having various trophic states. Ecol. Modelling 17: 129–145.

Bratring, F.W.A. 1799. Die Grafschaft Ruppin in historischer, statistischer und geographischer Hinsicht. Ein Beitrag zur Kunde der Mark Brandenburg. Berlin: G. Hayn. XIV + 618.

Burkhardt, A. 1960. Märkische Sagen und Märchen. Berlin: Altberliner Verlag Lucie Groszer. 284.

Casper, S.J. 1974. Die Aufgaben der limnologischen Forschung im Naturschutzgebiet Stechlin. Naturschutzarb. Berlin Brandenburg 10: 34–36.

Driescher, E. 1983. Historisch-geographische Veränderungen von Gewässereinzugsgebieten im Jungmoränengebiet der DDR. Geograph. Ber. 107: 103–118.

Fontane, Th. 1897. Stechlin. In: Über Land und Meer. 79. Band, No. 1. Stuttgart (First issue).

Fontane, Th. 1976. Wanderungen durch die Mark Brandenburg. Erster Teil. Die Grafschaft Ruppin. Berlin-Weimar: Aufbau-Verlag. XXX + 806. – The chapter 'Die Menzer Forst und der Große Stechlin' was written in 1873 and first published in 1874 in: Illustrierte Frauen-Zeitung. Ausgabe der Modenwelt und Unterhaltungsblatt. Lipperheide.

Haase, K. 1887. Sagen aus der Grafschaft Ruppin. Neuruppin. Verlag R. Petrenz.

Heitmann, M.-L. & Schubert, H. 1965. Lake Stechlin – A hydrometeorological research area. Budapest: Publ. IASH 66: 429–440.

Heitmann, M.-L., Richter, D. & Schumann, D. 1969. Der Wärme- und Wasserhaushalt des Stechlin- und Nehmitzsees. Abh. Meteorol. Dienst. DDR 96. Berlin: Akademie-Verlag.

Kloss, K. 1980. Pollenanalysen zur Vegetations-, Siedlungs- und Moorgeschichte am Südrand der ostmecklenburgisch-brandenburgischen Seenplatte (Kreis Gransee). Arch. Natursch. Landschaftsf. 20: 203–212.

Koschel, R. & Mothes, G. 1976. Change of the biological activity in a lake system through the inflow of cooling water from a nuclear power plant. Proc. Intern. Symp. Eutrophic. Rehabilitation of surface waters. Karl-Marx-Stadt. 219–233.

Krausch, H.-D. 1962. Nachrichten über Abfluß- und Seespiegelveränderungen des Stechlinsee-Gebietes. Limnologica 1: 211–215.

Krausch, H.-D. 1964. Die Pflanzengesellschaften des Stechlinsee-Gebietes. I. Die Gesellschaften des offenen Wassers. II. Röhrichte und Großseggengesellschaften, Phragmitetea Tüx. et Prsg. 1942. Limnologica 2: 145–203; 423–483.

Krausch, H.-D. 1968. Die Pflanzengesellschaften des Stechlinsee-Gebietes. IV. Die Moore. Limnologica 6: 321–380.

Krausch, H.-D. 1970. Die Pflanzengesellschaften des Stechlinsee-Gebietes. V. Wälder, Hecken und Saumgesellschaften. Limnologica 7: 397–454.

Krausch, H.-D. & Zühlke, D. 1974. Das Rheinsberg-Fürstenberger Seengebiet. (Werte unserer Heimat Bd. 25). Berlin: Akademie-Verlag.

Marcinek, J. 1981. Kleiner Exkursionsführer für das Rheinsberger Becken (Als Manuskript gedruckt). Leipzig.

Marcinek, J. & Nitz, B., 1973. Das Tiefland der DDR. Gotha/Leipzig: VEB Haack.

Müller-Stoll, W.R. & Krausch, H.-D. 1959. Verbreitungskarten brandenburgischer Leitpflanzen. Zweite Reihe. Wiss. Z. Pädagog. Hochsch. Potsdam Math.-Naturwiss. Reihe 4, 1958/59: 105–150.

Petersen, J. 1928. Fontanes Altersroman. Euphorion (Stuttgart) 29: 6–9, 51.

Retzdorff, W. 1907. Über das Vorkommen von *Aldrovandia vesiculosa* L. in der Mark Brandenburg. Verh. Bot. Verein Brandenburg 48: XXIII–XXVII.

Schmidt, R. 1918. Märkische Wassergeister. Ein Beitrag zur brandenburgischen Sagenkunde. Brandenburgia 26: 145–151.

Schräder, Th. 1962. Die Forschungsstelle für Limnologie der Deutschen Akademie der Wissenschaften zu Berlin. Limnologica 1: 1–19.

Schulenburg, W. von 1912. Innere Volkskunde. In: Landeskunde der Provinz Brandenburg. III. Band, Die Volkskunde. Ed. E. Friedel. Berlin.

Schultze, J.-H. 1955. Die naturbedingten Landschaften der DDR. Gotha: VEB Geogr.-Kartograph. Anstalt.

Schumann, D. 1968. Zur Definition, Verbreitung und Entstehung der Binnenentwässerungsgebiete. Geograph. Ber. 13: 22–32.

Thienemann, A. 1933. Vom Wesen der Limnologie und ihrer Bedeutung für die Kultur der Gegenwart (Eröffnungsansprache). Verh. Int. Verein. Limnol. 6: 21–27.

Thienemann, A. 1935. Der Stechlin-See als Präseismograph. Arch. Hydrobiol. 29: 346–347.

Thienemann, A. 1955. Die Binnengewässer in Natur und Kultur. Verständliche Wissenschaft 55. Heidelberg: Springer-Verlag.

Treichel, F. 1957. Die Haupt- und Nebenwasserscheiden in Mecklenburg. Diss. Math.-Nat. Fak. Univ. Greifswald.

Winter, H. 1870. Flora der Umgegend von Menz. Verh. Bot. Verein Brandenburg 12: 1–43.

PART TWO

The hydrographical, hydrometeorological and chemical background

2. The lakes of the Lake Stechlin area: aspects of their morphometry

L. Krey

All lakes treated are dead ice lakes virtually without natural surface runoff. The water level is regulated by precipitation on the lake surface, by ground water feed and runoff through the highly permeable sand layers surrounding the lakes, and by evaporation. Different aspects of their physiography are dealt with in Krausch (1962, 1964); Flössner (1964); Heitmann & Schubert (1965); Heitmann, Richter & Schumann (1969; in the following paragraphs cited as Schumann 1969), and Krausch in Krausch & Zühlke (1974).

Lake Stechlin

Lake Stechlin (Fig. 2.1) with its large number of bights was formed by deep melting of dead ice blocks and erosion of melt water channels after the Weichselian stage of the last glaciation ('Kombinierter Beckensee'; Marcinek & Nitz 1973). It is situated in the outwash plain immediately to the south of the terminal moraine of the Fürstenberger Staffel (Marcinek 1981). Formations of gravel and sand deposits reaching heights up to 84.5 m a. s. l. and approaching the shores are towards the north-east of the lake.

Lake Stechlin has an area of 4.25 km^2, a maximum depth of 68 m, and is stratified in summer with small hypolimnetic oxygen depletion. Towards the west and southwest, a sand plain follows at heights of 70–80 m a. s. l. bordered on the east by a ground moraine (Heitmann & Schubert 1965). Before 1750, surface in- or effluents were virtually absent (Krausch 1962). The natural real surface catchment area amounts to 12.36 km^2 (Fig. 2.2) of which 80% are covered by forests. The actual sub-surface catchment area amounts to 25.95 km^2 (Schumann 1969; see Richter p. 67). The actual water level is 59.67 m a. s. l. Before 1750, it was 60.7 m a. s. l. (Krausch 1974).

The lake is locally cut deeply into its environs with considerably steep shores on the northwest and northeast sides of the north bay. The Fenchelberg, the highest point (84.5 m a. s. l.) of the Lake Stechlin area, is situated immediately on its

Casper, S. Jost (ed.), Lake Stechlin. A temperate oligotrophic lake. ISBN 978-94-010-8930-2.

Figure 2.1 Lake Stechlin. View over the central basin from Fenchelberg

northeast shore. The deepest point of the lake basin (68 m) runs down to 8.5 m u. s. l. (cryptodepression) and lies at the point of intersection of two dead ice channels. They are the basis for the cross-like outline of the lake ('Kreuzsee' in German). The energy of relief amounts to 93 m vertical interval at nearly 500 m distance (Schumann 1969).

The morphometry of Lake Stechlin was described by Krausch (1964) and in detail by Schumann (1969). Morphometric data (Table 2.1), the bathymetric map (Fig. 2.3) and hypsometric curves and depth-volume relations (Fig. 2.4) were obtained from these studies and from our own measurements and calculations.

The lake is divided in four basins or bays: the north basin with the maximum depth of the whole lake (68 m); the relatively shallow central basin, 59 m deep; the west basin influenced directly by the cooling water circuit (see p. 58), 41 m deep and separated from the south basin (35 m deep) by a nearly 1.2 km long peninsula stretching from west to east (Fig. 2.3).

The relative depths of the basins are 3.5%, 5.4%, 3.1%, 5.2% (Table 2.1). Thus, the basins belong to the category of deep lakes with a relative small surface area.

During the stagnation phases the hypolimnion of the different basins is not changeable mutually. The shore line is 16.1 km long; the high value of shore line development (2.14) reflects the irregular shore line of the lake, i.e. it is rich structured by its bays. The volume was calculated to $96.88 \cdot 10^6$ m^3 and the mean depth to 22.8 m. The volumes of the four basins are: west basin $19 \cdot 10^6$ m^3, north

In the background, the nuclear power plant. Photo W. Scheffler.

basin $37 \cdot 10^6 m^3$, south basin $13 \cdot 10^6 m^3$, central basin $28 \cdot 10^6 m^3$.

The shore of Lake Stechlin is bounded by a more or less closed large zone of mixed forests mainly consisting of Pines, Beeches, Willows, and Alders (Krausch 1970; p. 402). The trees cannot prevent the lake from wind, as the prevailing and most effective wind directions are W, WSW, N and NNW, which are along the long axes (Richter 1968, 1984) of the lake. Holomixis from November to April and allochthonous input of organic matter – mostly leaves, twigs, pollen, and fruits from the surrounding trees (cf. p. 402) – are potentially significant for material balance and production in Lake Stechlin. Open land (settlements, meadows, baths) borders on the shores up to 4.8% only. Its influence is less.

Now the cooling water circuit of the nuclear power plant and the man-made surface run-in of water from Lake Dagow influence mainly the ecosystem.

Lake Nehmitz

Lake Nehmitz (Figs. 2.5, 2.8) is situated nearly 1 km west of Lake Stechlin (see Fig. 1.4). It was formed by deep melting of two separate dead ice blocks which, first of all, remained within the eroded melt water channels and were covered by sand and gravel. After melting and sinking they were filled with ground-water and coupled.

Lake Nehmitz has an area of $1.71 km^2$, a maximum depth of 18.6 m, and is

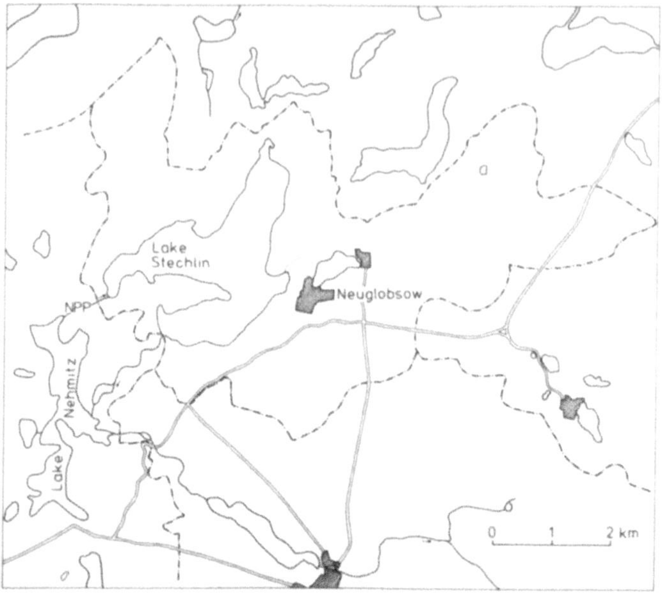

Figure 2.2 Lake Stechlin area and its watersheds.

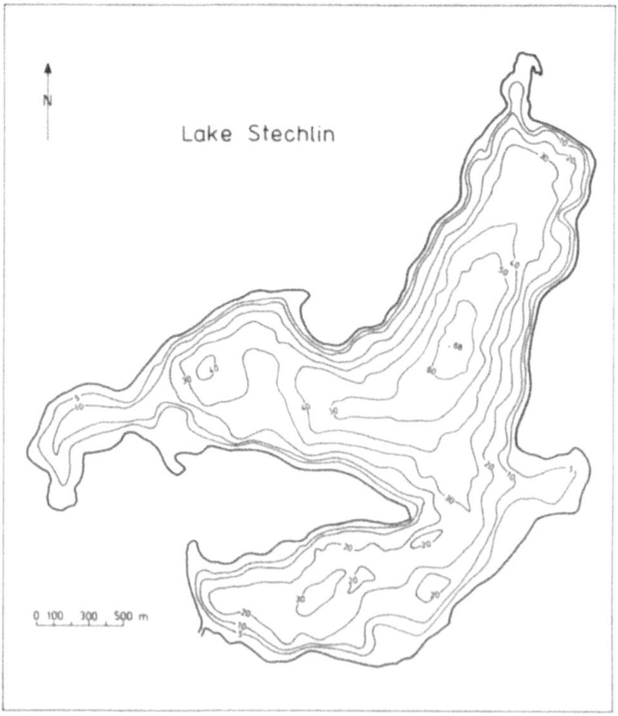

Figure 2.3 Lake Stechlin, bathymetric map.

Table 2.1 Morphometric data of some lakes of the study area (symbols according to Håkanson 1981).

	Stechlin					Nehmitz		Dagow	Große Fuchs-kuhle
	Total	W-basin	N-basin	S-basin	Central basin	Total	N-basin		
Surface area A (km^2)	4.3	1.1	1.3	0.9	1.0	1.7	1.0	0.3	0.02
Maximum depth D_{max} (m)	68	41	68	35	59	18.6	18.6	9.5	5.6
Mean depth $\bar{D}$ (m)	22.8	17.2	29.6	14.5	28	6.4	5.9	5.0	3.5
Depth of cryptodepression D_c (m)	8.1	–	8.1	–	–	–	–	–	–
Relative depth D_r (%)	2.9	3.5	5.3	3.3	5.2	1.3	1.6	3.4	1.6
Volume V ($1.0 \cdot 10^6$ m^3)	96.9	19	37	13	28	10.9	6.2	1.2	0.01
Maximum length L_{max} (km)	3.7	1.9	2.1	1.6	1.7	3.5	2.4	0.9	0.2
Maximum breadth B_{max} (km)	2.4	1.1	1.1	0.8	1.0	0.8	0.8	0.4	0.1
Mean breadth $\bar{B}$ (km)	1.2	–	–	–	–	0.5	0.4	0.3	0.08
Shore line length l_0 (km)	16.1	–	–	–	–	14.1	7.9	2.5	0.5
Shore line development F	2.1	–	–	–	–	3.9	2.2	1.4	1.2

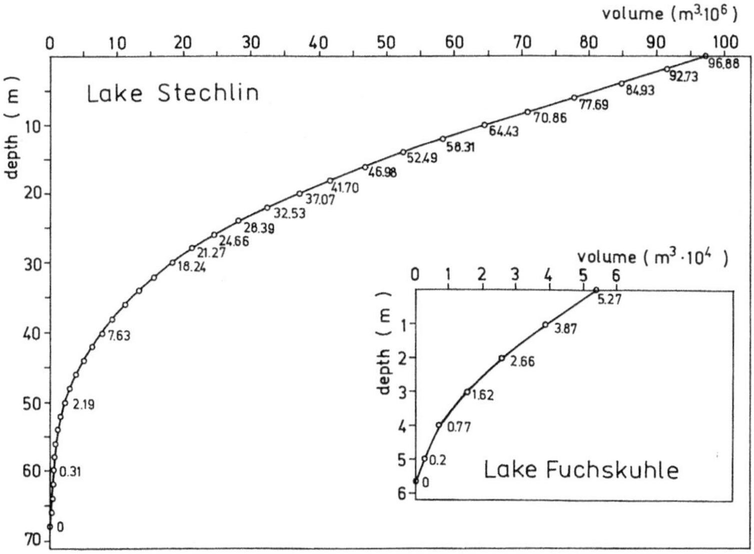

Figure 2.4 Lake Stechlin and Lake Fuchskuhle, hypsometric curves and water volumes

Figure 2.5 Lake Nehmitz. View over the north basin. In the foreground (right

stratified in summer, exhibiting hypolimnetic oxygen depletion. Before 1750, surface in- and effluents virtually were absent. At a high level of water only an insignificant above-ground outflow existed from the south basin by way of marshy lowland near Lake Zeuten into the rivulet Rhin (see Fig. 1.4). The Polzow canal connected Lake Nehmitz with Lake Stechlin and made an adjustable surface runoff towards the south into the River Havel System (Krausch 1962). The actual catchment area amounts to nearly 6.2 km^2 (Schumann 1969).

The actual water level is 59.48 m a. s. l. The shores, bounded by dense mixed forests (Krausch 1970), cut deeply into their environs.

Morphometric data (Table 2.1), the bathymetric maps (Figs 2.6, 2.9), and hypsometric curves and depth-volume relations (Fig. 2.7) were obtained from the studies of Schumann (1969) and from our own measurements and calculations.

The lake is divided into two basins by a shallow (mean water level 0.5 m) narrow underwater ridge: the north basin (in the following paragraphs named Nehmitz-N) with the maximum depth of the whole lake (18.6 m) and two small islands, and the south basin (in the following paragraphs named Nehmitz-S), 16 m deep. The existence of two nearly separate basins is of interest as the cooling water of the nuclear power plant circulates through the north basin only. The relative depth of the two basins is 1.25%.

The shore line is 14.1 km for the whole lake; its volume was calculated to $10.9 \cdot 10^6\ m^3$ and its mean depth to 6.4 m. The volume of the north basin was calculated to $6.15 \cdot 10^6\ m^3$ and the mean depth to 5.9 m.

ıe inflow-canal to the nuclear power plant. Photo W. Scheffler.

Lake Dagow

Lake Dagow (Fig. 2.10) is situated nearly 500 m east of Lake Stechlin, immediately in front of the terminal moraine of the Fürstenberger Staffel. It was formed by deep melting of a dead ice block, lying within a melt water channel.

Lake Dagow has an area of 240 500 m^2, a maximum depth of 9.5 m, and is stratified in summer, exhibiting hypolimnetic oxygen depletion. Surface inflow exists from the fen area of the Traden-Wiesen (Traden-meadows) east of Lake Dagow, whose waters from time to time are dammed up. It drains through the former Lake Kleiner Dagow into Lake Stechlin. The actual catchment area amounts to nearly 7 km^2. The actual water level of the lake is 60.1 m a. s. l. Bordered by the settlements, gardens, and parks of Neuglobsow and the former village Dagow, and by small woods, it has virtually no steep shores. A small island is the only conspicuous element of the well-rounded kettlehole-like basin with a shore development of 1.4 (Table 2.1).

Morphometric data (Table 2.1), the bathymetric map (Fig. 2.11), and hypsometric curves and depth-volume relations (Fig. 2.7) were obtained from our own measurements and calculations. The shore line is 2.5 km long; the small value of shore line development reflects the above-mentioned regular shore line of the lakes, i. e. there are few bights and no bays. The volume was calculated to $11.89 \cdot 10^5$ m^3 and the mean depth to 4.95 m.

Lake Dagow is mainly influenced by the waste waters of Neuglobsow and Dagow.

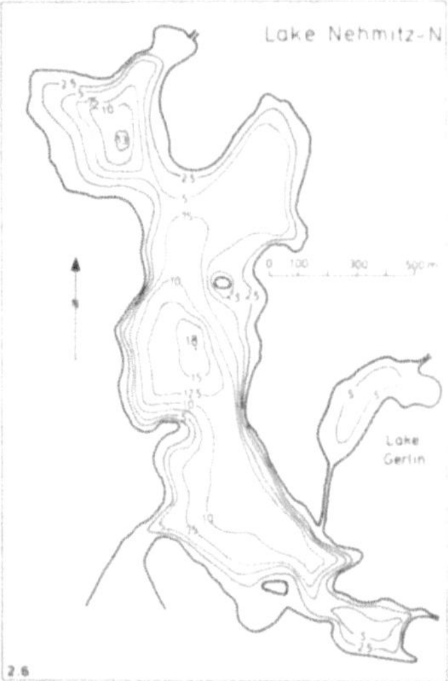

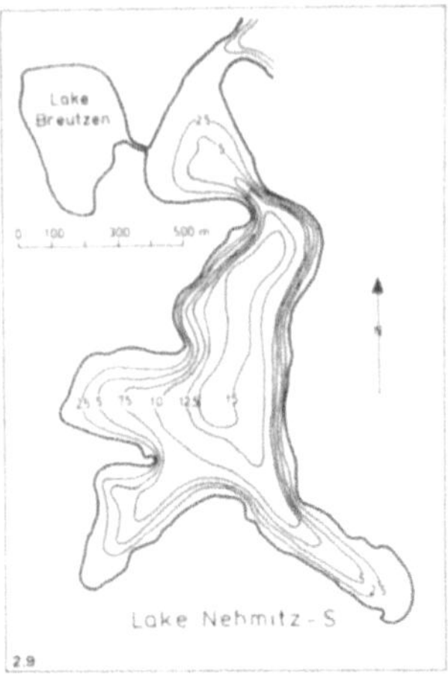

Figure 2.6 Lake Nehmitz, north basin, bathymetric map.
Figure 2.9 Lake Nehmitz, south basin, bathymetric map.

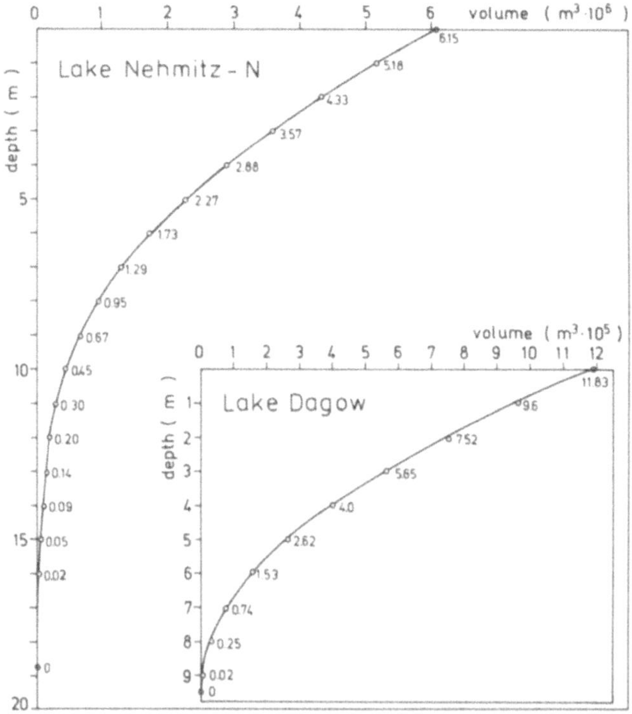

Figure 2.7 Lake Nehmitz and Lake Dagow, hypsometric curves and water volumes.

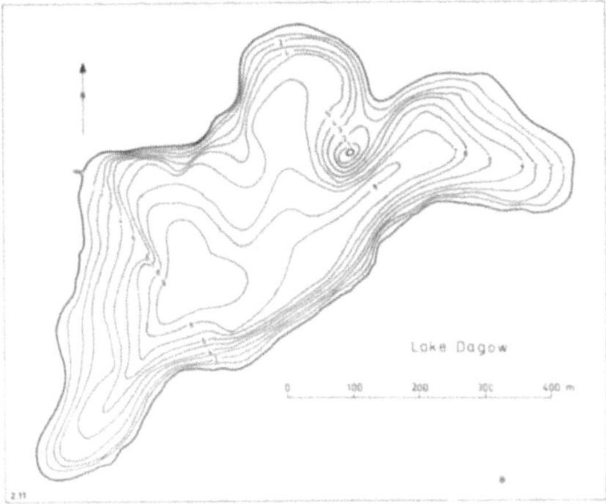

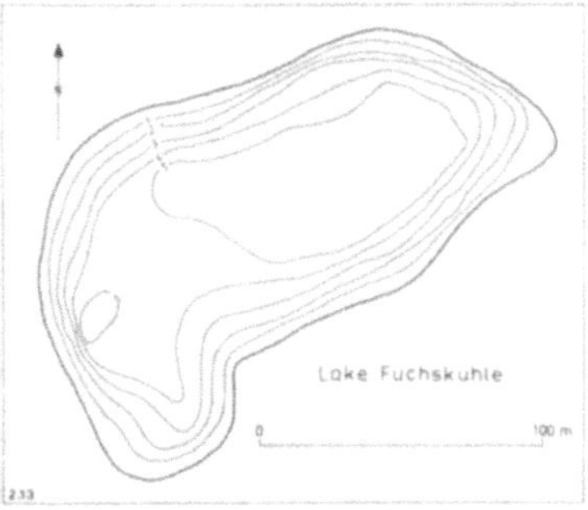

Figure 2.11 Lake Dagow, bathymetric map.
Figure 2.13 Lake Große Fuchskuhle, bathymetric map.

Lake Große Fuchskuhle

Lake Große Fuchskuhle (Fig. 2.12) is situated 1 km to the south of Lake Nehmitz far away from any habitations and sunk deeply into its woody environs. It is the residual lake of a *Sphagnum*-bog. It has an area of 15 000 m^2, a maximum depth of 5.5 m, and is stratified in summer, exhibiting hypolimnetic oxygen depletion. Its water is clear. Surface in- or effluents are absent. The actual catchment area amounts to 0.25 km^2. The actual water level of the lake is 59 m a. s. l. It is bordered by dense Pine-forests.

Morphometric data (Table 2.1), the bathymetric map (Fig. 2.13), and hypsometric curves and depth-volume relations (Fig. 2.4) were obtained from own measurements and calculations. The shore line is 0.5 km long; the small value of shore line development (1.19) reflects the regular shore line of the lake, i. e. it remembers a well-rounded kettlehole-like basin. The volume was calculated to 53 000 m^3 and the mean depth to 3.5 m.

Lake Großer Barsch

Lake Großer Barsch (Fig. 2.14) lies not far from the south point of Lake Nehmitz, within an undrained hollow of the ground moraine. Like Lake Große Fuchskuhle, it is the residual lake of a *Sphagnum*-bog (see Fig. 2.12). Its acid water is clear.

Morphometric data are not available.

Figure 2.8 Lake Nehmitz. View over the south basin to the north.
Photo W. Scheffler.

Figure 2.10 Lake Dagow. View over the lake to the west. Note the small island.
Photo W. Scheffler.

Figure 2.12 Lake Große Fuchskuhle. View over the lake to the northern shore. Photo W. Scheffler.

Figure 2.14 Lake Kleiner Barsch. View over the lake. Note the narrow *Sphagnum*-belts. Photo W. Scheffler.

References

Autorenkollektiv. 1962. Atlas der Bezirke Rostock, Schwerin und Neubrandenburg. Schwerin: Topographischer Dienst.

Flößner, D. 1964. Zur Cladocerenfauna des Stechlinsee-Gebietes II. Limnologica 2: 35–103.

Håkanson, L. 1981. A manual of lake morphometry. Berlin-Heidelberg-New York: Springer-Verlag.

Heitmann, M.-L. & Schubert, H. 1965. Lake Stechlin – A hydrometeorological research area. Budapest: Publ. IASH 66: 429–440.

Heitmann, M.-L., Richter, D. & Schumann, D. 1969. Der Wärme- und Wasserhaushalt des Stechlin- und Nehmitzsees. Abh. Meteorol. Dienst. DDR 96. Berlin: Akademie-Verlag.

Krausch, H.-D. 1962. Nachrichten über Abfluß- und Seespiegelveränderungen des Stechlinsee-Gebietes. Limnologica 1: 211–215.

Krausch, H.-D. 1964. Die Pflanzengesellschaften des Stechlinsee-Gebietes. I. Die Gesellschaften des offenen Wassers. Limnologica 2: 145–203.

Krausch, H.-D. 1970. Die Pflanzengesellschaften des Stechlinsee-Gebietes. V. Wälder, Hecken und Saumgesellschaften. Limnologica 7: 397–454.

Krausch, H.-D. & Zühlke, D. 1974. Das Rheinsberg-Fürstenberger Seengebiet (Werte unserer Heimat Bd. 25). Berlin: Akademie-Verlag.

Marcinek, J. 1981. Kleiner Exkursionsführer für das Rheinsberger Becken, Leipzig (manuscript).

Marcinek, J. & Nitz, B. 1973. Das Tiefland der DDR. Gotha/Leipzig: VEB Haack.

3. Hydrometeorology of the Lake Stechlin area

D. Richter & R. Koschel

Introduction

This chapter contains a detailed description of the hydrometerological conditions of Lake Stechlin and its catchment area. The description covers the climate in the lake's region, the water temperature, heat balance, amount of evaporation and light penetration of Lake Stechlin, and the water balance of the catchment area. Of particular interest are the changes caused since 1967 due to the operation of a nuclear power plant affecting the water temperature and heat budget of the lake. The results of long-term investigations are primarily suggested for use as background information to the wide-ranging ecological processes within the lake.

Climatic conditions (D. Richter)

General climatic conditions in Lake Stechlin catchment area

The macro-climate of the Lake Stechlin area belongs to the humid-moderate climates of Europa (Köppen 1931). Within the GDR classification, the region is allocated to the Mecklenburg-Brandenburg transitional climatic zone (Klima-atlas 1953). The region is characterized by the prevailing maritime influence with moderately warm summer seasons and relatively mild winter seasons. These general conditions are even better known due to the meteorological station in the immediate vicinity of Lake Stechlin. The measuring site selected for the meteorological measurements is located in a forest clearing of about 300 · 100 m near the southern shore of the lake (Fig. 3.1), at a distance of 40 m from the shore line. Since on the part of the shore zone of the clearing there are also trees and brushwood growing, the measuring site is rather well protected from every direction.

The meteorological data obtained there are representative for the prevailingly wooded region abundant with further numerous minor lakes. As compared with

Casper, S. Jost (ed.), Lake Stechlin. A temperate oligotrophic lake. ISBN 978-94-010-8930-2.

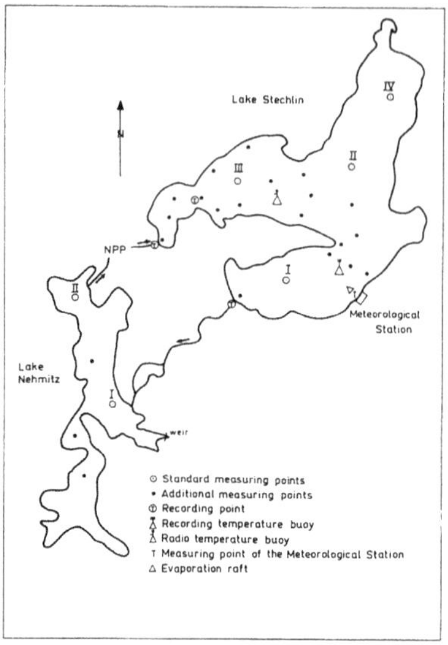

Figure 3.1 Water temperature measuring points of Lake Stechlin and Lake Nehmitz.

the neighbouring meteorological stations outside this region there are, especially in low-wind weather, higher values of humidity and generally lower air temperatures. Some of these differences can be recognized even in the long-term averages. Comparative measurements made on Lake Stechlin from June to November 1963 did not reveal any substantial differences from the values obtained at the meteorological measuring site.

At the Neuglobsow meteorological station, measurements have been made since 1958. The description of the climatic conditions is therefore based on the 23-year period 1958–1980. From a comparison of the meteorological stations Neustrelitz, Zehdenick and Neuruppin – all of them at an average distance of about 25 km – the norm values have been determined of air temperature, vapour pressure, sunshine duration and cloud cover for the period 1951–1980. In addition, air temperature and sunshine duration have also been reduced to the long-term averages of 1901–1980, for which the Potsdam and Waren series were used. These stations are about 80 km to the south and 50 km to the north of Lake Stechlin.

This background material is sufficient for the description of the climate prevailing in the Lake Stechlin area. Average and extreme values of a selection of meteorological parameters are compiled in Tables 3.1 and 3.2. Unless the observational period has been otherwise specified, the data refer to the period 1958–1980. Statements on precipitation are given when describing the water budget (see p. 70).

Air temperature

Detailed information on air temperature is contained in Table 3.1. This meteorological element is quite closely related to the water temperature and, therefore, also responsible for the radiation and heat balance of a lake. The long-term average (1901–1980) of the air temperature at Lake Stechlin is 7.9° C. The warmest month is July, with an average of 17.3° C, whereas January with an average of −0.9° C is the coldest month. Apart from January, only the February temperature is negative. The highest values of monthly means vary between 3.7° C in February and 19.8° C in August; the lowest are between −11.7° C, also in February, and 14.3° C in July. These data illustrate the quite well-balanced climatic conditions.

A comparison of the average values of 1901–1950 with the norm recommended by the World Meteorological Organization for 1951–1980 does not reveal any differences from the annual mean values. However, the annual variation shows some changes, i.e., the monthly mean values of the 1951–1980 series from January to May are permanently below and from August to December permanently above those of the 1901–1950 period. On account of the higher air temperatures in June and August and the lower July values, the summer seasons of the past 30 years have in general been somewhat more balanced.

Air temperatures above 30° C are on the average to be expected only on 4 days per annum, i.e., usually between June and August. A maximum of 25° C normally occurs on 28 days between April and October, i.e., mainly between June and August. The highest air temperature measured has been 36.9° C. On the average, air temperatures below 0° C occur between October and May, with July and August, so far, having been the only frost-free months. However, between December and March temperatures below −10° C may well occur, in a few rare cases even below −20° C. The hitherto lowest measured value was −25.0° C. The most days with frost occur in January and February, i.e., more than 20 days, but even in December and March frost should be expected for more than half of the month. Days with a maximum air temperature below 0° C occur from November to March, most frequently in January, with more than 10 days on the average. The 1962/63 winter was extremely cold, with 63 days having a temperature below zero, whereas in the rather mild winter of 1974/75 this occurred on only one day.

Atmospheric humidity

For a description of atmospheric humidity the data of vapour pressure, relative humidity and of the saturation deficit can be used. The average annual course of all three is reflected in Table 3.2.

The monthly average values of vapour pressure are between 5.3 hPa in January and February and 14.8 hPa in July. The deviations of the individual months are quite small, i.e., only 1–2 hPa. The variability of vapour pressure is, therefore,

Table 3.1 Mean and extreme values of the air temperature in °C at the Neuglobsow meteorological station.

	Jan	Feb	Mar	Apr	May	Jun	Jul	Aug	Sep	Oct	Nov	Dec	Year
Average 1901–1980	–0.9	–0.5	2.4	6.7	12.1	15.6	17.3	16.2	12.7	8.2	3.6	0.6	7.9
Highest average 1901–1980	4.9	3.7	6.8	9.8	15.3	19.2	19.4	19.8	16.3	11.2	7.2	5.0	9.6
Lowest average 1901–1980	–10.1	–11.7	–1.8	3.5	8.5	11.0	14.3	13.4	8.9	4.2	–1.1	–6.4	5.8
Average 1901–1950	–0.8	–0.4	2.6	6.8	12.2	15.4	17.4	16.1	12.6	8.0	3.4	0.5	7.9
Average 1951–1980	–1.2	–0.8	2.2	6.5	11.9	16.0	17.1	16.4	12.7	8.5	4.0	0.8	7.9
Mean maximum	6.9	9.7	15.1	21.0	26.3	29.8	30.3	30.1	25.9	21.0	13.3	9.4	32.0
Highest maximum	11.1	15.6	23.5	30.4	31.5	33.2	36.9	34.2	29.6	26.4	19.6	15.0	–
Mean minimum	–13.8	–12.7	–8.9	–3.6	–0.6	3.9	6.5	6.1	2.3	–1.1	–5.1	–11.0	–16.3
Lowest minimum	–25.0	–21.7	–17.7	–6.6	–4.8	–0.1	4.2	2.6	–0.3	–5.3	–14.6	–22.0	–
Mean number of days with $t_{max} \geq 25°C$	–	–	–	0.3	2.2	7.6	8.5	7.4	1.8	0.0	–	–	27.8
Mean number of days with $t_{max} \geq 30°C$	–	–	–	0.0	0.1	0.9	1.7	1.3	–	–	–	–	4.0
Mean number of days with $t_{min} < 0°C$	22.1	20.2	16.2	8.5	1.4	0.0	–	–	0.0	2.9	8.4	17.8	97.5
Mean number of days with $t_{max} < 0°C$	10.3	6.7	1.8	–	–	–	–	–	–	–	1.1	7.2	27.1

Table 3.2 Mean monthly and annual values of meteorological elements at the Neuglobsow meteorological station.

Element	Dimension	Jan	Feb	Mar	Apr	May	Jun	Jul	Aug	Sep	Oct	Nov	Dec	Year
S (80)	h/d	1.5	2.4	4.0	5.5	7.5	7.8	7.2	6.7	5.5	3.4	1.6	1.2	4.5
S (max)	h/d	3.0	4.0	6.4	8.4	10.6	11.0	10.1	9.3	8.2	7.1	3.0	2.0	5.6
S (min)	h/d	0.3	0.9	1.9	2.8	4.0	3.2	3.7	3.5	2.7	1.2	0.6	0.4	3.7
S (50)	h/d	1.6	2.4	3.9	5.4	7.6	7.7	7.1	6.7	5.4	3.3	1.7	1.3	4.5
S (30)	h/d	1.4	2.5	4.1	5.5	7.2	8.0	7.2	6.8	5.6	3.4	1.5	1.1	4.5
e_a (30)	hPa	5.3	5.3	5.9	7.3	9.9	13.0	14.8	14.6	12.3	9.8	7.4	6.1	9.3
C (30)	Tenths	7.7	7.3	6.6	6.4	6.2	5.9	6.3	6.1	5.9	6.7	7.7	7.9	6.7
RH	%	89.0	86.0	81.0	76.0	73.0	73.0	77.0	79.0	81.0	87.0	89.0	90.0	82.0
Δe_a	hPa	0.6	0.9	1.8	2.9	4.4	6.1	5.5	4.9	3.0	1.7	1.1	0.7	2.8
w_2	m/s	2.2	2.2	2.3	2.2	2.3	2.5	2.5	2.1	2.1	2.4	2.4	2.3	2.3

S (80) Average of sunshine duration 1901–1980.
S (max) Highest average of sunshine duration 1901–1980.
S (min) Lowest average of sunshine duration 1901–1980.
S (50) Average of sunshine duration 1901–1950.
S (30) Average of sunshine duration 1951–1980.
e_a (30) Mean vapour pressure 1951–1980.
C (30) Mean total cloud cover 1951–1980.
w_2 Mean wind speed.
h/d Hours per day.
hPa Hectopascal.
m/s Meters per second.
RH Mean relative humidity.
Δe_a Mean saturation deficit.

much less than that of the air temperature. The hitherto lowest and highest monthly mean values have been 3.1 and 16.8 hPa.

The annual variation of the relative humidity, i.e., between 90 percent in December and 73 percent in May, is still more balanced. Maximum and minimum values show, as compared with vapour pressure, a contrary pattern to the seasons, each occurring one month earlier. Even in the individual months the deviation from the long-term averages is merely 5 percent in winter and 10 percent in summer. The recorded extreme values of monthly averages are 94 percent and 62 percent.

The saturation deficit is a rather significant element for hydrometeorology. It is most appropriate as an indicator of the atmospheric dryness index, and there is under the prevailing climatic conditions a rather close relation to the potential amount of evaporation. With its saturation deficit of 6.1 hPa, June is relatively the driest month, whereas maximum saturation occurs in January with 0.6 hPa deficit. The highest and lowest individual monthly averages are 8.5 and 0.3 hPa. The dispersion is between 0.5 and 1.5 hPa.

Sunshine duration and total cloud cover

Due to its close relation to global radiation, the knowledge about sunshine duration is very important for the study of the radiation balance of a water body. Because measurements of sunshine duration are usually available for much longer periods of time, investigation can be extended over rather long periods. A comparison of the periods 1901–1950 and 1951–1980, on the average, does not reveal any differences; and the annual figures show but minor changes. The normal values of 1951–1980 are in the summer season somewhat higher, May excepted, whereas in the period November to January the sunshine duration is shorter than for 1901–1950. The sunshine duration, with mean values between 1.4 per day in December and 8.0 per day in June, is, therefore, respresentative even for long-term considerations. The lowest and the highest monthly average of 0.3 and 11.0 per day occurs also in these two months.

The mean total cloud cover is inversely proportional to sunshine duration. Its variability, however, is lower, December, with its 7.9 tenths is the month with the greatest cover, whereas the cover is lowest in June and September, with 5.9 tenths.

Wind speed

Measurements of wind speed are made at the meteorological station on an 18 m mast which exceeds the surrounding forest, and at a height of 2 m above the lake's surface. At the latter position the measurements were made by a contact anemometer, which was mounted on a raft that was at the same time used for evaporation measurements with floating evaporation pans (see p. 64). Wind speeds measured at the height of 2 m above the lake are shown in Table 3.2. Temporary gaps of measurements, and interruptions of measurements in winter, are complemented on the basis of comparisons with standard measurements in the height of 18 m. The wind speed does not reveal any typical annual variation, and the monthly averages of 2.1–2.5 $m \cdot s^{-1}$ deviate rather little from the annual average. Their hydrometeorological significance is primarily found in the application of daily values to the development of evaporation formulas and for the calculation of the amount of evaporation in case of measuring failures.

Water temperature (D. Richter)

Survey of the water temperature measurements

Thermal conditions are of primary importance to the heat budget and to the ecological system of a lake. Quite a number of relevant measuring programmes were carried out at Lake Stechlin, which include regular long-term routine investigations and other special investigations, the latter mainly serving the determination of the thermal pollution originating from the heated water coming from the nuclear power plant. A great member of results of all these investigations are described by Richter *et al.* (1979). A survey of the position of the various temperature measuring points is given in Fig. 3.1. Lake Nehmitz was also included into the investigations. The measurements of the vertical temperature distribution were initiated in 1958 at four measuring points on Lake Stechlin (I . . . IV) and at two measuring points on Lake Nehmitz (I, II). They were made from the surface down to a depth of 20 m at distances of 2 m, in Lake Stechlin even lower with distances of 10 m down to the bottom of the lake. Up to 1970 measurements were made every five days, and from 1970–1976 three times in a month. From 1967–1976 additional measuring points were operated serving the detailed investigation of the thermal influence, i.e., 19 additional points on Lake Stechlin and one more on Lake Nehmitz. In 1971 two extra measuring points were installed in the south basin of Lake Nehmitz where there is no thermal influence. After those special measurements, there have been only monthly checks of the vertical temperature distribution at the 4 standard measuring points of Lake Stechlin and at the 5 points in both basins of Lake Nehmitz.

Gaps in the measuring programme have occurred only in winter when the ice cover was not firm enough. In the beginning tilting thermometers were used, after that electrical temperature probes (Knoll 1976).

Since 1961 daily measurements of the water temperature at a depth of 30 cm have been made from a landing-stage extending 20 m into the lake nearby the meteorological station. This special measuring point is kept ice-free in winter.

For the investigation of a number of special problems, the following temperature recording systems were operating temporarily for several years:

- A recording temperature buoy in the south bay of Lake Stechlin with 12 measuring points arranged between 2 m above the lake's surface and down to 20 m below the water surface. Data transmission was made by electric cable to the station 400 m away, always in the summer months from 1964–1976.
- A radio temperature buoy, also with 12 measuring points, from which data were transmitted by radio signals to the meteorological station. This buoy is moored in the west bay of Lake Stechlin, just where – with the incoming warm water – average values of the water surface temperature are hardly representative for the total surface of the lake.
- Two temperature-recording points each with a temperature sensor, one in the inflow canal to measure the heated water coming from the nuclear power plant and one at the outflow to Lake Nehmitz.

Temperature of the water surface

Mean annual course. Figure 3.2 shows the average annual course of the prevailing temperature at the water surface under natural conditions (1958–1966) and under thermal discharge (1967–1976). For the calculation of the thermally influenced surface temperature there are included the weighted mean temperatures measured at each of the 23 measuring points and their systematic deviations from the temperature values obtained at the landing-stage near by the meteorological station.

During the winter months the percentage of ice and snow cover was also taken into account (Richter 1977). Under natural conditions the maximum water temperature of Lake Stechlin is, on the average, 18.9° C in July. In the winter months, January and February, the values are below zero, meaning that partial or complete ice cover may well be expected.

Since the beginning of the operation of the nuclear power plant the water temperature of the Lake Stechlin has been higher throughout the year. The increases are particularly remarkable in January/February and in July/August.

Distribution of the incoming heated water at the surface. The influence originating from the nuclear power plant may best be recognized from the distribution of the temperature of the water surface. For an analysis of this distribution, a total of 48 temperature measurements were made at more than 100 points in Lake Stechlin under different weather conditions. For these measurements a highly sensitive temperature probe was operated from a moving boat, along predetermined profiles. The isolines reveal the effective temperature increase relative to the natural water surface temperature.

The distribution of temperature depends primarily upon the prevailing wind

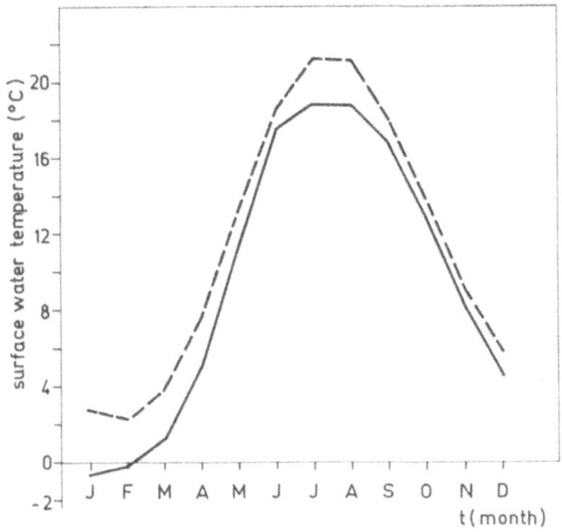

Figure 3.2 Mean annual course of the surface water temperature of Lake Stechlin before (——) and after (----) the thermal discharge.

conditions (Richter *et al.*, 1979). With westerly winds, the warm-water flow is found to be somewhat nearer to the northern shore of the peninsula than can be seen in Fig. 3.3. However, with easterly winds, the warm water is retained more within the west bay. The water temperature isolines then run in a diagonal direction through the bay.

Under strong-wind conditions the heating effect is caused by the prevailing,

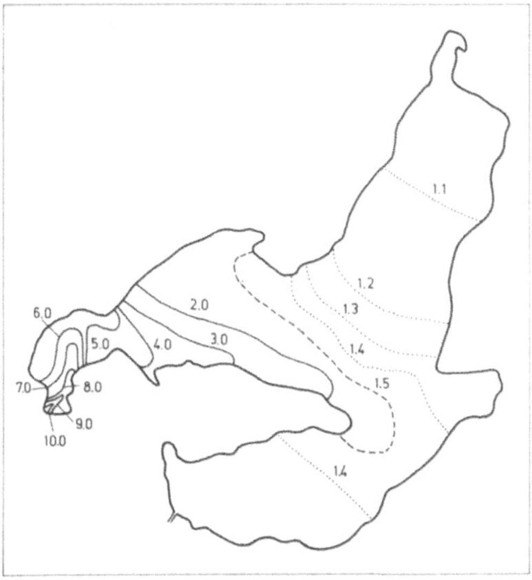

Figure 3.3 Mean increase of the surface water temperature by thermal discharge (in °C).

rather early, mixing processes well recognized only in the west bay, while, with only light wind prevailing, the warm-water flow can be tracked far into the north and the south bay. From measurements of the vertical temperature distribution we can see that even at a depth of 2 m there is a still well recognizable horizontal temperature distribution. The increase of the temperature about 4 K at the inflow and 1.4 K on the average is, however, much smaller than at the surface. At a depth of 4 m a direct influence from the warm-water is not present any more.

Cooling circuit and cooling process. The flow of the cooling water is shown in Fig. 3.3. The cooling water is, when heated up by the nuclear power plant, fed into the west bay of Lake Stechlin, from where it usually passes by the peninsula up to the south bay of Lake Stechlin and then via a canal through Lake Gerlin to Lake Nehmitz (Fig. 3.1). The flow of the cooling water inside Lake Stechlin is predominantly influenced by the wind. Inside Lake Nehmitz, the flow turns to the north. In the northernmost bay the cooled water is again taken by the nuclear power plant. There is no exchange of water with the south basin of the Lake Nehmitz, since the existing connection is merely rather shallow and narrow and, furthermore, in the summer month, quite closed by growing water plants (see p. 34).

The heated water is cooled mainly in Lake Stechlin. The basis for the calculation of the cooling process is the equation of the additional heat balance

$$Q_{npp} = \Delta Q_A + \Delta Q_E + \Delta Q_H + \Delta Q_{\Delta W} \tag{1}$$

which includes all those parameters immediately affected by the additional heat (Q_{npp}), i. e., upward terrestial radiation (ΔQ_A), evaporation (ΔQ_E), flux sensible heat (ΔQ_H) and heat storage in the water body ($\Delta Q_{\Delta W}$). Assuming stationary conditions and, furthermore, that the temperature of the incoming water (t_{WA}) is equal to the temperature (t_{WO}) at the beginning of the cooling process (cooling area F = O), the cooling equation reads

$$(t_{WO} - t^n_{WO}) = (t_{WA} - t^n_{WO}) \exp \left\{ \frac{K_L \cdot F}{c \cdot m \cdot \varrho} \right\} \tag{2}$$

where t^n_{WO} = natural or equilibrium temperature, and K_L = coefficient of the additional heat loss (Kuhn 1972; Günneberg 1973)

$$K_L = \frac{dQ_A}{dt_{WO}} + \frac{dQ_L}{dt_{WO}} + \frac{dQ_H}{dt_{WO}} \; . \tag{3}$$

For the calculation of K_L the difference quotient of equation (3) with temperature intervals of $\Delta t_{WO} = 1\,K$ may be used. The following relations can be applied to the Lake Stechlin conditions to calculate the daily loss of heat ($J \cdot cm^{-2}$):

$$Q_A = 0.94 \sigma t_{WO}{}^4 \tag{4}$$

$$Q_E = (39.5 + 49.4 v_2{}^{0.5}) (E_{(tWO)} - e_L) \tag{5}$$

$$Q_H = (24.1 + 30.1 v_2^{0,5})\,(t_{WO} - t_L) \quad (6)$$

where σ = Stefan-Boltzmann constant,
v_2 = wind speed at a height of 2 m ($m \cdot s^{-1}$),
$E(t_{WO}) - e_L$ = vapour pressure gradient (hPa) and
$t_{WO} - t_L$ = temperature gradient (K) between the water surface and a height of 2 m.

A comparison between the theoretical cooling values and the measured cooling values shows the actual cooling amount to be much faster at the beginning and slower than the theoretical curve later. The differences between the calculated and the measured cooling may be explained by the mixing of the incoming heated water with the cooler water of Lake Stechlin. Such a fact is, generally known as initial mixing (World Meteorological Organization, 1972). In Lake Stechlin, this kind of mixing is observed mainly at the first 0.06 km^2 of the overlaid area, where the cooling of the incoming heated water is about 4 K, although the heat supply into the atmosphere is calculated to be only 1.2–1.6 K. After 0.3–0.4 km^2, the cooling amount is smaller than would be possible in the absence of mixing. After 1.5–2 km^2 have been overlaid, the heated water obtains a temperature of only about 2 K above the natural temperature of Lake Stechlin. The difference in density of water becomes so small that, with the effect of wind, a complete mixing occurs. Therefore, in the summer the epilimnion and in winter the Lake Stechlin basin as a whole is heated by 1 K on the average.

The water then leaves Lake Stechlin with this slightly increased temperature, although if no mixing takes place the area of 4.25 km^2 is quite sufficient for a complete cooling. The effect upon Lake Nehmitz is much smaller but still recognizable.

Vertical distribution of the water temperature

The vertical distribution of the water temperature was measured under natural conditions in five-day cycles. The resulting monthly averages for 1958–1965 are contained in Table 3.3. A detailed description of these results is given by Heitmann *et al.* (1969).

On average the lowest water temperatures are generally measured in February, where, in accordance with the density of the water, the absolute minimum is observed at the surface.

Maximum temperatures occur in August in the epilimnion between depths of 0 and 8 m. In the layer of 10–14 m the highest values are measured in September or October, whereas in the layer below the highest temperatures are attained in November.

The annual development of the vertical temperature distribution in the upper 20 m layer of Lake Stechlin is shown in Fig. 3.4. From December until April the

Table 3.3 Monthly averages of the vertical temperature distribution in Lake Stechlin 1958–1965 (Heitmann *et al.* 1969).

Depth (m)	Jan	Feb	Mar	Apr	May	Jun	Jul	Aug	Sep	Oct	Nov	Dec
0	2.83	1.87	2.37	4.97	10.35	16.86	18.69	18.76	16.89	12.82	8.49	5.03
2	2.98	2.10	2.47	4.83	10.14	16.56	18.45	18.64	16.78	12.81	8.49	5.07
4	3.09	2.25	2.53	4.70	9.80	16.00	18.10	18.51	16.71	12.80	8.49	5.11
6	3.18	2.33	2.57	4.60	9.19	14.23	17.35	18.25	16.63	12.79	8.49	5.13
8	3.24	2.40	2.59	4.48	8.23	11.53	14.63	17.06	16.47	12.77	8.49	5.15
10	3.30	2.48	2.62	4.37	7.20	9.19	10.71	13.03	14.67	12.70	8.48	5.16
12	3.34	2.55	2.65	4.26	6.43	7.57	8.30	9.32	10.62	11.89	8.47	5.18
14	3.38	2.61	2.68	4.17	5.89	6.63	7.00	7.51	7.82	8.83	8.40	5.19
16	3.43	2.68	2.71	4.10	5.56	6.10	6.35	6.65	6.76	7.03	7.79	5.19
18	3.46	2.74	2.74	4.05	5.37	5.79	6.02	6.22	6.30	6.41	6.97	5.20
20	3.50	2.79	2.77	4.00	5.23	5.59	5.79	5.93	5.97	6.07	6.47	5.20
30	3.61	2.97	2.88	3.86	4.71	4.93	5.05	5.13	5.13	5.15	5.22	5.04
40	3.66	3.15	2.94	3.76	4.52	4.69	4.78	4.86	4.86	4.87	4.91	4.75

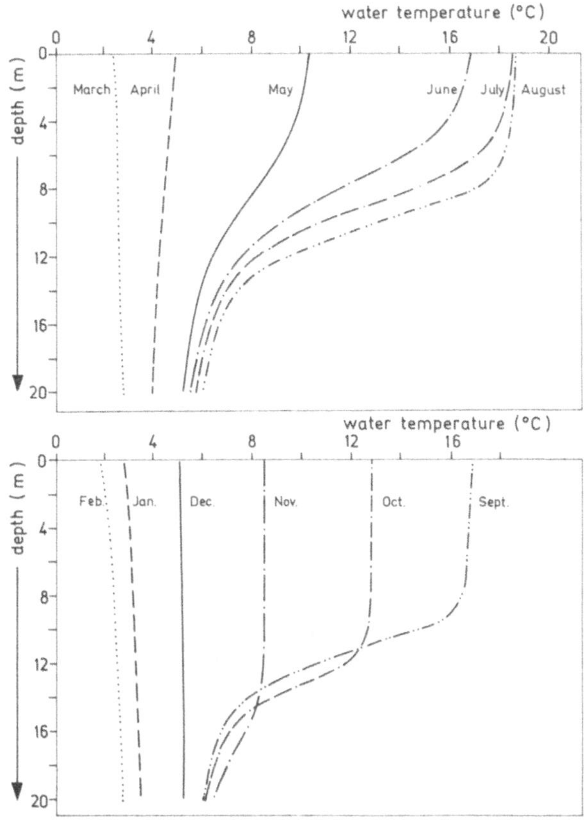

Figure 3.4 Monthly means of the vertical temperature distribution in Lake Stechlin, 1958–1965.

temperature differences between the surface and the bottom of the lake are only rather smal. May and November are transitional months with respect to the development and the destruction of a temperature stratification. From June to October a stable stratification prevails, with nearly isothermal conditions in the epilimnion. In the thermocline the temperature decreases rapidly, whereas in the hypolimnion there is only a minor temperature decrease. Since the operation of the nuclear power plant, the temperatures have been increased at all depth levels. As an example, in Fig. 3.5 is illustrated the average vertical temperature distribution which was measured on 1 August before and after the thermal discharge at the measuring point III. At that time, the temperature stratification is most stable. The maximum heating occurs in the epilimnion, but in the layer below the heating effect is still well recognized, and is also evident from the average annual course of the water temperature in the hypolimnion at a depth of 20 m (Fig. 3.6). Even at this depth the temperature is higher by 0.5 to 1.3 K as compared with the natural conditions. The lowest heating rate occurs from June to October when there is a thermocline.

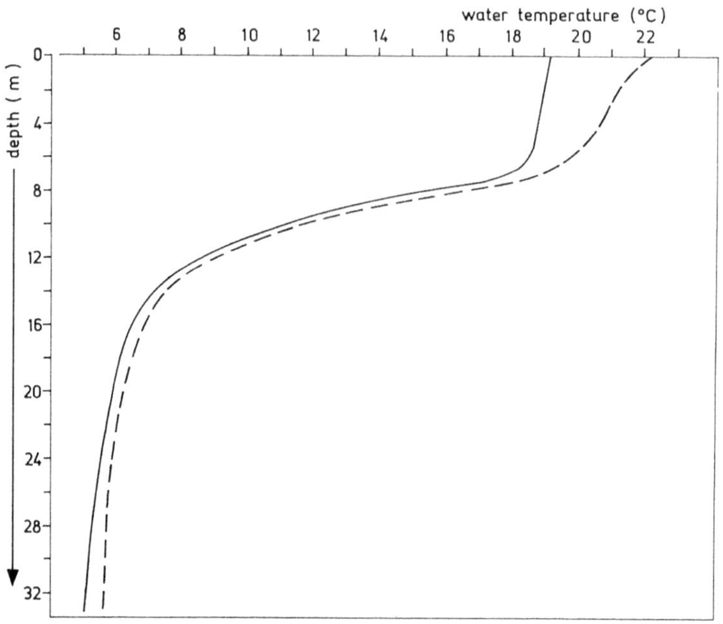

Figure 3.5 Mean vertical temperature distribution at August 1st before (——) and after (----) the thermal discharge, measuring point III.

Figure 3.7 is a comparison between the vertical temperature distributions of some lakes in the Lake Stechlin region. It presents the results of special measurements at the beginning of August 1982, with a maximum of stability of the temperature stratification in the lakes. The limit of the thermocline depends on the mixing effect of the wind-spead, whereas the depth of epilimnion is increasing

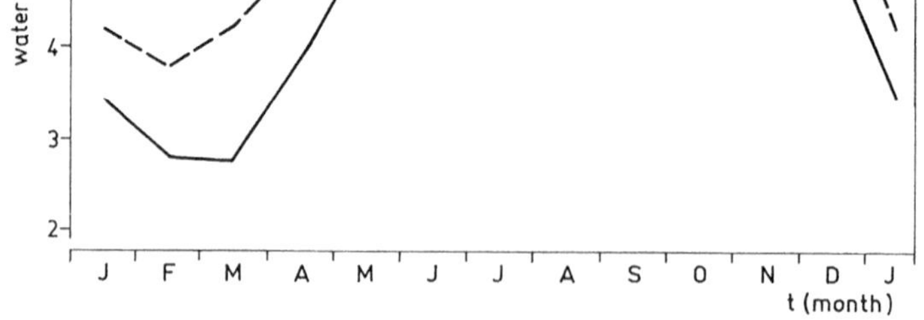

Figure 3.6 Mean annual course of the water temperature in a depth of 20 m (in °C; symbols as in Fig. 3.5).

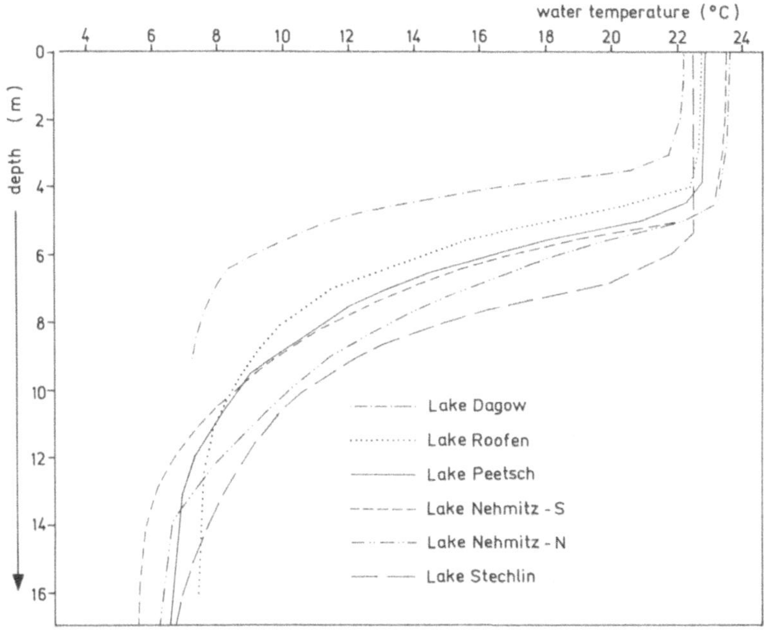

Figure 3.7 Vertical temperature distribution in different lakes of the Lake Stechlin area. August 9–13, 1982.

with the extent of lake surface. The different temperatures of the epilimnion are caused by the weather development during the period of measurement.

Circulation, stagnation and the development of the thermocline

Lake Stechlin is one of those lakes that regularly have a vertical temperature stratification every year. Following Hutchinson & Löffler (1956), it belongs to the type of dimictic lakes. In spring and autumn there develops a full-circulation period during which equal temperatures prevail between the surface and the bottom of the lake.

These unstable conditions are responsible for the complete mixing inside the lake caused by wind effects during those periods. Under natural conditions this period continues in spring from mid-March to mid-May. The density maximum at a water temperature of 4.2° C is, on the average, measured in early April. Along with an increasing heating, a thermal stratification is gradually developed with the character of a stable thermocline ($\Delta t_{w_z} \geqq 1$ K) in the second half of May. Then the phase of summer stagnation begins, with its maximum stability occurring from late July to early August. From mid-August the destruction of the thermal stratification begins along with the thermocline continuously moving downward. From November the period of autumnal circulation begins. The maximum density occurs in late December. From January to early March there is the winter stagnation phase. If a wind-caused mixing is prevented due to an ice cover, rather long-lasting inverse temperature stratification may occur during that time.

Since the operation of the nuclear power plant, some distinct changes of the vertical temperature distribution in Lake Stechlin have been observed. While under natural conditions the formation of the thermocline begins around mid-May at a depth of 8–10 m, under thermal load it certainly forms in late April at the much shallower depth of 2–4 m.

Throughout the summer stagnation phase a somewhat higher position of the thermocline continues. Due to the additional heating of the epilimnion and the somewhat more upward movement of the lower limit of the thermocline, the temperature gradients of the thermocline are, therefore, generally greater. While under natural conditions the thermocline temperature decreases only by 1.5–2.0 $K \cdot m^{-1}$, the corresponding values are, under thermal load, between 2.0 and 2.5 $K \cdot m^{-1}$, on the average. The maximum gradients are also remarkably greater.

Before the operation of the nuclear power plant, the measured maximum value was 4.2 $K \cdot m^{-1}$, while after it gradients of 7–8 $K \cdot m^{-1}$ occurred rather frequently. The destruction of the thermocline followed by the autumnal full-circulation is, under thermal discharge, to be expected in mid-November. The temperature then decreases until late January down to about 4° C, in the process of which the autumnal full-circulation period practically passes into the spring full circulation, whereas there is no characteristic period of winter stagnation. Only during rather cold winter seasons, when a complete ice cover forms in spite of the supply of warm water, an inverse temperature stratification may still occur in February – restricted, however, to a rather short period. The annual variation and the average level of the thermocline before and after the thermal discharge into Lake Stechlin are reflected in Fig. 3.8.

An example of the thermocline development in the years 1975 and 1976 is also given in this figure. Both of these years are characterized by relatively warm and sunny summer months, but 1975 was without thermal discharge. Differences in the level of the thermocline are in general agreement with the mean courses.

The depths where the vertical temperature decrease is equal to 1 K are indicated as the thermocline.

Heat balance components (D. Richter)

Radiation balance

A substantial part of the heat exchange at the water surface is caused by the radiation balance, which can be determined nearly exactly by measurements. For deep lakes, however, difficulties arise in regard to the fixed installation of a radiation balance meter. Auxiliary support, such as a raft or buoy, has to be used

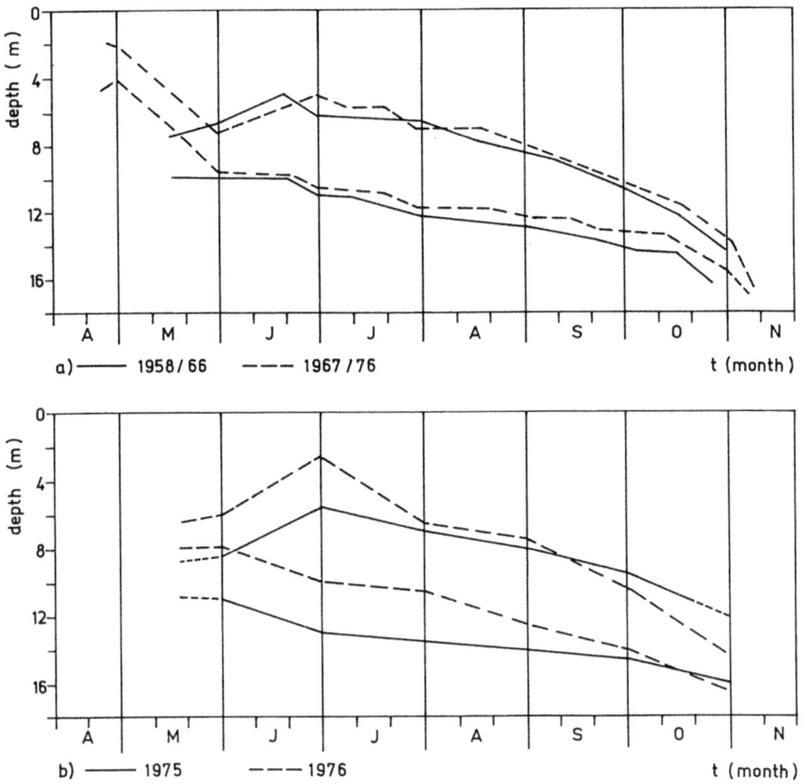

Figure 3.8 Mean annual development of the thermocline before (——) and after (----) the thermal discharge (a) and annual development of the thermocline in 1975 and 1976 (b).

to mount the instrument, which then entails oscillations of the carrier system or influences upon the radiation balance. At Lake Stechlin these potential sources of error have, at last, successfully been removed after the construction of a subsurface raft to be used as the carrier system. A triangular raft, each side 4 m long, is quite firmly moored at a depth of 2 m below the water surface. Only a thin mast, painted black, to which the radiation balance meter and a downward directed pyranometer for the measurement of the reflected global radiation are mounted, rises above the water surface. This measuring system is shown in Fig. 3.9. Above the radiation balance meter (Sonntag 1963), 0.15 mm fishing lines are used as a protection from seagulls. The measurement of the global radiation has been made since 1971 on land at a mast of the meteorological station about 100 m away from the position of the radiation balance meter.

Hence, for the purpose of balancing the radiation components back to 1958, calculated values had to be used. For the global radiation a relation had to be derived to the available long-term measuring series of the meteorological station at Potsdam. The following relation given by Angström (1915)

Figure 3.9 Sub-surface radiation balance measurement system.
Photo W. Scheffler.

Figure 3.10 Measuring of evaporation. Raft-based GGI-3000 evaporation pans.
Photo W. Scheffler.

$$Q_G = Q_T\,(a + b\,R) \tag{7}$$

served as a reference, where Q_G = global radiation, Q_T = Rayleigh radiation, R = relative sunshine duration, a and b = constants. Assuming equal Rayleigh radiation for the Lake Stechlin (LS) and Potsdam (Po), we obtain:

$$Q_G\,(LS) - Q_G\,(Po) = k\,Q_T\,[R\,(LS) - R\,(Po)]. \tag{8}$$

From the available measurements at both stations, the empirical coefficient k = 0.412 was obtained. The shortwave reflexion (Q_R) from the water surface was determined as a function of season and cloud amount, following Dirmhirn (1953). These albedo values are mainly in agreement with recent results obtained by Kirillova (1970).

When the lake was covered by ice or snow, constant albedo values of 20 percent were used for ice cover and of 50 percent for snow cover.

For the calculation of the net terrestrial radiation (Q_N) quite a variety of empirical relations are available in scientific bibliographies. They should not be utilized until their applicability to the individual local conditions has been checked. It should be noted that, owing to the depth of Lake Stechlin, rather great differences may occur between the temperature of the water surface and of the air. Therefore, such empirical formulas can only be suitably used if they are based upon the independent calculation of the upward and downward terrestrial radiation, for which relevant examples are given by Bolz and Fritz (1950), Anderson (1952) or Kirillova (1970). The latter includes the differences between water and air temperature using a correction term. The selection of an empirical formula applicable to the Lake Stechlin conditions was made with a view to the overall optimum adaptation to the heat balance – which was achieved by the Kirillova formula in case an average Asklöf (1920) cloud coefficient of 0.765, which is more appropriate for Central Europe, is used:

$$Q_N = 0.91\sigma T_a^4\,(0.41 - 0.05\,\sqrt{e_a})\,(1 - 0.765\,B) + 0.0034\,(T_{WO} - T_a) - 0.0065 \tag{9}$$

where T_{WO} = temperature of the water surface, T_a = air temperature, and e_a = vapour pressure (hPa).

Incoming heat from the nuclear power plant

The determination of the gain of heat caused by the operation of the nuclear power plant (npp) was made after measurements of the incoming amount of water (M) and the temperature difference (Δt) between the incoming and outgoing water. Assuming a constant water density $\varrho = 1$ and required amount of heat $c = 4.187$ J for the heating of 1 cm^3 of water by 1 K, the additional gain of heat (Q_{npp}) can be calculated from the product $M \cdot \Delta t$ with references to the area (A) of Lake Stechlin

$$Q_{npp} = \frac{\varrho\, c\, M \Delta t}{A} \qquad [\text{in J} \cdot \text{cm}^{-2}]. \tag{10}$$

Along with some operative information related to the nuclear power plant, this equation may be used for continuous calculation of the additional gain of heat.

Heat storage in the lake

For the calculation of the heat storage in Lake Stechlin and its changes, the measurements of the vertical temperature distribution and the division of the lake volume into equal layers each 2 m thick are required. Under the same conditions as are used for the gain of heat by the nuclear power plant (i.e. $\varrho = 1$; $c = 4.187$ J), we can now calculate the heat content (Q_W) from the product of layer volumes and the related water temperature (t_{Wi}) integrated over the number (z) of the layers:

$$Q_W = \sum_{i=1}^{z} V_{z_i} t_{w(z_i)}. \tag{11}$$

The diurnal mean value of the change of heat content (Q_W) (in J · cm^{-2}) we then obtain with reference to the number (n) of the days of a month and the lake's area (A) by the difference of the heat content at the beginning (Q_{W1}) and at the end (Q_{W2}):

$$Q_{\Delta W} = \frac{Q_{W2} - Q_{W1}}{n\, A}. \tag{12}$$

Sensible and latent heat

After the balancing of the hitherto described heat budget components there remains a residual amount of heat, which is mainly used by the exchange of sensible (Q_H) and latent (Q_E) heat:

$$Q_G - Q_R - Q_N \pm Q_{\Delta W} + Q_{npp} = Q_H + Q_E. \tag{13}$$

This equation of the energy balance neglects further components, such as transport of energy by precipitation and the exchange of ground-water, exchange of heat at the bottom of the lake, gain of heat by biochemical processes, etc. The contribution of these processes to the total heat exchange is, as a whole, negligible.

The division of the residual heat (13) can be made using the Bowen ratio. There is, however, one difficulty for Lake Stechlin. Owing to the slow heating in spring, negative temperature gradients frequently occur and the application of the

Bowen ratio will fail. Further, due to the thermal load there is some influence from the heated surface of the lake on the temperature and humidity conditions above the lake. To find a solution to the problem, reference was made to the amounts of evaporation measured with evaporation pans on the raft. If the amount of evaporation is known, we obtain from the Bowen ratio:

$$Q_H = Q_E \cdot 0.61\,(t_{WO} - t_a)/(e_o - e_a). \qquad (14)$$

The resulting flux of sensible heat is rather closely related to the temperature gradient $(t_{WO} - t_a)$ (Richter *et al.* 1979). We obtain the empirical relation

$$Q_H = 72.9\,(t_{WO} - t_a)^{0.85} \qquad (15)$$

with an average daily total of Q_H in $J \cdot cm^{-2}$. When this empirical formula is applied to negative temperature gradient, a direct division of the residual heat, obtained from (13), can be obtained.

The heat balance of Lake Stechlin

The heat balance of Lake Stechlin was studied with a particular view to the changes resulting from the thermal load caused by the nuclear power plant. Tables 3.4 and 3.5 contain the average annual course and the annual totals of the various components of the heat budget equation, covering the undisturbed period 1958–1966 and the first 10 years under influence, i. e., 1967–1976. It should be noted that the whole of 1966 was still considered undisturbed, even though already in autumn there was some minor temporary heat supply produced by the test operations of the power plant. The results of the heat balancing contained in the two tables are clear evidence of the changes produced by thermal load. Particularly, during the winter months there is a remarkable reduction of short-wave reflexion, which results from the average lesser amount of ice cover on the lake.

A survey of the ice cover of Lake Stechlin and for comparison also of Lake Nehmitz is given in Fig. 3.11 as an ice calendar. In addition to the minor ice cover after thermal discharge, there is also an earlier beginning and a gradual melting of the ice on Lake Stechlin.

There are also changes of the annual course of the heat content variation. So the heating begins in February and the cooling in August – both starting one month earlier than should occur under natural conditions. The greatest alterations are determined in the heat balance components which depend on the surface temperature. That is particularly evident from the average annual totals (Table 3.5). Of the additional gain of heat, 57 per cent contributes to the increase of the flux of latent heat (evaporation), 31 per cent to the increase of the flux of sensible heat, and 12 per cent to the increase of net terrestrial radiation.

Table 3.4 Mean daily totals of the heat balance components of Lake Stechlin ($J \cdot cm^{-2}$) under natural conditions (1958–1966) and with thermal discharge (1967–1976).

		Jan	Feb	Mar	Apr	May	Jun	Jul	Aug	Sep	Oct	Nov	Dec
1958–1966													
Global radiation	Q_G	218	435	812	1248	1666	1985	1696	1440	1151	599	239	159
Short-wave reflexion	$-Q_R$	54	109	172	117	138	167	134	126	121	63	25	21
Net terrestrial radiation	$-Q_N$	306	301	314	306	343	419	373	381	448	414	352	339
Change of heat content	$-Q_{\Delta W}$	−352	−75	234	791	766	502	205	38	−247	−498	−699	−641
Sensible heat flux	$-Q_H$	84	33	−38	−147	−50	75	159	188	234	230	264	243
Latent heat flux	$-Q_E$	126	67	130	181	469	822	825	724	603	414	306	218
Additional gain of heat	$-Q_{npp}$	–	–	–	–	–	–	–	17	8	24	9	21
1967–1976													
Global radiation	Q_G	197	385	829	1227	1629	1934	1805	1574	996	490	243	138
Short-wave reflexion	$-Q_R$	29	71	109	113	130	155	147	147	96	50	25	17
Net terrestrial radiation	$-Q_N$	327	331	373	385	377	419	414	456	440	402	381	364
Change of heat content	$-Q_{\Delta W}$	−301	25	331	603	620	536	268	−8	−343	−481	−620	−599
Sensible heat flux	$-Q_H$	197	105	75	50	50	142	180	201	276	285	314	314
Latent heat flux	$-Q_E$	214	176	297	373	611	850	1030	955	766	528	431	318
Additional gain of heat	$-Q_{npp}$	269	323	356	297	159	168	234	177	239	294	288	276

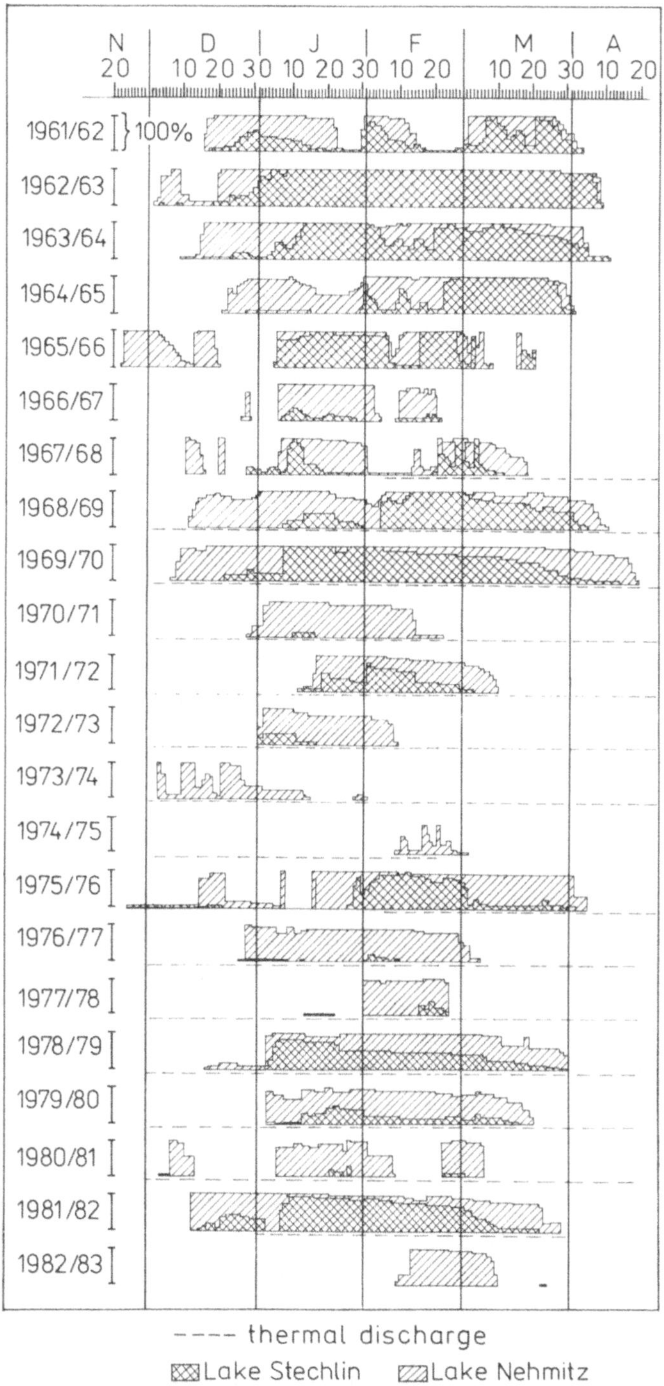

Figure 3.11 Ice calendar of Lake Stechlin and Lake Nehmitz, 1961–1983.

Table 3.5 Annual totals of the heat balance components of Lake Stechlin (KJ · cm^{-2}).

Year	Q_G	$-Q_R$	$-Q_N$	$-Q_{\Delta W}$	Q_{npp}	$-Q_H$	$-Q_E$
1958	349.8	37.8	115.8	8.0	–	40.1	148.1
1959	405.5	43.0	150.6	– 3.7	–	40.0	175.6
1960	341.4	36.7	126.2	4.1	–	37.8	136.6
1961	353.7	33.4	130.1	– 10.5	–	46.2	154.5
1962	340.3	33.1	125.1	– 4.3	–	41.6	144.8
1963	370.4	46.4	137.2	3.6	–	32.1	151.1
1964	362.1	38.7	138.8	5.7	–	31.4	147.5
1965	336.4	40.6	134.7	– 7.4	–	41.4	127.1
1966	337.3	31.0	121.6	8.0	26.7	45.6	157.8
$\bar{x}$	355.2	37.8	131.1	0.4	3.0	39.6	149.3
1967	360.4	31.9	141.4	2.4	93.7	62.8	215.6
1968	360.7	33.6	146.4	– 1.9	105.9	71.6	216.9
1969	347.8	37.1	148.0	– 5.5	113.8	73.2	208.8
1970	354.2	43.7	135.5	16.7	112.9	60.1	211.1
1971	351.0	31.5	144.4	11.1	94.8	64.9	193.9
1972	321.6	30.1	146.7	– 19.8	91.2	73.8	182.0
1973	345.4	30.6	140.0	– 1.6	83.1	75.4	184.1
1974	324.3	27.3	129.5	19.6	120.4	64.2	204.1
1975	366.7	32.9	142.2	– 22.0	18.6	54.8	177.4
1976	360.4	35.1	146.7	2.8	99.0	67.7	207.1
$\bar{x}$	349.3	33.4	142.0	0.2	93.3	66.9	200.1

Amount of evaporation (D. Richter)

Measuring and calculation methods

A major concern of the studies at Lake Stechlin was the amount of evaporation from the free water surface, the determination of which is important for both the heat budget and the water budget. The most important basis of the investigations are the daily evaporation amounts from 3 raft-based GGI-3000 evaporation pans which are available except during ice cover in the winter months. The pans are mounted at one side opposite the mooring point of a triangular raft of sheet-iron tubes. Because of the one-point mooring the raft always moves according to the wind direction while the pans are in leeward position. Measurements made are volumetric (Fig. 3.10).

The three evaporation pans are made of sheet-iron painted with silver-bronze, enamelled sheet-iron and piacryl. This assembly of different pans was originally intended to serve the investigation of the influence of the material upon the amount of evaporation. The results show that for diurnal and monthly totals the average for all three pans is in rather close agreement with the actual amount of evaporation from a lake (Richter 1977).

When measurements are made from raft-based evaporation pans failures and interruptions cannot be avoided. With strong wind, from about 7–8 $m \cdot s^{-1}$, the pans are filled by the wave motion of the water.

Ducks and seagulls can also interfere with results. In view of the ice cover, measurements are discontinued in winter. Missing measurements are complemented using empirical relations between the amount of evaporation (E) and the influencing meteorological elements. The most appropriate way for the calculation of diurnal values is the equation

$$E = (0.16 + 0.20\, v_2^{0.5})\,(e_o - e_a) \tag{16}$$

where v_2 = wind speed at a height of 2 m, and $(e_o - e_a)$ = difference of vapour pressure (hPa) between the water surface and the height of 2 m.

The influence of wind can be set aside for monthly values. The most suitable simulation is given by taking into account the average diurnal total of the global radiation (Q_G), using

$$E = 0.337\,(e_o - e_2) + 0.000566\, Q_G - 0.036 \tag{17}$$

where Q_G is in $J \cdot cm^{-2}$.

Both empirical formulas can be applied only if the lake's surface is sufficiently exposed to wind. For lakes protected from wind reductions are required.

Results of the investigation of evaporation

Evaporation values, measured and complemented by calculations, from Lake Stechlin have been available since 1958. A comparison of the average annual course under natural conditions and under thermal influence is shown in Fig. 3.12. The increase of evaporation continuing over the whole of the year is well recognized. On the average, the annual total increased from 633 mm (1958–1966) to 843 mm (1967–1976). The highest increase of evaporation is found in the summer months June/August, when the share in the total increase is about 40 per cent.

The variation of the annual totals of the evaporation amounts from 1958–1976 is shown in Fig. 3.13. For comparison, the balanced evaporation amounts (from p. 64) have also been included. Both methods of evaluating the evaporation clearly show the increase of evaporation that has occurred since the beginning of the thermal load. On the average, the amount of evaporation is lower by 4 per cent when the heat balance method is applied.

Some special investigations were made to test the validity of the raft-based evaporation measurements made under thermal load for the whole of the lake's surface. Therefore, between 6 and 16 August 1974 daily temperature measure-

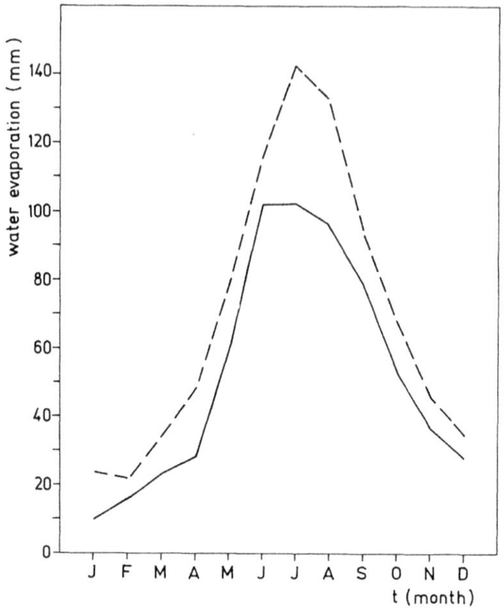

Figure 3.12 Mean annual course of the evaporation amount before (——) and after (----) the thermal discharge.

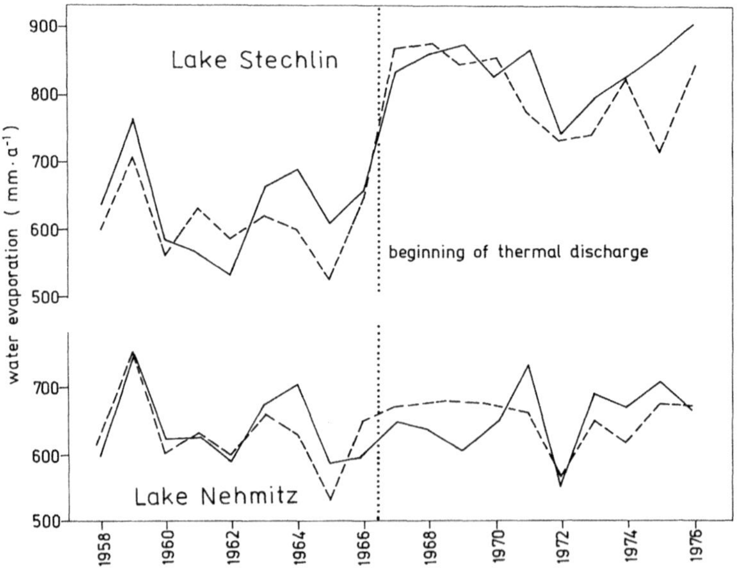

Figure 3.13 Course of the yearly evaporation amount of Lake Stechlin and Lake Nehmitz; (——) measured and completed by the empirical method, (----) calculated by the heat balance method.

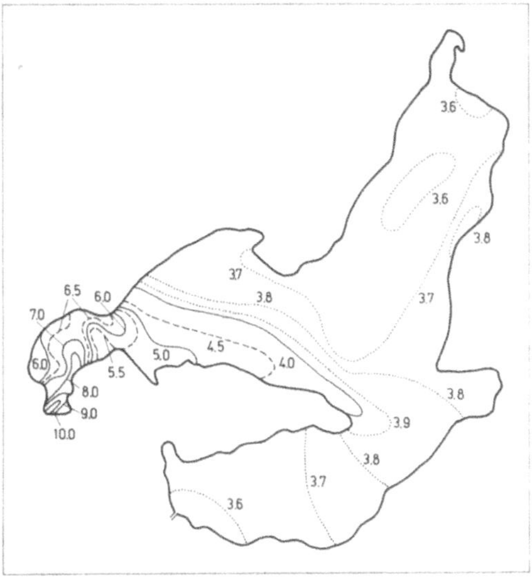

Figure 3.14 Distribution of the evaporation amount of Lake Stechlin, depending upon the incoming heated water, August 6–16, 1974.

ments were made at more than 100 measuring points of the lake using a sensitive temperature probe. The climatic data of the meteorological station and an evaporation formula like (16) were used to calculate the average distribution of evaporation as shown in Fig. 3.14, resulting in a daily amount of evaporation of 3.9 mm while during the same period 3.7 mm were measured. Since for the empirical calculation the reaction of the heated lake surface upon the vapour pressure above the lake was not taken into account, this amount of evaporation is suggested to be too high by 0.1–0.2 mm (Richter *et al.* 1979). Hence, there is good agreement between the measured and calculated amount of evaporation over the whole of the lake's surface.

Water balance components (D. Richter)

General remarks

Lake Stechlin and its catchment area are part of the major Havel-river System. Up to the middle of the 18th century there was no surface runoff from Lake Stechlin. But after the Polzow canal had been built from the south bay of Lake Stechlin to Lake Nehmitz and then through the Polzow valley, there was a connection available to the Havel river (see p. 10).

The water budget of Lake Stechlin is, apart from precipitation and evaporation, mainly controlled by the subsurface inflow and runoff. Surface inflow is only temporary from Lake Dagow with high water level. Runoff depends upon the water level of Lake Nehmitz, which is controlled by a weir at the outflow to the Polzow canal in the southeastern bay of the north basin of this lake (Fig. 3.1). Owing to the rather important role ground-water exerts on the water budget of Lake Stechlin, a quantitatively precise determination of the whole range of water budget components is not feasible; this particularly applies to the subsurface inflow and runoff.

The measurement of all the water budget components is rather difficult, even for the catchment area, since the surface and the subsurface catchment areas are not identical (Heitmann *et al.* 1969). Owing to the location of Lake Stechlin, there is ground-water inflow only from the northwest and the southeast, while the subsurface runoff is directed northwestward to the Havel river and southwestward to Lake Nehmitz. Due to the influence of the ground-water, the water level of Lake Stechlin varies only by 20–30 cm. The rather close relation to the ground-water is also evident from 10 minor lakes in the catchment area whose normal water level varies only between 59.5 and 60.8 m above m. s. l., although the height difference over the area is more than 40 m with a rather varied relief (see p. 9).

With regard to existing conditions, the determination of the water balance appears to be useful only if the whole catchment area of Lake Stechlin is included. Some components require calculations. When establishing the so-called water budget equation for Lake Stechlin, the following elements are to be included: areal amount of precipitation (P_A), losses due to evaporation from the free water surfaces (E_W), losses due to evaporation from overgrown land surfaces (ET_L), surface runoff (R_O), subsurface runoff (R_U), changes of water storage in the lakes (ΔW_L), changes of water storage in ground-water (ΔW_G) and changes of water storage in the soil (ΔW_S).

From these elements the storage below the surface can be assessed only by long-term balancing:

$$P_A - R_O - R_U - E_W - ET_L \pm \Delta W_L \pm \Delta W_G \pm \Delta W_S = O. \qquad (18)$$

Amount of precipitation

Daily measurements of the amount of precipitation are made at the meteorological station and 8 additional measuring points in the catchment area using rain Hellmann-type gange. Furthermore, there are 3 totalizers installed to improve the network density, with data read twice a month. Comparisons show that there are no significant differences between the monthly and annual totals of precipitation between the catchment area and the meteorological station. Hence, for an assessment of the precipitation regime in the Lake Stechlin area, the station data available since 1958 are sufficient. A survey of the measured average monthly totals and of the hitherto measured maximum and minimum monthly totals is reflected in Table 3.6. With a mean annual total of 592 mm, the monthly values of precipitation are, on the average, between 32 mm in February and 76 mm in July. Maximum and minimum monthly totals are 151.6 and 1.0 mm. There will usually be high monthly totales when there occur individual events of heavy or continuous precipitation, which have frequently yielded daily totals of 50–60 mm. Even extreme totals of 100 mm on one day are possible. These data show that the amounts of precipitation are highly variable, indicating that precise research of the spatial and time-related conditions demands long-term measurements.

Since the measured amount of precipitation may be falsified by wind speed, wetting and other effects (Karbaum 1969; Sevruk 1982), further investigations of the water budget are based on corrected values.

Using a GDR correction standard, +10 per cent are applied to liquid, and +30 per cent to solid precipitation. The water balance values presented hereafter include the higher precipitation amounts corrected as described.

Catchment area evaporation

The catchment area evaporation is mainly composed of the amount of evaporation from free water surfaces and wooded land areas. The percentages, in the region, of lake surfaces is 16 per cent; 95 per cent of the land surfaces are covered by Pine and Beech woods. The methods used for the determination of the amount of evaporation from free water surfaces are described on pages 64–65. For the calculation of the evapotranspiration from the wooded land surfaces the Kortüm equation (1961)

$$ET_L = P_A\,(1 - e^{-wQ_G/rP_A}) \tag{19}$$

was applied, which combines elements of the heat and the water budget, and where P_A is the areal precipitation, Q_G is the global radiation, r is the evaporating heat and w is a factor resembling the evaporational activities in the area.

This factor is obtained if ET_L is known from a simple water budget equation

$$ET_L = P_A - R_O \tag{20}$$

covering a rather long period of time.

The amounts of evaporation for individual years can, following Glugla (1970), be calculated as the difference (Δ) from the long-term average:

$$\Delta ET_L = \Delta P_A\ [\,1 - e^{-w\bar{Q}_G/\gamma\bar{P}_A}\,(1 + \frac{w\bar{Q}_G}{r\bar{P}_A} - \frac{w}{r}\,\frac{\Delta Q_G}{\Delta P_A})\,]. \tag{21}$$

If required, monthly values can be obtained using monthly percentages (Kusmin 1968) or the vapour saturation deficit (Richter 1964). The total catchment area

Table 3.6 Mean, maximum and minimum monthly and annual totals of precipitation (mm) measured at the Neuglobsow meteorological station, 1958–1981.

Amount of precipitation	Jan	Feb	Mar	Apr	May	Jun	Jul	Aug	Sep	Oct	Nov	Dec	Year
Average	40.9	31.7	34.4	44.8	58.0	60.0	76.1	57.0	44.6	40.9	50.8	53.0	591.9
Maximum	131.2	75.6	69.4	82.1	144.6	130.2	151.6	112.7	141.0	105.0	94.7	112.3	779.4
Minimum	15.6	3.2	6.6	6.3	26.5	16.0	11.0	9.2	1.0	8.7	10.9	10.7	379.5

Table 3.7 Water balance of the Lake Stechlin catchment area, mean annual course and annual totals, 1958–1970 in mm.

Mean annual course	Jan	Feb	Mar	Apr	May	Jun	Jul	Aug	Sep	Oct	Nov	Dec	Year
Amount of precipitation	43	42	37	51	70	59	77	63	46	44	57	54	643
Catchment evaporation	6	10	22	42	71	101	89	79	49	26	14	7	516
Total runoff	12	11	13	12	12	12	9	6	6	7	6	9	115
Storage in the lakes	2	2	0	1	1	− 7	− 1	− 2	− 2	1	4	2	1
Storage in the groundwater	1	2	3	3	4	2	1	− 2	− 1	0	− 3	1	11
Soil moisture storage	22	17	− 1	− 7	− 18	− 49	− 21	− 18	− 6	10	36	35	0
Annual totals	**1958**	**1959**	**1960**	**1961**	**1962**	**1963**	**1964**	**1965**	**1966**	**1967**	**1968**	**1969**	**1970**
Amount of precipitation	630	437	640	760	587	566	588	639	846	694	743	579	656
Catchment evaporation	489	480	515	550	475	484	483	498	581	557	576	499	520
Total runoff	89	61	48	163	161	78	93	83	151	183	139	102	147
Storage in the lakes	20	− 7	29	− 21	− 15	9	− 4	21	11	− 23	− 7	7	− 5
Storage in the groundwater	22	− 31	− 16	44	25	− 29	− 16	2	57	56	4	16	13
Soil moisture storage	10	− 66	64	24	− 59	24	32	35	46	− 79	31	− 45	− 19

evaporation of the Lake Stechlin environs (ET_A) is then obtained from the percentages of land and lake surfaces:

$$ET_A = 0.84\ ET_L + 0.16\ E_W. \qquad (22)$$

Results of related studies are contained in the survey of the water balances on page 72.

Surface and subsurface runoff

Surface runoff from the Lake Stechlin area is regulated by the weir at the outflow point of Lake Nehmitz. Since the canal connecting Lake Stechlin and Lake Nehmitz was built in 1960–1961, i.e., two years after the beginning of the investigations, nothing can be said about the natural runoff. From a study concluded in 1970, regulated runoff was, on the average, $95\,l \cdot s^{-1}$ or $71\,mm \cdot a^{-1}$, when related to the catchment area. In individual years, between 0 and $142\,mm \cdot a^{-1}$ were measured.

Much more difficult to determine is the subsurface runoff. For this, precise data on the ground-water level must be available. These data can be obtained from monthly measurements made at more than 30 ground-water pipes in the catchment and observations of the water level at 17 lakes within and in the vicinity of the area.

Ground-water maps permit a fairly precise survey of ground-water runoff conditions (Schumann 1970). Following the Darcy law, the subsurface runoff can be calculated from the ground-water level gradient and the permeability of the soil. Related to the whole area, there are well balanced values of $59.0\,l \cdot s^{-1}$ or $44\,mm \cdot a^{-1}$ on the average, with variations between 35 and $48\,mm \cdot a^{-1}$. The monthly variation of the total mean runoff of $115\,mm \cdot a^{-1}$ is shown in Table 3.7.

Variations of the water storage

Any variation of water storage in the Lake Stechlin area is controlled by a change of the water content of the lakes, in the ground-water and in the soil between the surface and ground-water level. As the changes of the water level in the lakes are rather small, the annual storage usually also remains below 20 mm in respect to the whole catchment area. The long-term storage is nearly balanced. The storage of ground-water could be more than doubled. It is calculated from the change of the ground-water level and the available hollows in the soil (Heitmann *et al.* 1969). The investigations made between 1958 and 1970 reveal substantial increases caused by abundant precipitation during a number of years, resulting in a mean annual storage of 11 mm. Precise information on the changes of the water storage in the soil is not feasible. Since the ground-water level lies partly up to

25 m below the land surface, there is thought to be a huge storage volume whose humidity content can, practically, not be determined. Only from the balancing of all of the remaining water budget components can some indications of the order of magnitude of the changes be obtained in advance. From experience, the maximum of storage variation in the soil may be more than 400 mm, corresponding to the amount of precipitation in a relatively dry year.

Water balance of the Lake Stechlin catchment area

Table 3.7 shows the survey of the water balance of the Lake Stechlin area regarding the mean annual course and also the individual years of the period 1958–1970. The calculated catchment area evaporation is composed of the amount of evaporation from the free water surfaces (626 mm $\cdot$ a^{-1}) and from the wooded land surfaces (495 mm $\cdot$ a^{-1}). The variation of the water content in the soil was obtained in each case as a constant value. The most important influence upon the water balance of the area is exerted by the amounts of precipitation and evaporation, where the catchment area evaporation is highest in years with abundant precipitation.

There are large variations of the runoff. On account of the regulation, the connection with other elements is changed. For example, there was only subsurface runoff in 1960 when the weir for the Polzow canal was built, and the water level of the lakes rose considerably. In subsequent years the water level was lowered again, caused by an increased runoff. Altogether, the variations of storage in the lakes are the smallest in comparison to the other water balance components. It should be noted that there is a steady increase of the ground-water content after a number of years with abundant precipitation, which is a sufficient evidence of the rather high storage capacity of the soil at the ground-water level. The annual course of water content of the soil is determined by draining during the summer months and increasing soil moisture in winter. The significant influence of evaporation is well recognized. It should, finally, be stated that owing to the partly rather low position of the ground-water level, wet or dry periods frequently influence the ground-water variation with a delay of up to more than one year. Corresponding reactions are observed upon the water level of lakes without runoff.

Light penetration (R. Koschel)

Surface incident radiation and photosynthetically active radiation

The annual values and monthly averages of total incoming global radiation and of albedo in the Lake Stechlin area are collected on pages 62 and 64 (Tables 3.4 and

Table 3.8 Monthly averages ($J \cdot cm^{-2} \cdot d^{-1}$) and annual values ($KJ \cdot cm^{-2} \cdot a^{-1}$) of global radiation in the study area (measurements of Institute of Hydrometeorological Research, Lake Stechlin station).

Year	Jan	Feb	Mar	Apr	May	Jun	Jul	Aug	Sep	Oct	Nov	Dec	TAR*
1970	214	360	829	967	1687	2286	1712	1633	1130	419	239	134	354.2
1971	255	348	816	1223	1775	1432	2102	1658	966	641	234	92	251.0
1972	180	343	850	1043	1365	1825	1712	1444	1013	620	205	163	321.6
1973	163	339	544	1030	1591	1901	1712	1742	921	553	301	155	345.4
1974	172	285	883	1344	1507	1608	1382	1470	1038	373	226	130	324.3
1975	222	499	812	1143	1746	2114	1871	1758	1063	423	260	109	366.7
1976	222	481	796	1382	1645	2211	1955	1541	871	335	226	159	360.4
1977	159	343	641	1001	1909	1687	1524	1210	1068	486	222	126	314.8
1978	167	461	729	1424	1880	1775	1583	1394	678	502	197	130	333.7
1979	172	381	607	1172	1817	1704	1340	1465	1026	741	197	134	328.2
1980	197	348	808	1122	2031	1671	1474	1344	1038	507	230	142	333.5
1981	193	423	528	1357	1670	1499	1398	1310	1017	423	230	176	311.9
1982	193	469	837	1394	1721	1788	2152	1645	1248	519	285	121	377.4
Mean	193	391	745	1200	1721	1808	1686	1509	1006	503	235	136	340.0
SD**	28.5	67.7	122.2	164.4	179.8	260.8	270.1	170.9	134.9	114.5	31.2	22.5	20.7

* TAR = Total annual global radiation. ** SD = Standard deviation.

3.5). The monthly variation of global radiation and the total annual global radiation during the period 1970–1982 is given in Table 3.8. The main deviation from the average of annual global radiation varied between +11% and −8.3%, whereas the main deviation from the average of monthly global radiation varied between +47% and −33%. The daily variation in radiation can be greater than 100% (Fig. 3.15). For example, the global radiation can amount to 2000 $J \cdot cm^{-2} \cdot d^{-1}$ in March or 3100 $J \cdot cm^{-2} \cdot d^{-1}$ in June on clear and cloudless days, whereas the values amount to only 150 $J \cdot cm^{-2} \cdot d^{-1}$ and 550 $J \cdot cm^{-2} \cdot d^{-1}$, respectively, in the same months on days with total cloud coverage.

The total photosynthetically active radiation (PAR) is only part of incoming global radiation and enters the waterbody. PAR of natural daylight radiation varies from 43.5% (cloudy day) to 53% (cloudless day, sun elevation 65%) with an average of 46%–48% (Vollenweider 1969). Measurements of PAR are difficult and require special spectroradiometers. 46% of incoming global radiation (corrected for surface reflection losses) were used in our calculations of energy fluxes. All calculations of PAR were made on the basis of energy. Despite the fact that photosynthesis is a quantum process, the relations of photosynthesis to irradiance in natural waters can be studied on the basis of energetic values as well as of quanta without significant error. The ratio between energy and quanta varies by no more than ±10% and is $2.5 \pm 0.25 \cdot 10^{18}$ quanta $\cdot s^{-1} \cdot W^{-1}$ (Morel &

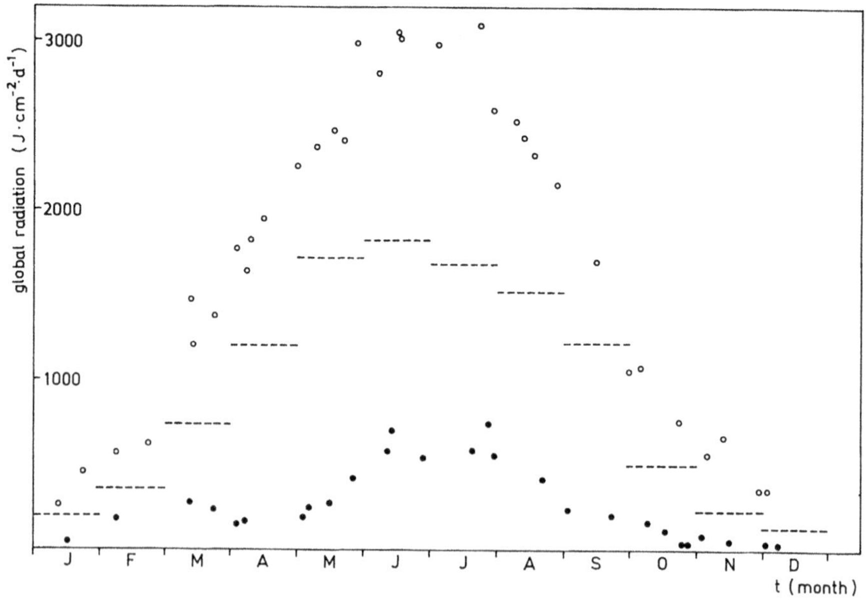

Figure 3.15 Global radiation in Lake Stechlin area (measurements from Institute of Hydrometeorological Research, Lake Stechlin station). Broken lines: monthly mean daily radiation (1970–1982); open circles: clear and cloudless days; solid circles: rainy and very cloudy days.

Table 3.9 Conversion factors for energy, quanta (illuminance) units in Baltic lakes (according to Krambeck 1982).

General conversion factors:
4.187 J = 4.187 WS = 1 cal.
1 E = $6.023 \cdot 10^{23}$ quanta.

Conversion factors including the spectral distribution (400 nm–700 nm):
1 W = 5 $\mu E \cdot s^{-1}$ (above the surface of Baltic lakes).
1 W ≈ 5–0.09 d_e $\mu E \cdot s^{-1}$ (d_e = depth of the epilimnetic layer ≥5 m; 1000 lux ≈21 $\mu E \cdot m^{-2} \cdot s^{-1}$).

Smith 1974). The energy can be corrected for other calculations on the basis of quanta values. Some conversion factors between the different units are given in Table 3.9.

Underwater light climate

The underwater light analyses were made with a selenium photocell (Sauberer 1962, Westlake 1965). Intensity was measured in four different spectral ranges through standard Schott-filters BG 12 (435 nm, blue), VG 9 (525 nm, green), OG 18/BG 12 (590 nm, orange) and RG 2 (660 nm, red). Measurements were carried out around noon (±2 h) on clear or cloudy days. The vertical extinction coefficient is defined according to

$$\varepsilon_\lambda = \frac{1}{z_2 - z_1} \cdot (\ln i_1 - \ln i_2) \tag{23}$$

(i_1 = intensity measured at depth z_1, i_2 = intensity measured at depth z_2, ε = vertical extinction coefficient, λ = represents the sensivity centre of filter-photocell combination). Correspondingly, the transmission (T) is given by

$$T_{(z2 - z1)} \% = i_2 \cdot (i_1 \cdot 100)^{-1}. \tag{24}$$

The light penetration of visible spectral components can deviate about ±10% per metre from the average during the year in Lake Stechlin (Fig. 3.16). Correspondingly, the relative intensity fluctuated at different depths (Fig. 3.17). Green and orange light are the most penetrating spectral components. The light intensity of these wavelengths fluctuates between 45% and 65% of surface intensity at 2.5 m depth, 10% and 25% at 7.5 m depth, 1% and 6% at 15 m depth (Fig. 3.17). The light attenuation in the blue and red regions is smaller than for green and orange light. The relative intensity varies between 26% and 43% (blue light) and 19% and 32% (red light) of surface intensity at a depth of 2.5 m, for example. These are principally results of a relatively clear-water lake without the influence of larger amounts of algae, detritus material or colouring matter. Similar conditions

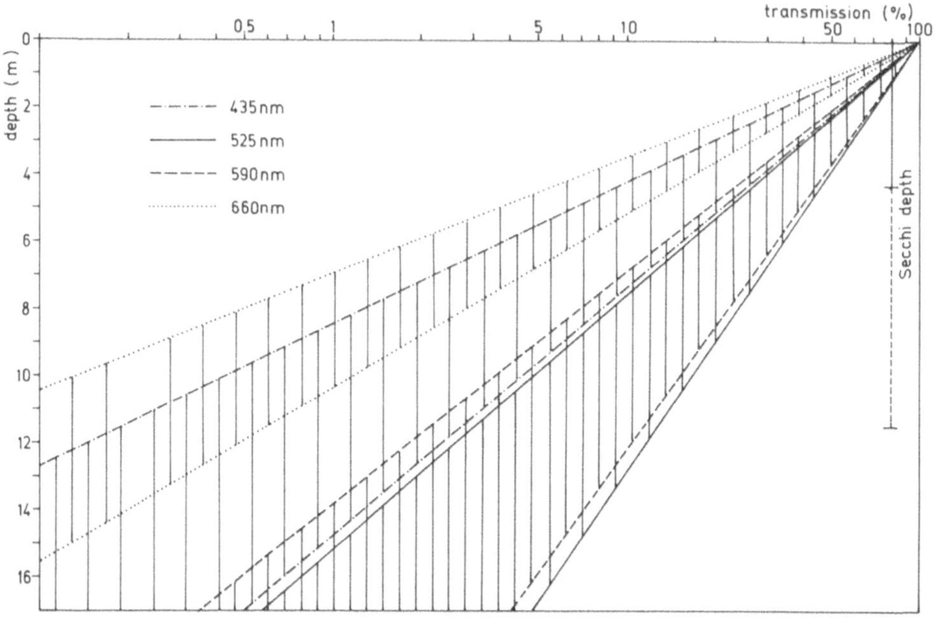

Figure 3.16 Range of percentage transmission of different light wavelengths at various depths in Lake Stechlin.

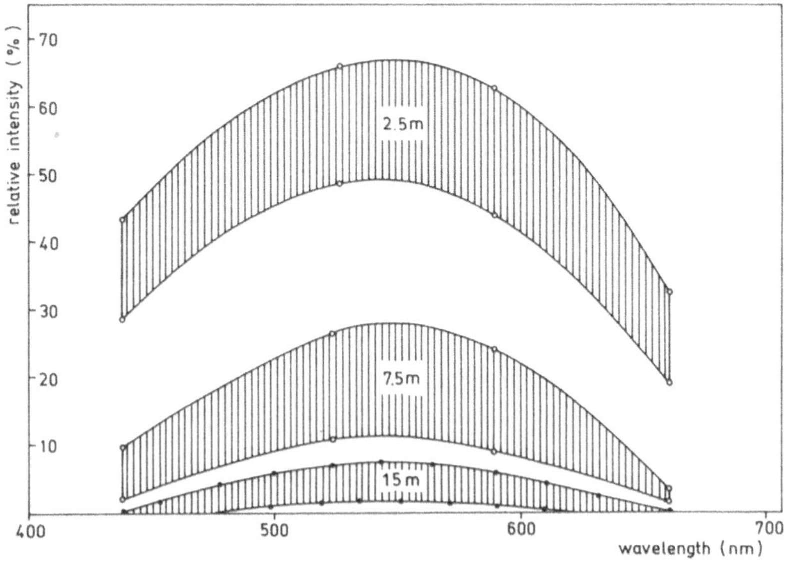

Figure 3.17 Range of relative intensity of PAR at various depths in Lake Stechlin.

exist in Lake Nehmitz (Fig. 3.18). In our study area the light attenuation of eutrophic lakes behaves differently, of course, from values of oligotrophic Lake Stechlin or mesotrophic-oligotrophic Lake Nehmitz. Mass-developments of planktonic algae leads necessarily to higher transmission and extinction coefficients in eutrophic (Lake Dagow) or hypertrophic lakes (Lake Haussee) (Figs. 3.18, 3.19; Table 3.10). In addition, the relations between the transmissions of single wavebands are different. The importance of red light especially increases in the whole euphotic zone (Fig. 3.19; Table 3.10).

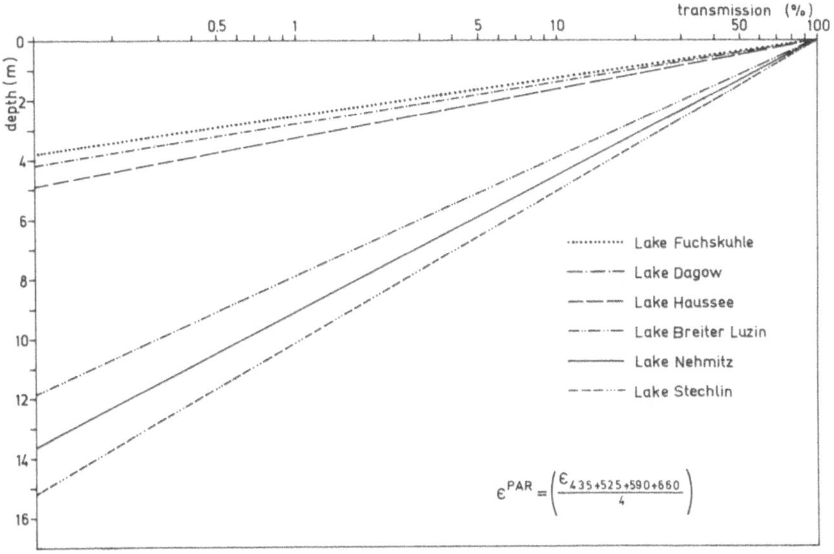

Figure 3.18 Percentage transmission (annual mean) of PAR at differing depths in various lakes of the areas of Lake Stechlin and Lake Breiter Luzin.

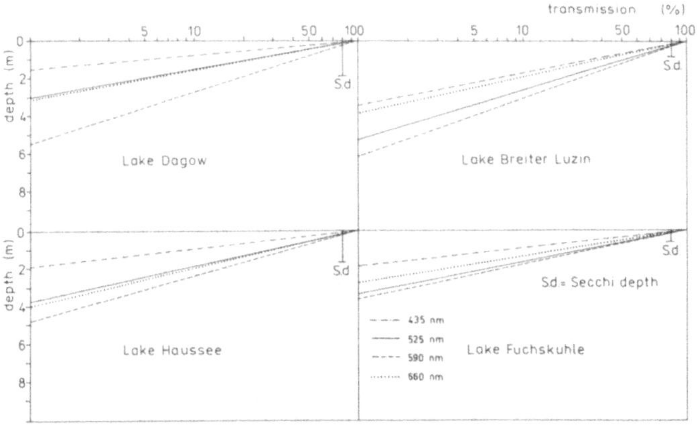

Figure 3.19 Percentage transmission of different light wavelengths (annual means) in two eutrophic lakes (Lakes Dagow and Haussee), during a 'whiting' phase (Lake Breiter Luzin) and during a mass-development of an ultraplanktonic green alga (Lake Große Fuchskuhle).

Table 3.10 Vertical extinction coefficient of different wavebands in lakes of the areas of Lake Stechlin and Lake Breiter Luzin.

Lake	Observation time	Vertical extinction coefficient ($ln \cdot m^{-1}$)											
		435 nm			525 nm			590 nm			660 nm		
		Min	Max	Mean	Min	Max	Mean	Min	Max	Mean	Min	Max	Mean
Lake Stechlin	1970–1982	0.31	0.76	0.46	0.19	0.31	0.25	0.20	0.38	0.29	0.45	0.69	0.54
Lake Nehmitz	1972–1978	0.40	0.85	0.65	0.21	0.55	0.40	0.25	0.55	0.38	0.44	0.75	0.62
Lake Dagow	1972–1978	1.66	4.27	2.46	0.69	2.07	1.13	0.56	1.11	0.95	0.79	1.86	1.16
Lake Große Fuchskuhle	1982	0.60	3.00	1.85	0.52	1.46	0.89	0.49	1.36	0.87	0.79	1.66	1.10
Lake Breiter Luzin	1978–1982	0.49	1.36	0.77	0.28	0.89	0.45	0.27	0.78	0.40	0.42	1.15	0.67
Lake Haussee	1978–1980	1.64	3.94	2.42	0.92	1.84	1.24	0.71	1.59	1.00	0.83	2.04	1.16

Figs. 3.18, 3.19 and Table 3.10 also indicate values of light penetration during extreme phases of light climate in two other lakes. The representative percentage transmission of Lake Breiter Luzin, a deep mesotrophic lake, was measured during an autochthonous 'whiting'-phase (see p. 413). The values of Lake Große Fuchskuhle, a shallow acid lake, are due to mass-development of an ultraplanktonic green alga. Diameters of calcite crystals and of ultraplanktonic algae are about the same (2–3 μm). The Secchi depth was 0.8 m in Lake Breiter Luzin and 0.6 m in Lake Große Fuchskuhle during these phases. Wavelength between 525 nm remains the dominating spectrum in both lakes too. But the transmission during the mass-development of ultraplanktonic green algae is disproportionally smaller than the transmission during calcite precipitation. Nevertheless the red light penetrates relatively deeper than blue light into algae-rich water.

Secchi disc transparency (Secchi depth)

The Secchi disc transparency was measured over a period of more than 20 years, from 1958–1982 (Richter & Wiedrich 1967; Heitmann 1976; Koschel 1974, 1981; Koschel & Mothes 1976). Usually one measurement per month was carried out. In some years measurements were made at intervals of 2 weeks. A white Secchi disc of 30 cm diameter was used. The Secchi disc transparency was measured through an area of shaded water surface. Reflection and glittering were eliminated by using a bathyscope.

The Secchi depth in Lake Stechlin is given in Fig. 3.20 for the 9 years from 1970–1982. The mean, maximum and minimum values of Secchi depth are presented in Table 3.11. Five periods can be distinguished generally in yearly periodical fluctuation of the Secchi depth. The highest transparency values of the year occurred in Period I (January–February). The observed maximum lay between 9 m and 12 m. This period is characterized generally by low values of seston concentration. Period II (March–April) is the phase of increase and spring minimum of the Secchi depth. At the beginning of March (in some years earlier) the Secchi depth increased more or less rapidly. The transparency diminishes to an average of 4.5–6 m for about 4 weeks in April/May. The minimum was recorded at the beginning of May in 1972 with 4.0 m. In this period the spring population of phytoplankton was built. Period III (June/beginning of July) is the phase of summer maximum of Secchi depth. At the end of May the Secchi depth increases rapidly and rises to summer maximum between 8 m and 11 m. Period IV (July–September) is the phase with the summer minimum. After the summer maximum, Secchi depth increases rapidly and obtains the summer minimum between July and August with values of about 5 m. At the end of September (or earlier) the Secchi depth decreases and passes into phase I with highest values of transparency at the end of November and in December (Period V).

This characteristic oscillation was observed in all years from 1972–1982. Only

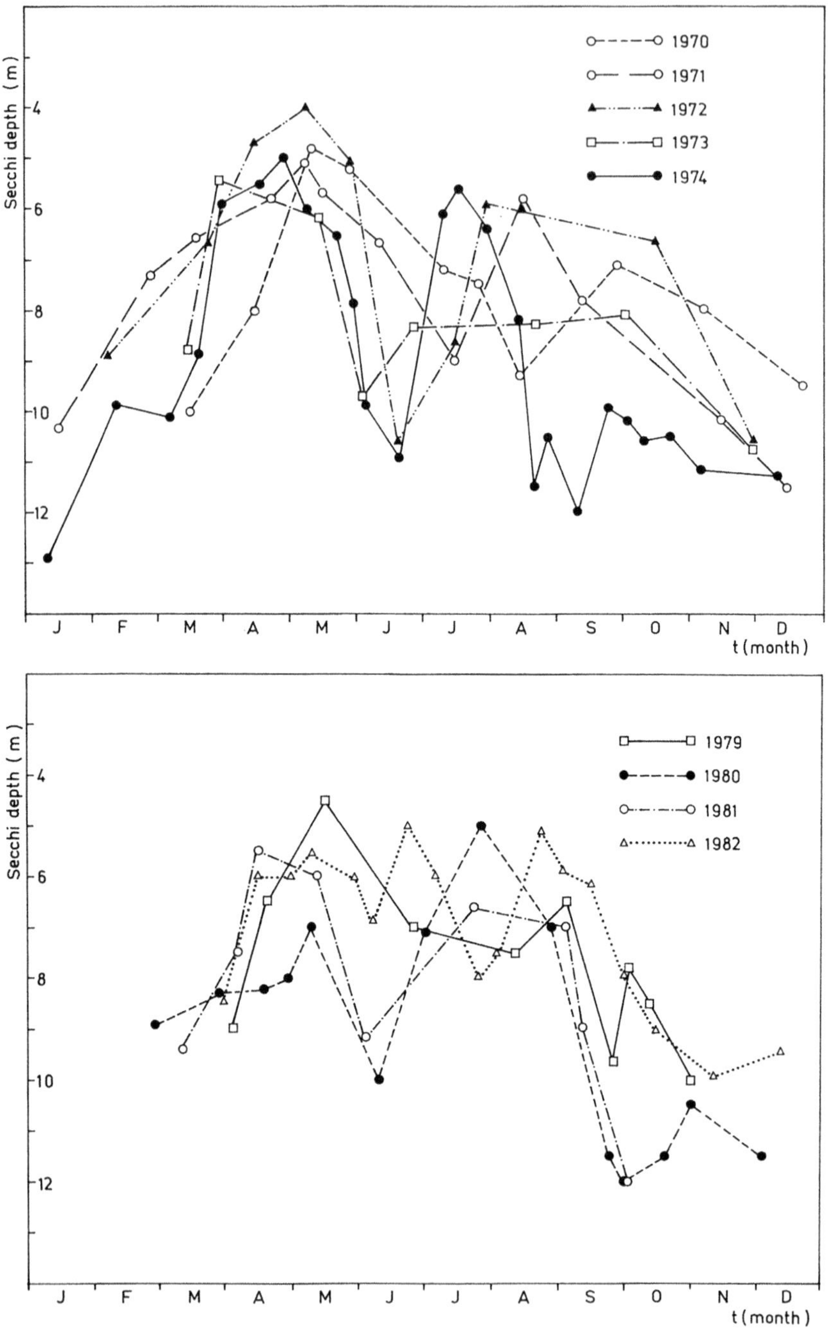

Figure 3.20 Secchi disc transparency in Lake Stechlin during 1970–1982.

Table 3.11 Recorded mean, maximum and minimum values of Secchi depth (m) in Lake Stechlin from 1970–1982.

Year	Mean (annual)	Mean (period Apr–Sep)	Max winter	Min spring	Max summer	Min summer
1970	8.2	7.1	10.5	4.8	9.3	7.2
1971	8.1	6.8	11.5	5.1	9.0	5.8
1972	7.6	6.4	10.5	4.0	10.8	5.9
1973	8.6	7.9	10.8	5.5	9.8	–*
1974	9.4	8.0	12.9	5.0	11.5	5.7
1979	8.4	6.7	10.0	4.5	9.7	6.5
1980	8.9	7.8	11.5	7.0	10.0	5.0
1981	8.4	7.5	12.0	5.5	9.0	6.6
1982	7.7	6.4	9.9	5.6	8.0	5.2
Mean	8.4	7.2	11.1	5.2	9.7	6.0

* Not observed.

the time intervals and absolute values are different from year to year (Table 3.11, Fig. 3.20).

From 1958–1965 (before the power plant came into operation) the Secchi depth was generally higher, on the average 2 m (Richter & Wiedrich 1967; Koschel & Mothes 1976; see p. 428). But periodical oscillation was about the same. The highest value was observed in January 1961 with a Secchi depth of 15.5 m.

A similar fluctuation of Secchi depths can be observed in other lakes of the Lake Stechlin area (Fig. 3.21): high transparency in winter, decrease in spring with a spring minimum, increase in late spring or summer with a summer maximum following a decrease in summer with a summer minimum and a subsequent increase to the winter maximum. The mean annual values of Secchi depths move between 1.9 m (Lake Dagow) and 7.4 m (Lake Nehmitz-N) in other lakes of the study area (Table 3.12).

Table 3.12 Recorded mean, maximum and minimum values of Secchi depth (m) in other lakes of the study area.

Lake	Year	Mean (annual)	Mean (period Apr–Sep)	Max winter	Min spring	Max summer	Min summer
Lake Dagow	1974	1.9	1.4	4.0	1.0	1.5	1.3
Lake Große Fuchskuhle	1982	2.6	2.0	4.7	0.5	3.8	2.0
Lake Nehmitz-N	1974	7.4	6.4	11.0	4.0	8.2	5.6
Lake Nehmitz-S	1974	6.5	6.2	8.2	3.7	7.7	5.6

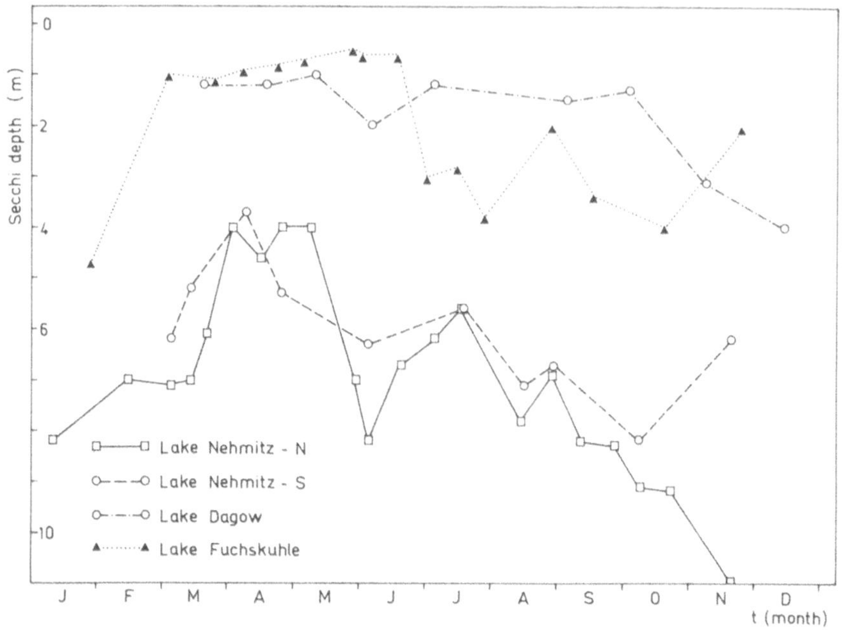

Figure 3.21 Secchi disc transparency in Lake Nehmitz (1974), Lake Dagow (1974) and Lake Große Fuchskuhle (1982).

The Secchi depth in Lake Nehmitz-N increased after operation of nuclear power plant, on the average, 1 m to 2 m.

Relationships between Secchi depth and vertical extinction coefficients

An inverse proportionality exists between Secchi transparency and extinction coefficients for visible light. This relatively close correlation was observed by various authors in different waters (e. g. Vollenweider 1960; Schanz 1982). This correlation is determined primarily on the basis of phytoplankton concentration. According to Megard *et al.* (1980), the vertical extinction coefficient for PAR depends linearely upon the concentration (c) of chlorophyll of phytoplankton:

$$\varepsilon^{PAR} = \varepsilon_w \cdot \varepsilon_c \cdot c \tag{25}$$

ε_w = attenuation of light by water and substances others than algae; ε_c = attenuation per unit concentration of chlorophyll-a in water.

Equation (25) is valid if ε_w and ε_c are constant. However, this can not always be expected (Schanz 1982). Local conditions – for example, humic fractions, turbidity (calcite precipitates), allochthonous matter – affect ε_w and ε_c and partly influence the vertical extinction coefficient too. Thus it is better to analyse the

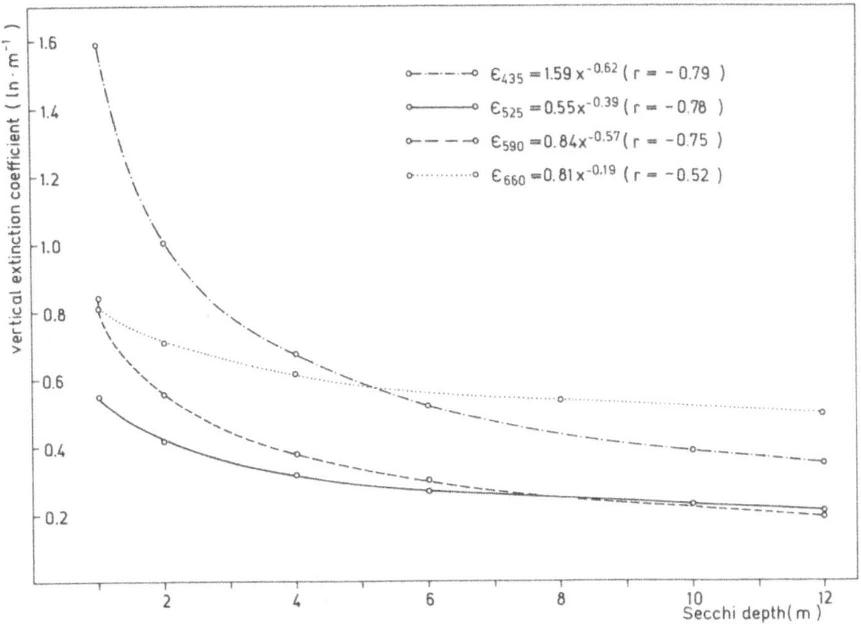

Figure 3.22 Relationships between the Secchi transparency and vertical extinction coefficient of different wavelengths in Lake Stechlin.

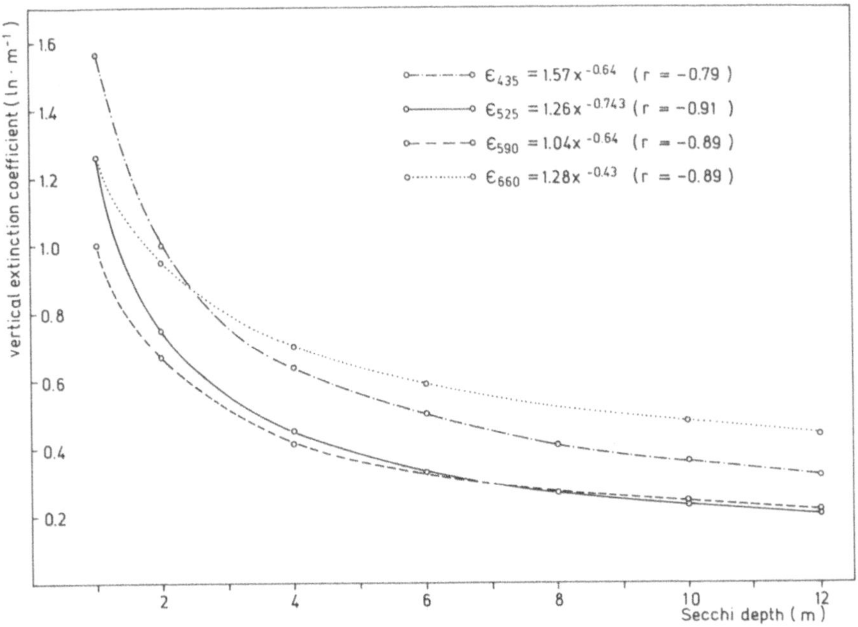

Figure 3.23 Relationships between the Secchi disc transparency and vertical extinction coefficient of different wavelengths in lakes of the study area.

relation between the Secchi depth and the extinction coefficient in every lake or in every lake area.

The connection between Secchi depth and vertical extinction coefficients in Lake Stechlin and its area are presented in Figs. 3.22–3.25. The Secchi depths are connected with the vertical extinction coefficient for individual wavelengths in Lake Stechlin in Fig. 3.22. Fig. 3.24 shows relationships between Secchi depth and approximated vertical extinction coefficient for all visible light in Lake Stechlin. In this procedure the assumption is made that the vertical extinction coefficient for visible light in the euphotic zone is made up in two ways:

$$\varepsilon_3^{PAR} = 1/3\ (\varepsilon_{435\ nm} + \varepsilon_{525\ nm} + \varepsilon_{660\ nm})\ \text{(Vollenweider 1969)} \quad (26)$$
$$\varepsilon_4^{PAR} = 1/4\ (\varepsilon_{435\ nm} + \varepsilon_{525\ nm} + \varepsilon_{590\ nm} + \varepsilon_{660\ nm}) \quad (27)$$

where the ε's were determined as averages of the whole euphotic zone.

The same relations which are demonstrated in Fig. 3.22 and 3.24 for Lake Stechlin are plotted in Fig. 3.23 and 3.25 for the whole lake area (except acid lakes).

The given relationships were expressed by simple potential equations. Only the equation of the connections between the Secchi depth and the extinction coefficient of blue light is identical for Lake Stechlin and for the lakes of its area (Figs. 3.22, 3.23). The other equations are different for Lake Stechlin and its lake area.

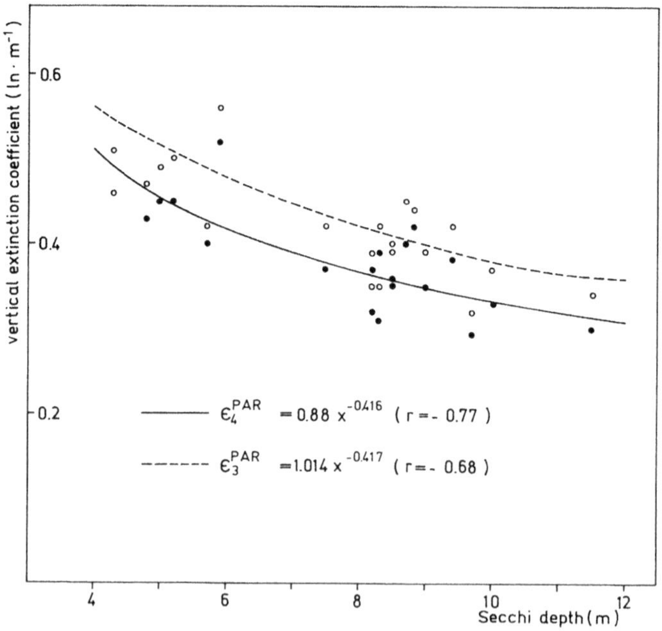

Figure 3.24 Relationship between the Secchi disc transparency and the vertical extinction coefficient for PAR in Lake Stechlin; solid circles = (ε (equation 26); open circles = ε (equation 27).

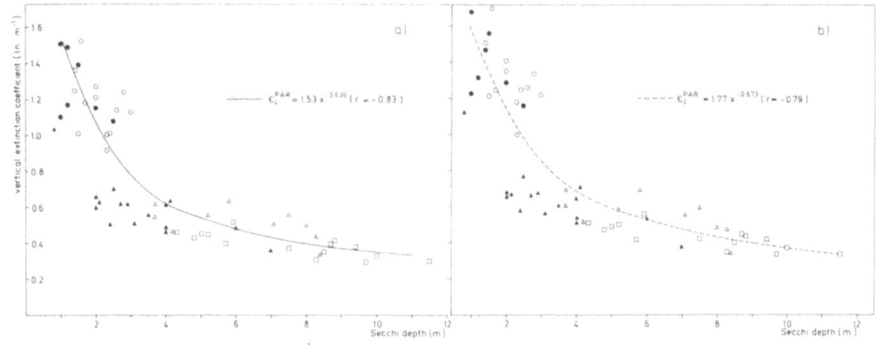

Figure 3.25 Relationships between the Secchi disc transparency and the vertical extinction coefficient for PAR in lakes of the study area. (a) – ε (equation 26); (b) – ε (equation 27); □ Lake Stechlin, △ Lake Nehmitz, ▲ Lake Breiter Luzin, ○ Lake Dagow, ● Lake Haussee.

But a satisfactory statistical certainty was established for all the other relations too, except for the connection between the Secchi depth and the extinction coefficient for red light in Lake Stechlin (Fig. 3.22). The calculated extinction coefficient on basis of 4 wavelengths of visible light (equation 27) is about 10% lower than the same on basis of 3 wavelengths (equation 26). The statistical satisfactory correlation between Secchi depth and the vertical extinction coefficient for visible light is a little better on the basis of 4 wavelengths.

References

Anderson, E.R. 1952. Energy budget studies. In: Waterloss investigations, Vol 1, Lake Hefner studies. Techn. Report, Geol. Survey Circular 229: 71–119.

Angström, A. 1915. A study of the radiation of the atmosphere. Smithonian Inst. Misc. Coll. Washington 65, No. 3.

Asklöf, S. 1920. Über den Zusammenhang zwischen der nächtlichen Wärmestrahlung, der Bewölkung und der Wolkenart. Geogr. Ann. 2: 253–259.

Bolz, H.M. & Fritz, H. 1950. Tabellen und Diagramme zur Berechnung der Gegenstrahlung und Ausstrahlung. Z. Meteor. 4: 314–317.

Dirmhirn, I. 1953. Einiges über die Reflexion der Sonnen- und Himmelsstrahlung an verschiedenen Oberflächen. Wetter und Leben 5: 86–94.

Glugla, G. 1970. Zur Berechnung des aktuellen Wassergehaltes und Gravitationswasserabflusses im Boden. Diss., K.-M.-Univ. Leipzig, Sekt. Physik.

Günneberg, F. 1973. Abwärme in Gewässern. Dt. Gewässerkdl. Mitt. 17: 14–21.

Heitmann, M.–L. 1976. Die thermische Belastung des Stechlinsees durch ein Kernkraftwerk. Materialien des Internationalen Symposiums über Eutrophierung und Sanierung von Oberflächengewässern. 20. – 25. 9. 1976 Karl-Marx-Stadt, DDR, Bd. III: 133–146.

Heitmann, M.-L., Richter, D. & Schumann, D. 1969. Der Wärme- und Wasserhaushalt des Stechlin- und Nehmitzsees. Abh. Meteor. Dienstes DDR Nr. 96 (12).

Hutchinson, G.E. & Löffler, H. 1956. The thermal classification of lakes. Proc. Nat. Acad. Sci. Washington 42.

Karbaum, H. 1969. Der Niederschlag als Wasserhaushaltsgröße. Abh. Meteor. Dienstes DDR Nr. 86 (11).

Kirillova, T.B. 1970. Radiacionnyi Režim ozer i vodochranilišč. Leningrad: Gidrometizdat.

Klimaatlas, 1953. Klimaatlas für das Gebiet der Deutschen Demokratischen Republik. Berlin: Akademie-Verlag.

Köppen, W. 1931. Grundriß der Klimakunde. Berlin, Leipzig: De Gruyter.

Kortüm, F. 1961. Über den Zusammenhang zwischen dem Energie- und dem Niederschlagsangebot und der Gesamtverdunstung. Z. Meteor. 15: 189–192.

Koschel, R. 1974. Einfluß der physikalischen und chemischen Umweltfaktoren auf die Primärproduktion des Phytoplanktons im Stechlinsee. Diss. A, Techn. Univ. Dresden, 1–72.

Koschel, R. 1981. Lichtklima. In: Mothes, G., Physikalische und chemische Parameter der Wasserbeschaffenheit in Seen des Stechlinseegebietes. Limnologica 13: 3–6, 15, 21.

Koschel, R. & Mothes, G. 1976. Änderung der biologischen Produktivität in einem Seensystem durch den Kühlwasserzufluß eines Kernkraftwerkes. Materialien des Internationalen Symposiums über Eutrophierung und Sanierung von Oberflächengewässern. 20.–25.9.1976 Karl-Marx-Stadt, DDR, Bd. III: 219–233.

Knoll, W. 1976. Ein neues Gerät für die Messung von Wassertemperaturen. Wasserwirtschaft-Wassertechnik 26 (2): III–IV.

Krambeck, H.-J. 1982. Solar energy and quanta in Baltic lakes. Arch. Hydrobiol. 95: 197–206.

Kuhn, W. 1972. Physikalisch-meteorologische Überlegungen zur Nutzung von Gewässern für Kühlzwecke. Arch. Meteor. Geophys. Bioklimatol. A, 20: 95–122.

Kusmin, P.P. *et al.* 1968. Intra-annual distribution of evaporation from land in the UdSSR. In: Sowj. Hydrol., Selected papers 3: 5–11.

Megard, R.O., Settles, J.C., Boyer, H.A. & Combs, W.S. 1980. Light, secchi discs and trophic states. Limnol. Oceanogr. 25: 373–377.

Morel, A. & Smith, R.C. 1974. Relation between total quanta and total energy for aquatic photosynthesis. Limnol. Oceanogr. 19: 591–600.

Richter, D. 1964. Sättigungsdefizit und Verdunstung. Acta Hydrophys. 20: 73–109.

Richter, D. 1977. Zur einheitlichen Berechnung der Wassertemperatur und der Verdunstung von freien Wasserflächen auf statistischer Grundlage. Abh. Meteor. Dienstes DDR, Nr. 119 (16).

Richter, D. & Wiedrich, R. 1967. Die Sichttiefen im Stechlin- und Nehmitzsee. Limnologica 5: 299–305.

Richter, D., Neubert, W. & Klämt, A. 1979. Temperatur und Wärmehaushalt des thermisch belasteten Stechlin- und Nehmitzsees. Abh. Meteor. Dienstes DDR Nr. 123 (16).

Sauberer, F. 1962. Empfehlungen für die Durchführung von Strahlungsmessungen an und in Gewässern. Mitt. Int. Verein. Limnol. 11.

Schanz, F. 1982. Light conditions in Lake Zurich 1979–1981. Part I: Secchi disc transparency. Vierteljahrsschr. Naturforsch. Ges. Zürich 127: 357–367.

Schumann, D. 1970. Erste Ergebnisse einer Wasserhaushaltsuntersuchung im Norddeutschen Jungmoränengebiet. Wasserwirtschaft-Wassertechnik 20: 198–202.

Sevruk, B. 1982. Methods of correction for systematic error in point precipitation measurement for operational use. WMO-No. 589, Geneva.

Sonntag, D. 1963. Ein Pyranometer bzw. Effektivpyranometer mit galvanisch erzeugter Thermosäule. Z. Meteor. 17: 49–56.

Vollenweider, R.A. 1960. Beiträge zur Kenntnis optischer Eigenschaften der Gewässer und Primärproduktion. Mem. Ist. Ital. Idrobiol. 12: 201–244.

Vollenweider, R.A. 1969. A manual on methods for measuring primary production in aquatic environments. IBP Handbook No. 12. Oxford: Blackwell, pp. 158–171.

Westlake, D.F. 1965. Some problems in the measurement of radiation under water. Photochem. Photobiol. 4: 849–868.

World Meteorological Organization, 1972. Pollution of the environment. Cooling water discharges to lakes and seas. WMO Regional Association VI (Europe), WMO-No. 329, Luzern.

4. The chemical environment

G. Mothes, R. Koschel & G. Proft

Introduction

Changes in the chemical features of Lakes Stechlin, Nehmitz and Dagow as a result of the cooling water circuit of the nuclear power plant must be expected both spatially and temporally. In the following sections we will give basic data about the phosphorus, carbon, nitrogen, silicium and oxygen content of the water characterizing the chemical state of the lake system before and during the operation of the nuclear power plant.

The results give an insight into the temporal and spatial concentration gradients, the long-term changes and the external loads of these elements (import and export). This is intended to contribute to a deeper understanding of the main features of lake metabolism, such as primary production, secondary production, microbial activities, and sedimentation which will be described in other chapters of this volume.

The phosphorus budget (R. Koschel & G. Mothes)

Phosphorus occupies a key position in the budget of substances in our study area. Orthophosphate concentrations in all the lakes analysed in the euphotic zone are partially far below the chemical detection limit (Mothes 1974, 1981; Koschel 1981b, additionally unpublished material). From spring to autumn bioproduction is mainly limited by phosphorus (Koschel 1974a, b, 1976, 1980, 1981a; Mothes 1974, 1981a, 1981b). Since phosphorus is decisive for the budget of substances in our study area, this nutrient is shown in detail.

The annual course of pelagic phosphorus

All phosphorus analyses were carried out by the molybdenum-sulphuric acid

Casper, S. Jost (ed.), Lake Stechlin. A temperate oligotrophic lake. ISBN 978-94-010-8930-2.

method (Vogler 1966; cf. Mothes 1974, 1981a; Proft, unpublished). Dissolved orthophosphate (DIP), the sum of the dissolved phosphates (DP) and the total phosphate (TP) were analysed. The difference between TP and DP results in the particular phosphate (PP).

In the chemical analysis of phosphorus concentrations <2 mg P · m^{-3} there may be considerable overestimation of the real phosphorus concentration (Rigler 1966, 1968; Koschel 1974a, 1981b; Mothes 1981a). Since reagents are added in the chemical analysis, PO_4-P from the DP fraction is set free. Moreover, there may be excretion of cellular phosphates during the filtration process, which (particularly in the lower concentration areas) can strongly influence the DIP values.

Calculations of the DIP concentration on the basis of analyses with the P-32 method (Koschel 1974a, 1980, 1981b; Mothes 1981a) showed that in an extreme case values may be measured by chemical analysis which are 100 times above normal (Fig. 4.1).

The chemically ascertained phosphorus concentrations <2 mg P · m^{-3} must, therefore, be treated with caution. In our discussion they mainly serve to characterize dynamic aspects and do not reflect absolute phosphorus concentrations.

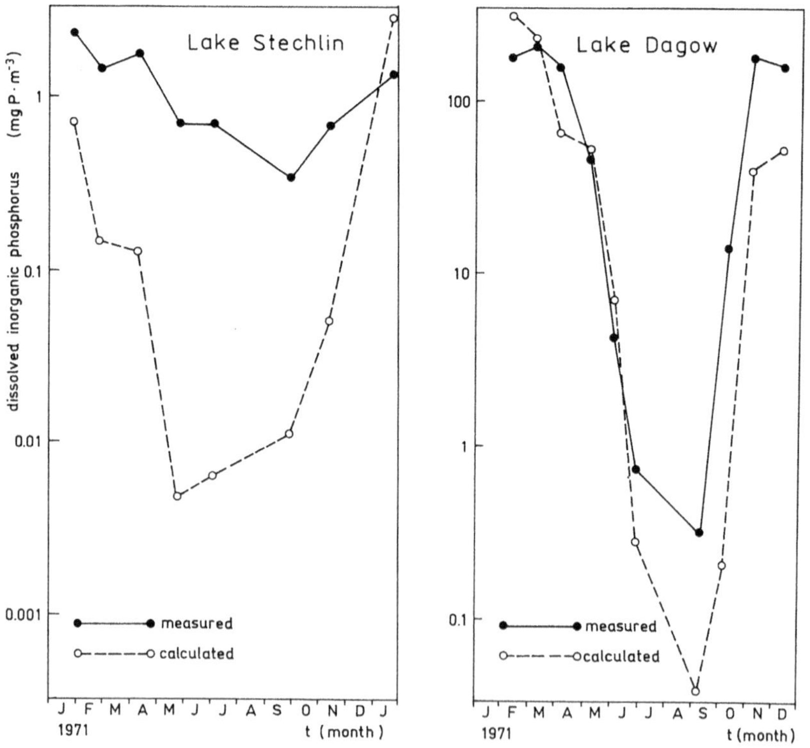

Figure 4.1 Comparison of chemical analysis values and PO_4-P-concentration in the surface water of Lakes Stechlin and Dagow calculated by P 32-analyses.

Lake Stechlin

The DIP concentration in the euphotic zone in Lake Stechlin during the year lies below 4 mg $P \cdot m^{-3}$; as a rule there are values of or below 1 mg $P \cdot m^{-3}$ (Fig. 4.2).

Throughout the winter an average of 2 mg $P \cdot m^{-3}$ is found. In some years, however, DIP values fall below the detection level as early as in February/March. This decrease is due to algae development in late winter (cf. p. 173).

In general one can say that the DIP concentration in the 0 m zone lies above comparable values in deeper strata of the epilimnion and the upper hypolimnion. Only in late autumn does the DIP concentration in the euphotic zone again reach 2 mg $P \cdot m^{-3}$ (Fig. 4.2).

During the summer stagnation period a distinct but, in absolute terms, minor phosphate accumulation takes place in the hypolimnion. At a depth of 60 m there are DIP concentrations between 6 and 15 mg $P \cdot m^{-3}$ towards the end of the summer stagnation period (late November, early December) (Figs. 4.2, 4.6).

DP shows a similar behaviour as DIP in its spatial and temporal distribution. Absolute values for DP in the euphotic zone vary between 8 mg $P \cdot m^{-3}$ at the beginning of the year and 3 mg $P \cdot m^{-3}$ later in the year. In extreme cases DP values fall below 1 mg $P \cdot m^{-3}$ (e.g. summer 1970; Fig. 4.3). At a depth of 60 m in the hypolimnion DP reaches values between 10 and 18 mg $P \cdot m^{-3}$ at the end of the summer stagnation period (Fig. 4.3).

The difference of DP and DIP varies in the euphotic zone between 1 and 8 mg $P \cdot m^{-3}$ and in the hypolimnion between 7 mg $P \cdot m^{-3}$ (beginning of summer stagnation) and 4 mg $P \cdot m^{-3}$ (end of stagnation).

The TP concentration in the free water of Lake Stechlin may lie between 8 and 18 mg $P \cdot m^{-3}$. In the euphotic zone maximum values are found at the end of the spring full circulation period. They range from 9 to over 16 mg $P \cdot m^{-3}$, on an average 10 mg $P \cdot m^{-3}$ (Figs. 4.4, 4.5).

At a depth of 60 m in the hypolimnion maximum values of 11–18 mg $P \cdot m^{-3}$ are to be found at the end of the summer stagnation period (late November, early December), medium values 13 mg $P \cdot m^{-3}$ (Fig. 4.6).

The share of PP in TP in the euphotic zone amounts to 50 per cent on an average (maximum 90 per cent). In the deeper strata of the hypolimnion its share is mostly below 50 per cent.

Lake Nehmitz

Phosphate concentrations in the euphotic zone of Lake Nehmitz are almost the same as those in Lake Stechlin. The DIP concentration lies below 4 mg $P \cdot m^{-3}$ throughout the year, from February–September often below 1 mg $P \cdot m^{-3}$.

In the hypolimnion, however, there is a strong phosphate accumulation during the summer stagnation period (cf. p. 395). At the end of the summer stagnation

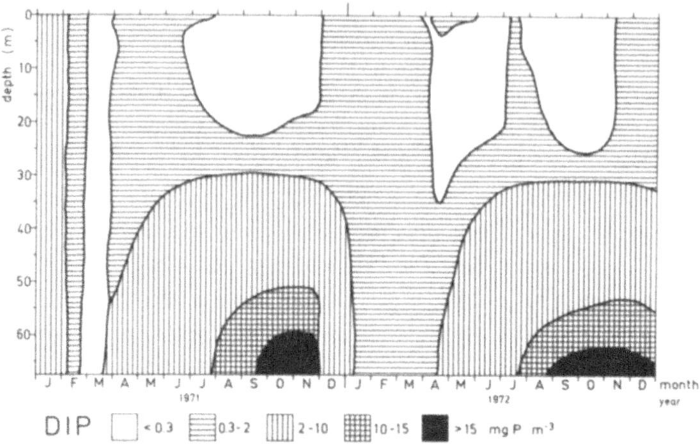

Figure 4.2 Temporal and spatial distribution of DIP in Lake Stechlin (deepest point) in 1971 and 1972.

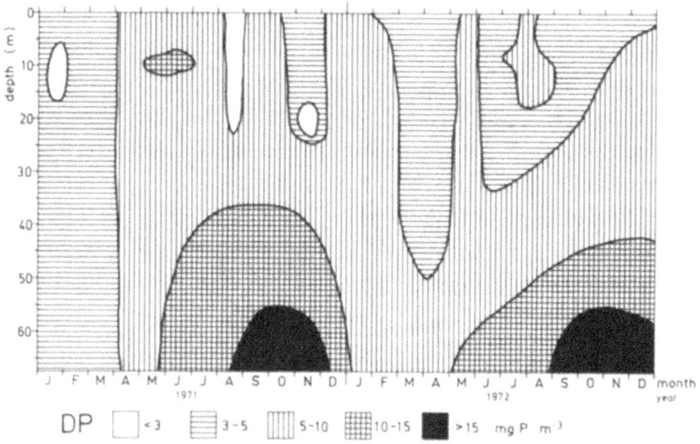

Figure 4.3 Temporal and spatial distribution of DP in Lake Stechlin (deepest point) in 1971 and 1972.

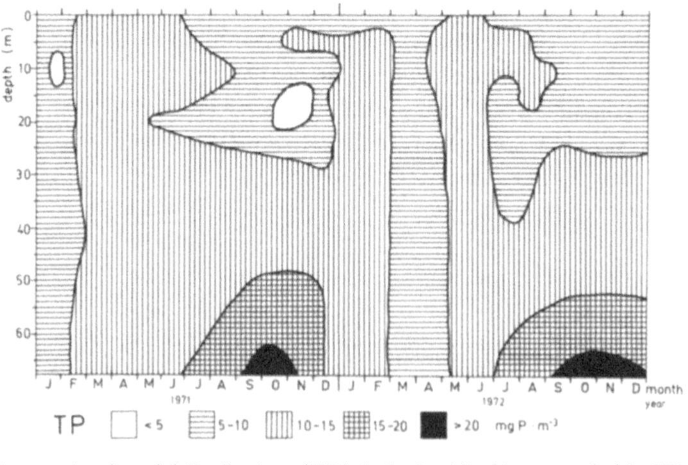

Figure 4.4 Temporal and spatial distribution of TP in Lake Stechlin (deepest point) in 1971 and 1972.

period (October), medium maximum DIP concentrations can be measured at 150 mg $P \cdot m^{-3}$ at a depth of 15 m, and 530 mg $P \cdot m^{-3}$ at a depth of 17.5 m. The values at a depth of 15 m vary greatly during the years of analysis (Table 4.1).

The DP concentration in the euphotic zone is 3–15 mg $P \cdot m^{-3}$. In the hypolimnion DP is basically determined by DIP.

The TP concentration reaches its maximum in the euphotic zone at the end of the spring full circulation period, similar to Lake Stechlin. Maximum values of 17 mg $P \cdot m^{-3}$ were measured (1973, 1975). In other years maximum values were lower (10–15 mg $P \cdot m^{-3}$). In the hypolimnion maximum values of 270 mg $P \cdot m^{-3}$ at a depth of 15 m and 560 mg $P \cdot m^{-3}$ at a depth of 17.5 m are found at the end of the summer stagnation period.

PP represents 50 per cent of TP on an average (maximum 80–90 per cent in spring and summer).

Lake Dagow

The eutrophic Lake Dagow shows medium maximum DIP concentrations of 115–190 mg $P \cdot m^{-3}$ and medium maximum TP concentrations of 165–335 mg $P \cdot m^{-3}$ at the beginning of the year (Table 4.2). The highest concentrations were found in the late nineteen sixties and early nineteen seventies. At that time large duck and Carp populations constituted a heavy load on the lake (cf. p. 472).

During the year the separate phosphorus shares are highly differentiated. The DIP concentration in the euphotic zone has a range of over two decimal exponents (Fig. 4.5) during the year. During the stagnation period concentrations of <3 mg $P \cdot m^{-3}$ in the upper zone of the epilimnion are not rare. At a depth of 7.5 m, concentrations at the end of August/beginning of September reach max-

Table 4.1 Mean maximum phosphorus concentration at the end of the summer stagnation period in the lower ranges of the hypolimnion of Lakes Stechlin, Nehmitz and Dagow.

Lake	Year	Depth (m)	DIP (mg $P \cdot m^{-3} \pm SD$*)	TP (mg $P \cdot m^{-3} \pm SD$*)
Stechlin	1970–1982 (Nov/Dec)	60	7.2 ± 2.5	13.0 ± 2.8
Nehmitz-N	1970–1982 (Sep/Oct)	15	105.0 ± 63.0	152.0 ± 75.0
Nehmitz-S	1970–1975 (Sep/Oct/Nov)	15	72.0 ± 35.0	145.0 ± 52.0
Dagow	1971 (Oct)	7.5	1340.0	1510.0
Dagow	1972–1982	7.5	490.0 ± 340.0	860.0 ± 510.0

* SD = standard deviation.

Table 4.2 Mean maximum phosphorus concentration at the end of winter in the free water of Lakes Stechlin, Nehmitz and Dagow.

Lake	Year	DIP (mg $P \cdot m^{-3} \pm SD$*)	TP (mg $P \cdot m^{-3} \pm SD$*)
Stechlin	1971–1982 (Feb/Mar)	2.2 ± 1.0	10.7 ± 1.1
Nehmitz-N	1971–1980 (Feb/Mar)	1.0 ± 0.5	13.4 ± 4.4
Nehmitz-S	1971–1975 (Mar/Apr)	0.3 ± 0.3	11.7 ± 1.3
Dagow	1971 (Mar)	190.0	335.0
Dagow	1972–1982 (Feb/Mar)	114.0 ± 34.0	164.0 ± 40.0

* SD = Standard deviation.

imum values of 1.4 g $P \cdot m^{-3}$ (1971) (Table 4.1). In the thermocline, during periods low in wind, there can be DIP gradients of 200–700 mg $P \cdot m^{-3}$ per metre of depth.

DP exceeds DIP on an average by 20 mg $P \cdot m^{-3}$. During the summer stagnation period there may be, however, more than 50 mg $P \cdot m^{-3}$ in the epilimnion and 500–1000 mg $P \cdot m^{-3}$ of dissolved non-orthophosphate concentrations in the hypolimnion. Such high hypolimnetic non-orthophosphate concentrations occur at the beginning of anoxia. In the further course of the summer stagnation period, the dissolved non-orthophosphate is converted to DIP, so that values approximate again the average difference of 200 mg $P \cdot m^{-3}$ between DP and PO_4-P.

During the summer stagnation period the TP concentration reaches minimum values of 30–55 mg $P \cdot m^{-3}$ in the epilimnion and maximum values of 500–1700 mg $P \cdot m^{-3}$ (7.5 m) in the hypolimnion.

The PP concentration lies between 10 mg $P \cdot m^{-3}$ (low plankton development) and 200 mg $P \cdot m^{-3}$ (mass development of algae). At times there can be up to 300 mg $P \cdot m^{-3}$ in the hypolimnion too, analysable as particular phosphate.

The pelagic phosphorus content during the investigation period (1971–1982) in Lakes Stechlin, Nehmitz and Dagow

The maximum phosphorus concentrations that occur for DIP at the beginning of the spring full circulation and for TP at the end of the spring full circulation in Lake Stechlin differ only to a small degree between 1971 and 1982 (Fig. 4.5). The medium value of the maximum DIP concentration is 2.2 mg $P \cdot m^{-3}$, that of TP 10.7 mg $P \cdot m^{-3}$ (Table 4.1). The maximum DIP concentration and the maximum TP concentration are not identical in the individual years of investigation (Fig. 4.5). High TP concentrations, as a rule, coincide with likewise high DIP values (e.g., 1971, 1973, 1975, 1979). If the TP concentrations are lower, there is no visible correlation between TP and DIP (e.g., 1976, 1981, 1982). The height of the maximum TP value at the end of the spring full circulation seems to depend on its

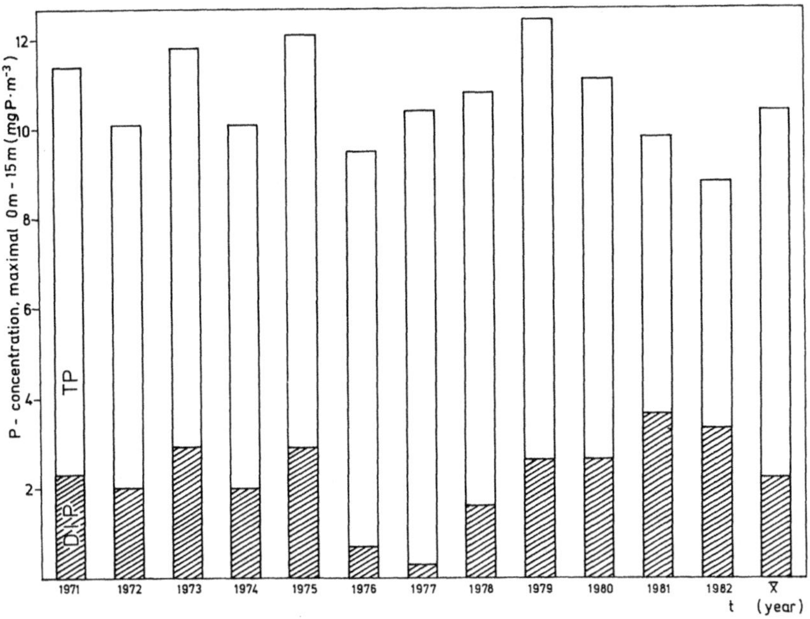

Figure 4.5 Maximum DIP-concentration at the beginning of the spring full circulation and maximum TP-concentration at the end of the spring full circulation from 1971 to 1982 in Lake Stechlin.

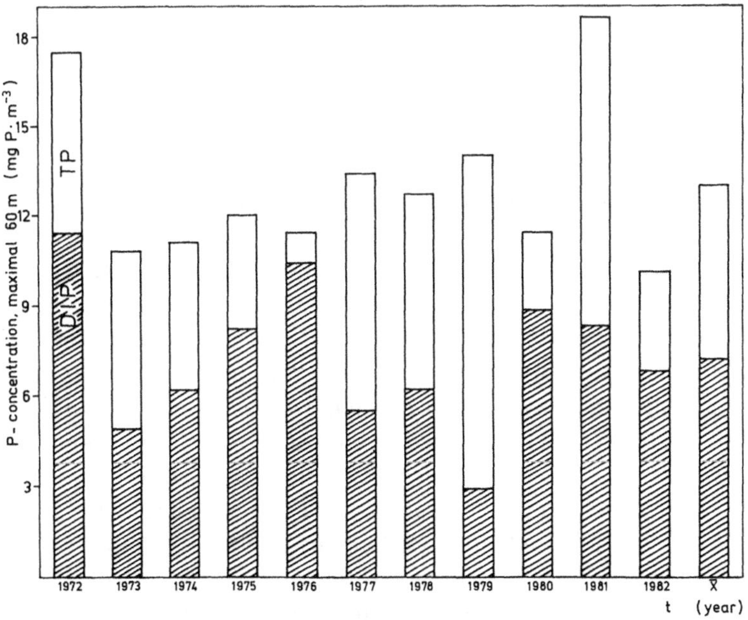

Figure 4.6 DIP- and TP-concentrations at the end of the summer stagnation period (November/ December), depth of 60 metres, in Lake Stechlin (1972–1982).

duration (Mothes 1981a). Prolonged circulation periods favour TP accumulation.

The hypolimnetic phosphorus accumulation in Lake Stechlin is low (Fig. 4.6). On an average, at a depth of 60 m, the TP concentration exceeds the values of early spring by only 20 per cent or so at the end of the summer stagnation period. The share of DIP in TP in the hyplimnion increases, however, in the course of the stagnation period. At the beginning of the stagnation period, at a depth of 60 m, it runs to a medium 20 per cent and at the end, to 60 per cent. Particularly high values of a hypolimnetic phosphorus accumulation were measured in 1970, 1971, 1972 and 1981. They possibly relate to the similarly high phosphate imports from Lake Nehmitz and Lake Dagow in those years (there were, however, no measurements of Lake Dagow imports between 1970 and 1972).

In Lake Nehmitz the DIP concentration is very low in late winter, on an average 1.0 mg $P \cdot m^{-3}$ (Lake Nehmitz-N) and 0.3 mg $P \cdot m^{-3}$ (Lake Nehmitz-S; Table 4.2). The TP values at that time are roughly similar to those of Lake Stechlin, with differences in the individual years of investigation being greater in Lake Nehmitz-N than in Lake Nehmitz-S (Table 4.2).

The DIP and TP concentrations at the end of the summer stagnation period, at a depth of 15 m, show remarkable differences in the individual years of investigation (Table 4.1). However, the 15 m values are possibly strongly influenced by instabilities of stratification at the end of the summer stagnation period.

In Lake Dagow there are mainly the DIP concentrations, and to a lesser degree those of TP, that have receded since the early seventies (Table 4.1). In the middle and at the end of the seventies the DIP concentration at the beginning of the spring full circulation was only some 30 per cent compared with the 1971 value. In 1981 and 1982 values increased again. This is connected with various building operations temporarily worsening the situation in Lake Dagow.

Horizontal differences of phosphorus content in Lake Stechlin

The phosphorus concentration in the euphotic range of the pelagic zone shows only minor differences in the three main basins (north, south and west basin). If the cooling-water circulation is in operation the highest concentrations are found in the south basin, followed by the west basin. The lowest values are measured in the north basin. On an average, values in the south basin are around 14 per cent (DIP) and 24 per cent (TP), and in the west basin around 10 per cent (DIP) and 6 per cent (TP) above comparable values in the north basin. This is due to the cooling-water circulation. The cooling water goes through the west basin and turns into the south basin before reaching the east basin. There it absorbs input from Lake Dagow and, at the end of the south basin, leaves Lake Stechlin towards Lake Nehmitz. The measuring point in the north basin is hardly touched by the cooling-water current.

In the terminal cove of the west basin we find in the immediate range of the

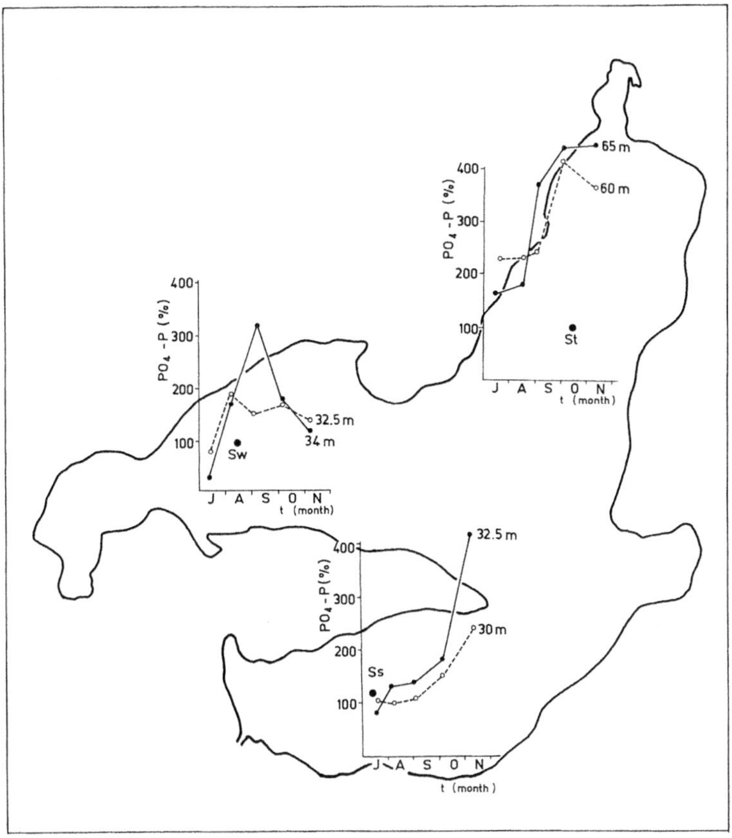

Figure 4.7 Comparisons of the phosphorus accumulation in the hypolimnion of the north, south and west basins of Lake Stechlin during the summer stagnation period (average values 1980–1981). Starting values in early spring = 100 per cent.

cooling-water current slowly increased DIP values over the thermally unloaded free-water regions. On an average, DIP values, at temperatures of over 28° C, are around 1.6 mg $P \cdot m^{-3}$ over comparable values in the uninfluenced pelagic zone (p. 425).

Also near the outlet of Lake Dagow in the east basin, there are at times increased phosphorus concentrations. In the depth horizons of the north basin directly above the lake bottom, there are, as expected, the highest phosphate values. In the south basin values at a depth of 32.5 m reach the same concentrations as in 65 m in the north basin (Fig. 4.7). A considerably lower phosphate enrichment occurs in the depth horizons of the west basin (Fig. 4.7). This is obviously an effect of the water flow in Lake Stechlin. In the west basin there are higher remineralization rates due to higher medium temperatures in the surface water region, so that the hypolimnic phosphate load is lower in this basin.

The 'import' of phosphorus

The phosphate import is composed of inputs from the atmosphere, from the contents of inflowing waters and from direct inputs of civilization.

Atmospheric phosphorus inputs

Initial investigations on the phosphorus input through precipitation were conducted in the early seventies (Mothes 1981a). In 1980 this problem was taken up again by an extended research programme taking account of the hard-to-measure 'dry' phosphorus input (dust, etc.).

The collecting recipients for continuous measurements were on a measuring raft some 50 m from the shore. The precipitated water was analysed immediately and in case of prolonged rain, no later than 24 hours after. The 'dry' phosphorus input was calculated from the difference between the sum total of the overall atmospheric phosphorus input and the phosphorus input of the precipitations. For the 'dry' phosphorus input we also distinguish between DIP, DP and TP. The DIP values are generally a bit too high, because part of the non-orthophosphate in the collecting recipients is converted to PO_4-P.

The phosphorus analyses of the precipitated water showed for 1980 medium DIP concentrations of 22.3 mg $P \cdot m^{-3}$ and a medium TP concentration of 31.7 mg $P \cdot m^{-3}$ (Table 4.3). Particularly high phosphorus concentrations were found in precipitated water following prolonged dry periods. In case of prolonged precipitations there is a rapid decrease in the phosphorus concentration in the precipitated water. The phosphorus concentration levels out at around 12 mg $P \cdot m^{-3}$ (DIP) and 16 mg $P \cdot m^{-3}$ (TP).

For the 'dry' input the medium for DIP was 20 mg $P \cdot m^{-2} \cdot a^{-1}$ and for TP 23 mg

Table 4.3 Medium, maximum and minimum phosphorus concentrations in the precipitated water of the Lake Stechlin area, 1980.

Phosphorus fraction	Number of measurings	Medium value ($mg\ P \cdot m^{-3}$)	Maximum value ($mg\ P \cdot m^{-3}$)	Minimum value ($mg\ P \cdot m^{-3}$)
Rainwater				
DIP	20	22.3	81	5
Dissolved non-DIP	15	2.8	5	1
PP	15	6.6	30	1
TP	15	31.7	104	13
Snow				
DIP	7	6.5	10.5	3.5
Dissolved non-DIP	7	1.8	3.0	1.0
TP	7	16.0	35.0	8.0

$P \cdot m^{-2} \cdot a^{-1}$ (Table 4.4). In prolonged dry periods particularly high 'dry' phosphorus inputs occur. In such phases maximum values of 0.14 mg $P \cdot m^{-2} \cdot d^{-1}$ for DIP and 0.18 mg $P \cdot m^{-2} \cdot d^{-1}$ for TIP were measured.

The ratio between precipitated input and 'dry' phosphorus input is about 1:1 for the Lake Stechlin area.

The overall atmospheric phosphorus input is 35.5 mg $P \cdot m^{-2} \cdot a^{-1}$ for DIP and 45.2 mg $P \cdot m^{-2} \cdot a^{-1}$ for TP (Table 4.4).

A comparison between the TP concentrations found between 1974 and 1980 in the precipitated water showed only minor differences. In 1974 the medium value was 28.3 mg $P \cdot m^{-3}$, in 1980 31.7 mg $P \cdot m^{-3}$. The phosphorus surface load through atmospheric inputs, compared with other regions, is relatively low for the Lake Stechlin area (Table 4.5). In the precipitated water concentrations are in the range of the values measured by Nehring & Wilde (1982) for the Darss, Zingst and Bodden areas. However, an increase in the phosphorus surface load through atmospheric inputs by only 50 per cent (cf. Table 4.5, in the south of the GDR about tenfold) would be sufficient to constitute a grave danger for the oligotrophic condition of Lake Stechlin, if only for these imports.

Table 4.4 Atmospheric phosphorus input into Lake Stechlin.

	DIP		TP	
	(mg $P \cdot m^{-2} \cdot a^{-1}$)	(kg $P \cdot a^{-1}$)	(mg $P \cdot m^{-2} \cdot a^{-1}$)	(kg $P \cdot a^{-1}$)
'Wet' input (rain, snow)	15.4	65.5	21.9	99
'Dry' input (dust, etc.)	20.1	85.3	23.3	93
Total	35.5	150.8	45.2	192

Table 4.5 Phosphorus concentration in precipitation and area-related values of atmospheric phosphorus input in different regions of the GDR.

Region	Precipitation	DIP (mg $P \cdot m^{-3}$)	TP (mg $P \cdot m^{-3}$)	TP, area-related (mg $P \cdot m^{-3} \cdot a^{-1}$)	Source
Southern GDR	Rain	332 (10–2000)			Peukert 1976
Lake Stechlin	Rain	22 (5–81)	32 (13–104)	23	This chapter
	Dust			22	
Darss-Zingst Bodden	Rain	6–29	7–49	24	Nehring & Wilde 1982
Baltic Sea	Rain	35 (2–124)	129 (59–412)	77	Nausch 1983

Phosphorus inputs through inflows

The phosphorus input through surface inflows into Lake Stechlin takes place via the cooling-water circulation of the nuclear power plant (on an average 10.5 million $m^3 \cdot a^{-1}$) and via the outflow ditch from Lake Dagow (from 1978–1982 an average of 0.3 million $m^3 \cdot a^{-1}$).

The cooling-water circulation of the nuclear power plant includes Lake Nehmitz. Part of the nutrient load in Lake Nehmitz, therefore, also goes to Lake Stechlin. With the cooling-water circulation the epilimnetic water stays in Lake Nehmitz some 15 days and in Lake Stechlin some 260 days. The nutrient load from surface inflows into Lake Nehmitz takes place via the sewage plant belonging to the nuclear power plant (0.04 million $m^3 \cdot a^{-1}$) and via the outflow ditch of the highly eutrophic Lake Wulwitz (0.2 million $m^3 \cdot a^{-1}$). The nutrient input from these two inflows has, however, largely been cut off in the last few years (cf. p. 474).

Table 4.6 summarizes the phosphorus loads (DIP, TP) of the inflows into the Lakes Stechlin and Nehmitz.

From Lake Dagow on an average 8.5 kg $P \cdot a^{-1}$ (DIP) and 25.6 kg $P \cdot a^{-1}$ (TP) go into Lake Stechlin. Phosphorus loads from Lake Dagow differ largely though (Fig. 4.8). In the years 1981 and 1982, which were rich in outflows, DIP loads of some 15 kg $P \cdot a^{-1}$ and TP loads of some 40 kg $P \cdot a^{-1}$ were measured. The highest phosphorus loads were measured in the winter months of January and February and in the summer months of July and August (Fig. 4.8). On the whole, however, phosphorus loads from Lake Dagow have greatly declined in the past few years. As late as 1974 a TP load of ca. 46 kg P from Lake Dagow was calculated for the period May–August (Mothes 1981b).

Table 4.6 Phosphorus input into Lakes Stechlin and Nehmitz through surface inflows.

Source	DIP (kg $P \cdot a^{-1}$)	TP (kg $P \cdot a^{-1}$)	DIP (mg $P \cdot m^{-2} \cdot a^{-1}$)	TP (mg $P \cdot m^{-2} \cdot a^{-1}$)
Phosphorus input into Lake Stechlin				
Lake Nehmitz	51.9	66.0	12.2	15.6
Warming up	60.4	0	14.2	0
Lake Dagow	8.5	25.6	2.0	6.0
Total	120.8	91.6	28.4	21.6
Phosphorus input into Lake Nehmitz				
Sewage plant	182.5	226.9	175.4	218.2
Lake Wulwitz	8.8	16.4	8.5	15.8
Total*	191.3	243.3	183.9	233.9

* In 1974–1980 the phosphorus input into Lake Nehmitz is only some 50 per cent, in 1980–1983 only some 40 per cent and after 1983 only some 1 per cent of the total indicated (cf. p. 99).

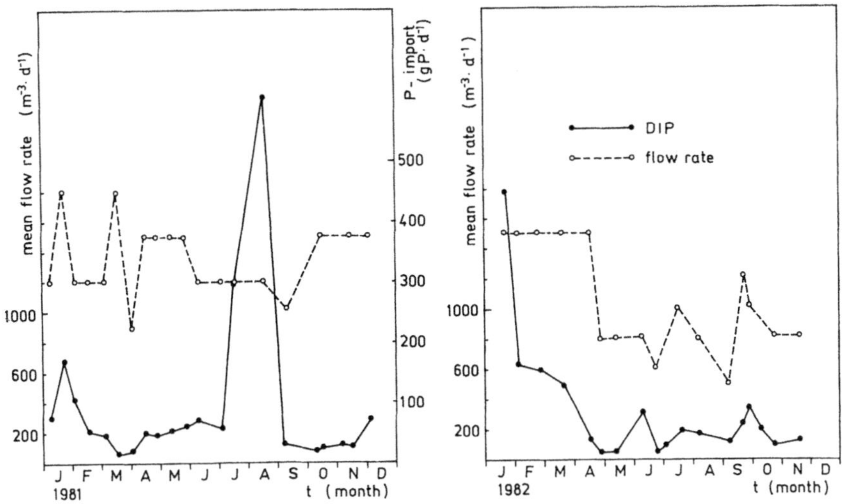

Figure 4.8 DIP load and outflow from Lake Dagow into Lake Stechlin in 1981 and 1982.

From Lake Nehmitz 51.9 kg P (DIP) · a^{-1} and 66 kg P (TP) · a^{-1} go into Lake Stechlin. In addition, during the months of summer, 60.4 kg P (DIP) are mobilized in the warm-water zone (T = 28° C) (cf. p. 425).

Until the mid-seventies Lake Nehmitz got 182.5 kg P · a^{-1} (DIP) and 226.9 kg P · a^{-1} (TP) from sewage plant outflow and 8.8 kg P · a^{-1} (DIP) and 16.4 kg P · a^{-1} (DIP) from Lake Wulwitz outflow. The phosphorus load has been gradually reduced up to the present moment (cf. p. 98; Table 4.6).

The phosphorus inflow through the ground-water is very low. The DP concentrations in the area near the lake, in the ground-water, run to 1 mg P · m^{-3}. Only in the local area of Neuglobsow-Dagow higher values were measured in the ground-water (DP on an average 40 mg P · m^{-3}). Overall, a phosphate load through ground-water of some 1 kg P (DIP) · a^{-1} can be calculated.

Phosphorus inputs by other sources

Phosphates are also introduced into Lake Stechlin by bathers. There is a phosphate import of some 0.1 g P (Schulz 1981, TGL 27885/01 1982) per bather/day. This input occurs mainly in the form of DIP.

For the calculation of the DIP input, 500 bathers per day were assumed for the months of July and August and 250 bathers per day for June and September. This alone totals a phosphorus input of 4.6 kg P · a^{-1}.

The foliage input in the months of October and November into Lake Stechlin is another source of phosphorus. Twenty-three t of dry mass per year in the shape of leaves can be calculated for Lake Stechlin (cf. page 406). With a phosphorus content of 0.04 per cent in the dry leaves mass, this amounts to a TP input of 9 kg P.

The export of phosphorus

The export of phosphorus from Lake Stechlin is considerably smaller than its import. There are three important ways of phosphorus export: phosphorus output through outflows, through hatching water insects and fishing.

As for the phosphorus export through outflows, there is an amount of 1.0 kg P (DIP) $\cdot a^{-1}$ and 6.0 kg P (TP) $\cdot a^{-1}$ for Lake Stechlin (medium outflow 490 000 $m^3 \cdot a^{-1}$, medium DIP concentration 2 mg P $\cdot m^{-3}$, medium TP concentration 12 mg P $\cdot m^{-3}$).

The phosphorus export through water insects was calculated at 18.1 kg P (TP) $\cdot a^{-1}$, with 4.3 kg P $\cdot a^{-1}$ coming from the profundal zone (medium biomass 2.8 g $\cdot m^{-2} \cdot a^{-1}$, and 1.9 mg P $\cdot m^{-2} \cdot a^{-1}$) and 13.8 kg P $\cdot a^{-1}$ from the littoral zone (medium biomass 10.2 g $\cdot m^{-2} \cdot a^{-1}$ and 6.9 mg P $\cdot m^{-2} \cdot a^{-1}$).

Fishing accounts for the highest phosphorus export. With a medium annual proportion of 5.7 t $\cdot a^{-1}$ (cf. p. 263) this results in 34.2 kg P (TP) $\cdot a^{-1}$. The specific phosphorus content of the fish biomass was assumed at 0.6 per cent (Barthelmes, personal communication). Overall, there is a P export of 58.3 kg P (TP) $\cdot a^{-1}$.

The 'external' phosphorus balance of Lake Stechlin

The present data on phosphorus import and export permit calculation of an 'external' phosphorus balance. To calculate this 'external' phosphorus balance the phosphorus export is deducted from the phosphorus input.

On the other hand, it is possible to establish an 'internal' phosphorus balance, too. The 'internal' phosphorus balance takes into account not only the phosphorus import and export but also the phosphorus flow between different zones, for example, between the pelagic and the profundal zones. At this point we shall only discuss the 'external' phosphate balance.

The balance values for the 'external' phosphorus balance of Lake Stechlin are contained in Table 4.7. With the cooling-water circulation in operation, there is a phosphorus surface load for Lake Stechlin of 65.2 mg P (DIP) $\cdot m^{-2} \cdot a^{-1}$ and 71 mg P (TP) $\cdot m^{-2} \cdot a^{-1}$. When the cooling-water circulation is out of operation, the DIP input is reduced to 38.8 mg P $\cdot m^{-2} \cdot a^{-1}$ and the TP input to 5.6 mg P $\cdot m^{-2} \cdot a^{-1}$. When the cooling-water circulation is in operation, the phosphorus load of Lake Stechlin attains values typical of mesotrophic conditions (cf. p. 421). With the cooling-water circulation not operating, the phosphorus load is in the oligotrophic range. Atmospheric inputs account for the major part of the phosphorus load. With 35.5 mg P $\cdot m^{-2} \cdot a^{-1}$ (DIP) the share of atmospheric inputs is 50

Table 4.7 'External' phosphorus balance of Lake Stechlin.

	DIP		TP	
	(kg $P \cdot a^{-1}$)	(mg $P \cdot m^{-2} \cdot a^{-1}$)	(kg $P \cdot a^{-1}$)	(mg $P \cdot m^{-2} \cdot a^{-1}$)
Inputs				
Precipitation	150.8	35.5	192.0	45.2
Lake Nehmitz	51.9	12.2	66.0	15.6
Warming up NPP[a]	60.4	14.2	0	0
Lake Dagow	8.5	2.0	25.6	6.0
Fallen leaves	0	0	13.0	3.1
Bathing	4.6	1.1	4.6	1.1
Ground-water	1.0	0.2	1.0	0.2
Total	277.2	65.2	302.2	71.2
Exputs[b]				
Outflow	0.8	0.2	6.0	1.4
Insects	0	0	18.1	4.3
Fishing	0	0	34.2	8.0
Total	0.8	0.2	58.3	13.7
Balance				
(Input–exput)	276.4	65.0	243.9	57.5

[a] NPP = nuclear power plant.
[b] Without exput by birds.

per cent when the cooling-water circulation operates and even 90 per cent when it does not.

The export of TP is some 20 per cent of the TP import. Interestingly enough, some 60 per cent of the phosphorus export is accounted for by fishing. Intensive fishing, therefore, can also have an enduring effect on the phosphorus balance of Lake Stechlin.

The carbon budget (G. Proft)

Carbon plays an important role in the metabolism of lakes. It exists in water in both inorganic and organic states as well as in particulate form. The subject of our contribution is the spatial and temporal distribution of the different carbon components. In addition to the quantitative relations of particulate to soluable organic carbon, we are especially interested in the pelagic biogenic calcite precipitation (cf. p. 412).

The dissolved inorganic carbon (DIC)

Lake Stechlin is a hard water lake with a content of 50 g Ca · m^{-3} and 2–3 g Mg · m^{-3}. Ca and Mg bind the main part of the inorganic dissolved carbon in the carbonate equilibrium.

The monthly examination of the vertical profile at the deepest point of Lake Stechlin shows (Table 4.8) that the DIC-content remains very constant. The standard deviation around the average value of 23.7 g C · m^{-3} is only 0.42 g C · m^{-3}. In comparison to neighbouring lakes with a high trophic level, the deviation is small. At the end of the stagnation period in the hypolimnion there is a maximum accumulation of DIC. The deviations in Stechlin (October 1978 – 117%) are lower than in the mesotrophic Lake Nehmitz-N (September 1978 – 143%; Table 4.9) and in the eutrophic Lake Dagow (September 1980 – 180%; Table 4.10; cf. also Ohle 1952; Heitmann 1967; Koschel *et al.* 1981; Mothes 1981).

The carbon turnover in the daily primary production has little influence on the DIC-value, because it is so low (<1%). In contrast to this, the decomposition of the organic material with the associated CO_2 release leads to a calcite solution and DIC-accumulation in the summer stagnation period.

The particulate inorganic carbon (PIC)

The carbon content of a lake is governed by the process of the pelagic biogenic calcite precipitation (Minder 1923, 1926, cf. p. 412); i.e., through the decalcification by assimilation connected with the extraction of CO_2 from the water and re-solution through respiration connected with the release of CO_2 (Kelts & Hsü 1978). This process also appears in the lakes subject to our research.

During the precipitation period the particulate inorganic carbon (PIC) and the CO_2 content in the oligotrophic Lake Stechlin, in the mesotrophic Lake Breiter Luzin and in the eutrophic Lake Dagow is governed by the differing depths of water. In addition an attempt was made to determine the quantity of precipitated calcite with the help of sedimentation traps (methods see Proft 1980, 1983). To characterize the $CaCO_3$-equilibrium conditions, a selective determination of H_2CO_3 in water samples is well suited (Liden 1982).

Calcite precipitation was only observed when the CO_2 values in water were less than 1 g CO_2 · m^{-3} (Figs. 4.9–4.11). High CO_2 values with a simultaneous reduction of PIC in the hypolimnion, especially at the end of the summer stagnation period, indicate the solution of sedimentating calcite crystals (Figs. 4.9–4.11).

In the eutrophic Lake Dagow after the spring circulation, as a result of the strong development of the phytoplankton, the free CO_2 is quite rapidly consumed; consequently the precipitation conditions for calcite are established here first (Fig. 4.11). In the epilimnion of Lake Breiter Luzin the $CaCO_3$ values between June and August are about 0.5–1 g (Fig. 4.10). In the oligotrophic Lake

Table 4.8 DIC-content (g C · m⁻³) in Lake Stechlin, 1978.

Depth (m)	Jan	Feb	Mar	Apr	May	Jun	Aug	Sep	Oct	Dec
0		24.5	23.5	26.2	22.5	24.2	23.5	20.8	22.0	24.5
2.5	19.2	23.5	23.5	24.2	22.7	23.5	23.7	20.8	22.0	23.5
5.0			23.5	24.0	22.7	23.8	23.5	21.8	21.5	23.5
7.5			23.5	23.0	22.5	23.8	24.5	21.5	22.5	24.0
10.0		22.5	24.0	23.7	22.7	24.2	24.5	20.8	22.5	23.5
15.0			23.5	25.5	23.0	24.2	24.7	23.8	24.7	23.0
20.0	22.0	22.5	23.0	22.8	22.5	24.2	26.7	25.0	24.7	23.0
40.0	21.0	22.5	26.2		22.7	24.5	25.5	23.5	25.0	23.5
60.0	27.2	23.0	24.8	24.2	22.7	24.8	24.5	24.8	25.7	24.0
$\bar{x}$	22.4	23.1	23.9	24.2	22.7	24.1	24.6	22.5	23.4	23.6

Table 4.9 DIC-content (g C · m⁻³) in Lake Nehmitz-N, 1978.

Depth (m)	Jan	Feb	Mar	Apr	May	Jun	Aug	Sep	Oct	Dec
0	26.8	23.5	23.8	25.3	23.5	24.2	22.0	21.0	22.2	24.2
2.5	24.0	21.0	25.0	25.5	21.7	24.5	22.5	21.5	22.5	22.7
5.0	24.8	21.0	25.5	25.5	22.5	23.8	21.2	21.0	21.7	22.5
7.5	26.8	23.5	23.8	25.3	23.0	24.3	22.6	22.0	21.7	23.3
10.0	24.0	25.0	23.3	25.5	22.7	25.5	26.5	23.5	22.0	24.0
15.0	24.0		22.5	25.7	24.7	27.5	28.0	30.0	28.5	23.0
$\bar{x}$	25.1	22.8	24.0	25.5	22.7	24.2	24.0	24.0	23.4	23.9

Table 4.10 DIC-content (g C · m⁻³) in Lake Dagow, 1980, 982.

Depth (m)	Apr	May	Jun	Jul	Aug	Sep	Oct	Nov	Dec
1980									
0		31		30	27.5	20	26.0	35	31
2.5		31		32	32.5	21	26.0	35	31
5.0		34	36	36.0	24	25.5	34	31	
7.5		35.5		38	39.0	36	26.5	35	31
$\bar{x}$		33		34	34	25	26	35	31
1982									
0	33	30	33	34	33	35.0	37	39	37
2.5	34	30	34	35	33	35.5	37	39	38
5.0	34	31	37	34	37.0	37	39	38	
7.5	35	34	41	42	40	46.0	37	–	38
$\bar{x}$	34	31	37	37	35	38	37	39	38

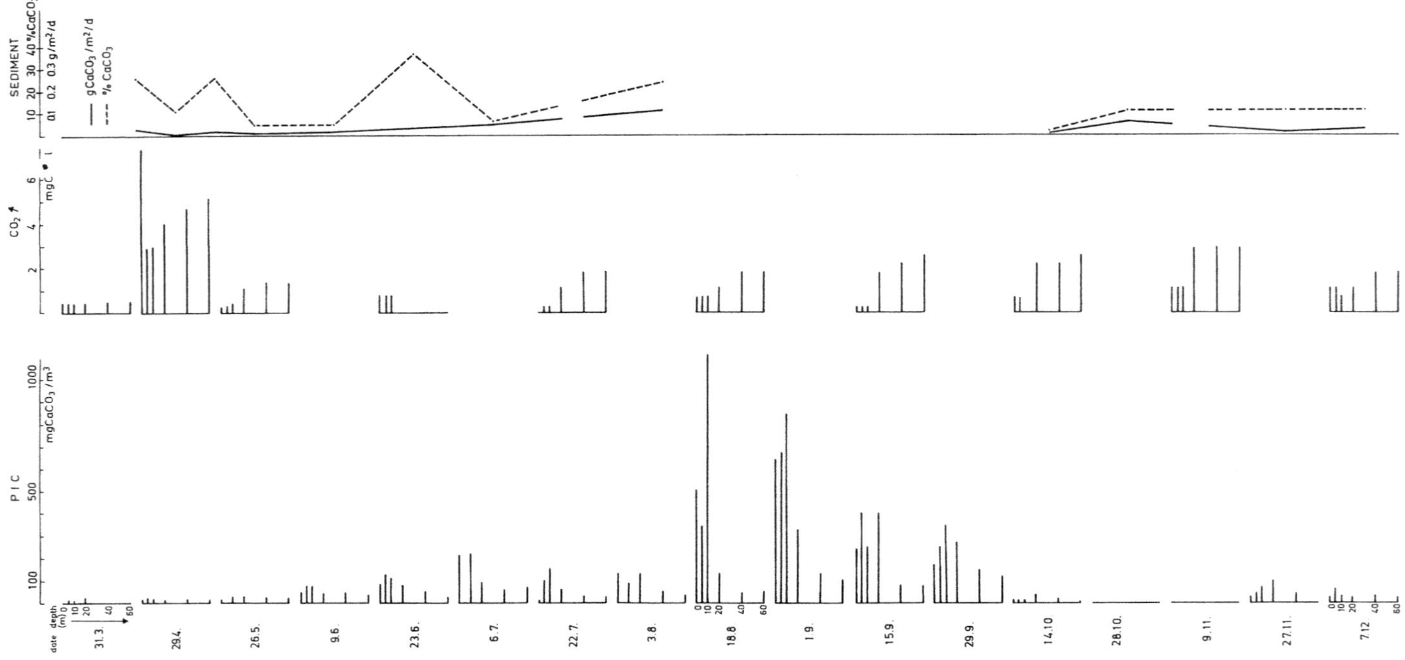

Figure 4.9 Lake Stechlin, PIC (mg $CaCO_3 \cdot m^{-3}$) and flushed CO_2 ($mg \cdot m^{-3}$) in various depths. Sedimentation (dry matter; $g \cdot m^{-2} \cdot d^{-1}$) and $CaCO_3$ (%) from sediment traps.

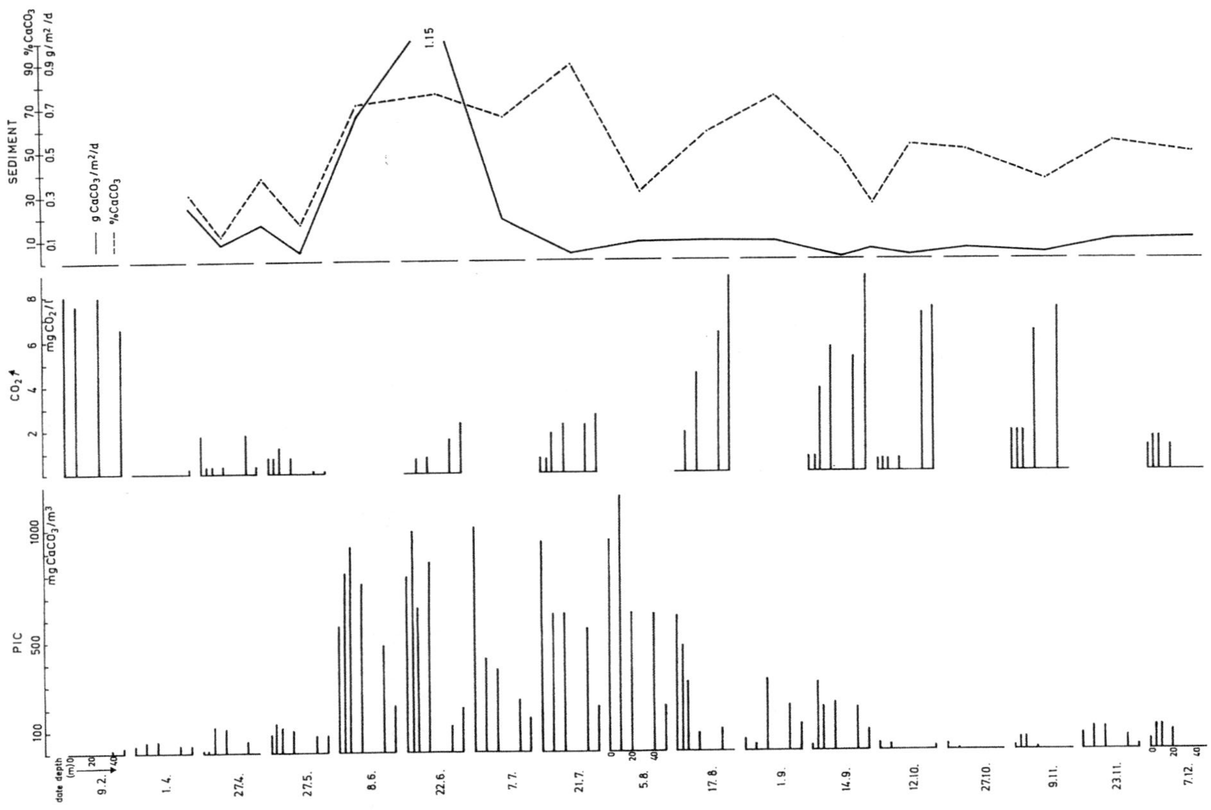

Figure 4.10 Lake Breiter Luzin, PIC and CO_2 in various depths. Sedimentation and $CaCO_3$ as Fig. 4.9.

Stechlin the maximum calcite precipitation (Fig. 4.9) begins only at the end of the summer stagnation during which, both before and after the maximum precipitation, weaker precipitation often occurs.

Low PIC-values in water (<0.1 g $CaCO_3 \cdot m^{-3}$) do not necessarily depend on calcite precipitation. Calcium-containing skeletons of organisms and resuspended sediment (e.g., littoral lake chalk) can also jointly determine the PIC-value.

High CO_2-values in the water at the end of the stagnation period exclude calcite precipitation. In deep water the PIC-level declines; the calcite crystals with a size up to 5 μm are dissolved (Kelts & Hsü 1978; Lehmann 1983).

In Lake Stechlin only an insignificant part of the calcite crystals reach the profundal sediment. In Lake Breiter Luzin, where the precipitation is stronger, the CO_2 released by the respiration cannot completely resolve the calcite crystals precipitated in the epilimnion. In Lake Dagow an anaerobic hypolimnion with a high CO_2 content indicates an increased chalk solution potentiality. The hypolimnion here is not very big, and consequently the time taken by the calcite precipitation is so low that a solution is not achieved (Fig. 4.11).

In the sediment traps directly under the epilimnion (Fig. 4.9–4.11; Koschel *et al.* 1983) between April and December 1982, levels of 5.5 g $CaCO_3 \cdot m^{-2}$ in Lake Stechlin, 83 g $CaCO_3 \cdot m^{-2}$ in Lake Breiter Luzin, and 147 g $CaCO_3 \cdot m^{-2}$ in Lake Dagow were established. However, these results must be interpreted with caution.

Thus in deep lakes between the spring and the autumn circulation a reduction in the calcium content of the epilimnion is observable. In Lake Breiter Luzin in spring only 56 g $Ca \cdot m^{-3}$ and in autumn only 40 g $Ca \cdot m^{-3}$ are found. In an epilimnion of ten metres depth the loss of 16 g $Ca \cdot m^{-3}$ is equivalent to a sedimentation of 400 g $CaCO_3 \cdot m^{-2}$. On the same basis Schröder *et al.* (1983) calculated for Lake Atter a calcite precipitation of 250 g $CaCO_3 \cdot m^{-2} \cdot a^{-1}$ and Müller (1971) for Lake Constance 250–380 g $CaCO_3 \cdot m^{-2} \cdot a^{-1}$.

The enormous differences between the calcite precipitation values obtained by the sediment traps and those obtained by calculation of the calcium reduction in the epilimnion still need to be explained.

The organic bound carbon in Lake Stechlin

Dissolved organic carbon (DOC) and particulate organic carbon (POC). In some lakes of the Mecklenburg Lake District the spatial and temporal distribution of dissolved organic and organic bound carbon was examined in 1978 and 1979. In them retention time of the water is very high. Because of this the autochthonous turnover in the lake can be established without much interference.

The DOC-content in Lake Stechlin at 3.8 g $C \cdot m^{-3}$ in comparison with the other lakes is low. The POC-portion at 0.29 g $C \cdot m^{-3}$ is about one tenth of the

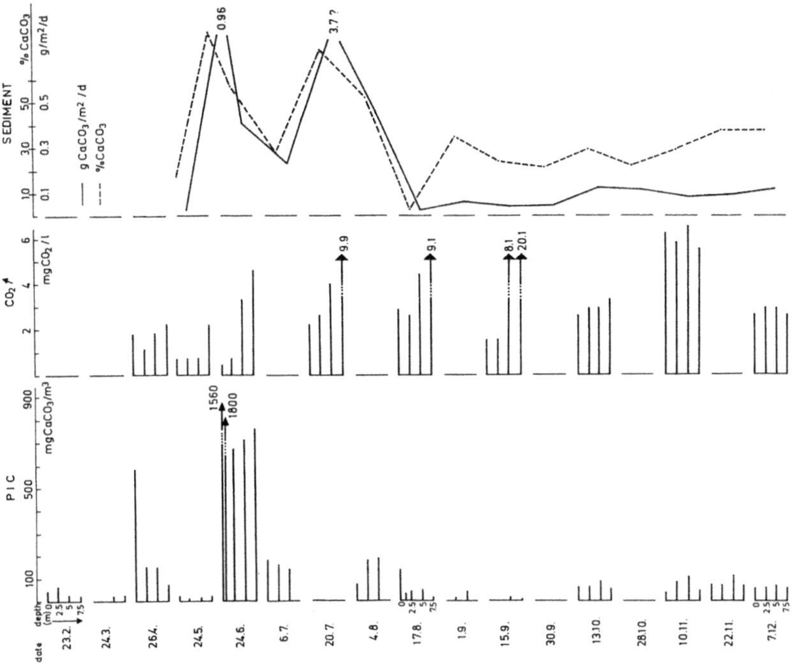

Figure 4.11 Lake Dagow, PIC and CO_2 in various depths. Sedimentation and $CaCO_3$ as Fig. 4.9.

DOC-portion. In the epilimnion the carbon content of the water is higher than in the hypolimnion (Fig. 4.12). This concerns both the dissolved and also the bound portion, the latter especially. The differences, averaged over a year, are 11% (DOC) and 31% (POC). This depends on the decomposition of the organic substance (Yanada & Maita 1976).

The carbon concentration in the course of a year varies between 2 and 5 g $C \cdot m^{-3}$ (DOC), and 0.1 and 0.6 g $C \cdot m^{-3}$ (POC). In comparison to the primary production and the plankton biomass, the deviation is low.

The ratio of dissolved organic carbon to particulate organic carbon (DOC/POC). With rising trophic levels the organic burden of the water increases. The ratio of bound carbon to dissolved carbon decreases strongly. The ratio DOC/POC shifts according to the trophic level. For still water (Wetzel 1976; cf. Lawacz 1977) gives a DOC/POC ratio of 10; it decreases in productive waters to 5 and occasionally to <1.

In Lake Stechlin the DOC/POC ratio is averaged at 16.8, with seasonal variations between 7 and 43 (average values of the depth profile). In times of increased plankton production it is mostly about 10; in periods of weak production it is noticeably higher (Fig. 4.13). Low DOC/POC values, which occur especially in winter as the result of storms or during circulation, can be caused by

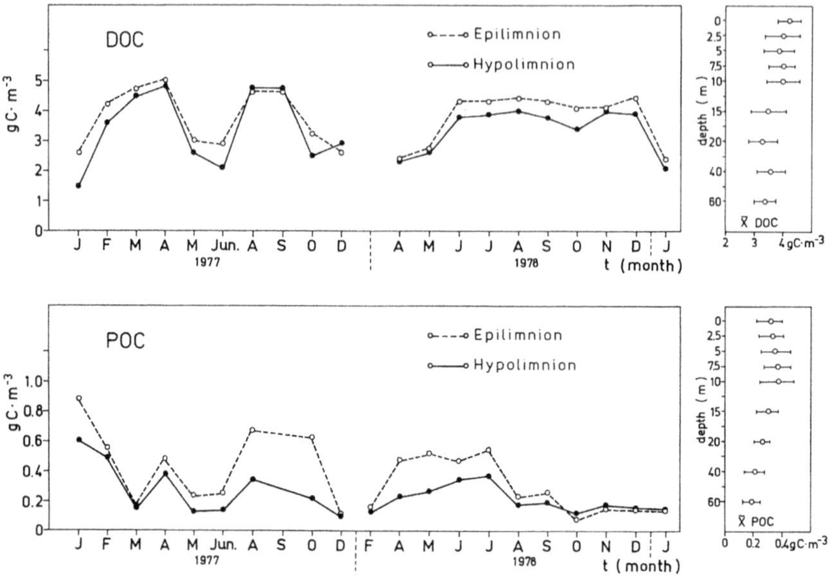

Figure 4.12 Lake Stechlin, DOC and POC in the epilimnion (0–10 m) and hypolimnion (15–60 m). Average of the various depths with standard deviation.

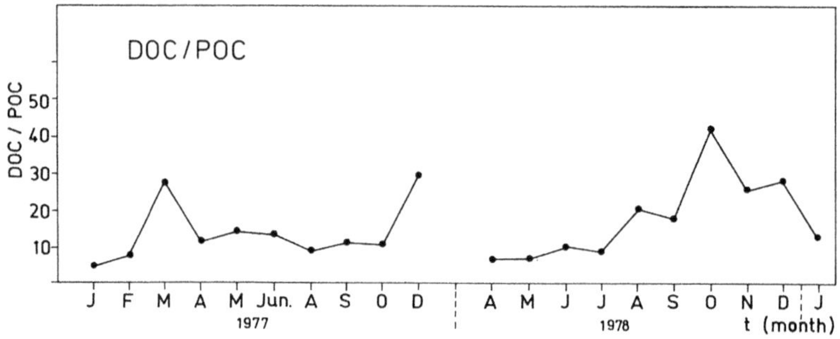

Figure 4.13 Lake Stechlin, ratio of DOC/POC 1977 and 1978; averages of the vertical profile samples.

resuspended sediment particles which lead to an increase in POC-values.

The lakes with a higher trophic level in the examined territory show values less than those of Lake Stechlin (Fig. 4.14). In Lake Große Fuchskuhle, an acid boglake, the DOC/POC ratio varies greatly. Here the algae mass development leads to DOC/POC values under 1; the particulate bound carbon quantity is greater than the soluable.

Size fractions of POC. To be able to answer the question about the composition of the particulate organic carbon, we used sieves (with meshes of 400, 200, 100 and

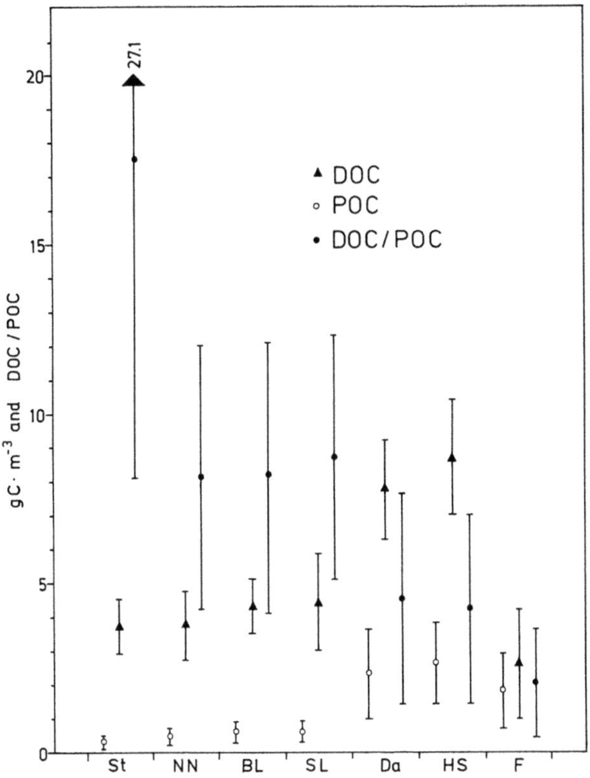

Figure 4.14 Ratio of DOC:POC, DOC and POC (g C · m^{-3}; with standard deviation) in Lakes Stechlin (St), Nehmitz-N (NN), Breiter Luzin (BL), Schmaler Luzin (SL), Dagow (Da), Haussee (HS) and Große Fuchskuhle (F).

50 μm) in order to fractionate the suspended particles into size groups (Sheldon *et al.* 1969, 1972). In this way we attempted to obtain results of the carbon content in plankton according to the sieve fraction (Proft & Ronneberger, in preparation). We proceed from the point that the carbon content in the large meshed nets corresponds essentially with the organisms obtained. In the small meshed nets, however, the carbon content of the seston in comparison with the plankton is of increased importance. Similar investigations into phosphorus were carried out by Peters (1983).

In Lake Stechlin the total POC-content determined using glassfibre paper is 290 mg C · m^{-3}. The individual sieve fractions produce portions between 1.5 and 3% (average values from 11 analyses). Their total accords roughly with those from a plankton net of 50 μm mesh producing 18 mg C · m^{-3} (~6% of the total POC-content). This value, compared with those from lakes with higher trophic levels, is low (Fig. 4.15).

In general changes in the plankton composition also lead to changes in the

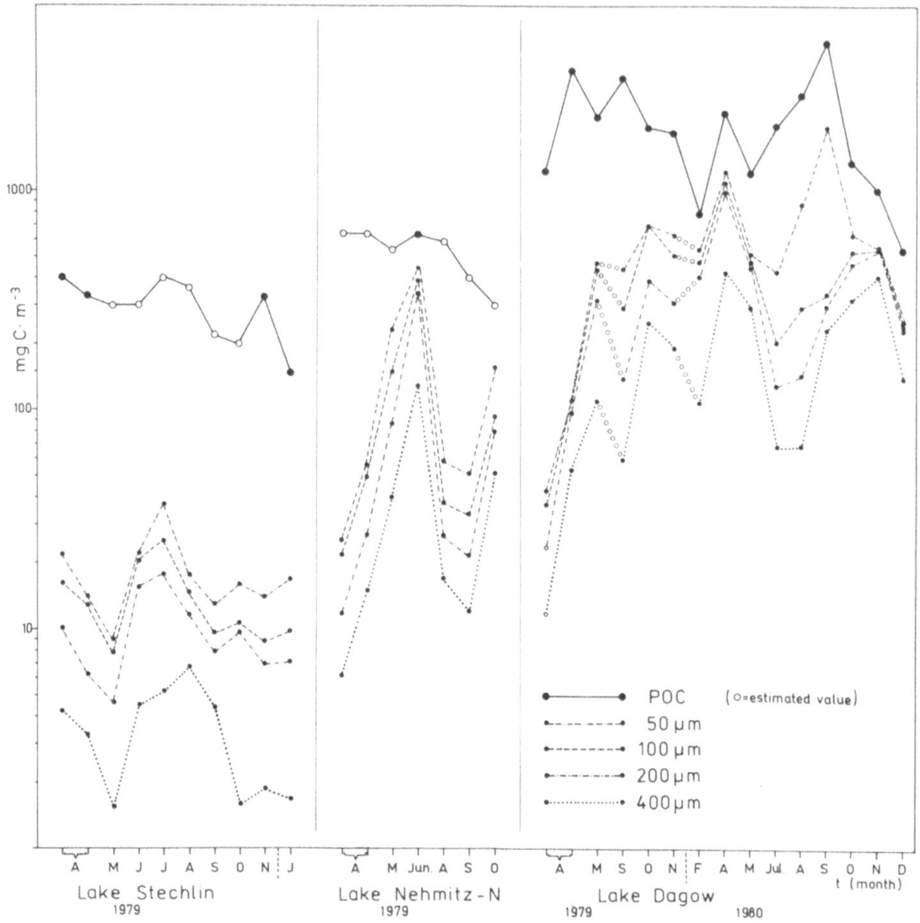

Figure 4.15 Carbon content of seston size fractions (cumulative 400 + 200 + 100 + 50 μm) and POC (mg C · m^{-3}) in Lakes Stechlin, Nehmitz and Dagow.

carbon content in the individual sieve fraction. However, the total POC-value hardly changes because the carbon is only shifted between the individual sizes or within the successive plankton communities.

The size of the POC value is dependent on the trophic condition of the water and is subject to seasonal variations (Fig. 4.12). It is quite independent of the species composition.

The nitrogen budget (G. Mothes & R. Koschel)

Introduction

All processes involving the transformation of substances in a body of water are linked to the existence of nitrogen compounds. Nitrogen enters the transformation in different oxidation forms (NO_3, NO_2, N_2, NH_3). In most cases it occurs in such a small quantity (except as for N_2) that we have to count it among the decisive minimum factors for autotrophic and in the last analysis also for heterotrophic production (Forsberg 1975). Concentration, temporal and spatial distribution as well as the balance of nitrogen compounds are, therefore, indispensable for a basic limnological assessment of our lakes.

Since 1972 continuous measurements of the nitrogen concentration in free water have been carried out in the Lake Stechlin area (Mothes 1981). These analyses mainly covered dissolved nitrate and ammonium nitrogen (Vogler 1976a, b) in order to characterize the concentration and dynamics of nitrogen compounds and to establish the interrelation with other nutrients such as phosphorus and carbon.

Nitrogen budget of Lake Stechlin

The temporal and spatial dynamics of the nitrate concentration in the pelagic zone of Lake Stechlin is depicted in Fig. 4.16. At the beginning of the year the nitrate concentration in the whole water column is some 90 mg $N \cdot m^{-3}$ (Table 4.11). During the spring full circulation period the nitrate concentration in the euphotic zone decreases rapidly. In May/June the measurable nitrate in the euphotic zone has been consumed. Nitrate concentrations below detection level (<2 mg $N \cdot m^{-3}$) are characteristic of the epilimnion until September.

During the summer stagnation period there is a nitrate accumulation in the hypolimnion with the nitrate concentration increasing with growing depth. At the end of the summer stagnation period nitrate values between 100 and 170 mg $N \cdot m^{-3}$, on an average around 150 mg $N \cdot m^{-3}$ (Table 4.11), can be measured at a depth of 60 m.

In autumn differences in concentration between the euphotic and the aphotic zones disappear slowly. By the end of the year the nitrate concentrations in the whole water column increase again to reach the above mentioned medium value of 90 mg $N \cdot m^{-3}$.

An exception was the year 1975, when an extreme depletion of nitrate (measuring values <2 mg $N \cdot m^{-3}$) was observed until November.

Ammonium shows temporal and spatial distribution different from nitrate (Fig. 4.17). In winter ammonium concentrations are generally below detection level (<8 mg $N \cdot m^{-3}$). At the beginning of spring (March/April) the ammonium

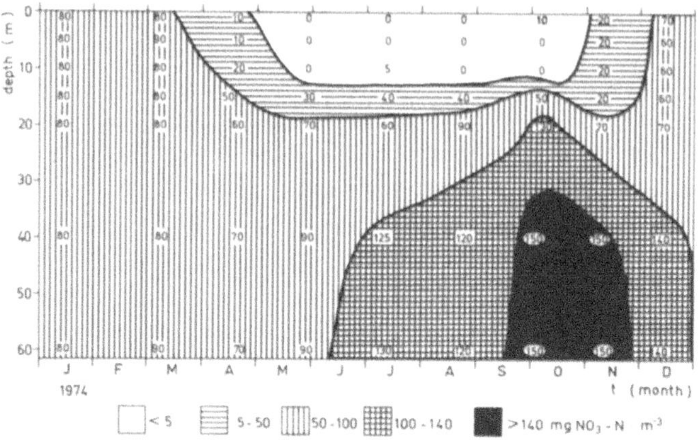

Figure 4.16 Temporal and spatial distribution of NO_3–N in the pelagic zone of Lake Stechlin (deepest point, 1974).

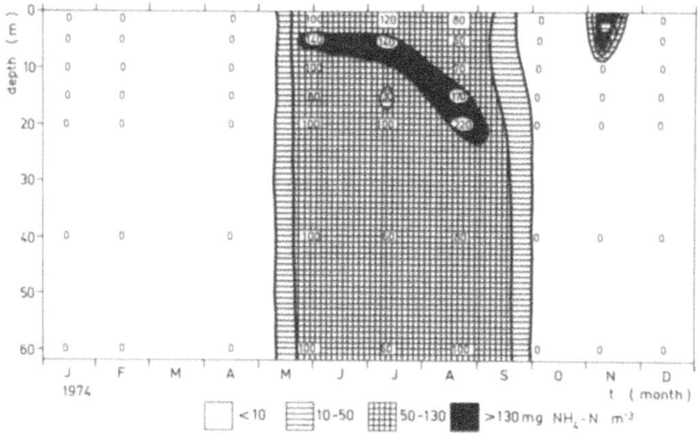

Figure 4.17 Temporal and spatial distribution of NH_4–N in the pelagic zone of Lake Stechlin (deepest point, 1974).

content increases and then fluctuates in ranges between 50 and 200 mg $N \cdot m^{-3}$ until September. Maximum values were measured particularly near the surface and in the metalimnetic and upper hypolimnetic regions (Fig. 4.17).

In the hypolimnion the medium ammonium values are somewhat below the values of the epilimnion and metalimnion. In autumn they fall relatively rapidly below detection level in the whole column of water. On October 15, 1975, for instance, no more free ammonium could be measured at a depth of 0–60 m.

Table 4.11 Maximum NO_3–N concentration at the beginning of the spring full circulation (sfc; 0–60 m) and at the end of summer stagnation period (ss; 60 m) in Lake Stechlin (1972–1982).

Date	NO_3–N concentration (sfc) (mg N · m^{-3})	Date	NO_3–N concentration (ss) (mg N · m^{-3})
14.3.72	68	26. 9.72	140
29.3.73	51	3.10.73	140
6.3.74	80	2.10.74	150
18.2.75	80	14.10.75	100
26.2.76	100	5.10.76	160
22.2.77	100	25.10.77	165
28.2.80	115	24.10.78	160
31.3.82	107		
$\bar{x}$	88	$\bar{x}$	145

Nitrogen budget of Lake Nehmitz

The nitrate concentration in the free water of Lake Nehmitz is about 70 mg N · m^{-3} ($\bar{x}$ 1975–1977) at the beginning of the year. Later in the year the nitrates in the euphotic zone are rapidly consumed. As early as April their amount may temporarily fall below detection level and remain unchanged during the whole summer stagnation period.

In the hypolimnion a nitrate accumulation is found at the beginning of the summer stagnation period. Further during this period, however, the largest part of the nitrates is consumed (anaerobic, cf. p. 120). In August/September values are mostly below detection level.

Ammonium mainly appears during the summer stagnation period. In the epilimnion concentrations are between 10 and 200 mg N · m^{-3}. In the hypolimnion the ammonium concentration increases with growing depth and duration of the stratification. In September/October it amounts to some 800 mg N · m^{-3} at a depth of 15 m ($\bar{x}$ 1973–1977).

The autumn full circulation removes the differences in concentration. Ammonium is oxidized but may still exist in measurable concentrations until the end of the year (100–8 mg N · m^{-3}).

The previous data mainly cover the north basin of Lake Nehmitz. The nitrate and ammonium values in the south basin are quite similar. At the beginning of the year, however, the nitrate values in the south basin are below comparable values of the north basin (e.g., January 16, 1974: 30 mg N · m^{-3} in the south basin; January 10, 1974: 70 mg N · m^{-3} in the north basin).

The ammonium values at the end of the summer stagnation period reach an average maximum concentration of 730 mg N · m^{-3} at a depth of 15 m in the south basin.

Nitrogen budget of Lake Dagow

At the beginning of the spring full circulation period Lake Dagow shows a nitrate concentration of 325 mg $N \cdot m^{-3}$ ($\bar{x}$ 1974–1981).

The nitrate is rapidly consumed in spring. From June to October only very low nitrate concentrations are measured in the whole body of water (mostly <10 mg $N \cdot m^{-3}$ or below detection level). A nitrate accumulation in the anaerobic hypolimnion does not occur.

The ammonium concentration is high at the beginning of the year (partly around 1,000 mg $N \cdot m^{-3}$). During the spring full circulation period values vary between 60 and 300 mg $N \cdot m^{-3}$. Late in spring and during the summer stagnation period values may fall below detection level for a short time (<8 mg $N \cdot m^{-3}$). As a rule, however, values vary between 10 and 200 mg $N \cdot m^{-3}$ with values around 200 mg $N \cdot m^{-3}$ being the rule. In the hypolimnion there is a marked ammonium accumulation during the summer stagnation period. The ammonium concentration increases with growing depth (at a depth of 7.5 m 1,000–3,000 mg $N \cdot m^{-3}$ from July to September). At times maximum values considerably exceed 3,000 mg $N \cdot m^{-3}$ (e.g. September 4, 1974: 3,390 mg $N \cdot m^{-3}$; August 26, 1976: 4,630 mg $N \cdot m^{-3}$; August 22, 1978: 3,970 mg $N \cdot m^{-3}$; October 5, 1981: 3,310 mg $N \cdot m^{-3}$). During the autumn full circulation period (October–December) too, often distinctly higher values can be measured at a depth of 7.5 m than in the water strata situated above.

Nitrogen to phosphorus ratio and limitation of bioproduction by nutrients

In the pelagic zone of Lakes Stechlin and Nehmitz the nitrogen supply is much more balanced than the phosphorus supply. The inorganic nitrogen concentration in the euphotic zone falls below detection level only for a short time, mainly in early summer and temporarily in autumn. As regards orthophosphate, however, concentrations above and below detection level occur from early spring until autumn (cf. p. 89).

A comparison of the annual concentrations of orthophosphate, nitrate and ammonium nitrogen shows, in addition (Figs. 4.2, 4.16 and 4.17), that the reduction of nitrate begins two to three months later than that of orthophosphate. Further, during the year, the nitrate is replaced by ammonium nitrogen so that the lack of dissolved nitrogen compounds is largely compensated.

The priority of P limitation in Lake Stechlin and Lake Nehmitz is underlined by the N : P ratio in the euphotic zone (Fig. 4.18). With an N : P ratio (by atoms) of over 26, one expects a predominance of the phosphorus limitation, and at a ratio of under 11 a nitrogen limitation (Forsberg *et al.* 1978; Uhlmann 1982). In the euphotic zone of Lake Stechlin, we found for the dissolved N and P compounds (nitrate and ammonium nitrogen and orthophosphate), an N/P ratio >30, with

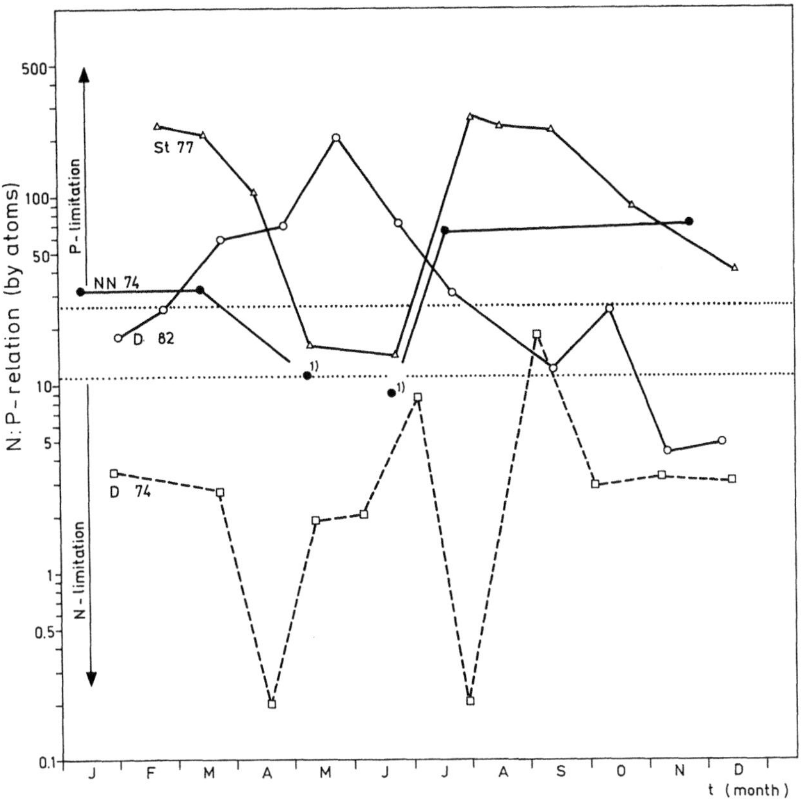

Figure 4.18 Seasonal distribution of N:P-ratio ($NO_3-N + NH_4-N : PO_4-P$, by atoms) in the upper euphotic zone of Lakes Stechlin, Nehmitz and Dagow. Lakes Stechlin and Nehmitz (deepest point, means 0–10 m), Lake Dagow (deepest point, means 0–2 m). ●[1)] = N: P-ratio values calculated.

maximum values of >200 from the beginning of the year until May and from July. Only late in spring and early in summer as well as for a short period in autumn May N : P ratios of <15 occur.

The dynamics of the N : P ratio in Lake Nehmitz are similar. In absolute terms, however, the N : P ratios in Lake Nehmitz fall below comparable values of Lake Stechlin.

Lake Dagow is quite different (cf. p. 91). Prior to 1975 the N : P ratios were normally under 10, with marked minima in April and July/August (N : P ratio ~0.5).

Different curative measures to restore Lake Dagow (cf. p. 474) have led to drastic changes. From January until June we now have N : P ratios ranging from 20–200. The N : P ratio increases from January to May (decrease in the orthophosphate concentration as far as below detection level). From May/June the N : P ratio decreases again and reaches values <25 at the end of July. In October/

November N : P ratios are around 5. Measured by the N : P ratios, Lake Dagow has developed from a nitrogen-limited lake which it had been for a long time (1974; Fig. 4.18) into a mainly phosphorus-limited lake (1982; Fig. 4.18).

Only in late summer/autumn can phases be derived from the N : P ratio which are more or less subject to a nitrogen limitation. In this period blue algae carrying heterocysts are on the increase and prevail. They cover the nitrogen deficit through N_2 fixation. In Lakes Stechlin and Nehmitz comparable phases for early summer can be derived from the N : P ratio.

The silicon budget (G. Mothes & R. Koschel)

A sufficient supply of silicic acid is of vital importance for the growth of various algae, particularly of diatoms and many Chrysomonadae. If there is only a minimum of silicic acid, their growth can be greatly reduced or totally stopped (Golterman 1975).

We carried out analyses of the silicon budget of the pelagic zone in order to give information on the annual silicon concentration as well as on the scope and time of maxima and minima and to contribute, along with other nutrient analyses, to a better understanding of diatom developments in our lakes. Measurements of the silicic acid content in the Lake Stechlin region have been carried out since 1971 (Mothes 1981).

Silicon budget of Lake Stechlin

The silicon concentration in Lake Stechlin showed a relatively constant temporal and spatial distribution in all the years under review (Figs. 4.19, 4.20).

At the beginning of the spring full circulation period, the medium values in the free water amount to 1.0–1.3 g $Si \cdot m^{-3}$ (Fig. 4.19). The silicon concentration decreases continuously later in the year in the euphotic region, but reaches minimum values as late as September/October, as a rule 0.5 g $Si \cdot m^{-3}$. On October 3, 1973, however, a considerably lower concentration was measured: 0.15 g $Si \cdot m^{-3}$ (medium value, depth of 0–10 m). From November on, values increased again.

In the hypolimnion a silicon accumulation occurs during the summer stratification period (Fig. 4.20). The silicon concentration increases with the depth and duration of the stratification. Maximum values at a depth of 60 m were 2.5 g $Si \cdot m^{-3}$ (measured in October and November). The highest value was measured at a depth of 60 m on November 11, 1976, with 2.9 g $Si \cdot m^{-3}$.

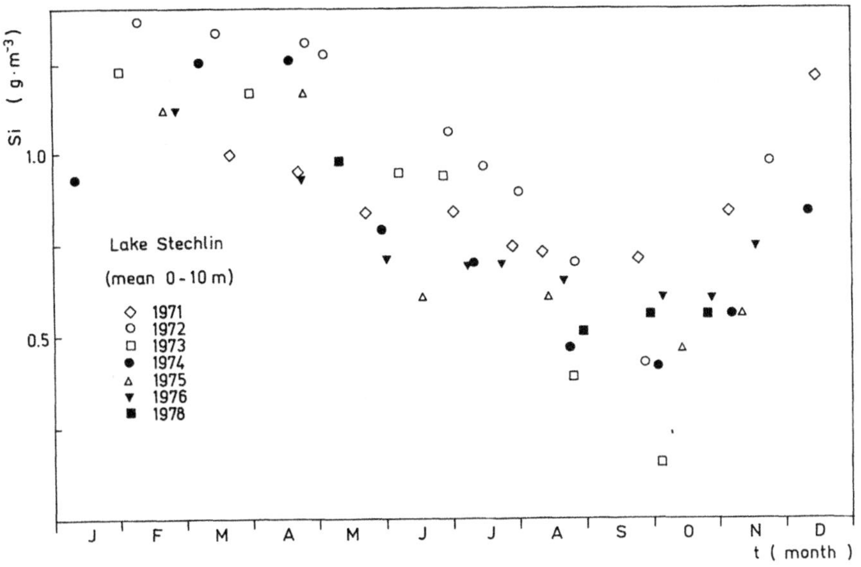

Figure 4.19 Silicon concentration in the upper euphotic zone of Lake Stechlin (means, 0–10 m).

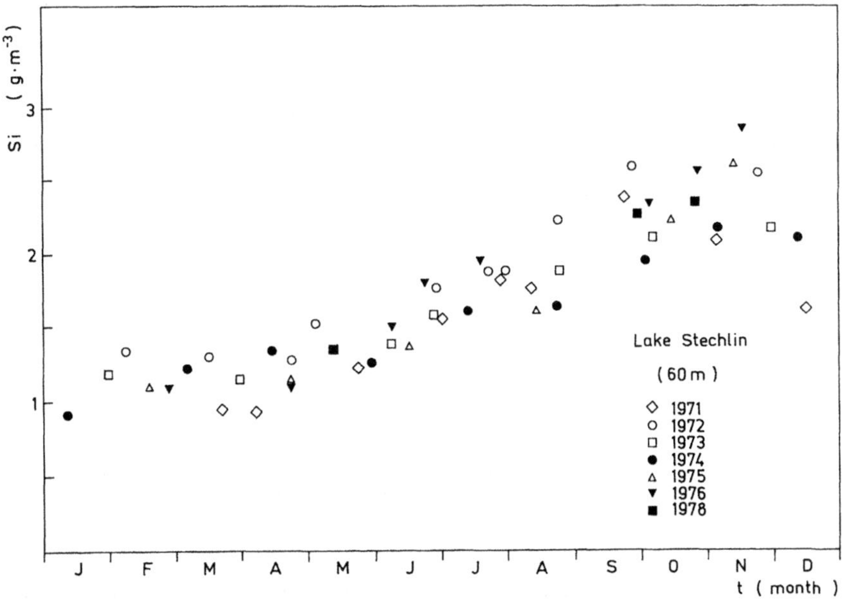

Figure 4.20 Silicon concentration in the lower aphotic zone (60 m) in Lake Stechlin.

Silicon budget of Lake Nehmitz

At the beginning of the year the silicon concentration in Lake Nehmitz is comparable to that in Lake Stechlin. Afterwards, however, the silicon consumption in Lake Nehmitz is faster. In May/June minima around 0.4 g Si · m^{-3} are observed in the euphotic region (e.g. June 18, 1975: 0.2 g Si · m^{-3}).

In summer the concentration slightly increases in the north basin, but decreases again in autumn (September/October; minimum values around 0.5 g Si · m^{-3}). In the hypolimnion there is a silicon accumulation in summer and autumn. At the end of the summer stagnation period values up to 4.5 g Si · m^{-3} are measured at a depth of 15 m.

In the south basin the minimum values fall below those of the north basin. In early June, as a rule, concentrations of as little as approximately 0.2 g Si · m^{-3} can be measured in the epilimnion. They are characteristic of the upper section of the epilimnion (0–5 m deep) from June to October. In October/November values increase again and approximate those of the north basin and those of Lake Stechlin. The silicon values in the hypolimnion of the south basin approximate those of the north basin.

Silicon budget of Lake Dagow

The silicon concentration in Lake Dagow is about 2.2 g Si · m^{-3} at the beginning of the year. A high silicon consumption starts below the ice so that early in March the concentration may fall below detection level. Values <0.1 g Si · m^{-3} are characteristic of the upper region (0–2.5 m of the euphotic zone) until the end of April, in some years (e.g., 1974, 1976) even until June. From July to September the silicon concentration reaches values between 0.4–1.1 g Si · m^{-3}, from October values >1.2 g Si · m^{-3} (in each case at a depth of 0–2.5 m).

A distinctive silicon accumulation occurs in the hypolimnion during the summer stagnation period. In August concentrations at a depth of 7.5 m mostly exceed 3.8 g Si · m^{-3}. The highest values at a depth of 7.5 m were measured in September 9, 1971: 6.4 g Si · m^{-3}; October 10, 1971: 7.2 g Si · m^{-3}.

Oxygen budget (G. Mothes & R. Koschel)

First data concerning the oxygen distribution in Lake Stechlin were given by Möller (1936). Regular analyses of the oxygen budget of Lake Stechlin and the adjacent Lake Nehmitz was begun in the late 1950s and early 1960s by the Research Institute for Hydrometeorology and the Research Department for Limnology of the German Academy of Sciences (Wagner 1961; Babenzien 1964; Heitmann 1967). From 1970 investigations were extended to other lakes in the area and conducted mainly by the Limnological Laboratory Stechlin (cf. Mothes 1981).

Lake Stechlin

Lake Stechlin is one the few deep stratified lakes of the Mecklenburg lakeland district, which is marked throughout the year by an aerobic pelagic zone (Fig. 4.21).

The highest oxygen concentrations in a year with values from 13 g $O_2 \cdot m^{-3}$ to 14 g $O_2 \cdot m^{-3}$ occur at the end of the spring full circulation period. Only in years with lengthy ice covers (cf. Richter *et al.*; 1979; p. 63) is 13 g $O_2 \cdot m^{-3}$ not exceeded. In 1959, e.g. (ice cover lasting 3 months) a maximum oxygen concentration of only 12.4 g $O_2 \cdot m^{-3}$ was measured at the end of the spring full circulation period (Wagner 1961).

Later in the year the oxygen concentration decreases with the stratification of the lake. This decrease is caused in the epilimnion mainly by the reduced ability of the water to solve oxygen, which is due to the rising temperature. In the hypolimnion the oxygen is used up by consumption processes.

In the metalimnion oxygen maxima and supersaturations can now be observed mainly in early summer (Fig. 4.21; cf. Mothes 1981). An oxygen saturation of 120 per cent is, however, hardly exceeded. In those horizons the water obviously warms up faster than the oxygen decreases. Biogenous processes (photosynthetic O_2 accumulation) play a minor part during this period (Mothes 1981). Before 1965 metalimnetic oxygen supersaturations were frequently observed in August too (Heitmann 1967; Babenzien 1964). They were possibly caused by the supply of

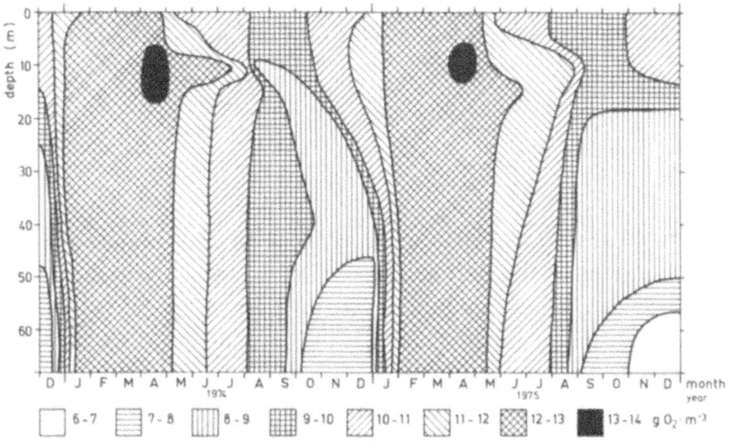

Figure 4.21 Oxygen distribution in Lake Stechlin (deepest point) in 1974 and 1975.

assimilation oxygen from the submersed macrophyte stocks to the metalimnion.

The hypolimnetic oxygen consumption during the summer stagnation period amounts to some 5 g $O_2 \cdot m^{-3}$; only in very small parts of the basin do values fall below 6 g $O_2 \cdot m^{-3}$. Directly above the sediment, values between 4 g $O_2 \cdot m^{-3}$ and 5 g $O_2 \cdot m^{-3}$ can be measured.

With the beginning of the autumn full circulation period the prevailing small oxygen deficit is slowly eliminated until the beginning of the next summer stagnation period.

Lake Nehmitz

The two basins of Lake Nehmitz show minor differences in their oxygen budgets. The maximum values of oxygen concentration occur during the spring full circulation and amount to some 13 g $O_2 \cdot m^{-3}$. Towards the end of the spring full circulation, differences between the surface and deep-water oxygen values can be detected. They are, however, only 1–2 g $O_2 \cdot m^{-3}$. Later in the year (summer stagnation period) the oxygen reserve in the deep parts of the basin is rather quickly used up. Late in June no oxygen can be measured below a depth of 15 m. Early in September (end of summer stagnation) the oxygen reserve of the whole hypolimnion is consumed.

In the epilimnion oxygen values are similar to those of Lake Stechlin. Supersaturation is low; an oxygen saturation of 120 per cent is rarely exceeded. Metalimnetic oxygen maxima can be detected in both basins, mainly in the southern part. They develop under the same conditions as in Lake Stechlin (cf. p. 119), however, the photosynthetic oxygen supply from the phytoplankton seems to play at least temporarily a greater part. In the metalimnetic horizons of Lake Nehmitz considerable developments of algae can take place (mostly blue algae, cf. p. 165).

The oxygen deficit is eliminated relatively fast as the autumn full circulation sets in.

Lake Dagow

The oxygen budget of Lake Dagow is one typical of a eutrophic lake. A short time after the beginning of the summer stagnation no more oxygen is to be found in deep water (7.5 m). Early in July the major part of the hypolimnion is free of oxygen. In the epilimnion oxygen saturation values of 150–200 per cent are frequently measured. On sunny days without wind oxygen gradients of up to 7 g $O_2 \cdot m^{-3} \cdot d^{-1}$ are measured in the chemocline. As the autumn full circulation (middle/end of September) starts, the deep horizons are again supplied with oxygen. By early January the oxygen deficit is offset again (from a saturation level of 50–60 per cent in mid-October to a hundred per cent in late December/early January). If there is an ice cover, however, there will be another oxygen decrease in deep water. In extreme cases even the whole body of water may become anaerobic (e.g., winter 1969/1970: ice cover lasting three months; high nutrient load of Lake Dagow).

Horizontal differences of oxygen content in Lake Stechlin

Heitmann (1967) detected differences in the oxygen distribution in the different basins of Lake Stechlin in the period 1958–1965. At the end of the summer stagnation period the lowest oxygen saturation in the south basin was measured. The highest values of oxygen saturation were found in the north basin (Table 4.12). These conditions remained after the starting up of the nuclear power station, too. As before, the lowest oxygen saturation values were measured first in the south basin at the end of the summer stagnation (November) and then in the west and the north basins (Table 4.12). Proportions have again shifted to the disadvantage of the south basin.

In absolute terms, the oxygen saturation concentrations at a depth of 34 m in the west basin and 32.5 m in the south basin, measured at the end of the summer stagnation period, are down 20 per cent over comparable values of 1958–1965. In the north basin differences are smaller. In 1963 and 1965, e.g., oxygen saturation values of 57 per cent were found in November at a depth of 60 m. In 1981/1982 the oxygen saturation at the same time and depth amounted to 54 per cent.

Table 4.12 Oxygen saturation at the end of the summer stagnation period (November) in the deep water of three basins of Lake Stechlin.

Basin	Depth (m)	Oxygen saturation (%)	
		Measured values 1958–1965[a]	Measured values 1981–1982
North basin	60	57	54
	65	–	46
West basin	32.5	68	53
	34	67	48
South basin	30	63[b]	44
	32.5	–	41

[a] According to Heitmann 1965.
[b] Depth of 28 m.

The pelagic oxygen content of Lake Stechlin before and during the operation of the nuclear power plant

Oxygen conditions in Lake Stechlin deteriorated after the cooling-water circulation system of the nuclear power plant began to function. This concerns mainly the oxygen saturation at the end of the summer stagnation period in the lower horizons of the hypolimnion. Conditions in the south and west basins changed markedly (Table 4.12). Before the nuclear power plant became operational, the oxygen saturation values in November amounted in the south basin to 60 per cent at a depth of 30 m, in the west basin to 70 per cent at a depth of 32.5 m. Comparable values in 1981 and 1982 were 44 per cent in the south basin and 53 per cent in the west basin. Differences are not so conspicuous in the north basin. Before 1965 the oxygen concentration below 60 m did not fall below 7 g $O_2 \cdot m^{-3}$ or a saturation level of 55 per cent. The values measured from 1970–1982 show a decrease here, too. At the end of the summer stagnation period (November) there is an average oxygen saturation level of 52 per cent at a depth of 60 m, with values varying between 49 and 59 per cent (Fig. 4.22). Even for the 40 m horizon of the north basin differences were found before and after the inauguration of the nuclear power plant. Heitmann (1967) indicates for this horizon a medium relative consumption of 21 per cent at the end of the summer stagnation period. After 1970 an average of 24 per cent was calculated.

The causes of this deterioration in the oxygen budget are due to an increased nutrient supply coming into Lake Stechlin, caused largely by the cooling-water circulation (cf. p. 100). This is not due to the waste heat of the cooling water, which rather counteracts any deterioration, but to the inflow of heated cooling water which reduces the ice cover period or in some parts totally prevents it. Thus the full circulation period is extended, and the lake can saturate to a maximum its body of water with oxygen.

Since the nuclear power plant became operational, the oxygen budget has been showing differences from year to year. From the hypolimnetic consumption rates, 'bad' and 'good' years can be derived. The variation scale of individual values, however, is relatively small. For the 60 metres horizon, for example, it amounts to 1 g $O_2 \cdot m^{-3}$ or 10 per cent of oxygen saturation at the end of the summer stagnation period (Fig. 4.22). There is no direct connection with variations in primary production measured during the same period (60 g $C \cdot m^{-2} \cdot a^{-1}$–138 g $C \cdot m^{-2} \cdot a^{-1}$).

Lake Stechlin offsets the increased nutrient load to be found through an increased circulation of materials in the epilimnion (among other things, a high share of nanoplanktonic primary production, high remineralization rates in the epilimnion and upper hypolimnion, high carbon turnover in the short-circuited circulation; cf. p. 423). The hypolimnetic oxygen consumption rates oscillate with another amplitude than, for example, the primary production of the phytoplankton. They indicate the process of increased production of materials in the epilim-

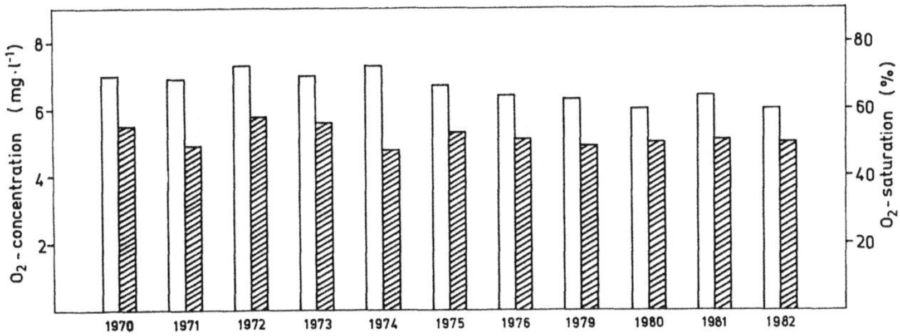

Figure 4.22 Oxygen saturation and concentration at the end of summer stagnation period (November) in deep water (60 m) of Lake Stechlin from 1970–1982. Hatched part – oxygen saturation; unhatched part – oxygen concentration.

nion only in a subdued form. The oxygen consumption in the hypolimnion of the two shallow lake basins (south and west basins) reacts to this most sensitively.

Comparison of investigated lakes

The four lakes dealt with here present different initial conditions for the oxygen budget, if only for their morphometric indices. Moreover, the oxygen budget is decisively determined by differing trophic conditions.

All the lakes have in common an oxygen maximum during or at the end of the spring full circulation period and a decrease in oxygen concentration in the hypolimnion which coincides with stratification. This decrease takes place with different speed coefficients. In the hypolimnion of Lake Stechlin there is a medium oxygen consumption of $0.025\ g\ O_2 \cdot m^{-3} \cdot d^{-1}$, in the hypolimnion of Lake Nehmitz of $0.115\ g\ O_2 \cdot m^{-3} \cdot d^{-1}$ and in the hypolimnion of Lake Dagow of $0.315\ g\ O_2 \cdot m^{-3} \cdot d^{-1}$.

The difference between Lake Stechlin and the two basins of Lake Nehmitz is mainly due to the different medium depths and thus the different sizes of the hypolimnion. In Lake Nehmitz a hypolimnion comparable in size to that of Lake Stechlin would also show high oxygen concentrations during the summer stagnation period. Conditions in Lake Dagow are different. Here the high rates of consumption are sufficient to reduce a substantially more sizable hypolimnion to anoxia.

In the epilimnion the medium summer O_2 deficits mainly depend on the water temperature and intermixing. Only in Lake Dagow can consumption processes in the lower region of the euphotic zone have a decisive, though mostly shortterm, share in the O_2 deficit. As a rule, this consumption is compensated, however, by supersaturation in the water horizons situated above.

The medium O_2 deficits in the epilimnion amount to $0.028\ g\ O_2 \cdot m^{-3} \cdot d^{-1}$ in

Lake Stechlin, to 0.032–0.056 in Lake Nehmitz and to 0.090 g $O_2 \cdot m^{-3} \cdot d^{-1}$ in Lake Dagow.

With the beginning of the autumn full circulation in Lake Dagow, a marked oxygen consumption in the pelagic zone sets in, caused by the reduction products accumulated in the hypolimnion, which is then followed by high O_2 increase rates, too. The increase in oxygen may amount in autumn to 1 g $O_2 \cdot m^{-3} \cdot d^{-1}$. These rates in Lake Stechlin and Lake Nehmitz vary from 0.3 and 0.4 g $O_2 \cdot m^{-3} \cdot d^{-1}$.

References

Babenzien, H.-D. 1964. Hydrobakteriologische Untersuchungen im Stechlinsee. Limnologica 2: 9–34.

Forsberg, C. 1977. Nitrogen as a growth factor in fresh water. Prog. Wat. Tech. 8: 275–290.

Forsberg, C., Ryding, S.O., Cleasson, A. & Forsberg, A. 1978. Water chemical analyses and/or algal assay? – Sewage effluent and polluted lake water studies. Mitt. Int. Verein. Limnol. 21: 352–363.

Golterman, H.L. 1975. Silicon and its compounds-chemistry, biochemistry and physiology of uptake by diatoms. In: Golterman, H.L.: Physiological Limnology. Amsterdam: Elsevier Scientific Publishing Company. 124–133.

Heitmann, M.-L. 1967. Die physikalisch-chemischen Verhältnisse im Stechlin- und Nehmitzsee. Limnologica 5: 307–329.

Kelts, K. & Hsü, K.J. 1978. Freshwater Carbonate Sedimentation. In: Lakes Chemistry, Geology, Physics (Ed. A. Lerman). New York-Heidelberg-Berlin: Springer-Verlag. 296–323.

Koschel, R. 1974a. Einfluß der physikalischen und chemischen Umweltfaktoren auf die Primärproduktion des Phytoplanktons im Stechlinsee. Diss. Techn. Univ. Dresden. 72.

Koschel, R. 1974b. Primärproduktionsuntersuchungen im Stechlinsee. Limnologica 9: 143–156.

Koschel, R. 1976. Der Einfluß des Phosphorangebotes auf die Primärproduktion des Phytoplanktons in einem geschichteten Klarwassersee (Stechlinsee, DDR). Limnologica 10: 325–346.

Koschel, R. 1980a. Untersuchungen zur Phosphataffinität des Planktons in der euphotischen Zone von Seen. Limnologica 12: 141–145.

Koschel, R. 1980b. Die Phosphataffinität von Mikroorganismengesellschaften im Stechlinseesystem und in anderen Seen. Limnologica 13: 129–145.

Koschel, R. 1981a. Die Bilanz der Primärproduktion des Phytoplanktons in den Gewässern des Stechlinseesystems. Limnologica 13: 55–81.

Koschel, R., Haubold, G., Kasprzak, P., Küchler, L., Proft, G. & Ronneberger, D. 1981. Eine limnologische Zustandsanalyse des Feldberger Haussees. Acta hydrochim. hydrobiol. 9: 255–279.

Koschel, R., Benndorf, J., Proft, G. & Recknagel, F. 1983. Calcite precipitation as a natural control mechanism of eutrophication. Arch. Hydrobiol. 98: 380–408.

Lawacz, W. 1977. Dissolved and particulate organic carbon (DOC and POC) in ecological studies. Acta hydrochim. hydrobiol. 5: 363–367.

Lehmann, R. 1983. Untersuchungen zur Sedimentation in einem oligotrophen Alpensee (Königssee) während der sommerlichen Schichtung. Arch. Hydrobiol. 96: 486–495.

Lidén, J. 1982. Equilibrium approaches to natural water systems – part 2. Schweiz. Z. Hydrol. 44: 393–404.

Maita, Y. & Yanda, M. 1976. Vertical distribution and seasonal variation of particulate organic matter in Funka Bay. Bull. Fac. Fish. Hokkaido Univ. 27: 152–159.

Minder, L. 1923. Die biogene Entkalkung im Zürichsee. Verh. Int. Verein. Limnol. 1: 20–32.

Minder, L. 1926. Biologisch-chemische Untersuchungen im Zürichsee. Z. Hydrol. 3: 1–69.

Möller, L. 1936. Stechlin-See und Sakrower See. Ein Beitrag zur Charakteristik eutropher und oligotropher Seetypen. Arch. Hydrobiol. 24: 137–156.

Mothes, G. 1974. Phosphatbestimmungen im Pelagial des Stechlinsees (1970–1971). Limnologica 9: 157–169.

Mothes, G. 1981a. Physikalische und chemische Parameter der Wasserbeschaffenheit des Stechlinseegebietes. Limnologica 13: 1–53.

Mothes, G. 1981b. Sedimentation und Stoffbilanzen in Seen des Stechlinseegebietes. Limnologica 13: 147–194.

Müller, G. 1971. Sediments of Lake Constance. In: Sedimentology of parts of Central Europe, Guidebook (Ed. G. Müller). Frankfurt/M.: Kramer. 237–252.

Nausch, G., 1983. Phosphoreinträge aus Niederschlägen im Küstenbereich der DDR. Acta hydrochim. hydrobiol. 11: 151–154.

Nehring, D. & Wilde, A., 1982. Untersuchungen über den atmosphärischen Nährstoffeintrag in die Ostsee. Acta hydrochim. hydrobiol. 10: 89–100.

Ohle, W., 1952. Die hypolimnische Kohlendioxyd-Akkumulation als produktionsbiologischer Indikator. Arch. Hydrobiol. 46: 153–285.

Peters, R. H., 1983. Size structure of the plankton community along the trophic gradient of Lake Memphrema Gog. Can. J. Fish. Aquat. Sci. 40: 1770–1778.

Peukert, V. 1976. Untersuchungen über den Einfluß atmosphärischer Verunreinigungen auf die Eutrophierung der Gewässer. Limnologica 10: 483–488.

Proft, G. 1980. Der Kohlenstoffgehalt und die Kohlenstoffracht der mittleren Saale. Limnologica 12: 195–206.

Proft, G. 1983. Automatische Bestimmung von Kohlendioxid und Carbonat in Wasser nach dem flow-stream-Prinzip. Acta hydrochim. hydrobiol. 11: 235–239.

Richter, D., Neubert, W. & Klämt, A. 1979. Temperatur und Wärmehaushalt des thermisch belasteten Stechlin- und Nehmitzsees. Abh. Meteor. Dienst. DDR 16 (No. 123): 1–40.

Rigler, F.H. 1966. Radiobiological analysis of inorganic phosphorus in lakewater. Verh. Int. Verein. Limnol. 16: 465–470.

Rigler, F.H. 1968. Further observations inconsistent with the hypothesis that the molybdenum blue method measures orthophosphate in the lake water. Limnol. Oceanogr. 13: 7–13.

Schröder, H.G., Windelph, H. & Schneider, J. 1983. Bilanzierung der biogenen Karbonatproduktion eines oligotrophen Sees (Attersee, Salzkammergut – Österreich). Arch. Hydrobiol. 97: 356–372.

Schulz, L., 1981. Nährstoffeintrag in Seen durch Badegäste. Zbl. Bakt. Hyg. I, Abt. B. 173: 528–548.

Sheldon, R. W., Prackash, A. & Sutcliffe, W. H., 1972. The size distribution of particles in the ocean. Limnol. Oceanogr. 17: 327–340.

Sheldon, R. W. & Sutcliffe, W. H., 1969. Retention of marine particles by screens and filters. Limnol. Oceanogr. 14: 441–444.

TGL 27 885/01. 1982. Fachbereichstandard Nutzung und Schutz der Gewässer. Wassergütebewirtschaftung See. Berlin: Ministerium für Umweltschutz und Wasserwirtschaft der DDR. 16.

Uhlmann, D. 1982. Evaluation of strategies for controlling eutrophication of lakes and reservoirs. Int. Revue ges. Hydrobiol. 67: 821–835.

Vogler, P. 1966. Zur Analytik der Phosphorverbindungen in Gewässern. Limnologica 4: 437–444.

Vogler, P. 1976a. Analysenautomation in Wasserlaboratorien mit flow-stream-Automaten. Teil 4. Die teilautomatische Bestimmung von organisch gebundenem Stickstoff und Gesamtphosphat. Acta hydrochim. hydrobiol. 4: 43–58.

Vogler, P. 1976b. Analysenautomation in Wasserlaboratorien mit flow-stream-Automaten. Teil 6. Die automatische Bestimmung von Nitrit und Nitrat im Wasser. Acta hydrochim. hydrobiol. 4: 227–238.

Wagner, G. 1961. Temperaturen und Sauerstoff im Großen Stechlinsee und Sacrower See. Eine Gegenüberstellung der Jahre 1958 und 1959. Z. Fischerei N. F. 10: 5–19.

Wetzel, R.G. & Rich, P.H. 1976. Carbon in freshwater systems. In: Carbon and biosphere. – Proc. 24th Brookhaven Symposium Biology. U.S. Atomic Energy Comm. Symposium Ser. CONF-720510. Nat. Techn. Inform. Serv. (Eds. G.M. Woodwelland & E.V. Pecan). Springfield.

PART THREE

The biological background

5. The plant communities

S.J. Casper, H.-D. Krausch & W. Scheffler

Aquatic macrophytes in the Lake Stechlin area (H.-D. Krausch)

The subsequent sections summarize some results of more or less continuous macrophytic vegetation studies between 1959 and 1983. Most of the facts, established before the nuclear power plant came into operation, already have been published (Krausch 1964, 1967, 1968a).

Historical aspects of the floristic and phytosociological research in the Lake Stechlin area

Floristic research began relatively late. In the 'Flora der Provinz Brandenburg' (Ascherson 1864) the little information recorded on the Lake Stechlin area was given by Albert Grunow (Bärenbusch) and Johann Friedrich Langmann (Neustrelitz); Johann Friedrich Sarcander (1865) published a floristic description of the surrounding of Fürstenberg (Havel).

From 1862 Hermann Winter (Menz) explored the area very intensively. He set down the results in his 'Flora der Umgebung von Menz', which includes 783 wild species of kormophytes and 175 species of mosses (Winter 1870). It gives an exact review of the flora of the area before the beginning of stronger influences of civilisation. Since that time floristic research has been essentially restricted to occasional excursions, which above all the Botanische Verein der Provinz Brandenburg undertook (p.e. 1928). The investigations about the bogs of Brandenburg by Kurt Hueck (1925) also included some bogs in the Lake Stechlin area.

At the time the most important floristic discovery was that of *Aldrovanda vesiculosa,* which was found in the ditch between Lake Zeuten and Lake Nehmitz (Winter 1870). This plant was last recorded in Lake Mehlitz and its surrounding in 1903 (Retzdorff 1907); now it is absent (Müller-Stoll & Krausch 1959). This is also true for *Najas marina* and *N. marina* subsp. *intermedia,* which Winter (1870) had found in Lake Nehmitz.

Casper, S. Jost (ed.), Lake Stechlin. A temperate oligotrophic lake. ISBN 978-94-010-8930-2.

During the vegetational research, carried out since 1959 by the Limnological Research Station Stechlin by H. Freitag (until 1961) and H.-D. Krausch (since 1962), many floristic discoveries and conformations of plants were made. The observations until 1964 are recorded in Fischer (1964), the later ones in the reports of the botanical journal 'Gleditschia' (Benkert 1973ff.).

A survey of all taxa of aquatic macrophytes found in the area with information on frequency and phytosociological connections is to be found in the chapter on general aspects of flora and fauna (p. 497).

The communities of aquatic macrophytes in the lakes of the Lake Stechlin area

Associations of Characeae (Figs. 5.1–5.2). In the oligo- to mesotrophic calcareous clearwater lakes of the Lake Stechlin area, Characeae take a leading part and often cover large planes of the bottom. The associations of Characeae have their chief unfolding in depths between 2 and 6 m, where they form mostly dense, now and then meadow-like stands. Altogether they grow at a depth of between 0.2 and 19–20 m.

The occurrence of Characeae-associations requires the presence of a certain minimum of soluble chalk in the water. Therefore, they are absent in the bog-lakes with acidic and non-calcareous water. In addition, Characeae need a certain quantity of light, differing from species to species. In years with above average insolation they are distinctly encouraged. On account of their high pretension of light Characeae are more or less tied to translucent and unpolluted waters. Some taxa such as *Chara aspera* and *Ch. filiformis* appear only in very clear lakes; other species also occur in mesotrophic, sometimes in less eutrophic reach, such as *Nitellopsis obtusa.* The centre of occurring of *Chara tomentosa* and *Ch. hispida* generally lies in the mesotrophic reach, and they die down in oligotrophic lakes. When oligotrophic lakes become mesotrophized such species increase in quantity, whereas the more sensitive species disappear. In typical eutrophic lakes such as Lake Dagow, Characeae are no longer to be found. Therefore the Characeae are important indicators of unpolluted chalk-oligotrophic to chalk-mesotrophic waters (Krause 1981).

The decline of Characeae in eutrophication and its absence in eutrophic lakes are also founded on their incompatibility with higher concentrations of NO_3 and PO_4. According to Forsberg (1964) the germination of Characeae is already stopped when the water contains more than 20 μg PO_4/l.

The increase of PO_4 in the water produces an increased growth of phytoplankton and periphyton, causing an adequate high absorbtion of light. The shoots of Characeae are often reduced on only one or two intact whorles and fructification becomes rare. In this way deficiency of light leads to a decline and disappearance of Characeae. The lack of O_2 and the increased sedimentation of mud may also be influencial. Therefore, the eutrophication of waters, which increased rapidly in

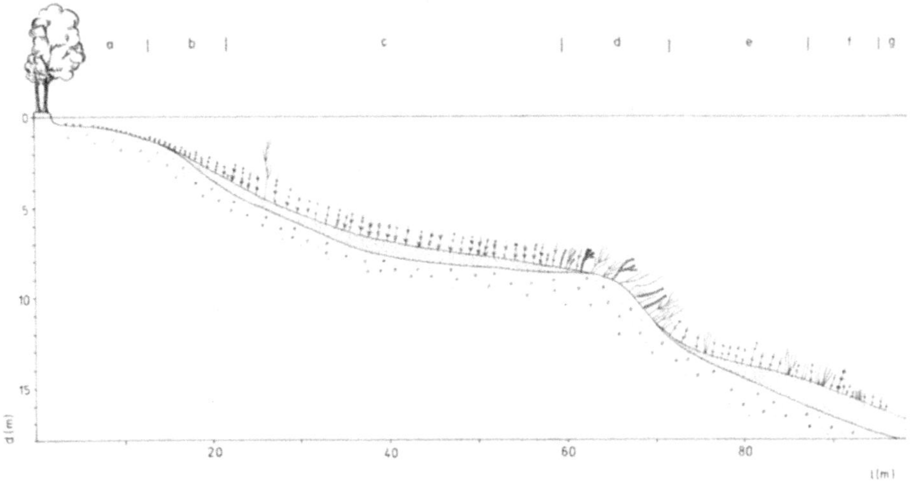

Figure 5.1 Vegetational transect through the north side of the peninsula of Lake Stechlin. a) sandy shore-bank, free from vegetation; b) Charetum asperae; c) Charetum filiformis; d) *Cladophora-Fontinalis* community; e) Nitello-Vaucherietum, facies of *Nitella;* f) Nitello-Vaucherietum, facies of *Vaucheria;* g) profundal, free from macrophytes.

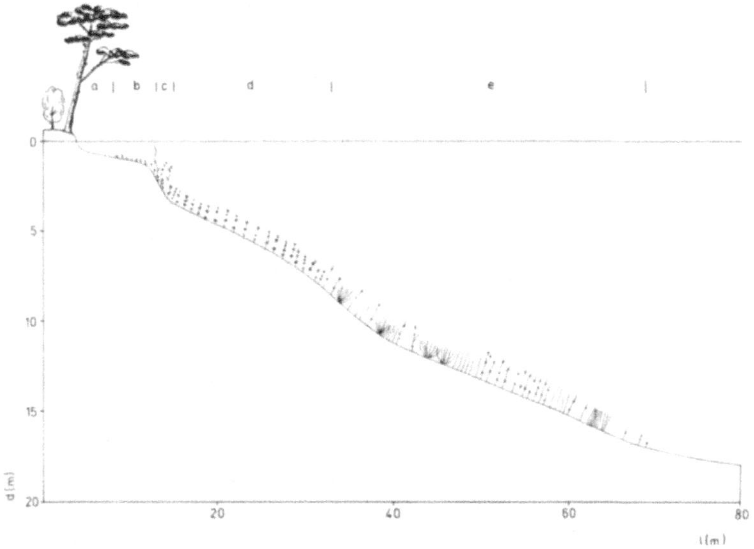

Figure 5.2 Vegetational transect through the north shore of the north bay of Lake Stechlin. a) sandy shore-bank, free from vegetation; b) Charetum asperae; c) Charetum filiformis, subassociation of *Potamogeton pectinatus;* d) Nitellopsidetum obtusae; e) Nitello-Vaucherietum.

the last few years, included a general retrogression of Characeae, which in Central Europe now belong to the endangered species of plants (Schmidt 1980). In some lakes of the Lake Stechlin area Characeae have already disappeared or became less frequent. In Lake Stechlin itself considerable displacements of the Characeae-vegetation, formerly wide-spread and rich in species, have been recorded. The following is a survey of the associations observed in the lakes of the Lake Stechlin area.

Charetum asperae settles at the wind-exposed sandy or partly stony, read-free shores of the oligotrophic to mesotrophic lakes at depths of between 0.2 and 1.3 m. It consists of sparse and low (sometimes only 2–3 cm high) stands of *Chara aspera.* In Lake Stechlin it was previously rather abundant, but has decreased very much (see Table 5.3).

Charetum filiformis grows at depths of between 1.2 and 5.0 m. In its dense, gray to greenishgray, chalk encrusted, middle high and nearly meadow-like stands occur mainly *Chara filiformis* and *Ch. aculeolata,* also frequently *Ch. fragilis* f. *hedwigii* and *Ch. aspera,* more infrequently *Ch. rudis* and some other species of *Chara.* It thrives at white-gray calcareous gyttja and covers large areas in the places with broad shore-banks. It is linked with oligotrophic to mesotrophic habitats and recedes quickly in eutrophication. In deeper water *Nitellopsis obtusa* also adds to this and forms a special variety of this association. In sheltered places the subassociation of *Potamogeton pectinatus* with *P. pectinatus* and single *P. gramineus, Myriophyllum spicatum* and *Ceratophyllum demersum* grows.

Charetum tomentosae, with *Chara tomentosa* as a principal ingredient, develops vast stands in mesotrophic lakes at depths of between 1 and 5 m. It is noticed at once due to its reddish-brown colour. In the lakes of the Lake Stechlin area it was rather rare, but has increased distinctly in the last years.

Nitellopsidetum obtusae replaces the last two mentioned communities in increasing depths of water. Its upper limit lies in proportion to the trophic state of the lake concerned at 2–6 m and its lower limit at 6–9 m. *Nitellopsis obtusa* forms high, dense and voluminous carpets, in which other species occur only singularly, mainly *Chara fragilis.* This association grows at deep graywhite calcareous gyttja. It is restricted not only to chalk-oligotrophic lakes, but is to be found in mesotrophic lakes and is also able to grow in eutrophic lakes, but with reduced vitality.

Nitello-Vaucherietum occupies the deepest zone of macrophytic vegetation. Forming dense stands or smaller tuffs, it is connected with *Nitellopsidetum obtusae* and advances to depths of nearly 20 m. Principal components are *Nitella flexilis,* which is able to tolerate lower temperatures (Corillion 1957), and the blackish-green *Vaucheria dichotoma.* Sometimes *Chara fragilis* and *Drepanocladus aduncus* are added to this. The association grows on brownish-black to deep black gyttja with transitions to sapropel. Where this layer of mud is thin, for instance on lacustric slopes, the subassociation of *Platyhypnidium rusciforme* with *Fontinalis antipyretica* and *Platyhypnidium rusciforme* develops. Closely related to this is a *Cladophora-Fontinalis*-community, which is to be found on slopes also and is perhaps identical with the *Aegagrophiletum holsaticae.*

Associations of pondweeds. ***Potametum lucentis*** is the most frequent pondweed-association. It appears in small strips between reed-zones and belts of Characeae, mainly at sheltered places and in the bays of the lakes. Conditions for its occurrence are the existence of a shallow shore-bank, wave-sheltered sites and a greater accumulation of sediments on the bottom. The subsoil consists mostly of a gyttja of great thickness. At abrupt steep banks the association developed is only fragmentary or is completely absent. ***Potametum lucentis*** mostly begins further in than the reeds at a depth of 1.20 m and descends to 6 m depth with an optimum between 2.5 and 4 m. Generally the pondweed-belt has a breadth of 3–4 m, but locally can extend as far as 20 m and more.

The characteristic and mostly predominating species is *Potamogeton lucens.* Its light-green submerged leaves are visible down to great depths in the clear water and form the aspect of the association (Fig. 5.3). Frequent components are *Myriophyllum spicatum, Ceratophyllum demersum* and *Elodea canadensis. Stratiotes aloides* f. *submersa, Ranunculus circinatus, Fontinalis antipyretica, Potamogeton natans, P. mucronatus, P. obtusifolius, Utricularia vulgaris, Myriophyllum verticillatum* and other plants occur sparsely.

The stands of the ***Potametum lucentis*** can be divided into three subassociations. The subassociation of *Nymphaea* forms the transition to the *Myriophyllo-Nupharetum* and contains already several floating plants such as *Nymphaea alba, Nuphar lutea* and *Potamogeton natans,* and pioneers of the reed-belt as *Phragmites australis, Typha angustifolia,* and *Equisetum fluviatile.* The typical subassociation forms the pith of the association. It can be divided into a richer variant of *Potamogeton perfoliatus* at eutrophic habitats and into a variant of *Chara fragilis* at meso- to oligotrophic habitats. Some species of Characeae (*Chara fragilis, Ch. hispida* subsp. *rudis, Ch. tomentosa, Nitellopsis obtusa*), which occur in this variant, form a more or less thick layer on the bottom. In Lake Stechlin and other oligotrophic lakes of the area, the subassociation of *Myriophyllum alterniflorum,* in which the submerged form of *Stratiotes aloides* occurs most frequently, can be found. The stands of *Stratiotes aloides* remain permanently on the bottom and also flower there. They are good indicators of very transparent calcareous lakes poor in nutrients. They are very sensitive for eutrophication and recede quickly when it occurs. Also here a poorer variant of *Chara fragilis* and a richer variant of *Potamogeton perfoliatus* can be divided.

Some plants of the pondweed-belt also form their own stands without *Potamogeton lucens.* In the Lake Stechlin area especially, there is a *Myriophyllum alterniflorum* community which still belongs to the pondweed-associations. But *Myriophyllum spicatum* also often forms pure stands at protected shallow sandy banks. *Potamogeton perfoliatus, Ceratophyllum demersum, Elodea canadensis* and *Ranunculus circinatus* are furthered by eutrophication, and then they may build up dense stands which, however, are often temporary. Abundant stands of

Figure 5.3 Potametum lucentis, subassociation of *Nymphaea,* at a sheltered place in Lake Stechlin. Photo: W. Scheffler.

Figure 5.4 Stand of sparse *Phragmites*-reed in Lake Stechlin. Photo: W. Scheffler.

Ranunculus circinatus are especially striking during the flowering period when they form white carpets on the surface of the water visible from a distance.

The ***Potametum lucentis*** enjoys its optimum conditions in the mesotrophic reach. There dense and luxuriant stands of *Potamogeton lucens* are to be found, for instance in Lake Nehmitz (Fig. 5.5). But it also settles in the oligotrophic reach and, on the other hand, it extends into the eutrophic reach. During eutrophication it is at first encouraged, but as the process of eutrophication continues it recedes quickly. Thus in Lake Dagow the ***Potametum lucentis*** was still found in 1960, but now it is absent and displaced by species-poor stands of *Ceratophyllum demersum.*

Potametum filiformis is another pondweed-association of the Lake Stechlin area. It grows in thin layers on shallow banks with sandy or sandy-muddy substrata. The optimum formations are situated on small bays of the shores or in protection of sparse reed-stands. There are sparse groups of *Potamogeton filiformis,* which contain few species, mostly of *Chara* and plants of the reed-belt. The association is linked with clear water. During eutrophication the plants of *Potamogeton filiformis* were soon encrusted with dense periphyton, which effects a progressive reduction of vitality and finally produces a decrease in the stand. In this way the association in the last years was pushed back locally in the Lake Stechlin area. The ***Potametum filiformis*** is a boreal association with its centre in the chalk-oligotrophic lakes of northern Europe.

Pondweed-associations of flowing water are nearly absent in the Lake Stechlin area. Only a *Potamogeton perfoliatus-Zannichellia palustris*-community in the Roofen-brook downstream of Menz with *Zannichellia palustris, Potamogeton perfoliatus, P. mucronatus* and *Elodea canadensis* on a sandy and gravel bed can be mentioned.

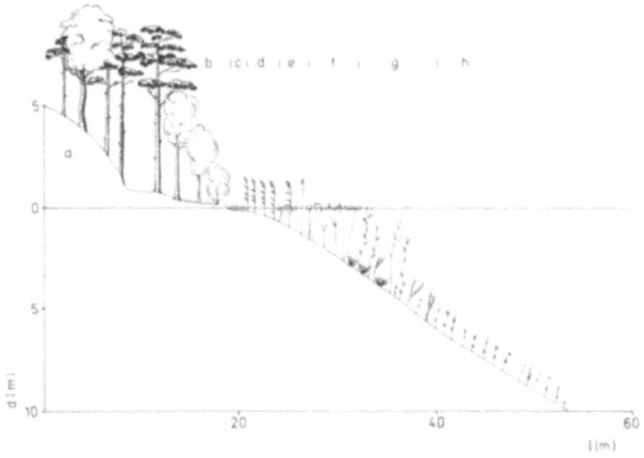

Figure 5.5 Vegetational transect through the east shore of Lake Nehmitz. a) Pine-Beech forest; b) Alder carr (Carici elongatae-Alnetum); c) Caricetum ripariae-acutiformis; d) Scirpo-Phragmitetum; e) Myriophyllo-Nupharetum, subassociation of *Schoenoplectus lacustris;* f) Myriophyllo-Nupharetum typicum; g) Potametum lucentis; h) Nitellopsidetum obtusae.

Associations of floating plants. In the deep lakes with precipitous slopes strongly affected by storms, associations of floating plants are rare and limited only to some places under the lee in the bays or on shallow banks. In the shallow lakes – for instance Lake Breutzen, Little Boberow Lake and Lake Mehlitz – they occupy wide areas and sometimes cover the surface of the lake.

Myriophyllo-Nupharetum is the most frequent association of floating plants. Its ecological spectrum extends from the oligotrophic to the eutrophic reach. In the Lake Stechlin area the majority of the stands belong to poorer varieties, which are indicated by the occurrence of Characeae, by high frequency of *Potamogeton natans* and *Utricularia vulgaris* as well as by the receding of *Ceratophyllum demersum.*

Myriophyllo-Nupharetum is an association with several layers of plants. On the surface of the water *Nuphar lutea, Nymphaea alba* and *Potamogeton natans* spread their floating leaves. A middle layer is formed by *Myriophyllum verticillatum, M. spicatum, Utricularia vulgaris, Ranunculus circinatus, Ceratophyllum demersum* and some *Potamogeton*-species, which also – with the exception of *Ceratophyllum demersum* – flower above water. There is still commonly a lower layer of Characeae on the bottom – *Stratiotes aloides, Fontinalis antipyretica,* sterile *Elodea canadensis* and *Potamogeton* species.

Myriophyllum verticillatum prefers pure and only slightly eutrophic water. It is not rare and occurs frequently in the shallow lakes especially, but is absent in the deep and cooler lakes, where *Myriophyllum spicatum* takes its place, and also in the stronger eutrophized lakes, for instance in Lake Dagow. There *Nymphaea alba* is developed luxuriantly, but the *Myriophyllum*-species and most of the other submerged plants with the exception of *Ceratophyllum demersum,* which forms dense stands, are absent. *Nuphar* occurs only sparsely.

In relation to water depth the ***Myriophyllo-Nupharetum*** of the Lake Stechlin area can be divided into three subassociations. The subassociation of *Stratiotes aloides* settles in habitats with a water depth of 1.8–3.5 m. It is characterized by the occurrence of the submerged form of *Stratiotes aloides.* The typical subassociation occupies the medium reach of depth of between 1.2 and 2.5 m. In shallower water the carpets of floating leaves are mixed with single blades or sparse growth of *Schoenoplectus lacustris, Phragmites australis, Typha angustifolia* or *Sparganium erectum* (subassociation of *Schoenoplectus lacustris*). In all of these subassociations besides typical variants there are variants of *Chara* with Characeae, which indicate mesotrophic to oligotrophic calcareous habitats.

In those habitats other species-poor communities of floating leaves-plants sometimes occur. In addition to a *Potameto-Nupharetum* with *Nuphar lutea, Potamogeton natans* and *Utricularia vulgaris* in some places there are stands of pure *Potamogeton natans.* On shallow sandy banks of chalk-oligotrophic lakes species-poor stands with undernourished forms of *Nymphaea alba* (*N. 'minor'*) grow. The acid bog-lakes contain only sparse stands of *Nuphar lutea* and/or *Nymphaea alba* mixed with *Drepanocladus fluitans.*

Species-poor stands of *Polygonum amphibium 'natans'* occur individually and as small groups in protected bays directly adjacent to the reed-belt or in gaps in the reeds, especially in some disturbed and slightly eutrophized places.

Associations of duckweeds. Duckweed-associations are restricted to shallow lakes and small waters.

Hydrocharitetum morsus-ranae is the most important association of this class. It grows in ditches and peat-cuttings, but also in very shallow and shrunken lakes as for instance in Lake Mehlitz, in Little Kessel Lake and Lake Glawke, covering here and there the whole surface of the residual lake. Its habitats are shallow water basins with muddy substrata, in which the water becomes warmer during the summer than it does in the large and deep lakes.

Characteristic species of the association are *Hydrocharis morsus-ranae* and *Stratiotes aloides*. Sometimes the latter forms dense stands mainly in shallow lakes, but also in the peat-cuttings (*Stratiotes*-facies). Shallow ditches are dominated mostly by *Hydrocharis morsus-ranae.* In addition *Lemna minor, L. trisulca, Utricularia vulgaris* and *Potamogeton natans* occur more or less regularly. The relatively frequent occurrence of *Utricularia vulgaris* and *Potamogeton natans* as well as the receding of *Spirodela polyrrhiza* and *Ceratophyllum demersum* indicate a poorer situated habitat in contrast to other areas. A typical subassociation here, *Sparganium minimum,* is frequent, which occurs in peat-ditches and peat-pools and forms a transition to the association of small bladderworts (see p. 137).

The proper duckweed-associations are represented in the Lake Stechlin area by the ***Spirodeletum polyrrhizae*** and the ***Lemnetum trisulcae.*** They sometimes form dense green covers in small waters such as ditches, ponds and pools. Its principal ingredients are *Lemna minor, L. trisulca* and *Spirodela polyrrhiza.* The ***Spirodeletum polyrrhizae,*** linked to eutrophic waters, is rare in the Lake Stechlin area and occurs often in impoverished forms (without *Spirodela*). The less present ***Lemnetum trisulcae*** settles mesotrophic clear waters and is more frequent, especially in former peat-cuttings. In the recent years it has formed extensive dense stands on inundated peat-meadows, Alder swamps and in the dyke between Lake Dagow and Lake Stechlin and was also found in former small fish-ponds near the east bay of Lake Stechlin.

Associations of small bladderworts. In ditches and flarks of the mires, but also in fens near the lakes, sometimes associations with bladderwort-species (*Utricularia minor, U. intermedia*) can be found.

Sparganietum minimi grows in shallow mire-ditches and flarks with a depth of 10–50 cm, more rarely in protected bays of meso- to oligotrophic lakes. When summer droughts lower the water-level, ditch and flark habitats sometimes dry out, causing *Sparganium minimum* and other components of this association to develop aerial forms. The characteristic and most dominating species is *Spar-*

ganium minimum. Other important plants of this species – poor association are *Utricularia intermedia, U. minor* and *Potamogeton gramineus.* In addition to the typical subassociation, in some richer habitats the subassociation of *Lemna minor* with *Lemna trisulca, Utricularia vulgaris* and *Hydrocharis morsus-ranae* occurs. The communities of *Utricularia minor* and *U. intermedia* can be summarized as *Scorpidio-Utricularietum.* It grows in shallow flarks of tall sedge- and reed-communities, especially in the Cladietum marisci in calcareous habitats of silted up bays and in mesotrophic fens. This association is characterized by some mosses of calcareous fens, for instance *Scorpidium scorpidioides.* In addition, *Chara vulgaris* is frequent. In recent years several former sites of this endangered association were destroyed by amelioration.

Reeds. In the Lake Stechlin area reeds generally do not take such a leading part as in eutrophic lakes with shallow shores situated in summerwarm areas. At the deep oligotrophic and mesotrophic lakes with steep slopes, reeds retreat markedly and are only to be found in bays and on the more shallow banks. Moreover, they often remain low and sparse (Fig. 5.4). On the smaller, shallow and mostly more eutrophic lakes the reeds are better developed. These lakes are surrounded by a dense and continuous reed-belt. In the small oligotrophic lakes the reeds also retreat and are often absent from the typical bog-lakes.

Scirpo-Phragmitetum with *Phragmites australis, Schoenoplectus lacustris, Typha angustifolia, T. latifolia, Sparganium erectum, Equisetum fluviatile* and *Sagittaria sagittifolia* is the most frequent reed-association. Often one of the species mentioned predominates, and these abundant stands are now considered by several authors as independent associations (e.g. *Sparganietum erecti* Roll 1938, *Phragmitetum australis* Schmale 1939, *Typhetum angustifoliae* Pign. 1953). Stands with predominating *Phragmites australis* are frequent in the Lake Stechlin area, also those with *Typha angustifolia. Schoenoplectus lacustris* forms sometimes sparse stands in shallow water. Here and there on shallow and muddy shores an *Equisetum fluviatile*-reed is found, for instance in the south bay of Lake Stechlin. On the other hand, stands of *Sparganium erectum* and of *Sagittaria sagittifolia* are very rare, and *Typha latifolia* seems to be completely absent.

In ***Scirpo-Phragmitetum,*** independent of the depth of the water, five subassociations can be established. The typical subassociation grows directly on the shore-line in only shallow flooded, temporarily dry habitats. Its reeds are tall and dense, but very species-poor, in which either *Phragmites australis* or *Typha angustifolia* is prevalent.

Reeds standing permanently in the water are wide-spread; they can advance to depths of 1.8 m. Shores with greater undulations are settled by the subassociation of *Potamogeton filiformis* and the subassociation of *Chara delicatula.* In places in the lee at depths of between 0.25 and 1.50 m the subassociation of *Nuphar lutea* occurs. Between reed plants which are not very dense, floating water plants such as *Potamogeton natans, Nymphaea alba* and *Nuphar lutea* commonly grow, but

Polygonum amphibium more rarely. In this subassociation the variant of *Carex rostrata* indicates poorer habitats, while the variant of *Glyceria maxima* indicates richer habitats with muddy substrata.

When the reed-peat becomes thicker by accumulation and the habitats dry out in summer, the subassociation of *Solanum dulcamara* develops. In addition to *Solanum dulcamara* several other plants of sedge-reeds and carrs such as *Lysimachia vulgaris, Galium palustre, Lythrum salicaria* and others grow. In the Lake Stechlin area this subassociation occurs only in the poorer variant of *Thelypteris palustris,* in which *Thelypteris palustris, Carex acutiformis* and *Acrocladium cuspidatum* are represented abundantly. In these stands *Phragmites australis* predominates. In some lakes with adjacent mires, in the course of miring they are invaded by *Sphagnum*-species, which lead to a *Phragmites*-rich mire-vegetation, as for instance in Lake Steutzen.

Cladietum marisci (as well as ***Scirpo-Phragmitetum***) is a frequent reed-association of the Lake Stechlin area. It is widespread in the calcareous oligotrophic and mesotrophic lakes and occupies in alternation with ***Scirpo-Phragmitetum*** large parts of the shores; at some lakes it is dominant. In the eutrophic lakes it occurs only occasionally (e.g., at Lake Dagow) or is completely absent. *Cladium mariscus* frequently forms 3–4 m broad belts, but on shallow banks or in bays it also can cover larger areas. The substrate is different and extends from pure sandy soils to mud and chalk-gyttja and to *Cladium*- and fen-peat.

Two subassociations can be distinguished. The typical subassociation on the shores of the lakes forms species-poor stands, containing besides the predominant *Cladium mariscus* often only some *Phragmites australis*. Richer in species is the subassociation of *Carex elata,* with numerous plants of sedge-reeds and carrs such as *Thelypteris palustris, Acrocladium cuspidatum, Carex elata* and others. It occupies the less wet habitats on land sites. Several variants of this subassociation contain plants from locally adjacent calcareous fens. *Utricularia minor, U. intermedia,* and *U. vulgaris* often grow in flarks.

Carex acutiformis-reed extends in nearly all lakes along the shore in often broken narrow bands, sometimes behind sparse *Phragmites*-reeds. The species-poor association is indicated by the dark-green blades of *Carex acutiformis*. Other species are rare. In habitats with a thicker peat-layer the variant of *Thelypteris palustris* grows. The subassociation of *Ranunculus repens,* growing near the lakes in fen-meadows and also containing meadow-plants, is very much richer in species.

A species-poor *Eleocharis palustris*-reed is occasionally found in protected spots of shallow sandy shores, for instance in the south bay of Lake Stechlin.

Juncus subnodulosus-reed occurs in chalk-oligotrophic lakes, extending as a narrow belt along the shores or behind the *Phragmites*- or *Cladium*-reeds. It only occupies small areas.

A sparse reed of *Carex lasiocarpa* is sometimes in evidence on sandy or sandy-muddy substrata on the shores of lakes extremely poor in nutrients. But also the poor bog-lakes are sometimes bordered by a narrow belt of *Carex lasiocarpa*.

Reeds of eutrophic habitats are almost absent in the Lake Stechlin area. Only *Glycerietum maximum* is sometimes found in small areas of eutrophized waters.

Changes in the macrophytic vegetation of Lake Stechlin

Before the starting up of the nuclear power plant Lake Stechlin was a typical calcareous oligotrophic lake with very clear water (Krausch 1964). With favourable light conditions the submerged vegetation could advance to depths of nearly 20 m. On the steep slopes of the shores the belts of macrophytic vegetation were often very narrow indeed, but in the bays they covered extensive areas (Fig. 5.6). As can be seen in Table 5.1, nearly 40% of the surface of the lake was covered with plant growth, and the belts of Characeae-communities with about 1.4 km^2 occupied a third of the total area of 4.2 km^2.

After the coming into operation of the nuclear power plant in 1966 as a result of the warming of the water, and the production of a fixed drift and supply of nutrients from Lake Nehmitz, the translucence of the water was reduced by the increase in plankton and periphyton. The changes in the vegetation of macro-

Table 5.1 Lake Stechlin. Areas of macrophytic plant-communities as part of the total area of the lake about 1960 [based on the map of vegetation by Freitag and Krausch in Krausch (1964) calculated by Krey].

Lake region	Area in ha	% of total area
Upper littoral (0–5 m)		
Alder carr (Alnetum)	0.3	0.1
Reed	9.6	2.3
Pondweed-communities	16.7	3.9
Macrophytes-free zones	7.5	1.8
	34.1	8.1
Lower littoral (5–20 m)		
Charetum filiformis and Charetum asperae	41.8	9.9
Nitellopsidetum obtusae	34.5	8.1
Cladophora-Fontinalis-community	10.9	2.9
Nitello-Vaucherietum	55.4	13.0
	142.6	33.6
Profundal (20–68 m)		
Macrophytes-free zones	247.5	58.3
Lake Stechlin	424.2	100.0

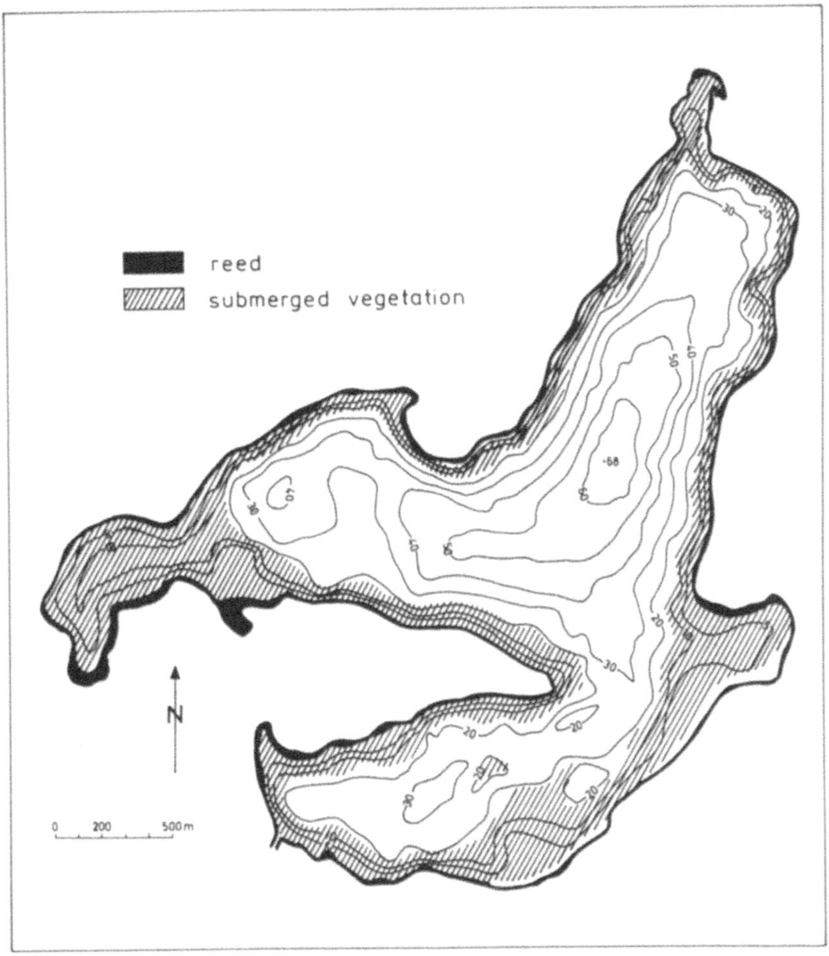

Figure 5.6 Occurrence of reeds and submerged vegetation in Lake Stechlin 1960–1963.

phytes were first seen in the Characeae. Nearly all communities of Characeae receded, and in many places desolated areas without macrophytic vegetation arose. Most sensitive species such as *Chara filiformis, Ch. aspera, Stratiotes aloides* and *Potamogeton rutilus* disappeared. *Ceratophyllum demersum, Potamogeton pectinatus, Ranunculus circinatus* and *Elodea canadensis,* which thrive in nutrient-richer water, increased suddenly and formed dense stands in the west bay. The eutraphent *Potamogeton crispus,* completely absent before, began to expand. An interesting phenomenon was the premature growth and flowering of water- and reed-plants. In the west bay *Nymphaea alba* and *Potamogeton natans* were flowering at the beginning of May when heated water flowed into the lake. The dependence of reed-growth on the temperature of water can be seen by the measurements taken on 11.5.1967 (Table 5.2).

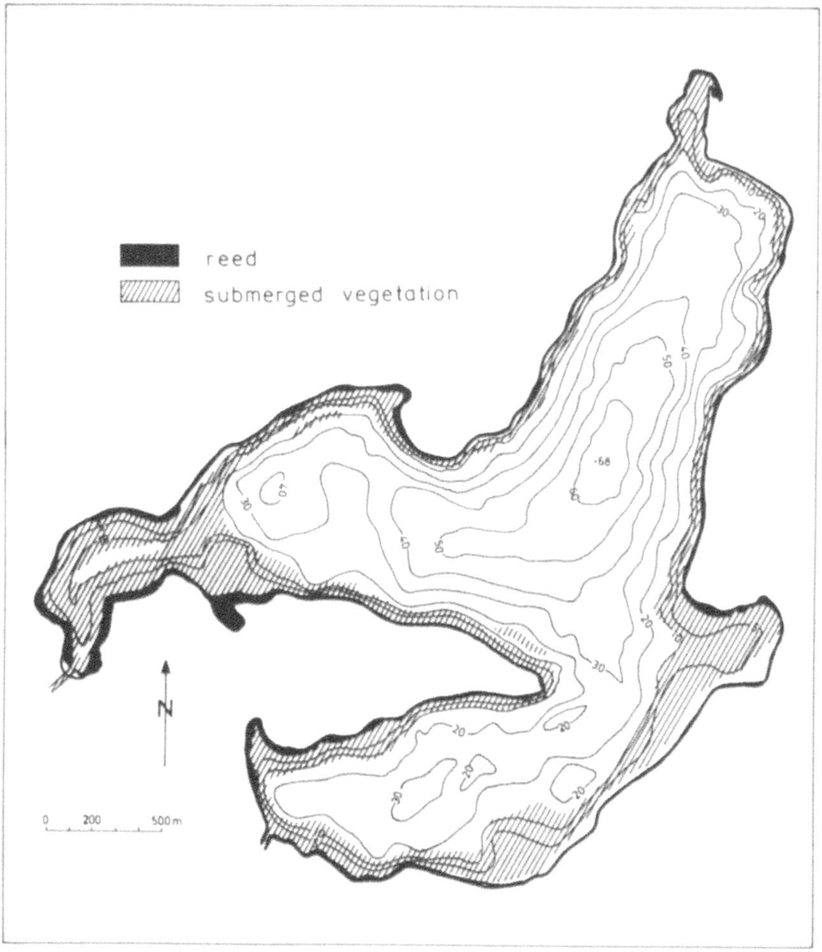

Figure 5.7 Occurrence of reeds and submerged vegetation in Lake Stechlin 1980–1983.

Table 5.2 Dependence of reed growth on water temperature in Lake Stechlin (11. 5. 1967).

Site of measurement	Temperature of surface water (°C)	Mean maximum growth above water level (cm)	
		Typha	Phragmites
End of south bay	16.5–16.6	39.4	Yet under water-level
East bay	17.8	–	25.4
Entrance of west bay (Ochsenkopf)	21.4	78.4	92.3
Middle of west bay	25.3	172.3	101.5
End of west bay	25.4	177.9	115.5

In the course of time a certain stabilization has taken place. The macrophytic vegetation has adapted under the new conditions. The displacement of the zonal vegetation seems to have ended (Fig. 5.7). The extent of some species has recedes. Some species which had disappeared, returned, but not more extensively than previously (e.g. *Chara filiformis*). Other new invading species (e.g. *Potamogeton crispus*) disappeared again. Some of former associations such as *Charetum asperae* and *Cladophora-Fontinalis antipyretica*-community disappeared, while the formerly rare *Charetum tomentosae* has spread. The changes of the individual communities from 1960/63 to 1980/3 can be seen in detail in Table 5.3.

Changes in the vegetation of Lake Nehmitz (north basin)

In 1960 the north basin of Lake Nehmitz (German named 'Flacher Nehmitz') had the vegetation of mesotrophic clearwater-lakes. The reeds – composed of *Phragmites australis, Typha angustifolia, Cladium mariscus* and *Schoenoplectus lacustris* – were frequently higher and denser than in Lake Stechlin, but on steep slopes here sparse stands were also to be found. On wind-exposed shores the *Charetum asperae* occurred locally on sandy substrata. The stands of the *Myriophyllo- Nupharetum* in the bays as well as those of *Potametum lucentis* were frequent and more extensive than in Lake Stechlin. The latter occurred only in the typical subassociation; *Myriophyllum alterniflorum* was absent. *Ceratophyllum demersum* and *Elodea canadensis* were much more frequent and more luxuriant than in Lake Stechlin, and here and there they formed dense stands. *Utricularia vulgaris* too occurred often and sometimes in immense (until 2.5 m long) specimens. At the bottom of the associations of floating plants and pondweeds in shallow water *Chara hispida* and *Ch. tomentosa* were often found. The pondweed belt reached to depths of 4.5 m. There, but occasionally at a depth of 2 m, joined associations of Characeae, mostly *Nitellopsidetum obtusae,* rarely also *Charetum filiformis.* They extended to depths of 4.5–6 m, individually to 6.5 m. At greater depths isolated spots of *Vaucheria dichotoma* and *Nitella flexilis* occurred down to about 7 m. Altogether the communities of Characeae covered much smaller areas than in Lake Stechlin. Following the circulation of the water from the nuclear power plant, a great quantity of oligotrophic water from Lake Stechlin came to the north basin on Lake Nehmitz, so that here too translucence increased and vegetation could advance to greater depths. *Myriophyllum alterniflorum* spread at first into the canal between Lake Stechlin and Lake Nehmitz, and then it migrated into Lake Nehmitz also. In the belts of reeds and floating plants changes were barely noticed. Of interest is the mass occurrence of *Ranunculus circinatus* in some places, where it forms remarkable white carpets on the surface when in flower.

Potametum lucentis often developed luxuriant and dense growth. *Cerato-*

Table 5.3 Changes in macrophytic vegetation of Lake Stechlin.

Community	1960–63	1980–83
Reeds (*Phragmites-, Typha angustifolia-, Cladium mariscus-, Carex acutiformis-, Eleocharis palustris*-reed)	In sheltered places only, mainly in bays, at depths of 0–1.2 (1.8) m	In the same places and of the same composition at depths of 0–1.2 m, locally a little more dense and higher than before. *Eleocharis palustris*-reed rather frequent as 1960–63.
Myriophyllo-Nupharetum (poorer stands)	In very sheltered places only, little groups, mostly fragmentary, at depths of 0.7–2.5 m	In the same places, barely changed. *Myriophyllum verticillatum* and *Stratiotes aloides* are missing.
Potametum lucentis (mostly in poorer stands with Characeae and *Myriophyllum alterniflorum*)	In sheltered places, mainly in bays, at depths of 2–4 m	In the same places, but frequently more luxuriant and more extended. Extension to other places. *Potamogeton mucronatus, P. rutilus* and *Stratiotes aloides* are missing; *Potamogeton perfoliatus* and *Ceratophyllum demersum* have increased.
Myriophyllum alterniflorum-community	Frequently in bays and in sheltered places at depths of 0.5–2.5 m	Nearly in the same areas, but *Stratiotes aloides* is missing.
Charetum asperae	On wind-exposed stony or sandy shores at depths of 0.2–1.3 m, often extensive stands	Mostly absent or only fragmentary stands
Charetum tomentosae	In few places, mostly fragmentary	Considerably increased, greater stands at various places at depths of 0.7–2.0 m
Charetum filiformis	Widely spread at depths of 1.2–5.0 m on gyttja	After heavy retrogression now stabilized, at depths of (0.5) 1.0–2.0 (3.0) m
Nitellopsidetum obtusae	Widely spread on calcareous gyttja especially in bays at depths of (3) 5–9 m	After retrogression now extensive and luxuriant stands at depths of (0.8) 2.0–8.0 (8.5) m
Nitello-Vaucherietum	Sparse to dense stands on dark gyttja at depths of 9–18 (−20) m	Only sparse stands at depths of 8–16 m, but sometimes 20 m
Cladophora-Fontinalis antipyretica-community	On sublacustric slopes often dense and luxuriant stands at depths of 7–13 m	Completely disappeared

phyllum demersum and *Elodea canadensis* still occurs abundantly, but there are also many *Potamogeton mucronatus* and *Stratiotes aloides,* which now are no longer found in Lake Stechlin. The communities of Characeae have increased since the sixties and reach depths of 9 m. In depths of 0.8–3.5 m, in gaps in the belt of floating plants and pondweeds, *Charetum filiformis, Charetum tomentosae* and *Charetum asperae* grow sometimes in dense stands. In addition to other species *Fontinalis antipyretica* also occurs. *Nitellopsidetum obtusae,* starting at 3.5 m, often forms pure, dense and luxuriant stands at depths of 5–6 m, reaching to about 6.5 m. Now *Nitella flexilis* and *Vaucheria dichotoma* grow at depths of 6–9 m. They form, partly singly, partly together, at depths of 7–8 m often dense and luxuriant stands.

Standing crop of macrophytes

Under 'standing crop' is to be understood that quantity of organic substance in the form of living substance, which (in proportion to unit of area and time) can be found at the date of investigation (biomass; related to plants, it is also called phytomass; Raspopov 1972). The biomass of waterplants is indicated as dry weight (DM) or as carbon (C).

Considering various conditions, biomass permits conclusions on the productivity of water-plants. In perennials a great part of the net primary production wanders into the subterranean biomass, for example in *Phragmites australis* up to 2/3. We have, as is general use, only ascertained the above ground biomass. According to Westlake (1965) water- and reed-plants lose 15% of annual net production in one growing season. To the loss of dead parts ('litter') and seeds are added grazing and damage by wind and waves. According to Raspopov (1972) the annual production of most emerged and submerged hydrophytes can be calculated with the formula: annual production = 1.2 × maximum (above ground and subterranean) biomass.

Methods. The sampling of the plants was carried out from the middle to the end of August, in the period of the development of maximum biomass. An area of 1 m^2 was staked out with a frame of sheet iron. The plants within the frame were torn away above the lake bottom; roots and rhizomes were not seized. The material was selected macroscopically, washed, air dried, oven dried at 105° C, and weighed. The calcite adhering to the leaves and stems, so far as it was not lost in the procedure of cleaning and drying, was also weighed.

Calculating the macrophytic biomass in the lake, the area of the individual water- and reed-plants was estimated on the basis of the abundance and dominance values in the samples of 1960–1963 (Krausch 1964; Table 5.4).

Macrophytic biomass of Lake Stechlin. The results of the biomass-determination are put together in Tables 5.4 and 5.5. All results are given as above ground biomass.

The mean biomass of all sampled material is about 390 g DM · m^{-2} (n = 40) in the submerged plants and about 900 g DM · m^{-2} (n = 6) in the reeds. Due to the large variation in the samples from the various places and due to the different sampling methods practised (samplings of individual species and of communities respectively) significant differences between the years 1963 and 1981/82 cannot be established.

Analyses of the C-content have shown that the average value of 45% of dry weight given in specialist literature (Kreeb 1983; [g] = 0.45 DM [g]) is not attained by most aquatic macrophytes except reeds in Lake Stechlin, due to the high calcite content. Most values are below (Table 5.5). The average C-content established by Kreeb (1983) corresponds to the content of organic carbon (OC). Our OC-values (Table 5.5) are related to biomass (DM) including $CaCO_3$ ($CaCO_3$ $\hat{=}$ inorganic carbon [IC] × 8.34). Calculated without $CaCO_3$ (DM–$CaCO_3$) our OC-values reach ~42 per cent, e.g. they can be compared with the value found by Kreeb (1983).

On the basis of the biomass values (Proft 1963, unpublished) of the plant communities, a total biomass of the aquatic macrophytes in Lake Stechlin of around 700 t DM or 160 t TC (107 t OC and 53 t IC $\hat{=}$ 440 t $CaCO_3$) is reached. The largest portions of the biomass are the Charagroups (270 t DM or 27 t OC) and *Nitellopsis obtusa* (175 t DM or 18 t OC; Table 5.5).

Comparing these values with those of other waters in the northern temperate zone reveals that the production of submerged plant-phytomass in Lake Stechlin related to its oligotrophic state is relatively high, which can be traced back to the high values of biomass of much Characeae-communities. Optimum stands of *Charetum filiformis* reach values up to 1000 g DM · m^{-2}; the mean value of our samples lies about 500 g DM · m^{-2}. Optimum stands of *Nitellopsidetum obtusae* show biomass-values of between 450 and 520 g DM · m^{-2}. On the other hand, the biomass-values of Characeae-stands in some Masurian lakes amount to 38–220 g DM · m^{-2} (Bernatowicz & Pieczynska 1965; Bernatowicz & Radziej 1964; Bernatowicz, Pieczynska & Radziej 1968). But the *Potametum lucentis* in Lake Stechlin, with a mean biomass of between 90–860 DM · m^{-2}, is within the usual bounds. The *Hydrilla verticillata-Potamogeton*-community of the comparable Lake Biwa in Japan has a mean biomass of 356 g DM · m^{-2} or 144 g OC (Ikusima 1975).

In the reeds of Lake Stechlin the stands of *Phragmites australis*, even at their optimum development, only reach biomass-values of 1080 g DM · m^{-2}. These values are significantly lower than those of *Phragmites*-reeds in eutrophic lakes. In Lake Müritz stands of *Phragmites australis* values of 1425–1575 g DM · m^{-2} were measured (Krausch, unpublished). In Lake Neusiedel in Austria the mean values of above ground biomass of *Phragmites* stands amount to about 1580 g

DM · m^{-2} (cut stands) respectively 1380 g DM · m^{-2} (uncut stands; Hübl 1966). On the other hand *Typha angustifolia* in its optimum stands in Lake Stechlin with 1380 g DM · m^{-2} reach values of biomass which can also be found in eutrophic waters. So in the region of Berlin biomass-values of 1073–1720 g DM · m^{-2} of stands of *Typha angustifolia* were registered (Sukopp & Markstein 1981). At Lake Stechlin it must be considered that optimum reed-stands only grow at a few places

Table 5.4 Standing crop of water plants, analysed August 1981 (submerged plants) and 1982 (reeds), Lake Stechlin.

Plant community and sample points	depth (m)	DM (g · m^{-2})
Potametum lucentis with *Chara*		
Near LL*	1.5	335
N-bay, Kreuzlanke	1.5–2.0	410
N-bay, Kreuzlanke (optimum stand)	1.5	650
S-bay, west of canal	1.5	90
S-bay	1.5	210
W-shore, Feenbucht	1.5	320
Neuglobsow bay	1.5	475
Neuglobsow bay	1.5	240
Charetum filiformis		
Near LL[1] (with *Potamogeton pectinatus*)	2.0	85
Near LL[1] (pure stand)	2.5	25
Middle of S-bay	2.5	185
N-bay, Kreuzlanke (optimum stand)	2.0	995
S-bay, west of canal (optimum stand)	1.0–2.0	995
S-bay, south shore of the peninsula (optimum stand)	2.0	1,000
S-bay	2.5	275
W-shore, Feenbucht	2.0	490
Nitellopsidetum obtusae		
Near LL[1]	4.5	365
Middle of S-bay	5.0	245
N-bay, Kreuzlanke	5.0–6.0	245
S-bay, west of canal	12.5	140
S-bay	5.0–6.0	450
Neuglobsow bay	2.0	450
Neuglobsow bay	4.0	520
Nitello-Vaucherietum		
N-bay, Kreuzlanke, *Vaucheria*-stand	10	2
Reeds		
S-bay, *Carex acutiformis*-reed		610
S-bay, *Typha angustifolia*-reed		1,380
S-bay, *Cladium mariscus*-reed		1,140
S-bay, *Phragmites*-reed		1,080
Neuglobsow bay, *Phragmites*-reed		1,080

* LL = Limnological Laboratory Stechlin.

Table 5.5 Biomass and C-content of water plants of Lake Stechlin (calculated according to Proft 1963, unpublished).

	Area (ha)	DM (g · m^{-2})	OC (g · m^{-2})	TC (% DM)	OC (% DM)	IC (% DM)	$CaCO_3$ (% DM)	DM (t/Lake)	TC (t/Lake)	OC (t/Lake)	IC (t/Lake)
Ceratophyllum demersum	0.2	400	54	21.5	13.6	7.9	66.1	0.8	0.17	0.11	0.06
Chara spp.	46.0	580	59	19.2	10.2	8.9	73.9	267.0	51.1	27.2	23.7
Elodea canadensis	0.5	500	183	37.0	36.7	0.3	2.7	2.5	0.93	0.92	0.01
Fontinalis antipyretica	7.5	334	134	40.5	40.2	0.2	2.0	25.0	10.1	10.1	0.06
Myriophyllum alterniflorum	5.0	380	75	26.1	19.8	6.3	52.5	19.0	5.0	3.8	1.2
Myriophyllum spicatum	5.0	400	83	26.5	20.8	5.6	46.7	20.0	5.3	4.2	5.3
Nitella flexilis	25.0	360	31	18.9	8.7	10.2	85.5	90.0	17.0	7.8	9.0
Nitellopsis obtusa	35.0	500	51	20.2	10.3	9.7	80.9	175.0	35.0	18.0	16.9
Platyhypnidium rusciforme	7.5	334	122	37.7	36.5	1.2	10.0	25.0	9.4	9.1	0.3
Polygonum amphibium	0.1	400	172	43.0	43.0	0.02	0.2	0.4	0.16	0.16	–
Potamogeton lucens	3.0	550	201	37.6	36.6	1.0	8.3	16.5	6.2	6.0	0.16
Potamogeton natans	1.0	550	236	48.8	43.0	1.0	8.0	5.5	2.4	2.4	0.05
Potamogeton perfoliatus	1.0	550	211	38.9	38.5	0.3	2.4	5.5	2.1	2.1	0.02
Stratiotes aloides	3.5	320	106	34.1	33.1	1.0	8.3	11.2	3.8	3.7	0.11
Vaucheria dichotoma	25.0	114	46	40.2	40.0	1.2	1.0	28.5	11.7	11.4	0.3
$\sum$	166.0	–	–	–	–	–	–	~ 692.0	~ 160.0	~ 107.0	~53.0
$\bar{x}$	–	418	118	33.0	29.0	3.7	30.0	–	–	–	–
SD	–	±124	68	9.7	13.1	3.9	33.3	–	–	–	–

in bays protected against wind and waves. The sparse stands of reeds, which are found on the other shores of Lake Stechlin, show a significantly lower biomass; in a sparse stand of *Typha angustifolia* (Table 5.4) a value of only 430 g DM $\cdot$ m^{-2} was measured.

The microphytobenthos (W. Scheffler)

Introduction

Thanks to the high transparency of its water, Lake Stechlin permits a production of periphyton up to a depth of about 12 m.

In the surf zone, which is poor in aquatic plants, predominantly stones (granite, gneiss, sandstone) and wood are prefered as growth substrata.

At 10 cm water depth periphyton also settles on aquatic plants like *Potamogeton, Myriophyllum* or *Chara.* In view of the paucity of reeds in the lake (see p. 138) the growth of rush is of relatively minor importance.

In research lasting five years biomass and primary production of periphyton were determined. For methodical reasons the totality of the community of microorganisms which constitutes the 'aufwuchs' was regarded as periphyton (Scheffler 1981). The qualitative analysis was restricted to the formation of a list of taxa of the benthonic diatoms on the basis of probes (see p. 491; Fig. 5.8).

Ecological questions – for instance diversity, abundance, or change of aspects – are not dealt with.

The biomass of periphyton in the littoral of Lake Stechlin

The biomass through the year. In the surf area on the sun-exposed north-west shore under the lee at measuring point Sl 1 the rates of biomass of periphyton increase (Fig. 5.9) after a winter minimum up to a first climax of development in spring (March–April). In April a short phase of depression occurs, immediately followed by the sharp rise to the maximum of annual development in the early summer. With one exception – autumn 1973 – the rates of biomass again decrease until the end of the year. These statements cannot be transferred to the whole littoral zone. On the wind-exposed and shadowed south-east shore, for instance, the development is definitely less clearly arranged. Here two phases of intensified production of periphyton alternate, but from year to year temporally displaced against one another. Regularly in August and September of each year a depression phase appears (Fig. 5.10). In the littoral of Lake Stechlin the following general tendencies of the production of biomass in the annual course can be stated.

In autumn and winter the production of periphyton stagnates. The active phase

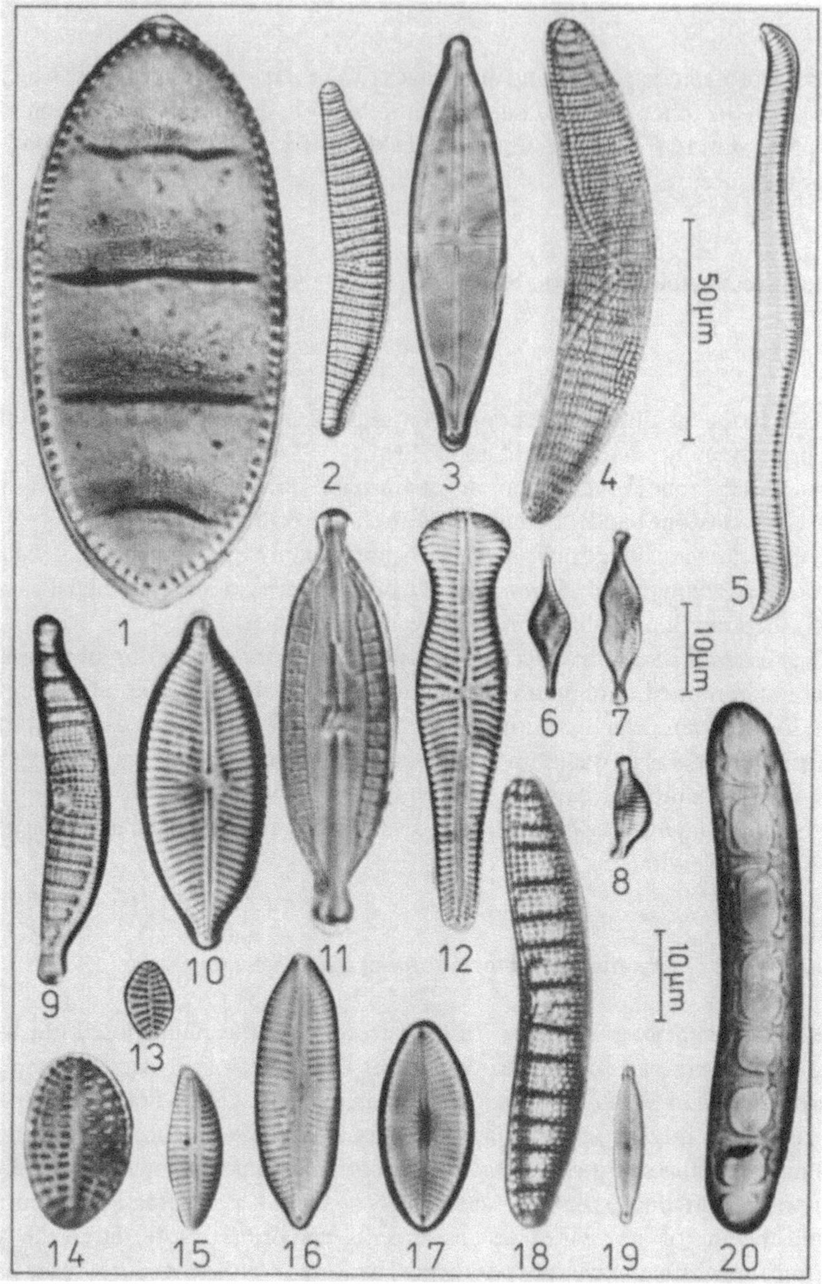

Figure 5.8 Benthonic diatoms of Lakes Stechlin and Nehmitz. 1) *Cymatopleura elliptica;* 2) *Epithemia turgida;* 3) *Stauroneis phoenicenteron;* 4) *Epithemia hyndmannii;* 5) *Rhopalodia gibba;* 6) *Synedra parasitica;* 7) *Synedra parasitica* var. *subconstricta;* 8) *Cymbella thumensis;* 9) *Epithemia zebra* var. *porcellus;* 10) *Cymbella obtusiuscula;* 11) *Mastogloia smithii* var. *amphicephala;* 12) *Gomphonema constrictum;* 13) *Cocconeis diminuta;* 14) *Cocconeis disculus;* 15) *Cymbella hustedtii;* 16) *Cymbella hybrida;* 17) *Navicula jentzschii;* 18) *Epithemia zebra;* 19) *Cymbella microcephala;* 20) *Epithemia mülleri.* Photos: W. Scheffler.

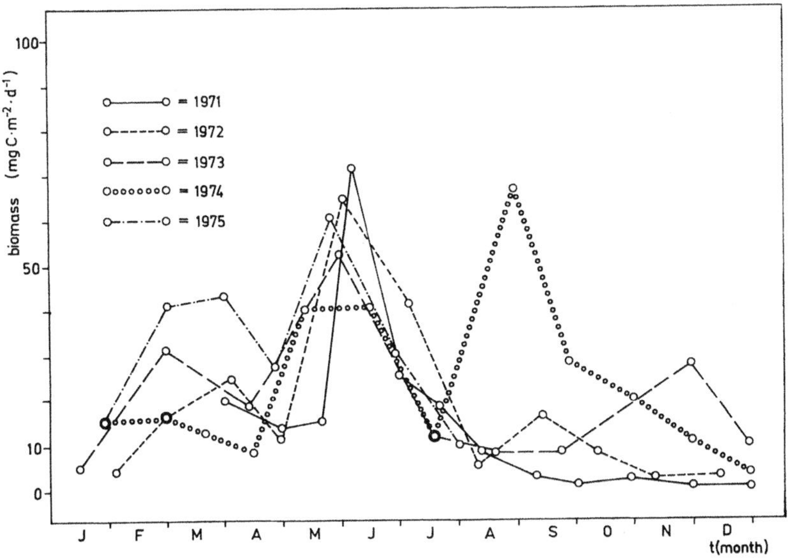

Figure 5.9 Seasonal periodicity of biomass in the littoral of Lake Stechlin (March 1971–August 1975; measurement point Sl 1; depth 0.05 m).

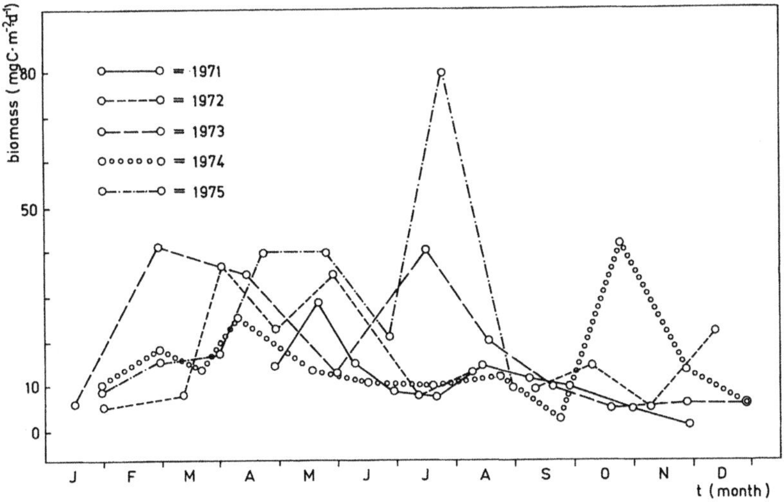

Figure 5.10 Seasonal periodicity of biomass in the littoral of Lake Stechlin (April 1971–August 1975; measurement point Sl 2; depth 0.05 m).

of production starts in February and finishes in mid-summer. In it there are two local and temporally displaced phases of intensified biomass production. Occasionally a small maximum in autumn appears between September and November.

The sums of the biomass. The figures of the biomass produced are summarized in Table 5.6. The years under study were adapted to the recognizable natural rhythm of the production of periphyton – February to January.

The annual sums vary between 3326 and 8743 mg $C \cdot m^{-2}$, and the average daily biomass between 9.1 and 24.0 mg $C \cdot m^{-2}$. On the strongly shadowed wind-exposed south-east shore (measuring point Sl 2 in Table 5.6) every year less biomass is produced than on the sunny, calm north-west shore (measuring point Sl 1 in Table 5.6), and as an average of all years under study, only 72 per cent. Obviously, the shadowed shore and a strong wash of the waves have – with the same substrata conditions – a negative influence on the production of periphyton.

The average of all five years studied and both research points show that from February–July 67.8 per cent and from August–January 32.2 per cent of the annual sum of the biomass are produced.

The summary of all rates shows an average medium annual sum of the biomass of 6059.4 mg $C \cdot m^{-2}$ and average daily biomass of 16.3 mg $C \cdot m^{-2}$.

Comparison with statements from other regions shows that Lake Stechlin belongs to the waters with a low periphyton production (Table 5.7).

The biomass in the thermally charged part of Lake Stechlin

The biomass through the year. Research on the influence of the thermal charge on the production of biomass of periphyton was carried out at two points. Measuring point Sk is situated immediately in the area of the outflow of the power plant cooling water. Here the water temperature is always about 10° C higher than in

Table 5.6 Mean daily biomass (d^{-1}) and annual sums of biomass (a^{-1}) in littoral at 0.05 m depth.

Year	Sl 1 (mg $C \cdot m^{-2}$)		Sl 2 (mg $C \cdot m^{-2}$)		m̄ Sl 1 + Sl 2 (mg $C \cdot m^{-2}$)	
	d^{-1}	a^{-1}	d^{-1}	a^{-1}	d^{-1}	a^{-1}
1971/72	12.6	4573.6	9.1	3325.8	10.9	3949.7
1972/73	17.3	6311.3	11.7	5262.4	14.5	5836.9
1973/74	23.7	8622.7	18.9	6893.1	21.3	7757.9
1974/75	24.0	8743.1	13.0	4743.5	18.5	6743.4
m 1971/75	19.4	7062.6	13.2	5056.2	16.3	6059.4

Table 5.7 Mean daily biomass of the periphyton in different lakes and reservoirs.

Waters	Depth (m)	Biomass ($mg\ C \cdot m^{-2} \cdot d^{-1}$)	Author/year
Lake Sodon (USA)	1.2 and 2.4	6.3	Newcombe 1949
Lake Sodon (USA)	0.2 and 1.2	19.9	Newcombe 1950
Lake Falls (USA)	0.4	78.4	Castenholz 1960
Lake Alkali (USA)	0.4	69.4	Castenholz 1960
Sedliče reservoir (CSSR)	1.0	135.7	Sládeček & Sládečková 1964
Lake Müggel (GDR)	0.2	75.3	Tscheu-Schlüter 1969
Lake Karutz (GDR)	0.2	9.5	Tscheu-Schlüter 1969
Lake Liebenberg (GDR)	0.2	61.5	Tscheu-Schlüter 1969
Lake Tiberias (Israel)	1.2	126.4	Dor 1970
Lake Stechlin (GDR)	0.05	16.3	Scheffler 1981

other uninfluenced parts of the lake; at 1 m depth a maximum of 30° C can appear.

Measuring point Swv is situated 500 m seaward of Sk. The temperature is increased, in comparison with the other parts of the lake, by from 4–7° C. Measuring point St in the middle of the lake was used as control. It is not influenced by the artificial warming. All-year research under conditions of artificial warming were not practicable, because the supply of the cooling water often was interrupted by the non-operation of the power plant.

The terminal development of the production of periphyton under continuously artificial warming at measuring point Sk, expressed as medium daily biomass ($mg\ C \cdot m^{-2} \cdot d^{-1}$), is shown in Fig. 5.11. Two periods of different development are discernible. From July–January relatively little growth occurs. From January–February a strong increase in biomass production starts and continues up to April, at least. It was not possible to observe this striking development to the end in any year because the supply of the cooling water was interrupted at the latest in April. The low summer rates make it possible to assume that even with a constant supply of warm water the strong biomass production would come to an end about May.

At measuring point Swv, in the area of less warming up, the development is only temporarily analogous (Fig. 5.12). The low biomass rates from November–January are similar. The increase of the biomass production in spring is less and is interrupted in three out of five years by a falling monthly rate. Contrary to measuring point Sk, a maximum in autumn appears from September–October.

At measuring point St the annual development is essentially more uniform (Fig. 5.13). The wintry rest phase continues from December–April. Then the phase of growth begins reaching its annual climax in August. From September the biomass rates decrease until the wintry rest phase is reached in November.

The artificial warming causes a terminal anticipation of the development of

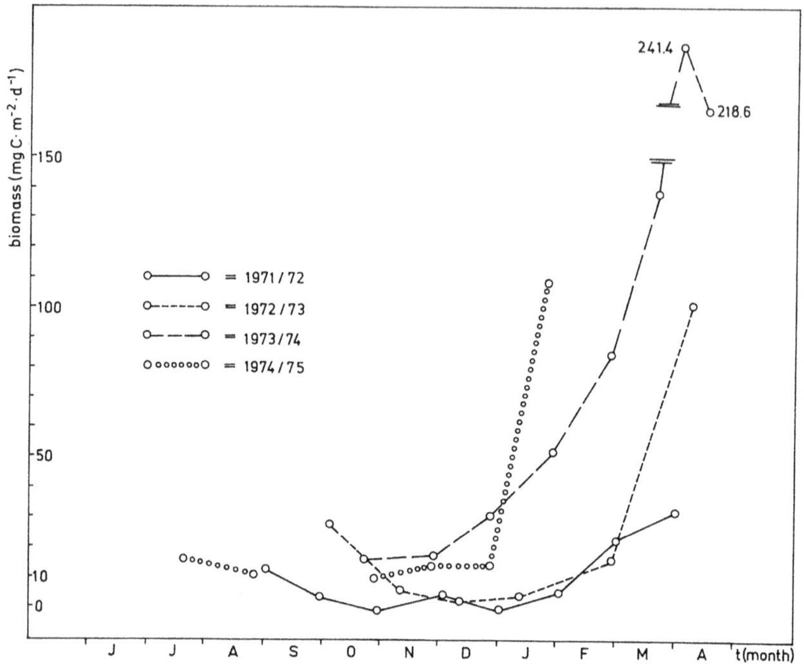

Figure 5.11 Seasonal periodicity of biomass in the littoral of Lake Stechlin under thermal load (1971–1975; measurement point Sk; depth 1.0 m).

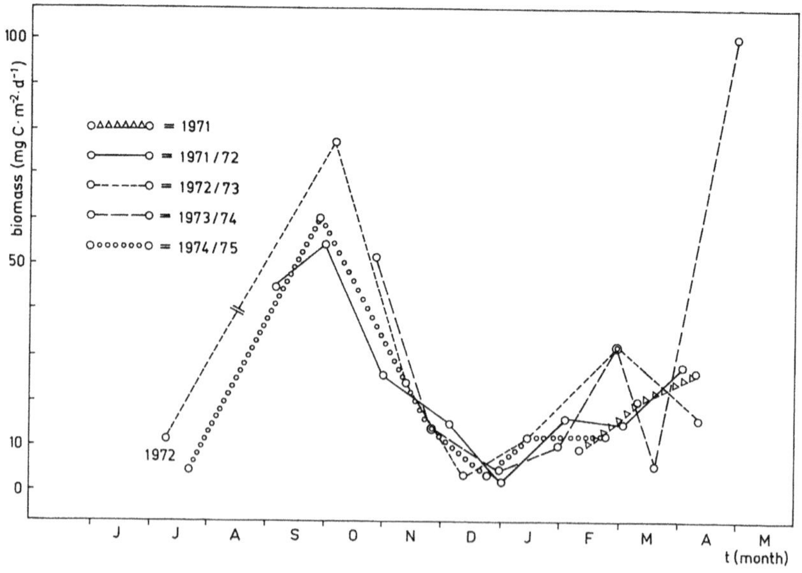

Figure 5.12 Seasonal periodicity of biomass in Lake Stechlin under thermal load (1971–1975; measurement point Swv; depth 1.0 m).

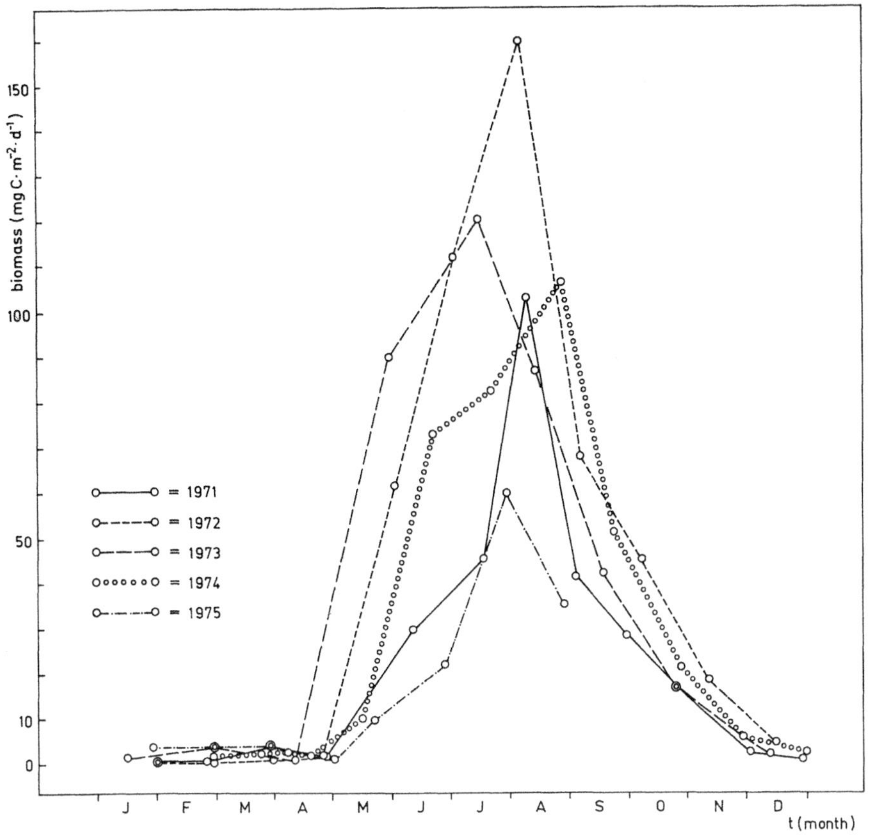

Figure 5.13 Seasonal periodicity of biomass in Lake Stechlin (January 1971–August 1975; measurement point St; depth 1.0 m).

periphyton in Lake Stechlin. The start of growth begins as early as January instead of April. The annual climax is reached in May–June, at measuring point St only in August–September. This corresponds to relations in the artificially heated Lake Wabamum (Canada; Klarer & Hickman 1975). An increased autumnal growth of periphyton takes place only in the area of artificial warming, but irregulary.

The sums of biomass. The power plant operates continuously only in the winter half-year. Directly comparable (because the results were obtained under the same external conditions) are the figures for the period 14th October–17th April (1971, 1972, 1973) and the period 14th October–13th February (1974/75) which are listed in Table 5.8. In each year a marked gradient is discernible between the three areas of the lake. Through artificial heating (Sk and Swv) in the winter half-year about three- to twentyfold biomass is produced in comparison to the uninfluenced point St. At measuring point Sk the largest biomass developed with 12.8 g C · m^{-2} followed by St with 10.2 g C · m^{-2} and Swv with 9.2 g C · m^{-2}. These statements do not answer the question of the influence of the warming on the bioproduction of periphyton unambiguously, because six months a year go by without any inflow of warm water.

But if the biomass in the winter half-year alone is considered (lower part of the diagram columns in Fig. 5.14), then the percentage of the period 10th October–17th April of the medium annual sums in the area of strongest warming (Sk) is 52.2 per cent (= 6.7 g C · m^{-2}), more than five times larger than the normal temperature measuring point St (9.2 per cent = 0.9 g C · m^{-2}). Measuring point Swv at an intermediate position has 38.8 per cent (= 3.6 g C · m^{-2}). In the winter half-year the thermal charge stimulates the production of biomass.

This phase of increased biomass production can be defined still further. From 14th October–14th January of the years under study, relatively low biomass rates still appear. The portion of the whole winter half-year is 17.3 per cent. In the following period, between 15th January and 17th April, the growth is comparably greater: 82.7 per cent of the biomass of the winter half-year is produced. With a high thermal charge the time from January–April is the phase with the most intensive growth of periphyton.

At measuring point Swv the portions of both quarters amount to 45 and 55 per cent, respectively of the whole biomass of the winter half-year. They are approximately equal.

Table 5.8 Sums of biomass and mean daily biomass of periphyton under thermal load 1971–1975.

Thermal load	Strong		Little		Without	
Station and depth	Sk 1.0 m		Swv 1.0 m		St 1.0 m	
Biomass	($\sum$mg C · m^{-2})	($\bar{m}$ mg C · m^{-2} · d^{-1})	($\sum$mg C · m^{-2})	($\bar{m}$ mg C · m^{-2} · d^{-1})	($\sum$mg C · m^{-2})	($\bar{m}$ mg C · m^{-2} · d^{-1})
14.10.1971–17.4.1972	2443.3	13.1	3618.1	19.5	828.0	4.5
14.10.1972–17.4.1973	4292.5	23.0	4358.0	23.4	1584.3	8.5
14.10.1973–17.4.1974	13036.1	70.1	5353.2	28.2	668.2	3.6
$\bar{m}$ 3 years	6590.6	35.4	4443.1	23.9	1026.5	5.5
14.10.1974–13.2.1975	4558.0	31.1	1789.9	14.6	846.7	6.9

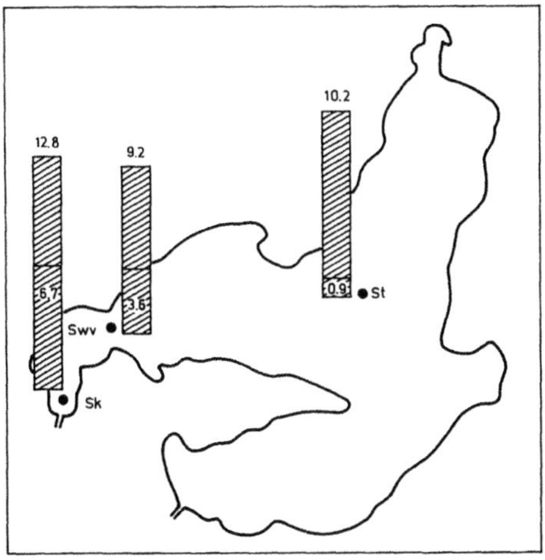

Figure 5.14 Mean annual sums of biomass in g C · m^{-2} in Lake Stechlin (November–October for 1971/72, 1973/74, 1974/75; measurement points Sk, Swv, St; depth 1.0 m). The lower part of the columns indicates the biomass sums from 14. 10.–17. 4. of the same years.

The phytoplankton (S.J. Casper)

Introduction

Phytoplankton studies have been carried out in the Lake Stechlin area as a whole, but especially in Lakes Stechlin and Nehmitz, more or less continuously, since the foundation of the Limnological Laboratory Stechlin. From 1959 to 1965 particular stress was laid on the qualitative aspects of the phytoplankton; species composition and temporal and spatial distribution of some dominant species were thoroughly investigated to establish the standing stock of the algal population before the nuclear power plant came into operation. In the period from 1966 to 1972 the investigations were continued, now with consideration given to possible changes caused by the influence of the cooling water circuit. Quantitative aspects such as seasonal and spatial periodicity, seasonal cycles of the individual species and of the total cell number, and seasonal cycles of the biomass and its components, were carefully studied. Special attention was given to the examination of various size fractions of the standing stock, especially of the nano- and ultraplankton, and of the vertical distribution of the dominant algal species.

This section deals with only selected aspects of the phytoplankton of the Lake Stechlin area. Firstly, the species composition and the seasonal periodicity of the phytoplankton during the whole period between 1959 and 1982 are surveyed. Secondly, the present situation is compared with that during the periods before and after the nuclear power plant was brought into operation. Thirdly, attention is paid to the vertical distribution of some dominant phytoplankton species.

Finally some results on the proportion of nanoplankton in the total phytoplankton and on the seasonal variation of this proportion are presented.

The present compilation summarizes the results of nearly 20 years of phytoplankton studies in the Lake Stechlin area done by several workers. The detailed and thorough investigations of Miss Gisela Busse (1965, 1972) from 1959 to 1972 were continued by the studies of L. Küchler (1981, 1982) from 1973 to 1978. I gratefully acknowledge their contributions. Their available data have permitted many correlations and generalizations to be made. My own contributions from 1981 to 1983 have mainly had the purpose of identifying some doubtful algal species, especially of the Chrysomonadales and Volvocales, to prove the quantitative data, and to continue the examination of the seasonal periodicity of the dominant phytoplankton species. Since 1981, W. Scheffler has been studying the diatoms.

The investigation of the phytoplankton is essential when studying a lake ecosystem. Being the primary producers of the system, the algae represent the key-link in the energy and material flow of lakes. The results presented here contribute to a better understanding of the relationships between standing crop biomass, primary production, and the food chain.

Material and methods

From 1959 to 1962 surface and deep water samples were taken in 15 waters of the area, but, more or less regularly once a month, only in Lakes Stechlin, Nehmitz and Gerlin. Phytoplankton catches for the investigation of qualitative aspects were collected with a coarse plankton net (mesh size ~55 μm, Müllergaze 25). The samples were killed for preservation with formalin to 3–4% solution. The collection of more than 500 vials provides material for species identification in connection with counting. During this first period the large species (microplankters) were probably over-emphasized, because most of the nano- und ultraplankton escaped attention.

In the following years particular stress was laid on the quantitative determination of the phytoplankton standing stock. For a short time a stable water volume unit (200 l) was collected by pumping, and filtered through a plankton net, but this inadequate method was soon replaced by collection with a Ruttner sampler. Subsamples were immediately preserved with Lugol's solution.

The numeric analysis of the standing stock of micro- (50–500 μm) and nanoplankton (5–50 μm) was carried out according to Utermöhl (1958), using a Zeiss inverted microscope. The ultraplanktonic species ('μ-algae'; <5 μm) were enriched by using Chemapol membran filters (pore-size 1.5 μm), sedimentated in a Thoma-chamber, and enumerated according to Overbeck (1962) and Braune (1975). Each sample was examined several times.

The counting score for each species was converted to cell concentration (cells per litre). Specific cell volumes were calculated for all species observed in the quantitative samples by using the data of Nauwerck (1963) and from measurement of the individual species in Lake Stechlin (Küchler 1981). Phytoplankton specific weight was taken as 1.0 for calculating fresh weight biomass ($10^9\ \mu m^3 \hat{=}$ 1 mg biomass) according to Schwoerbel (1966). For calculation of phytoplankton biomass related to 1 m^2 lake surface, only the plankters collected in the euphotic zone (Lake Stechlin 0–17.5 m; Lake Nehmitz-N 0–12.5 m; Lake Nehmitz-S 0–10 m; Lake Dagow 0–3.5 m) were considered.

These general outlines of the methods of sampling, counting and calculating have varied to some extent throughout the period of investigation. Nevertheless, the greater part of the available data is sufficiently reliable to allow at least qualitative evaluation. During 1963 to 1964, 1969 to 1972, and 1976 to 1978, the species composition and seasonal periodicity, including cell counts and biomass calculations, were studied in detail.

Species composition and seasonal periodicity 1963 to 1964

Cell number

Lake Stechlin. According to the classification by Hutchinson & Löffler (1956), Lake Stechlin is dimictic in years with a severe winter, e.g., if ice covers the lake (Heitmann, Richter & Schumann 1969; Richter p. 55). That was in the period from September 1962 to November 1964, before the nuclear power plant became operational.

During the years under consideration Lake Stechlin was covered by a long term ice layer up to 35 cm (cf. Heitmann, Richter & Schumann 1969; Richter p. 55; ice calendar Fig. 3.11; Fig. 5.15). Winter stagnation occurred, the water layers now having an inverse temperature difference of relatively low stability. The maximum inverse stratification was established on February 5th ($\bar{t}_w = 2.6°$ C). During winter 1962/63 this period lasted until the first 10 days of April (Heitmann, Richter & Schumann 1969).

Only from January to March 1963 are data of the phytoplankton growth available. Under ice cover a community poor in species and small in cell number occurred at water temperatures between 0.5 and 3.6° C (Fig. 5.15). *Mallomonas fastigata* (20.2.: 16)[1] and *Oscillatoria rubescens* (20.2. and 15.3.: 9) reach the highest cell numbers (Fig. 5.15). In holes broken into the ice cover red flocks of *Oscillatoria rubescens* float on the water surface. Further, the winter community consists of the following species of which not one is a typical winter form: *Cyclotella* spp., *Stephanodiscus minutus, Asterionella formosa, Synedra acus, Botryococcus braunii, Eudorina elegans, Tetraedron minimum, Gymnodinium* spec., and *Peridinium* spec.

[1] In the following sections all numbers without measure stand for cells or colonies $\times 10^3 \cdot l^{-1}$.

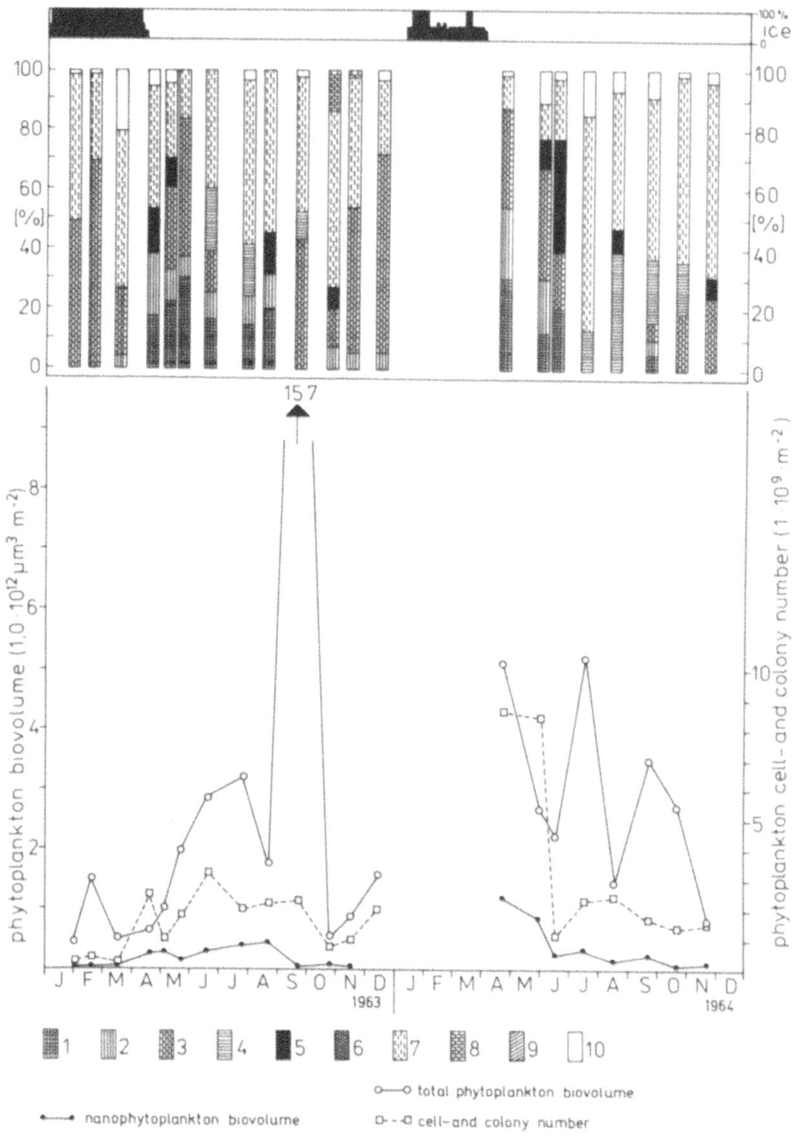

Figure 5.15 Seasonal periodicity of phytoplankton biovolume, cell and colony number, proportion of the main groups of the phytoplankton community, and ice cover in Lake Stechlin, 1963–1964 (according to Busse, unpublished data, altered and enlarged). 1) microplanktonic diatoms; 2) nanoplanktonic diatoms; 3) microplanktonic chrysomonades (mainly *Dinobryon, Mallomonas*); 4) dinomonades; 5) nanoplanktonic phytoflagellates; 6) nanoplanktonic green algae; 7) blue-green algae; 8) microplanktonic green algae (mainly *Volvox, Pandorina*); 9) ultraplanktonic green algae ('μ-algae'); 10) remainder.

About mid-March, after break-up and thaw of the ice cover, when solar radiation and daylength begin to increase, spring complete circulation starts reaching isothermy at the end of March ($\bar{t}_w = 3.1°C$) and maximum density on April 11th. This period lasts until late April, when thermal vertical heterogeneity begins (Heitmann, Richter & Schumann 1969; Richter p. 57).

In temperate lakes spring complete circulation is the phase of the so-called diatom spring maximum (Findenegg 1943). Due to mineralization in winter an adequate nutrient supply is combined with high turbulence.

In 1963 the situation was more complicated (Fig. 5.15). After the late ice break-up on April 6th spring phytoplankton community growth was delayed (17.4.; t_w = 5.5–4.6° C between 0 and 20 metres). Nanoplanktonic phytoflagellates increase to 205 000 cells per litre, but diatoms do not reach adequate amounts. Nanoplanktonic *Cyclotella* species (21), *Synedra acus* (8), *Synedra* spp. (6), and *Stephanodiscus minutus* (4) remain at very low cell numbers. *Oscillatoria rubescens* increases to 9000 cells per litre. At the beginning of May (2.5.; t_w = 7.4–5.0° C between 0 and 20 metres), the phytoflagellates had disappeared. The nanoplanktonic *Cyclotella* (11) – and *Synedra* (7) – species decrease. *Dinobryon cylindricum* (4) starts a moderate growth.

Phytoplankton growth during spring 1964 is a good example for the diatom spring maximum (Fig. 5.15). Ice cover broke on March 13th, and during April a rapid and impressive phytoplankton growth started (21.4.; t_w = 16.5–4.3° C between 0 and 30 metres) consisting mainly of nanoplanktonic *Cyclotella* species (325), *Synedra* spp. *'nano'* (197), *Asterionella formosa* (80), and *S. acus* (12; Table 5.9). Several other algae start growing: for instance, *Dinobryon cylindricum* (10), and *Tetraedron minimum* (43). At the end of May (29.5.) the diatom maximum gradually breaks down. The numbers of *Cyclotella* species (80), *Synedra* spp. *'nano'* (100), *Asterionella formosa* (16) decrease; only *Synedra acus* (12) and *Diatoma elongatum* (8) show low peaks. The diatom maximum is replaced by a considerable growth of nanoplanktonic phytoflagellates, which increases to 386 000 cells per litre, starting its development immediately after the *Cyclotella-Synedra* maximum (Table 5.9). Other species of the community in order of numeric importance are: *Scenedesmus costato-granulatus* (134), *Phacotus lenticularis* (26), *Dinobryon bavaricum* (23), *Oscillatoria rubescens* (14), *Tetraedron minimum* (12), *Chrysamoeba radians* (10), and *Pseudosphaerocystis schroeteri* (8). More than twenty other phytoplanktonic species not reaching more than 4000 cells per litre each were detected in Ruttner samples; cell densities of other ones were below the detection limit of the counting method.

It is noteworthy that nanoplanktonic phytoflagellates were detected in high cell numbers in depths of between 0 and 40 metres.

During the last 10 days of May, in the early stages of stratification, a stable thermocline develops at a depth of 4–7 metres ($\bar{t}_w = 6.8°C$). The summer stagnation phase begins. Epilimnetic water temperature increases to around 18° C in July and August, reaching temporarily maximum values of 22–23° C, depending on daily weather conditions. During the whole period which lasts until mid-August the thermocline steadily descends and is enlarged to a vertical distance on an average of 6–7 metres. In the second half of July (6–12 m) and the first half of August (7–13 m) it has its most stable position (Heitmann, Richter & Schumann 1969; Richter p. 57).

During the summer stagnation phase 1963 phytoplankton growth still suffers from the delay in the spring complete circulation (Fig. 5.15). Species and cell numbers do not reach values comparable to the 1964 ones. At the end of May (21.5.) *Dinobryon cylindricum* (44; Table 5.9) dominates, accompanied by *Synedra acus* (30), *Asterionella formosa* (12), and *Oscillatoria rubescens* (12) in deeper layers. During mid-June (18.6.) *Dinobryon cylindricum* (31), *D. bavaricum* (16), *Synedra acus* (38), *S.* spp. *'nano'* (25), and *Oscillatoria rubescens* (15) form the community together with some other species not reaching values higher than 4000 cells per litre each, among them *Ceratium hirundinella* and *Anabaena lemmermannii.* At the end of July (25.7.) *Synedra* spp. *'nano'* increases to 148 000 cells per litre after a steady growth which can be compared to the *Cyclotella* growth during July 1964. Nanoplanktonic *Cyclotella* species (40) and *Chrysamoeba radians* (12) increase to moderate amounts. *Anabaena lemmermannii* (4) ought to be mentioned too.

Table 5.9 Dominant phytoplankton species (maximum cell numbers $1.0 \cdot 10^4$ and more) in Lake Stechlin (sample point St) 1963–1964 (F = cell thread; K = cell colony).

Species	Date	Maximum cell number/depth ($\times 10^3 \cdot l^{-1}$/m)	Average cell number (0–40 m) ($\times 10^3 \cdot l^{-1}$)	Maximum biovolume ($\times 10^6 \mu m^3 \cdot l^{-1}$)	Average biovolume (0–40 m) ($\times 10^6 \mu m^3 \cdot l^{-1}$)
Anabaena lemmermanii (K)	17.09.63	46/0	10	2,300	500
Asterionella formosa	21.04.64	80/0 – 30	72	64	58
Chrysamoeba radians	21.08.63	104/0	51	52	26
Cyclotella kützingiana	28.09.64	38/0	24	13	8
Cyclotella spp *'nano'*	21.04.64	312/0	186	47	28
Dinobryon bavaricum	29.05.64	23/15	9	18	7
Dinobryon cylindricum	21.05.63	44/20	21	35	17
Mallomonas fastigata	17.09.63	211/0	32	2,532	382
Nanoplanktonic phytoflagellates (<10 μm)	29.05.64	386/5	219	19	11
Oscillatoria rubescens (F)	20.10.64	100/15	23	300	69
Scenedesmus costato-granulatus	29.05.64	134/30	43	26	5
Synedra acus subsp. *angustissimus*	18.06.63	39/20	15	20	8
Synedra spp. *'nano'*	21.04.64	197/20	182	39	36

In 1964 (16.6.) during the first half of summer stratification, spring maximum breaks down rapidly to small phytoplankton densities (Fig. 5.15). The nanoplanktonic phytoflagellates disappear nearly totally. The phytoplankton community, rich in species, is represented by a limited number of species in noticeable quantities. *Chrysamoeba radians* (16) occurs in the upper water column between 0 and 30 metres. Nanoplanktonic *Cyclotella* species (27), *Synedra* spp. *'nano'* (38), and *S. acus* (14) live in the layers between 30 and 40 metres together with *Scenedesmus costato-granulatus* (33) and *Oscillatoria rubescens* (10). During July (15. 7.) nanoplanktonic *Cyclotella* species reach a short term summer maximum (160) accompanied by *Anabaena lemmermannii* (8) in upper and *Oscillatoria rubescens* (35) and *Scenedesmus costato-granulatus* (60) in deeper layers.

After mid-August, the epilimnion becomes cooler, turbidity increases, the boundaries of the thermocline (7–13 m) begin gradually to sink to the bottom, e.g., summer stratification becomes more and more unstable; autumnal incomplete circulation sets in. This transitional period lasts until the end of November (Heitmann, Richter & Schumann 1969; Richter p. 57).

In 1963 (21.8.; $t_w = 18.3–5.2°$ C between 0 and 30 metres) *Chrysamoeba radians* runs up to 104 000 cells per litre (Table 5.9), accompanied by microplanktonic *Asterionella formosa* (80), by nanoplanktonic *Stephanodiscus minutus* (9), *Anabaena lemmermannii* (3), and *Oscillatoria rubescens* (8). In mid-September (17.9.) *Mallomonas fastigata* (211) reaches maximum cell numbers at the water surface (Table 5.9). The nanoplanktonic *Cyclotella* species (127), *Anabaena lemmermannii* (46; Table 5.9), and *Oscillatoria rubescens* (40) occur in considerable amounts. At the end of October (24.10.) nanoplanktonic phytoflagellates (32) and nanoplanktonic *Cyclotella* species (21) form an unstable community with small cell numbers (Fig. 5.15).

In 1964 (14.8.) an explosive growth (Fig. 5.15) of nanoplanktonic phytoflagellates occur, reaching the highest cell number of the whole community (216). It is accompanied by nanoplanktonic *Cyclotella* species (19) in upper layers and by *Scenedesmus costato-granulatus* (24) and *Oscillatoria rubescens* (8) in deeper layers. At the end of September (28.9.), the nanoplanktonic phytoflagellates decrease (18) and are replaced by a moderate growth of nanoplanktonic *Cyclotella* species (38) and by a considerable development of *Asterionella formosa* (80), *Scenedesmus costato-granulatus* (43) and *Oscillatoria rubescens* (71). During October (20.10.) the nanoplanktonic phytoflagellates increase moderately (36) and the diatoms decrease. *Oscillatoria* reaches its maximum cell number (100; Table 5.9).

After mid-November, caused by falling temperatures, increasing density of epilimnion and autumn storms, destratification has taken place and metalimnion has disappeared totally. The period of autumnal complete circulation reaching isothermy ($\bar{t}_w = 4.7°$ C) on mid-December and maximum density on December 27th lasts until the beginning of January (Heitmann, Richter & Schumann 1969; Richter p. 57).

During this period phytoplankton community is characterized by small cell numbers (Fig. 5.15). As during the period of autumnal incomplete circulation, the nanoplanktonic phytoflagellates dominate, with cell numbers up to 69 (12.12.1963) and 77 (20.11.1964). The cell numbers of the whole community in which nanoplanktonic diatoms, *Oscillatoria rubescens* – then in the upper layers between 0 and 15 metres – and *Mallomonas fastigata* stand out do not exceed 101 000 (1962) and 96 000 cells per litre (1964), respectively.

The description of the seasonal periodicity given above is based on cell numbers. The phytoplankton growth during 1963 and 1964 may be generalized as follows.

Phytoplankton growth usually starts as early as mid-March, after the ice break-up on Lake Stechlin, resulting in a spring maximum of diatoms and two species of Chlorophyceae in April and May when the cell numbers reach their maximum. At the onset of thermal stratification at the end of May, diatoms gradually disappear and a moderate growth of *Dinobryon* species starts. In summer the phytoplankton community consists mainly of diatoms, Chrysophyceae, Chlorophyceae and Cyanophyceae with low densities. Diatoms increase to a moderate maximum (*Synedra* spp. in 1963 and 1964, *Cyclotella* spp. in 1964) continuing during September. During August and September, when thermal stratification becomes more and more unstable, algal numbers of different species increase considerably. Nanoplanktonic phytoflagellates increase to high cell numbers, *Mallomonas fastigata, Chrysamoeba radians,* and *Oscillatoria rubescens* reach their annual maximum growth. In mid-November, when thermal stratification has broken down, algal numbers decrease rapidly. In winter under ice cover algal growth decreases to its lowest annual amount.

Although there are many simularities in the succession of algal populations in the course of the two years, the exact pattern differs from year to year. There is a variation of nearly four weaks in the timing of the spring maximum. In 1963 the spring complete circulation was delayed because the ice cover disappeared very late. A typical diatom spring maximum did not occur. Also maximal cell numbers may vary considerably; some species, present in relatively small numbers in one year, may exhibit massive growth in another one. Moreover, the number of species which are regular inhabitants of Lake Stechlin is far higher then the species mentioned above (see p. 485 ff.).

Under natural conditions, the yearly fluctuation in temperature and available sunlight combined with random meteorological fluctuations govern the seasonal periodicity of the phytoplankton community, which is, in term of phycology, an *Asterionella formosa–Synedra–Mallomonas–Dinobryon–Oscillatoria rubescens*–community. In 1926 Krieger (1927) took random net-samples from Lake Stechlin and established the dominance of *Asterionella formosa* and *Oscillatoria rubescens,* then accompanied by *Mallomonas tonsurata, Cyclotella comta* var. *quadrijuncta, Gloeococcus schroeteri* (= *Sphaerocystis schroeteri*), *Stephanodiscus astraea, Diatoma elongatum, Synedra acus, Centronella reicheltii,*

Eudorina elegans, Pandorina morum, Cosmarium phaseolus, Anabaena hassallii.

Therefore, we can establish that during the period 1926 to 1965 the phytoplankton community of Lake Stechlin has been kept from change, loss, or destruction. It was, in the terms of the nineteenth century algologists, a north-temperate diatom–*Mallomonas* lake.

Lake Nehmitz-North. In the undisturbed Lake Nehmitz-N the seasonal periodicity of the phytoplankton development follows the rise or fall of temperature and available sunlight according to the specific morphometric (p. 31), hydrometeorologic (p. 41), and nutrient features (p. 89). During the two years under consideration before the nuclear power plant came into operation, relatively similar figures (Fig. 5.16) occur. Under ice cover (19.2.1963) *Cyclotella kützingiana* (2), *Asterionella formosa* and *Peridinium* spec. dominate the community with very low cell numbers. In the period between the ice break-up in 1963 (24.4.) and mid-May (17.5.), *Dinobryon divergens* (30; Table 5.10), *D. sociale,* and *Mallomonas fastigata* (6), accompanied by *Cyclotella kützingiana* and *Gymnodinium* spec., develop remarkably. In mid-June the maximum cell numbers are reached: *Dinobryon sociale* amounts to 106 000 cells per litre. *Cyclotella kützingiana* (13), *Anabaena lemmermannii* (3.5 colonies per litre), and *Ceratium hirundinella* show good growth. In mid-summer (26.7.) blue-green algae (*Oscillatoria rubescens* and *Anabaena lemmermannii;* 25) increase to moderate maxima, accompanied by *Ceratium hirundinella* and *Mallomonas fastigata.* In late summer (12.9.) microplanktonic (*Fragilaria crotonensis,* 4; *Asterionella formosa,* 4) and nanoplanktonic diatoms (*Cyclotella kützingiana,* 4) gradually appear; *Ceratium hirundinella* increases (5) to its highest cell number, and *Mallomonas* stands by 7; but the dominant species is *Oscillatoria rubescens* with 48×10^3 cell-

Table 5.10 Dominant phytoplankton species (maximum cell numbers 1.0×10^4 and more) in Lake Nehmitz-N 1963/1964 (F = cell thread; K = cell colony).

Species	Date	Average cell number (0–12.5 m) ($\times 10^3 \cdot l^{-1}$)	Average biovolume (0–12.5 m) ($\times 10^6\ \mu m^3 \cdot l^{-1}$)
1 *Cyclotella kützingiana*	27.05.64	12	4.5
2 *Cyclotella* spp. *'nano'*	27.05.64	311	45
3 *Dinobryon divergens*	03.05.63	30	24
4 *Dinobryon sociale*	27.05.64	737	590
5 *Pseudokephyrion entzii*	26.04.64	100	5
6 *Mallomonas fastigata*	26.07.63	13	156
7 *Microcystis* spec. (K)	24.09.64	111	444
8 Nanoplanktonic phytoflagellates (<10 μm)	23.11.64	133	6.5
9 *Oscillatoria rubescens* (F)	12.09.63	48	144

threads per litre (Table 5.10). In autumn the cell number of the whole community decreases rapidly. The dominating species are *Mallomonas fastigata, Fragilaria crotonensis, Asterionella formosa, Cyclotella* and *Stephanodiscus* species.

Between the annual phytoplankton development in 1963 and 1964 there are no striking differences.

In 1964 the spring community at the end of April (26.4.) consists mainly of nanoplanktonic *Cyclotella* spp. (187), chrysophycean phytoflagellates of the *Kephyriopsis*-type (*Pseudokephyrion entzii* 100; Table 5.10) and of *Mallomonas fastigata*. The maximum cell and colony numbers amount to 1075×10^3 per litre at the end of May (27.5.). *Dinobryon sociale* (737; Table 5.10) and nanoplanktonic *Cyclotella* spp. (323; Table 5.10) are the dominant members. A second low autumn maximum follows (24.9.), distinguished by the considerable growth of nanoplanktonic phytoflagellates (71; 21.10.: 87; 23.11.: 133; Table 5.10), dinomonades, and blue-green algae mainly of the *Microcystis*-type (280). In late autumn nanoplanktonic (*Cyclotella* spp. 31) and microplanktonic diatoms (*Fragilaria crotonensis* 1; *Asterionella formosa* 1) accompanied by *Mallomonas* spp. (3) ought to be mentioned.

To sum up: Lake Nehmitz-N is, in algal-floristic terms, a chrysophycean-diatomean-dinomonadean-cyanophycean lake. Compared with Lake Stechlin, it is distinguished by the microplanktonic *Fragilaria crotonensis* (in Lake Stechlin then not observed) and by a broader spectrum of blue-green algae (*Microcystis*). The seasonal periodicity is – in contrast to that of Lake Stechlin – characterized by a typical late summer-autumn development of microplanktonic diatoms.

Lake Nehmitz-South. The two years investigated (1963, 1964) show relatively different figures (Fig. 5.17). The development will not be discussed in detail (Table 5.11).

In 1963 the numeric proportion of *Mallomonas-Dinobryon*-species seems to be higher than in 1964. The summer-autumn-development of dinomonades of 1964 is missed in 1963 (the dinomonades are then present in spring). The considerable growth of microplanktonic diatoms (1963 in the second half of the year, 1964 throughout the year) is characteristic, represented by *Fragilaria crotonensis, Asterionella formosa, Tabellaria flocculosa* and *T. fenestrata* var. *fenestrata*.

Biomass

Lake Stechlin. By calculating cell biovolume and fresh weight biomass, respectively, in the euphotic layer related to 1 m^2 lake surface, it is evident that at certain times nanoplanktonic diatoms (*Cyclotella* spp., *Synedra* spp. '*nano*') dominate. But some of the larger species – for instance, *Ceratium hirundinella* and *Mallomonas fastigata,* or cell-colonies like *Asterionella formosa, Uroglena* spp., *Botryococcus braunii,* and *Dinobryon* spp., or cell threads like *Oscillatoria rubescens* and *Anabaena lemmermannii* – occur at high biomass values and dominate the algal community too.

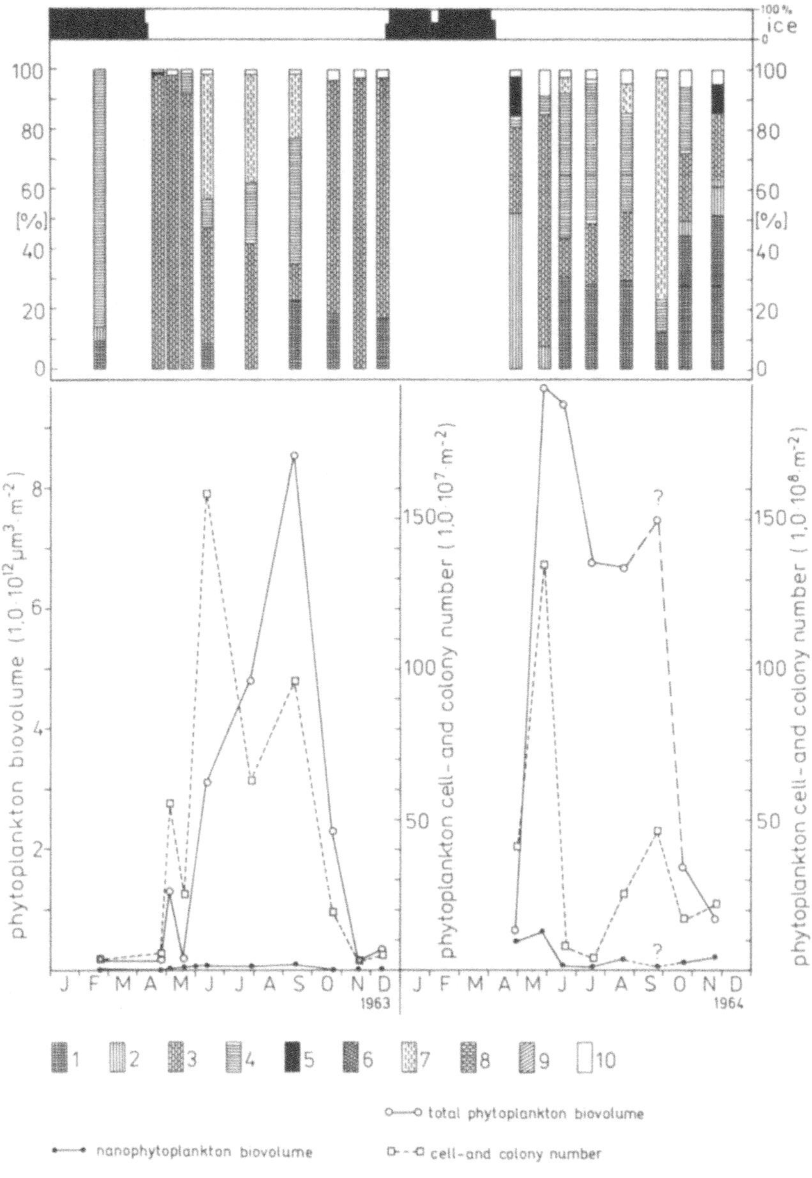

Figure 5.16 Seasonal periodicity of phytoplankton biovolume, cell and colony number, proportion of the main groups of the phytoplankton community, and ice cover in Lake Nehmitz-N, 1963–1964 (according to Busse, unpublished data, altered and enlarged; figures as in Fig. 5.15).

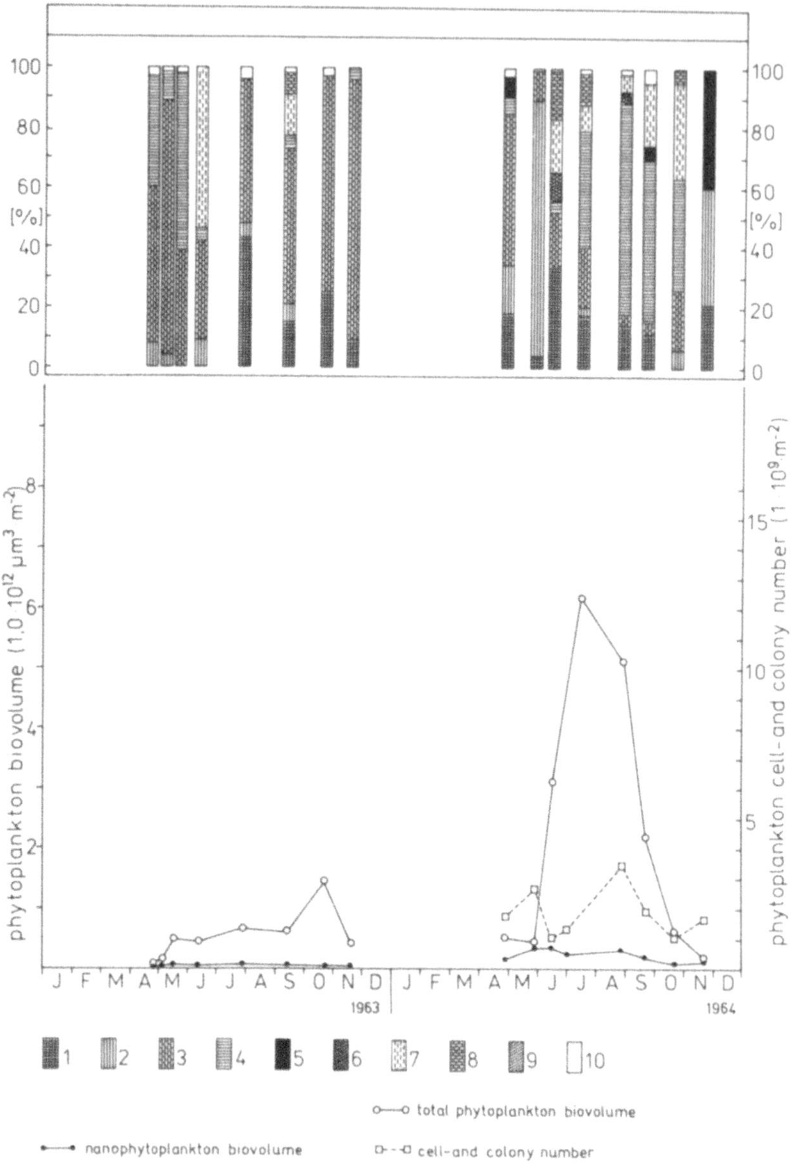

Figure 5.17 Seasonal periodicity of phytoplankton biovolume, cell and colony number, proportion of the main groups of the phytoplankton community in Lake Nehmitz-S, 1963–1964 (according to Busse, unpublished data, altered and enlarged, figures as in Fig. 5.15).

Table 5.11 Dominant phytoplankton species (maximum cell numbers 1.0×10^4 and more) in Lake Nehmitz-S 1964 (K = cell colony).

Species	Date	Average cell number (0–10 m) ($\times 10^3 \cdot l^{-1}$)	Average biovolume (0–10 m) ($\times 10^6 \mu m \cdot l^{-1}$)
1 *Ankistrodesmus* spec.	19.08.64	17	4
2 *Asterionella formosa*	16.06.64	128	102
3 *Cyclotella kützingiana*	16.07.64	54	19
4 *Cyclotella* spp. *'nano'*	27.05.64	254	38
5 *Dinobryon divergens*	16.06.64	35	28
6 *Dinobryon sociale*	26.04.64	12.5	10
7 *Pseudokephyrion entzii*	26.04.64	53	2.5
8 Nanoplanktonic phytoflagellates (<10 µm)	19.08.64	268	13.5
9 *Phacotus lenticularis*	19.08.64	17	4
10 *Sphaerocystis schroeteri* (K)	16.06.64	16	32
11 'Volvocale' (K)	16.07.64	22	66

During winter under ice cover the total algal biomass is rather small (0.4–1.5 $g \cdot m^{-2}$; Fig. 5.15). Microplanktonic phytoplankters such as *Mallomonas fastigata* (20.2.63: 70%) and *Oscillatoria rubescens* (15.3.63: 54%) constitute about 98% (January and February) and 74% (March) respectively, of the biovolumen of the winter community.

In 1963 immediately after ice break-up (17.4.) *Oscillatoria rubescens* stood at 44%. The nanoplanktonic (19%) and microplanktonic (17%) diatoms attained moderate peaks, notwithstanding the very high number of nanoplanktonic phytoflagellates (14%). By the beginning of May (2.5.) they had nearly disappeared. Microplanktonic diatoms (*Asterionella formosa,* 22%) and the chrysomonade *Dinobryon cylindricum* (28%) reached considerable biomass values. Nanoplanktonic diatoms decreased to 10%, whereas *Oscillatoria rubescens* stood at 24%.

In 1964 a diatom spring maximum amounted to 54% (21.4.; nanoplanktonic diatoms 24%, microplanktonic diatoms 30%) and 30% (29.5.; nanoplanktonic diatoms 18%, microplanktonic diatoms 12%), accompanied by *Dinobryon*-colonies (33% and 37% respectively) and *Oscillatoria rubescens* (11–12%). The high cell number of nanoplanktonic phytoflagellates (Fig. 5.15) represented only 10% of the total biomass.

In 1963 during the summer stagnation phase in May (21.5.) *Dinobryon cylindricum* accounted for about 46% of the algal biovolume, whereas *Asterionella formosa* (29%) and *Oscillatoria rubescens* stood at smaller values. Nanoplanktonic diatoms (*Synedra* spp.) had decreased to 8%, as they do also during mid-June (9.5%). Then (18.6.) the algal community was dominated by Cyanophyceae (*Oscillatoria rubescens* 22%, *Anabaena lemmermannii* 18%). *Dinobryon* spp. (14%) and *Asterionella formosa* (16%) disappeared gradually. Note the peak (19%) of *Ceratium hirundinella*. This species reached 17% in July (25.7.), when *Anabaena lemmermannii* amounted to 55% in the total biomass. *Asterionella formosa* (14%) and nanoplanktonic diatoms (10%) contributed considerably to the moderate total biomass value (3.2 g · m^{-2}).

In 1964 during the first half of summer stratification (16.6.) Chrysophyceae dominated the community (*Uroglena* spec. 33%, *Dinobryon bavaricum* 19%, *Chrysamoeba radians* 5%). *Oscillatoria rubescens* (9%) and *Anabaena lemmernammii* attained considerable values. Diatoms were 26% (*Asterionella formosa* 20%). During July (15.7.) the biomass of Cyanophyceae increased enormously (*Anabaena lemmermannii* 60%, *Oscillatoria rubescens* 12%); *Ceratium hirundinella* reached 13%, and the diatoms 12%

During August 1963 (21.8.) *Dinobryon* spp. almost totally disappeared; *Chrysamoeba radians* took their place with a biomass peak of about 15%. *Anabaena lemmermannii* (50%) dominated the community, whereas *Oscillatoria rubescens* only occurred in very small accounts (5%). *Asterionella formosa* (19%) and nanoplanktonic diatoms (12%) achieved only moderate values. In mid-September (17.9.) the phytoplankton community reached its highest biomass value of the whole year with 15.7 g · m^{-2}. *Mallomonas fastigata* (43%), *Anabaena lemmermannii* (39%), *Ceratium hirundinella* (9%), and *Oscillatoria rubescens* (6%) contributed considerably to the relatively enormous biomass. At the end of October (24.10.) an extremely small algal biomass (0.3 g · m^{-2}) was calculated, dominated by *Oscillatoria rubescens* (60%), *Mallomonas fastigata* (13%), and *Botryococcus braunii* (14%).

During August 1964 (14.8.) *Ceratium hirundinella* (39%), *Anabaena lemmermannii* (35%), and *Oscillatoria rubescens* (12%) dominated the biomass. Nanoplanktonic phytoflagellates amounted to 7%. In September (22.9.) *Oscillatoria rubescens* (35%), *Ceratium hirundinella* (21%), and *Anabaena lemmermannii* (19%) kept their rather high biomass, then accompanied by small amounts of *Mallomonas fastigata* (6%), *Asterionella formosa* (5%), and *Cyclotella kützingiana* (5%). During October (20.10.) the diatoms nearly disappeared. *Oscillatoria rubescens* reached its highest biovolume (62%), *Ceratium hirundinella* stood at 17%, *Dinobryon* spp. and *Mallomonas fastigata* amounted to 10% and 9%, respectively.

In late autumn 1963 (15.11.) *Mallomonas fastigata* (48%) and *Oscillatoria rubescens* (44%) dominated the community, whereas in 1964 (20.11.) *Mallomonas fastigata* attained a mere 24%, but *Oscillatoria rubescens* amounted to 65%, whereas nanoplanktonic phytoplankters reached only 7%.

On a biomass basis the most important species in Lake Stechlin in 1963 and 1964 were microplanktonic phytoplankters such as *Mallomonas fastigata* (late summer to winter), *Dinobryon* spp. (mid-spring to early summer). *Anabaena lemmermannii* (mid-summer to early autumn), *Oscillatoria rubescens* (mid-autumn to late spring), and *Asterionella formosa* (late spring to summer). Nanoplanktonic diatoms occurred in spring and late summer at high biovolumes. Some other species such as *Uroglena* spp., *Chrysamoeba radians,* and *Botryococcus braunii* amounted to noticeable biomass for short periods only. Nearly all other species mentioned in the count-tables accounted, on an average, for 8% of the total phytoplankton volume.

The mean total biomass per annum in 1963 was 2.6 $g \cdot m^{-2}$, in 1964 3 $g \cdot m^{-2}$ (Table 5.13); the maximum in 1963 15.7 $g \cdot m^{-2}$ (17.9.) and in 1964 5.2 $g \cdot m^{-2}$ (15.7.); the minimum in 1963 0.4 $g \cdot m^{-2}$ (31.1.) and in 1964 0.8 $g \cdot m^{-2}$ (20.11.).

Lake Nehmitz-North. The mean total biomass per annum in 1963 was 2.1 $g \cdot m^{-2}$ and in 1964 5.4 $g \cdot m^{-2}$ (Table 5.13); the maximum in 1963 8.5 $g \cdot m^{-2}$ (12.9.) and in 1964 9.7 $g \cdot m^{-2}$ (17.5.), and the minimum in 1963 0.1 $g \cdot m^{-2}$ (24.4.), and in 1964 0.7 $g \cdot m^{-2}$ (26.4.; cf. Fig. 5.16).

Taking into account some uncertainty in calculating the biomass in 1963, the values can, nevertheless, be considered as representative. In February under ice cover *Peridinium* spec. dominated a community poor in species. No distinct spring maximum followed. *Dinobryon divergens* (24.4.; 58%, 3.5.: 68%, 16.5.: 92%), *D. sociale* (24.4.: 10%, 3.5.: 7%), and *Mallomonas fastigata* (24.4.: 16%, 3.5.: 14%) reached the highest values (Fig. 5.16).

During the summer stagnation biomass amounted to a considerable level. At that time *Dinobryon sociale* occupied a good position (12.6.: 34%), but was surpassed by *Anabaena lemmermannii* (12.6.: 42%). *Ceratium hirundinella* stood at 9%. In mid-July (26.7.) *Mallomonas fastigata* (42%) dominated the community, accompanied by *Ceratium hirundinella* (20%), *Anabaena lemmermannii* (19%) and *Oscillatoria rubescens* (17%). When the development reached its highest peak (12.9.; Fig. 5.16) *Ceratium hirundinella* stood at 42%, *Oscillatoria rubescens* at 21%, *Mallomonas fastigata* at 13%, and *Fragilaria crotonensis* reached its annual maximum with 18% (Fig. 5.16).

In autumn *Mallomonas fastigata* dominated (21.10.: 78%, 18.11.: 97%, 11.12.: 81%). Only *Fragilaria crotonensis* (21.10.: 18%) and *Asterionella formosa* (11.12.: 16%) amounted to values which were worthy of serious attention.

In 1964 (Fig. 5.16) there were no striking differences in the biomass development in Lake Nehmitz-N compared with those of 1963. *Mallomonas fastigata* (24.4.: 29%, 27.5.: 76%, 23.11.: 18%), *Ceratium hirundinella* (16.6.: 49%, 16.7.: 35%, 19.8.: 28%, 21.10.: 22%), *Dinobryon sociale* (27.5.: 76%, 19.8.: 12%), and *Fragilaria crotonensis* (16.6.: 23%, 16.7.: 27%, 19.8.: 25%, 24.9.: 69%, 21.10.: 44%, 23.11.: 41%) were the dominant taxa. Noteworthy were nanoplanktonic diatoms (24.4.: 52%) and phytoflagellates (23.11.: 10%) and the Cyanophyceae (19.8.: 11%, 24.9.: 75%), which could not be exactly calculated (*Microcystis* spp., *Anabaena lemmermannii, Chroococcus turgidus*).

In both 1963 and 1964 the phytoplankton biomass of Lake Nehmitz-N was distinguished by the dominance of *Mallomonas-, Dinobryon-, Ceratium-, Peridinium-,* and Cyanophyceae-species. In contrast to Lake Stechlin, *Fragilaria crotonensis* played an important role. This may be the source of the *Fragilaria* population of Lake Stechlin which started after 1966.

Lake Nehmitz-South. The mean total biomass per annum in 1963 was 0.5 g · m^{-2} and in 1964 2.3 g · m^{-2} (Table 5.13); the maximum in 1963 1.5 g · m^{-2} (21.10.) and in 1964 6.2 g · m^{-2} (16.6.); and the minimum in 1963 0.03 g · m^{-2} (24.4.) and in 1964 0.2 g · m^{-2} (23.11.; Fig. 5.17).

Taking into account some uncertainty in calculating the biomass in 1963, the values demonstrate different figures between 1963 and 1964, which cannot be verified beyond doubt. Therefore, we shall only discuss the figures for 1964 (Fig. 5.17).

At the end of the spring circulation (26.4.) *Dinobryon sociale* (20%), *Mallomonas fastigata* (18%), nanoplanktonic *Cyclotella* species (16%), *Synedra* spec. (9%), and *Tabellaria fenestrata* var. *fenestrata* (10%) dominated the community. On May 27th the *Cyclotella* development reached its highest peak (85%), accompanied by *Mallomonas fastigata* (10%). During the summer stagnation *Asterionella formosa* (16.6.: 33%, 16.7.: 13%), *Fragilaria crotonensis* (19.8.: 13%, 24.9.: 11%), *Anabaena lemmermannii* (16.6.: 18%, 16.7.: 9%), *Pandorina-Eudorina* (16.6.: 16%, 16.7.: 11%), *Sphaerocystis schroeteri* (16.6.: 10%), *Ceratium hirundinella* (16.7.: 41%, 19.8.: 70%, 24.9.: 55%, 21.10.: 37%), and *Mallomonas fastigata* (16.6.: 10%, 16.7.: 18%, 24.9.: 9%, 21.10.: 20%) dominated the community. On October 21st an undetermined blue-green alga (*Microcystis*?!) stood at 30%.

During autumn circulation the community consisted mainly of nanoplanktonic *Cyclotella* and *Stephanodiscus* species (40%), *Asterionella formosa* (19%), and nanoplanktonic phytoflagellates (35%).

In contrast to Lake Nehmitz-N and Lake Stechlin, Lake Nehmitz-S was distinguished by a more complicated phytoplankton community. Remarkable were the high proportions of *Ceratium hirundinella,* green algae and nanoplanktonic diatoms and phytoflagellates. The considerable growth of microplanktonic diatoms (1963 in the second half of the year, 1964 throughout the year) ought to be mentioned, represented by *Fragilaria crotonensis, Asterionella formosa, Tabellaria flocculosa,* and *T. fenestrata* var. *fenestrata.*

Species composition and seasonal periodicity 1969 to 1972

Cell number

Lake Stechlin. During this period the nuclear power plant operated more or less continuously. In 1970 it did not work on 85 days, in 1971 on 143 days (on 125 days from 17.4. to 19.8.), and in 1972 on 123 days (on 65 days from 15.4. to 19.8., and on 46 days from 22.7. to 5.9.; cf. Figs. 5.18, 5.19). The phytoplankton sample point St was situated in the minor thermally loaded part of the lake (mean increase of temperature in 0–2 m depth: 1.2–1.4 K; Richter *et al.* 1979) east-north-east from the centre of temperature.

During the operation of the nuclear power plant a real winter stagnation phase does not set in. The summer stagnation phase begins during May and is completely developed between June and December. The thermocline lies between 10 and 15 m depth. During winter and early spring (December to April) the water temperature is ~4° C (p. 52).

In 1969 the ice covered 50 per cent of the lake area until March 20th. In 1970 the ice cover was 100 per cent between January 5th, and February 5th, and melted to 50 per cent by March 20th. During the winter periods 1970/71 and 1971/72, the lake was ice-free (Richter *et al.* 1979; see Fig. 3.11).

From December to April the phytoplankton community consists of the following species in order of numeric importance: nanoplanktonic *Cyclotella* species (e.g. *C. comta, C. comta* var. *oligactis, C. glomerata, C. operculata*), nanoplanktonic phytoflagellates smaller than 10 μm (e.g. *Rhodomonas minuta* var. *nanoplanktonica,* and *Erkenia subaequiciliata*), *Tabellaria fenestrata* var. *asterionelloides* (only 1969 and 1970), *Scenedesmus costato-granulatus, Chrysococcus* spec., *Tetraedron minimum, Fragilaria crotonensis, Pseudokephyrion entzii, Kephyrion* ssp., *Asterionella formosa, Eudorina elegans, Coelastrum microporum, Pediastrum tetras, P. duplex, Oocystis* spec., *Gymnodinium* spec., *Botryococcus braunii, Stephanodiscus minutus, Cosmarium* ssp., *Cymatopleura solea, Diatoma elongatum, Dinobryon* ssp., and *Ceratium hirundinella.*

Only the nanoplanktonic *Cyclotella* species (February 1972: 552 000 cells per litre; Table 5.12), and the nanoplanktonic phytoflagellates (March 1972: 450) are present in high numbers. More than 10 000 cells per litre reach *Scenedesmus costato-granulatus* (March 1971: 34), and *Chrysococcus* spec. (April 1969: 28).

The other above-mentioned species are present in very small numbers (0.5 to 7.5), on average, not amounting to more than 4000 cells per litre. Often they are detectable only in net-samples but not in Ruttner-samples.

The late winter – early spring maximum (February-March) of nanoplanktonic *Cyclotella* species, together with the considerable growth of nanoplanktonic phytoflagellates, was favoured by the absence of any ice cover and the intensive solar radiation on many sunny days.

When the summer stagnation phase set in (April to June) the nanoplanktonic *Cyclotella* growth came to an end (May 1971: 20, April 1972: 12), but in 1969 and 1972 during this phase a second minor peak of growth was reached (June 1969: 126, May 1972: 78; cf. Figs. 5.18, 5.19).

During this transitional period a considerable short-term growth of *Pseudokephyrion entzii* was observed (May 1971: 48, May 1972: 210; Table 5.12). It was absent in 1969 and 1970. The other small phytoflagellates increased cell numbers by 68 (1969), 118 (1971), and 2160 (1972; Table 5.12). The microplanktonic *Asterionella formosa* attained 104 000 cells per litre (1972). Species such as *Scenedesmus costato-granulatus* (52), *Chrysococcus* spec. (34), *Eudorina elegans* (29),

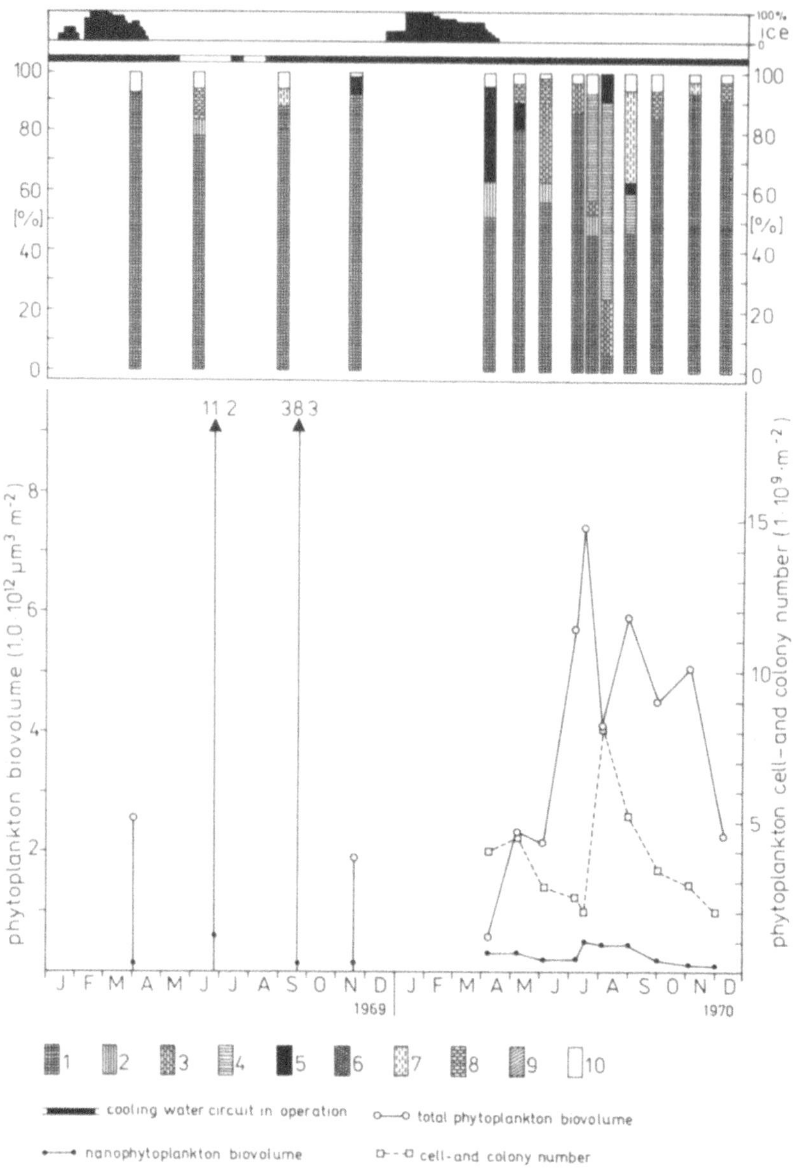

Figure 5.18 Seasonal periodicity of phytoplankton biovolume, cell and colony number, proportion of the main groups of the phytoplankton community, cooling water circulation, and ice cover in Lake Stechlin, 1969–1970 (according to Busse, unpublished data, altered and enlarged; figures as in Fig. 5.15).

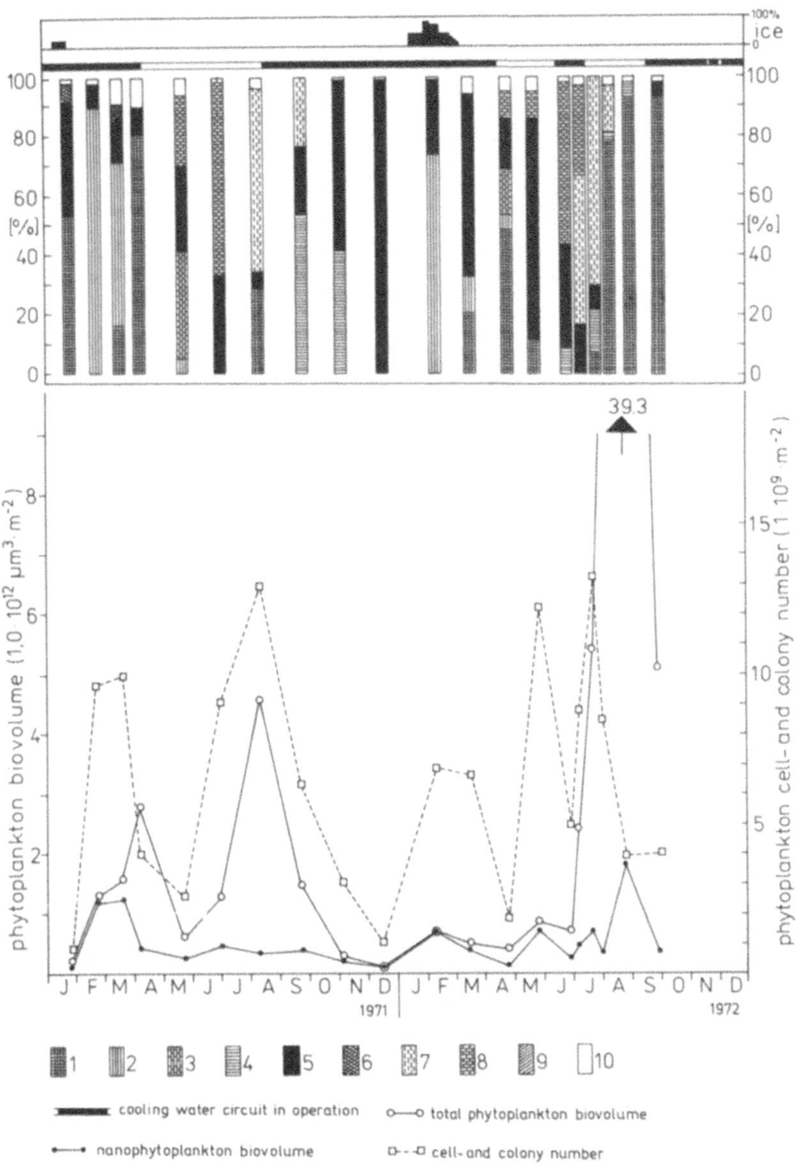

Figure 5.19 Seasonal periodicity of phytoplankton biovolume, cell and colony number, proportion of the main groups of the phytoplankton community, cooling water circulation, and ice cover in Lake Stechlin, 1971–1972 (according to Busse, unpublished data, altered and enlarged; figures as in Fig. 5.15).

Table 5.12 Dominant phytoplankton species (maximum cell numbers 1.0×10^4 and more) in Lake Stechlin 1969–1972 (F = cell filament; K = cell colony).

Species	Date	Maximum cell number/depth ($\times 10^3 \cdot l^{-1}$/m)	Average cell number (0–40 m) ($\times 10^3 \cdot l^{-1}$)	Maximum biovolume ($\times 10^6 \mu m^3 \cdot l^{-1}$)	Average biovolume (0–40 m) ($\times 10^6 \mu m^3 \cdot l^{-1}$)
Asterionella formosa	01.09.70	120/2.5	40	96	26
Chrysococcus spec.	07.04.70	38/0	28	2	1.5
Cyclotella kützingiana	01.09.70	64/5	8	22	3
Cyclotella spp. *'nano'*	23.02.71	552/2.5	349	83	52
Dinobryon sertularia	22.07.72	400/5	80	140	28
Eudorina elegans (K)	04.07.72	31/0	10	93	30
Fragilaria crotonensis (F)	24.08.72	149/10	31	4,470	930
Nanoplanktonic phytoflagellates (<10 μm)	23.05.72	2,150/0	499	108	25
Nitzschia spec.	25.06.69	398/15	43	138	11
Phacotus lenticularis	24.08.72	53/2.5	18	13	4
Pseudokephyrion entzii	23.05.72	210/0	49	10	2.5
Tabellaria fenestrata var. *asterionelloides*	22.09.69	680/10	128	2,040	384
Tetraedron minimum	04.08.70	32/0	6	1	0.2

Dinobryon crenulatum (27), *Tetraedron minimum* (27), and *Fragilaria crotonensis* (21) reached cell numbers worthy of mention. In June 1969 a nanoplanktonic *Nitzschia* species grew to 398 000 cells per litre.

During summer (July to September) nanoplanktonic phytoflagellates dominate: September 1969: 163, July 1971: 1000; July 1972: 151. *Fragilaria crotonensis* (August 1972: 149) and *Eudorina elegans* (July 1971: 21, July 1972: 31) reach their greatest annual growth quantity continuing their development during the transitional period (Table 5.12). Considerable cell numbers were found in *Phacotus lenticularis* (August 1972: 53), *Tetraedron minimum* (August 1971: 31), *Dinobryon sertularia* (July 1972: 20), *D. bavaricum* (August 1971: 11), and *Anabaena lemmermannii* (July 1972: 10).

During autumn the development of various diatoms took place. The 1969 September-maximum of the microplanktonic *Tabellaria fenestrata* var. *asterionelloides,* then together with *Fragilaria crotonensis* and nanoplanktonic phytoflagellates, is mentioned below (p. 177). *Phacotus lenticularis* (10) and *Scenedesmus costato-granulatus* (9) are noteworthy too. A similar great growth of nanoplanktonic phytoflagellates occurred in both September 1971 (598) and 1972

(498), the latter accompanied by nanoplanktonic *Cyclotella* species (30) and *Fragilaria crotonensis* (10).

The phytoplankton development in 1970 cannot be compared directly with that in 1969, 1971 or 1972 (Figs. 5.18, 5.19). Apparently due to the long ice cover, the nanoplanktonic *Cyclotella*-population started its growth in April (37) and reached its maximum in June (125). The small phytoflagellate-population started its development in March, and reached a first peak in May (333). Keeping up its high level throughout June and July, it increased to 786 000 cells per litre in August. After a short depression at the beginning of September the 200 000 cells level was reached again in autumn. During the whole period some other species reached considerable growth: *Chrysococcus* spec. (38, in April), *Tetraedron minimum* (32, in August), *Scenedesmus costato-granulatus* (22, in June), *Asterionella formosa* (120, in September), *Fragilaria crotonensis* (14, in July), *Cosmarium* spec. (19, in July), *Phacotus lenticularis* (40, in August and September), *Cyclotella kützingiana* (64, in September), and *Tabellaria fenestrata* var. *asterionelloides* (46, in December).

The most striking and unique phenomenon in the phytoplankton growth in the Lake Stechlin area during this period was the enormous growth of *Tabellaria fenestrata* var. *asterionelloides*. Starting with 11 000 cells per litre in April 1969, it increased to 149 000 during June and reached at the end of September its greatest quantity (680; Table 5.12), then accompanied by *Fragilaria crotonensis* (118), and nanoplanktonic phytoflagellates (163). This surprising development[2] was continued in 1970 at a lower level. Occurring throughout the whole year, it increased to 67 000 cells per litre at the end of September, and in December it was represented by 46 000 cells. Since its break-down in January 1971 (3) and its small pseudo-regeneration in April 1971 (39), the *Tabellaria fenestrata* var. *asterionelloides*-population has not reached any comparable cell numbers. Now it belongs to the quasi dormant stock of the species-rich phytoplankton in Lake Stechlin.

To the quasi-dormant stock of species without the considerable growth mentioned above (p. 176) belong all the taxa which also occur later on during summer and autumn: *Ankistrodesmus pfitzeri, Acanthoceras zachariasii, Bitrichia chodatii, Ceratium cornutum, Chroococcus turgidus, Chrysamoeba radians, Coelastrum reticulatum, Dinobryon cylindricum, D. sociale, D. suecicum, Elakatothrix gelatinosa, Mallomonas crassisquama, M. fastigata, M. tonsurata, Melosira* spec., *Nitzschia sigmoidea, Oscillatoria rubescens, Pandorina morum, Pediastrum boryanum, Peridinium* spp., *Rhizosolenia longiseta, Staurastrum paradoxum, Synedra acus, Tabellaria fenestrata* var. *fenestrata,* and *Trachelomonas* spec.

[2] The taxon had been never seen before in Lake Stechlin; so far *Tabellaria fenestrata* was represented by its type-variety and occurred only in small cell numbers. Krieger (1927) reported the taxon from Lake Dagow (11.4.1926). Perhaps Lake Stechlin was inoculated by a *Tabellaria fenestrata* var. *asterionelloides* population from Lake Dagow.

Concerning the cell number during the 1969 to 1972 period, it can be established that, with the exception of the extraordinary two-year development of the microplanktonic *Tabellaria fenestrata* var. *asterionelloides*, nanoplanktonic diatoms and phytoflagellates dominated the phytoplankton population in Lake Stechlin. These two groups reached cell numbers exceeding 100 000 per litre during the greatest part of the vegetation period. Only four further taxa surpassed the 100 000 cell limit temporarily: the microplanktonic *Fragilaria crotonensis* and *Asterionella formosa*, and the nanoplanktonic *Nitzschia* spec. and *Pseudokephyriopsis entzii*. The other species did not contribute considerably to the cell number of the phytoplankton population.

Biomass

Lake Stechlin. Only four samples are at our disposal for 1969 (Fig. 5.18). On April 1st *Tabellaria fenestrata* var. *fenestrata* dominated the phytoplankton community (88%), and on May 25th it consisted mainly of *Fragilaria crotonensis* (73%) together with *Tabellaria fenestrata* var. *asterionelloides* (7%) and 'Volvocales' (9.4%). On September 20th microplanktonic diatoms stood at 89% of the total biomass (*Tabellaria fenestrata* var. *asterionelloides* 27%, *Fragilaria crotonensis* 62%), which reached values never before calculated for Lake Stechlin with about 38 $g \cdot m^{-2}$ (Fig. 5.18).

During 1970 microplanktonic diatoms dominated throughout the year. The biovolume maximum (15.7.: 7.5 $g \cdot m^{-2}$) cannot be compared with that of 1969, but the mean biovolume of 3.4 $g \cdot m^{-2}$ per annum was nevertheless considerable (Table 5.13). It is noteworthy that the cooling water circulation operated throughout the year. After the ice break-up nanoplanktonic phytoflagellates reached 30% of the total low biomass (7.4.).

During the transitional phase between April and June microplanktonic algae governed the community (5.5.70: 81%: *Fragilaria crotonensis* 67%: 2.6.70: 98%: *Fragilaria crotonensis* 49%, *Dinobryon* spec. 24%, *Eudorina elegans* 9%).

During the summer stagnation the proportion of microplanktonic diatoms was still important (7.7.: *Fragilaria crotonensis* 70%, *Asterionella formosa* 15%; 14.7.: *Fragilaria crotonensis* 43%, *Asterionella formosa* 7%, *Tabellaria fenestrata* var. *asterionelloides* 3%), but from mid-July onward *Ceratium hirundinella* (14.7.:

Table 5.13 Yearly average phytoplankton biomass values ($g \cdot m^{-2} \cdot a^{-1}$) in Lakes Stechlin (S), Nehmitz-N (NN) and Nehmitz-S (NS); calculated according to the available data of Busse (1965, for 1963–1972) and Küchler (1982, for 1973–1978, partly altered).

Year	1963	1964	1970	1971	1972	1973	1974	1975	1976	1977	1978
S	2.6	3.0	4.1	1.4	6.7	4.5	4.2	9.1	12.9	3.1	5.1
NN	2.1	5.4	–	2.0	3.2	10.1	4.5	8.6	3.6	3.5	17.5
NS	0.5	2.3	–	7.0	3.6	–	–	–	–	–	–

35%; 4.8.: 64%) dominated the community, first accompanied by *Dinobryon* spp. (7.7.: 10%, 14.7.: 6%, 4.8.: 17%, 1.9.: 13%), later on by *Anabaena lemmermannii* (1.9.: 30%), whereas nanoplanktonic phytoflagellates played a subordinate role (Fig. 5.18).

From September onward an important long-term growth of *Tabellaria fenestrata* var. *asterionelloides* (80–90%) occurred dominating the whole late autumn – early winter community (Fig. 5.18).

In 1971 mean biomass values per annum attained only 1.3 $g \cdot m^{-2}$, e.g. 40% of the value of 1970 (Table 5.12). The maximum biomass amounted to 4.5 $g \cdot m^{-2}$ (10.8.). From April to August the cooling water circulation of the nuclear power plant did not operate.

From January to the end of March 1971 (Fig. 5.19) nanoplanktonic diatoms, mostly *Cyclotella* spp., dominated the small biovolume of the algal community (23.2.: 89%; 19.3.: 55%) accompanied by nanoplanktonic phytoflagellates (23.2.: 9%; 19.3.: 27%) and *Tabellaria fenestrata* var. *asterionelloides* (26.1.: 47%; 19.3.: 17%), apparently the last remainder of the long-term *Tabellaria*-growth during late autumn 1970. It was the typical late winter – early spring aspect of phytoplankton of an icefree temperate lake.

In 1972 similar phenomena were established. Note that nanoplanktonic phytoflagellates dominated the community on March 14th (60%; nanoplanktonic diatoms 12%), whereas in February (8.2.) nanoplanktonic diatoms accounted for 75% of the total biomass.

During April 1971 (Fig. 5.19) microplanktonic phytoplankters dominated (6.4.: 82%, *Tabellaria fenestrata* var. *asterionelloides* 47%, *Fragilaria crotonensis* 7%, *Cymatopleura solea* 27%; 20.5.: 50%, *Dinobryon* spp. 30%, Volvocales 19%), in 1972 32% (25.4.: *Fragilaria crotonensis*). Nanoplanktonic phytoflagellates were mostly subordinated, but reached 32% in 1971 and in 1972 76% on May 23th accompanied by Volvocales (*Eudorina elegans* 56% on June 28th). In 1971 the total biomass during this period was very small (up to 0.6 $g \cdot m^{-2}$), whereas in 1972 the corresponding values reached 2.7 $g \cdot m^{-2}$.

During the summer stagnation period 1972 (Fig. 5.19) nanoplanktonic phytoflagellates (4.7.: 19%; 19.7.: 10%), *Anabaena lemmermannii* (4.7.: 50%; 19.7.: 67%; 31.7.: 15%), *Eudorina elegans* (4.7.: 30%), and *Ceratium hirundinella* (19.7.: 12%) governed the community. But as early as mid-July the growth of microplanktonic diatoms started (19.7.: 10%) and later reached considerable proportions (31.7.: 78%: *Asterionella formosa* 63%, *Fragilaria crotonensis* 15%; 24.8.: 95%: *Asterionella formosa* 7%, *Fragilaria crotonensis* 88%; 26.9.: 92%: *Fragilaria crotonensis*). The mean biomass per annum in 1972 amounted to 6.7 $g \cdot m^{-2}$ (Table 5.13), the highest value observed between 1963 and 1974.

The period 1969–1972 under consideration (Fig. 5.18, 5.19) is distinguished by the extremely high dominance of microplanktonic algae, especially of diatoms. In 1969 and 1970 *Tabellaria fenestrata* var. *asterionelloides* and *Fragilaria crotonensis* played the first role, in 1972 *Fragilaria crotonensis* and *Asterionella formosa*.

There seemed to be a retreat of *Tabellaria fenestrata* var. *asterionelloides*. The period between June 1970 and June 1971 showed its own features. Nanoplanktonic algae dominated in periods of small total biovolumes of the community during phases, when working and non-working periods of the nuclear power plant alternated. Noteworthy was the appearance of greater quantities of green algae of the *Eudorina–Pandorina–Botryococcus*–group.

Lake Nehmitz-North. The mean total biomass per annum in 1971 was $2.0\,g \cdot m^{-2}$ and in 1972 $3.2\,g \cdot m^{-2}$ (Table 5.13); the maximum in 1971 $7.5\,g \cdot m^{-2}$ (7.9.) and in 1972 $14.1\,g \cdot m^{-2}$ (10.8.); the minimum in 1971 $0.2\,g \cdot m^{-2}$ (1.12.) and in 1972 $0.6\,g \cdot m^{-2}$ (10.4.; Fig. 5.20). The values do not differ greatly from those of the period 1963 to 1964.

Comparing the biomass development of Lakes Nehmitz-N and Stechlin, the conspicuous conformity of the figures is worthy of attention. Small biomass values between March and July 1972 in Lake Nehmitz-N (Fig. 5.20) correspond exactly to those in Lake Stechlin (Fig. 5.19). Biomass peaks in late summer (7.9.1971; 10.8.1972) in Lake Nehmitz-N coincide with those in Lake Stechlin (10.8.1971; 24.8.1972). In 1971 they were dominated (89%) by blue-green algae (*Pseudanabaena catenata, Microcystis* spec.) in Lake Nehmitz-N, but in 1972 by *Mallomonas fastigata* (65%), accompanied by *Fragilaria crotonensis* (18%) and blue-green algae (9%). The maxima in Lake Stechlin in 1971 were formed by blue-green algae (60%) and *Fragilaria crotonensis* (29%), whereas in 1972 *Fragilaria crotonensis* (88%) alone dominated the community.

Apart from trifling differences, the distribution of the biomass and the proportion of the main groups of phytoplankters are surprising similar. Here and there the enormous growth of nanoplanktonic phytoflagellates and diatoms in phases of small biomass values is obvious (Figs. 5.19, 5.20). The late autumn or early spring development of nanoplanktonic diatoms can be seen. Special features are the greater growth of green algae in Lake Stechlin and the striking dominance of *Mallomonas fastigata* in Lake Nehmitz.

To sum up: biomass periodicity, value, and dominating taxa coincide very well; that means, in the period 1971 and 1972, the phytoplankton communities of Lakes Nehmitz-N and Stechlin form a unit due to the cooling water circulation.

Lake Nehmitz-South. The mean total biomass per annum in 1971 was $7.0\,g \cdot m^{-2}$ and in 1972 $3.6\,g \cdot m^{-2}$ (Table 5.13); the maximum in 1971 $17.5\,g \cdot m^{-2}$ (4.10.) and in 1972 $6.9\,g \cdot m^{-2}$ (15.9.); the minimum in 1971 $1.5\,g \cdot m^{-2}$ (4.6.) and in 1972 $1.6\,g \cdot m^{-2}$ (26.10.; Fig. 5.21).

The dominant taxa were *Fragilaria crotonensis* (1971: 4.5.: 60%; 1972: 5.7.: 26%), *Tabellaria flocculosa* (11.6.1971: 20%), *Tabellaria fenestrata* var. *asterionelloides* (1971: 1.12.: 60%; 1972: 26.10.: 49%, *Asterionella formosa* (1971: 1.12.: 15%; 1972: 25.5.: 38%), *Dinobryon sertularia* (22.3.1971: 73%), *Ceratium hirundinella* (1971: 4.8.: 70%; 10.8.: 86%), *Oscillatoria rubescens* (1971: 3.11.: 27%; 1972: 15.9.: 22%).

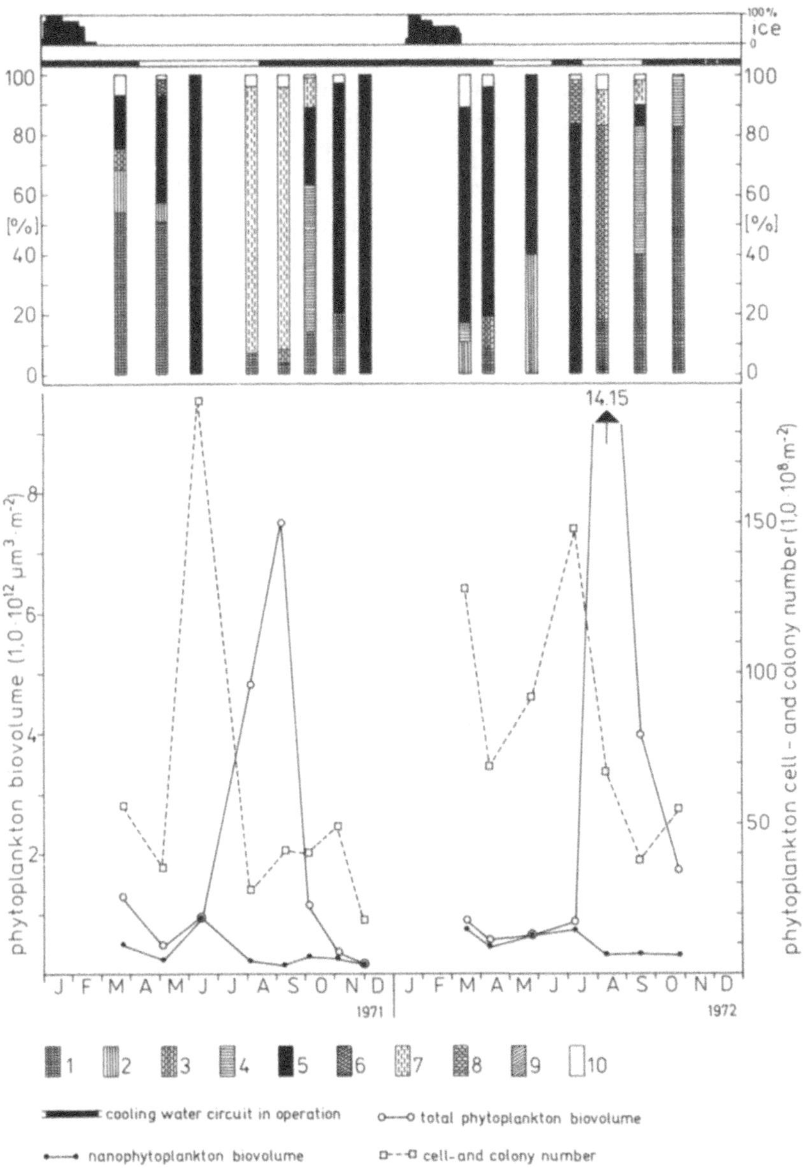

Figure 5.20 Seasonal periodicity of phytoplankton biovolume, cell and colony number, proportion of the main groups of the phytoplankton community, cooling water circulation, and ice cover in Lake Nehmitz-N, 1971–1972 (according to Busse, unpublished data, altered and enlarged; figures as in Fig. 5.15).

Microplanktonic diatoms, *Ceratium hirundinella*, and blue-green algae dominated the community more or less throughout the year. There were only a few similarities between the period 1971–1972 and 1963–1964, in which the *Mallomonas–Dinobryon* species played the important role, accompanied by the above mentioned groups. The periodicity of the biomass development by no means coincides with that of Lake Nehmitz-N or Lake Stechlin. Lake Nehmitz-S is not influenced by the cooling water circulation.

Species composition and seasonal periodicity 1976 to 1978

The phytoplankton development in Lakes Stechlin and Nehmitz between 1963 and 1972 has been discussed in detail in the preceding sections on the basis of the never published investigations by Busse (1965, 1972). For the period between 1973 to 1978, the results of the studies of Küchler (1981, 1982) are at our disposal. They will be summarized in the following paragraphs[3], especially the years 1976–1978.

Our Figs. 5.22 (Lake Stechlin) and 5.23 (Lake Nehmitz-N) pertain to the seasonal periodicity of cell or colony numbers, respectively, and the biomass of the main groups of the phytoplankton community from 1976 to 1978. During this time the nuclear power plant operated more or less continuously. The cooling water circulation made the two lakes – to a certain degree – a hydrologic unit. Cell numbers and biovolumine reached great values, the greatest ones ever observed in our study area (Table 5.13).

Cell number

Lake Stechlin. The average cell or colony numbers per annum were: in 1976 $65 \cdot 10^9 \cdot m^{-2}$; in 1977, $23 \cdot 10^9 \cdot m^{-2}$; in 1978, $167 \cdot 10^9 \cdot m^{-2}$. The maxima were: in 1976, $121 \cdot 10^9 \cdot m^{-2}$ (28.10.); in 1977, $29 \cdot 10^9 \cdot m^{-2}$ (13.9.), and in 1978, $493 \cdot 10^9 \cdot m^{-2}$ (5.12.; cf. Fig. 5.22).

The three years under consideration differ strongly from each another. 1976 can be called the year of blue-green algae (cf. Fig. 5.22); almost throughout the year they play a dominant role. *Oscillatoria* spp., *Anabaena lemmermannii* and *Chroococcus* spp. governed the community. The numbers of the colonies counted cannot be compared directly with those of the single cells of nanoplanktonic *Cyclotella* spp. (1.6.: 1524[4]) or ultraplanktonic green algae (28.10.: 6940), but they enable us to qualify the phytoplankton development in 1976 as an extraordinary one. Nanoplanktonic phytoflagellates (20.7.: 1292), *Dinobryon* spp. (30.6.: 1116),

[3] Küchler's (1981, 1982) lists of taxa were, after correction, incorporated in our own list (see Appendix pp. 485–496). The phytoplankton development in Lake Nehmitz-S, not influenced by the cooling water circulation, will not be regarded.

[4] In the following sections the numbers stand for $1.0 \cdot 10^3 \cdot l^{-1}$.

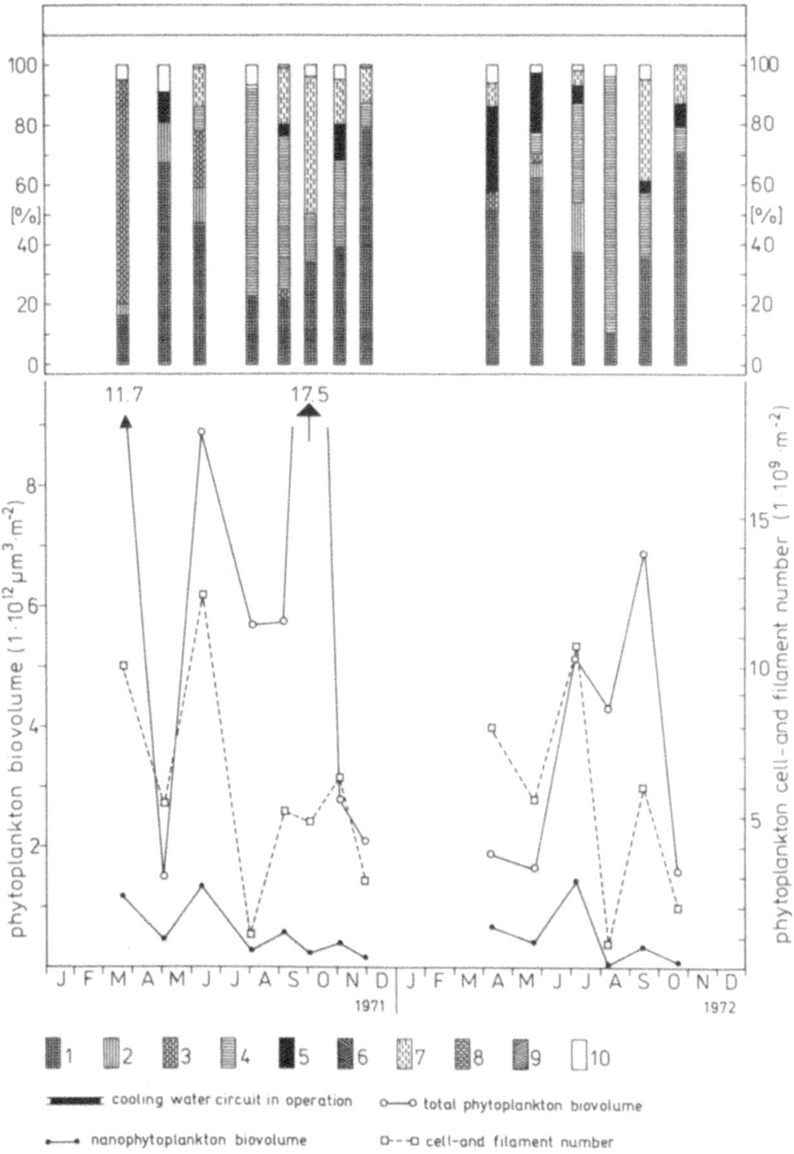

Figure 5.21 Seasonal periodicity of phytoplankton biovolume, cell and colony number, proportion of the main groups of the phytoplankton community in Lake Nehmitz-S, 1971–1972 (according to Busse, unpublished data, altered and enlarged; figures as in Fig. 5.15).

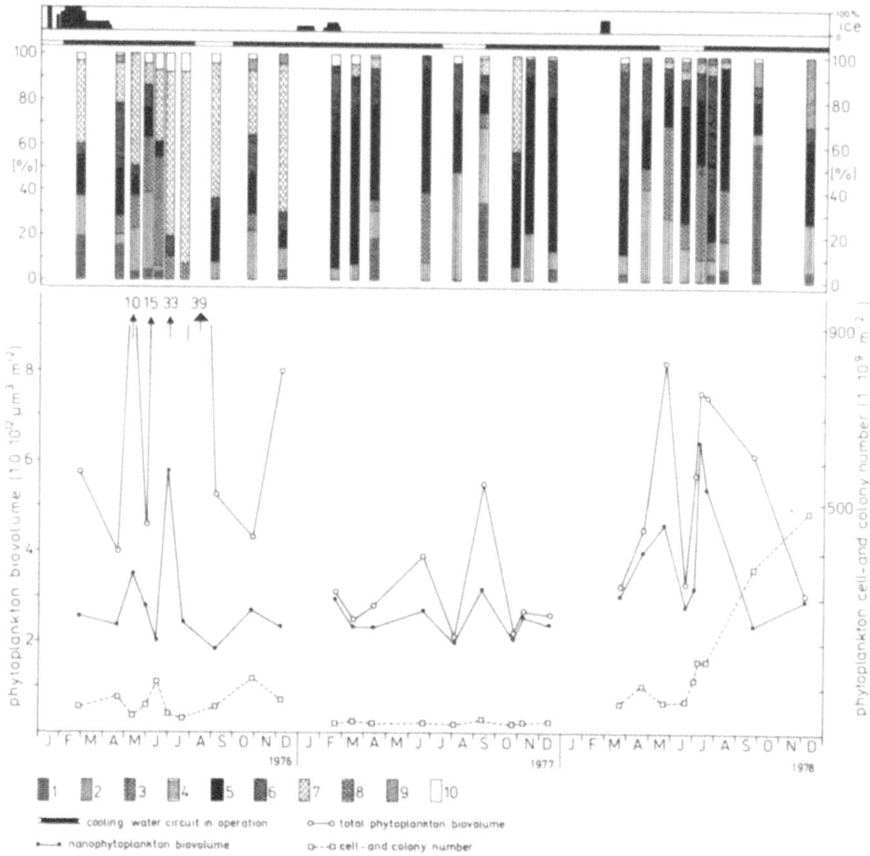

Figure 5.22 Seasonal periodicity of phytoplankton biovolume, cell and colony number, proportion of the main groups of the phytoplankton community, cooling water circulation, and ice cover in Lake Stechlin, 1976–1978 (according to Küchler 1982; altered strongly and enlarged; figures as in Fig. 5.15).

Ankistrodesmus spec. (30.6.: 528) were also present in high cell numbers. Microplanktonic diatoms played no comparable role. As a whole, nano- and ultraplanktonic green algae seemed to be in progress.

In 1977, the cell or colony numbers, respectively, reached only small quantities on a surprisingly even level throughout the year compared with those of 1976 (Fig. 5.22). The blue-green algae – with the exception of a short period in autumn – have disappeared to a great extent. Their place has been taken by nanoplanktonic phytoflagellates and coccoid green algae which reached high cell numbers throughout the year (21.6.: 386; 2.8.: 346; 25.10.: 872). In mid-June *Dinobryon* spp. amounted to 88 (21.6.), and diatoms attained 716 (13.9.: *Cyclotella 'nano'*) or 106 (13.9.: *Asterionella formosa*), respectively.

In 1978, the cell numbers increased enormously again (Fig. 5.22). The ultraplanktonic green algae amounted to 21 313 · 10^3 (5.12.), the coccoid green algae to

$823 \cdot 10^3$ (10.7.) or $348 \cdot 10^3$ (26.9.) cells per litre. The nanoplanktonic phytoflagellates attained 863 (20.6.) and 708 (18.10.). Similar high numbers were reached by nanoplanktonic *Cyclotella* spp. (23.5.: 865), whereas *Dinobryon* spp. stood at $240 \cdot 10^3$ cells per litre (23.5.). Noteworthy is the great growth of *Fragilaria crotonensis* in autumn (26.9.: 269).

In the period under consideration (including the data available for 1973 to 1975), the long period of cooling water circulation apparently generated a series of changes in the phytoplankton community. Apart from the quantitative properties which will be discussed below the dominance of species differed markedly from year to year. The striking phenomena are:

- the enormous development of ultraplanktonic green algae from 1973 onward[5]);
- the 'blooming' of blue-green algae (for instance *Anabaena lemmermannii*) during 1973, 1974, and 1976;
- the remarkable growth of nanoplanktonic diatoms (*Cyclotella* spp., *Stephanodiscus* spp.) during the whole period;
- the rapid retreat of *Tabellaria fenestrata* var. *asterionelloides;*
- the 'regeneration' of microplanktonic diatoms (*Asterionella formosa, Fragilaria crotonensis*) and chrysomonades (*Dinobryon* spp., *Mallomonas* spp.).

Even so, vernal and autumnal diatom maxima, in spring accompanied by chrysomonades, together with blue-green algal and dinomonadinean maxima in summer, have remained a characteristic of the phytoplankton annual cycle. In the early eighties this tendency has become more and more distinct, corresponding to the increasing discontinuity of the cooling water circulation.

Lake Nehmitz-North. From Fig. 5.23 it is obvious that the phytoplankton development discussed on the basis of cell numbers does not coincide with that of Lake Stechlin. To a certain degree the years 1976 and 1977 are exchangeable; the low level of cell numbers in 1977 of Lake Stechlin corresponds to that of Lake Nehmitz-N in 1976. Only in 1978 are high cell numbers counted in both the lakes.

The average cell or colony numbers were: in 1976, $21.8 \cdot 10^9 \cdot m^{-2}$; in 1977, $25.3 \cdot 10^9 \cdot m^{-2}$, and in 1978, $559 \cdot 10^9 \cdot m^{-2}$. The maxima were: in 1976, $33 \cdot 10^9 \cdot m^{-2}$ (21.4.); in 1977, $75 \cdot 10^9 \cdot m^{-2}$ (22.6.), and in 1978, $875 \cdot 10^9 \cdot m^{-2}$ (Fig. 5.23).

The dominant taxa in 1976 were nanoplanktonic phytoflagellates (21.4.: 1200), coccoid green algae (21.4.: 570), nanoplanktonic *Cyclotella* spp. (21.4.: 805), *Fragilaria crotonensis* (21.4.: 43), and blue-green algae (*Chroococcus* spec. and *Oscillatoria* spp.: 57, on September 9th).

In 1977, *Dinobryon* spp. attained 348 (22.6.), *Asterionella formosa* stood at 22 (26.10.), and *Fragilaria crotonensis* at 39 (14.9.). Nanoplanktonic *Cyclotella* spp. did not exceed 249 (22.6.), *Oscillatoria* spp. amounted to 154 (14.9.), but

[5] It must be taken into account that the existence of these forms may have been neglected in the preceding period of investigation.

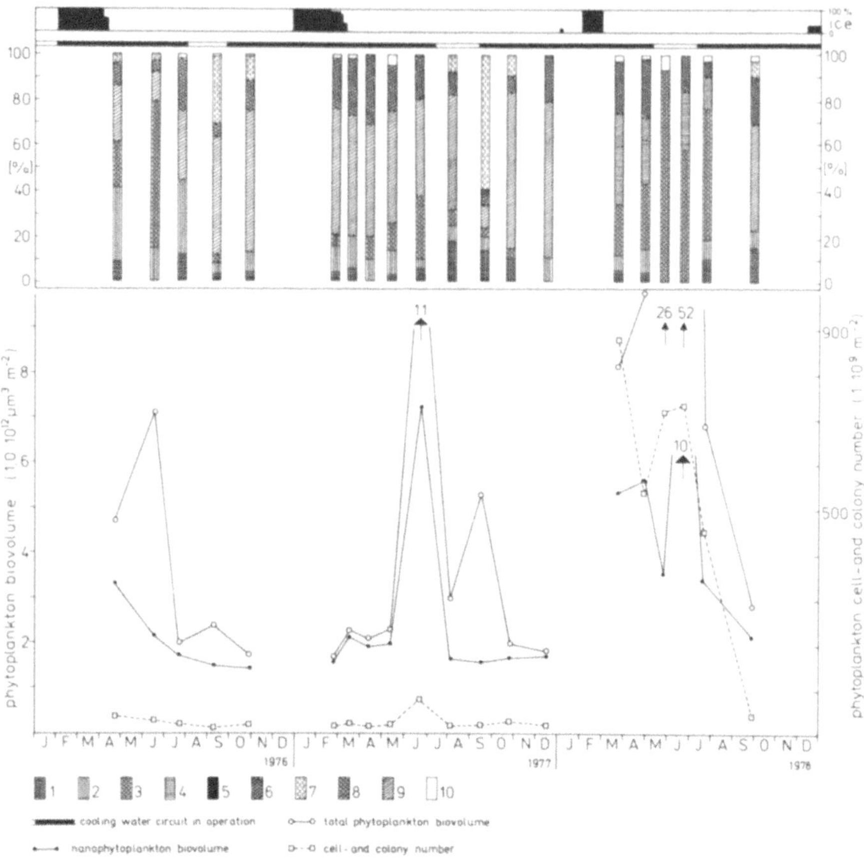

Figure 5.23 Seasonal periodicity of phytoplankton biovolume, cell and colony number, proportion of the main groups of the phytoplankton community, cooling water circulation, and ice cover in Lake Nehmitz-N, 1976–1978 (according to Küchler 1982; altered strongly and enlarged; figures as in Fig. 5.15).

throughout the year nanoplanktonic phytoflagellates dominated with cell numbers of about 1000.

In 1978, the highest cell numbers have been counted. Ultraplanktonic green algae attained on March 22nd to incredible – with regard to the conditions in our study area – 65 000 · 10^3 cells per litre (26.4.: 36 000; 24.5.: 40 000; 21.6.: 41 000; 19.7.: 52 000). The coccoid nanoplanktonic green algae with 13 700 (21.6.) and *Dinobryon* spp. with 2900 (21.6.) amounted to numbers not known until now in our study area. Nanoplanktonic phytoflagellates and *Cyclotella* spp. stood at 1600 (26.4.) and 577 (26.4.), respectively. *Synedra* spp. increased to 54 (22.3.), and *Fragilaria crotonensis* reached its maximum in mid-summer at about 66 (19.7.).

The striking increase in cell number must be regarded as the result of a process of fertilization under the influence of the cooling water circulation. But the phytoplankton community of Lake Nehmitz-N did not react in the same way as that of Lake Stechlin. The range of possibilities on the basis of the existing standing stock is apparently enormous. There is no single way to push the balance in favour of minor or larger algal species – of ultra-, nano-, or microplanktonic phytoplankters.

There are distinct signs of an increase in nanoplanktonic species, but Küchler (1982), in our opinion, overemphasized this trend (see p. 185). Already in the period between 1960 and 1964 nanoplanktonic diatoms played a dominant role in the community (p. 171).

Biomass

The calculation of the biovolume in the period between 1973 and 1978 is very difficult, because the data published by Küchler (1981, 1982) are partly based on inadequate and different methods – that is, with regard to the nanoplanktonic *Cyclotella* species, the ultraplanktonic and nanoplanktonic green algae and, perhaps, to the blue-green algae. There are signs that insufficient care had been given to *Ceratium hirundinella*. We have tried to calculate anew the biovolumina on the basis of the available data in Küchler's diaries, but failed in many cases due to the lack of information for exact interpretation[6].

Lake Stechlin. The average total biomass per annum was: in 1976, 12.9 $g \cdot m^{-2}$; in 1977, 3.1 $g \cdot m^{-2}$, and in 1978, 5.1 $g \cdot m^{-2}$ (Table 5.13). The maximum in 1976 was 38.6 $g \cdot m^{-2}$ (30.6.), in 1977 5.5 $g \cdot m^{-2}$ (13.9.), and in 1978 8.2 $g \cdot m^{-2}$ (23.5.); the minimum in 1976 was 4 $g \cdot m^{-2}$ (26.2.), in 1977 2.1 $g \cdot m^{-2}$ (2.8.), and in 1978 3.1 $g \cdot m^{-2}$ (5.12.).

In February 1976 blue-green algae (36%), nanoplanktonic phytoflagellates (18%), and *Cyclotella* species (19%) together with microplanktonic diatoms (18%) dominated the community. During April coccoid green algae accounted for about 28%; the other groups stood at 17% (blue-green algae) or 20% (phytoflagellates and diatoms). In mid-May an *Anabaena lemmermannii* maximum (48%) appeared, accompanied by the growth of *Dinobryon* spec. to 11%. Nanoplanktonic *Cyclotella* species attained 20% and amounted to about 35% on June 1st, when *Dinobryon* spp. stood at 25%. Then the vernal diatom – chrysomonadean maximum was reached. Whereas the *Dinobryon* spp. in mid-June still formed 48% of the total biomass, the *Cyclotella* species disappeared, and *Anabaena lemmermannii* together with *Oscillatoria* spp. developed considerably (36%). At the end of June *Anabaena lemmermannii* (71%) governed the community. *Dinobryon* spp. stood at 9%. The blue-green algal summer maximum lasted until September (20.7.: 86%, 7.9.: 57%), whereas nanoplanktonic phytoflagellates reached 23%. But in December blue-green algae anew amounted to 66% of the total biomass. In Lake Stechlin, 1976 was the year of

[6] For measurement, counting, and calculation I thank heartily my wife, Rosmarie.

blue-green algae on the basis of biomass as well as on cell number.

1977 can be called the year of nanoplanktonic phytoflagellates and green algae. Throughout the whole year their proportion in the total biomass did not sink to a level as lower as 18% (13.9.). Normally it stood at 50% and more. *Dinobryon* spp. reached 31% on June 21st, and *Cyclotella* spp. amounted to 13% on April 12th and to 48% on August 8th. In mid-September microplanktonic diatoms (*Asterionella formosa, Fragilaria crotonensis*) accounted for 26% (13.9.). Other groups of phytoplankters played no noticeable role.

In 1978, the situation altered to a certain degree. Indeed, nanoplanktonic phytoflagellates and green algae played an important role. But nanoplanktonic diatoms (*Cyclotella* spp., for instance 21.3.: 9%; 23.5.: 28%; 25.4.: 33%; 20.6.: 15%; 4.7.: 10%; 18.7.: 13%; 5.12.: 22%) held a remarkable position through great parts of the year. *Fragilaria crotonensis* and *Asterionella formosa* reached their maximum on September 26th with 60%, and the ultraplanktonic green algae amounted to 11%. *Dinobryon* spp. stood at 42% on July 4th. The most striking feature has been the low level of blue-green algal growth.

Lake Nehmitz-North. The average total biomass per annum was: in 1976, 3.6 $g \cdot m^{-2}$; in 1977, 3.5 $g \cdot m^{-2}$, and in 1978, 17.5 $g \cdot m^{-2}$ (Table 5.13). The maximum was: in 1976, 4.7 $g \cdot m^{-2}$ (21.4.); in 1977, 11.2 $g \cdot m^{-2}$ (22.6.), and in 1978, 51.6 $g \cdot m^{-2}$ (21.6.). The minimum was: in 1976, 1.7 $g \cdot m^{-2}$ (27.10.); in 1977, 1.7 $g \cdot m^{-2}$ (23.2.), and in 1978, 2.8 $g \cdot m^{-2}$ (27.9.)

In 1976, nanoplanktonic phytoflagellates (27.10.: 51%), nanoplanktonic *Cyclotella* spp. (21.4.: 32%; 21.7.: 32%), *Dinobryon* spp. (16.6.: 64%), microplanktonic diatoms (21.7.: 12%), and blue-green algae (8.9.: 30%) were the dominant members of the community.

In 1977, nanoplanktonic phytoflagellates (11.5.: 79%) and coccoid nanoplanktonic green algae (13.4.: 31%) governed the community during the greater parts of the year. Blue-green algae (14.9.: 59%), *Dinobryon* spp. (22.6.: 31%), nanoplanktonic *Cyclotella* spp. (15.3.: 14%), *Fragilaria crotonensis* (3.8.: 13%), and *Asterionella formosa* (26.10.: 11%) formed the community too.

Also in 1978 nanoplanktonic phytoflagellates and coccoid nanoplanktonic green algae were dominant members of the community, but only at the end of September did they exceed more than 50%. Surprising was the good development of *Dinobryon* spp. (24.5.: 89%; 21.6.: 56%; 19.7.: 39%). *Ceratium hirundinella* (21.6.: 24%), and *Peridinium* spec. (22.3.: 24%). A similar remarkable growth of *Peridinium* had, so far, been observed in 1963 (19.2., under ice cover; cf. p. 171) during a period when the nuclear power plant was not yet in operation. The ultraplanktonic green algae stood at a maximum of 20% (22.3.), *Fragilaria crotonensis* reached 12% (27.9.), but blue-green algae played only a subordinate role.

Some aspects of the vertical distribution of the phytoplankton in Lake Stechlin

Normally the vertical distribution of the phytoplankton community is characterized by maximum values of cell numbers and biomass in depths between 0 and 7.5 m, e.g., the epilimnetic layer of the lake is the typical phytoplankton horizon (cf. the similar distribution of the primary production activity, p. 289). Light inhibition near the lake surface during strong illumination, reduced utilization of light during low light intensity periods, different adaptation to temperature or pH-values, different settling or migration rates, different growth and mortality rates can contribute to a more complicated distribution pattern, to a distinct stratification within the phytoplankton community.

Lake Stechlin has a vertical temperature stratification every year (p. 55), with a stable summer stagnation phase from May to October. Under natural conditions, the thermocline lies in depths of between 5 and 16 m, but under the influence of the cooling water circulation in depths of between 4 and 16 m (Fig. 3.8, p. 57). During summer it lies generally in depths of between 5 and 12 m. Although there is no typical metalimnetic phytoplankton horizon, some species of the community prefer this water layer. The phenomenon that in clear water lakes an increase of cell numbers and biomass can occur in the metalimnetic water layer is well known (Nauwerck 1966; Brettum 1972). It may be the result of adaptation to medium light intensities and is well demonstrated by the phytoplankton community in Lake Stechlin. Nano- and microplanktonic diatoms (for instance *Cyclotella* spp.; Fig. 5.25), ultra- and nanoplanktonic green algae (for instance *Oocystis* spp., *Scenedesmus* spp., *Tetraedron* spp.) prefer the water layers of the epi and metalimnetic horizons (cf. Küchler 1981). But *Mallomonas fastigata,* said to be a member of the 'thermocline community', prefers the upper water layers in Lakes Stechlin and Nehmitz.

As many other filamentous blue-green algae, *Oscillatoria rubescens* (Fig. 5.26) is said to be a taxon adapted to cold water (Huber-Pestalozzi 1938). In winter under ice cover it forms red clouds in Lakes Stechlin and Nehmitz-N (p. 159). During summer it occurs in greater depths, often directly under the thermocline (Busse 1965 – for Lake Stechlin; Küchler 1981 – for Lake Nehmitz-N). Between 10 and 30 m it represents nearly the total biomass of a deepwater community accompanied by small amounts of *Tetraedron minimum, Scenedesmus costato-granulatus* and *Chroococcus limneticus.*

Long living microplanktonic diatoms such as *Fragilaria crotonensis* (Fig. 5.28) or *Tabellaria fenestrata* var. *asterionelloides* (Fig. 5.27) can be found in depths of between 0 and 20 m during their main development in late summer and early autumn.

Asterionella formosa (Fig. 5.24), also known as a taxon often living in deeper water layers (Ruttner 1929/30), prefers distinctly these horizons in Lake Stechlin (1963: 5–30 m; 1964: 5–40 m). During late autumn, winter, and early spring it is distributed evenly from the surface to the bottom. At the end of May it disappears

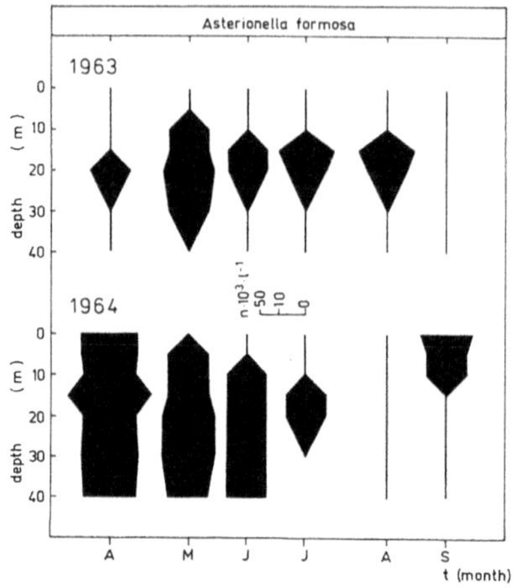

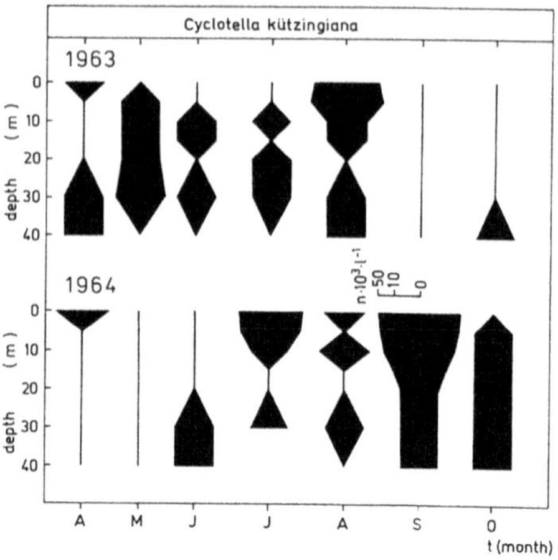

Figure 5.24 Vertical distribution of *Asterionella formosa* in Lake Stechlin, 1963–1964 (according to Busse, unpublished data, altered).

Figure 5.25 Vertical distribution of *Cyclotella kützingiana* in Lake Stechlin, 1963–1964 (according to Busse, unpublished data, altered).

from the epilimnetic layer and develops in great cell numbers in depths of between 10 and 30 (40) m.

Conclusions

The species composition of the phytoplankton community of the Lakes Stechlin and Nehmitz-N affected by the cooling water circulation after 1965 did not alter fundamentally.

Some striking phenomena, which cannot, so far, be explained in detail, should be mentioned. *Oscillatoria rubescens,* known as one of the main members of the phytoplankton community since 1926 (Krieger 1927), regularly growing in the deeper layers of the metalimnion and upper layers of the hypolimnion in great numbers during the summer stagnation period between 1960 and 1964 (see p. 164), seems to have been disappeared completely. The species has not been mentioned by Küchler (1981, 1982; but see his 'Gattung *Oscillatoria*' in his paper of 1981, p. 96). After the bringing into operation of the nuclear power plant microplanktonic diatoms such as *Tabellaria fenestrata* var. *asterionelloides* (Koschel & Mothes 1976) and *Fragilaria crotonensis* appeared in considerable numbers in all lakes flown through by the cooling water (see p. 177). *Tabellaria fenestrata* var. *asterionelloides* had not been detected before 1967, neither in Lake Stechlin nor, and this is puzzling, in Lakes Nehmitz-N or Gerlin, but it was mentioned by Krieger (1927) in connection with Lake Dagow.

The existing lists of taxa presented by Busse (1965, 1972) and Küchler (1981,

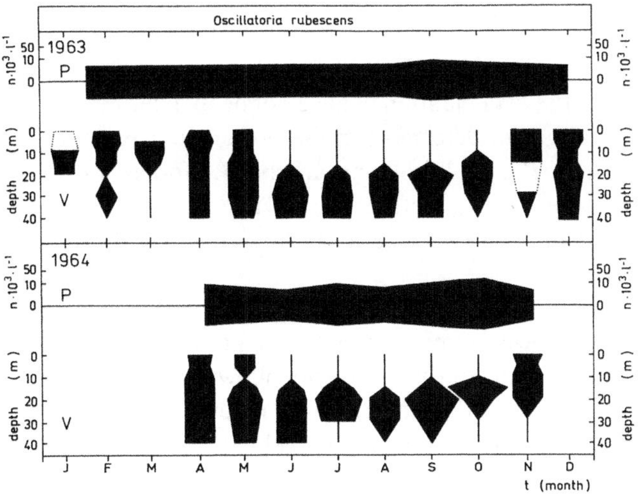

Figure 5.26 Seasonal periodicity (P) and vertical distribution (V) of *Oscillatoria rubescens* in Lake Stechlin, 1963–1964 (according to Busse, unpublished data, altered).

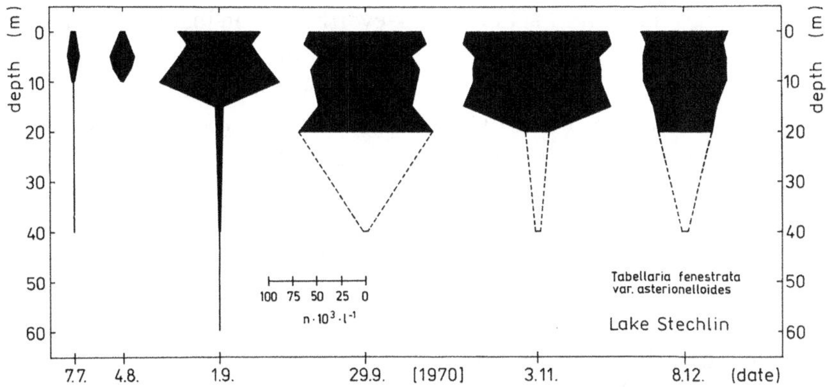

Figure 5.27 Vertical distribution of *Tabellaria fenestrata* var. *asterionelloides* in Lake Stechlin, summer to winter 1970, under the influence of the cooling water circulation (according to Busse, unpublished data, altered).

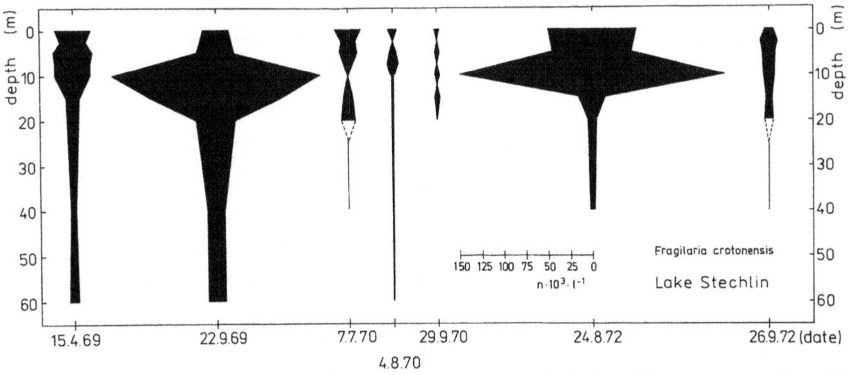

Figure 5.28 Vertical distribution of *Fragilaria crotonensis* in Lake Stechlin on various dates 1969, 1970, 1972 (according to Busse, unpublished data, altered).

1982), differing considerably from each other, cannot be compared directly because many of the taxa mentioned have not been determined beyond doubt or in other cases, although detectable in the preserved vials, have not been mentioned at all. So the data published by Koschel & Mothes (1976, Table 2, p. 221) are partly based on a misunderstanding of the data from Busse. To a certain degree the given numbers of species reflect the then actual level of knowledge of the researchers.

Several authors (Järnefelt 1956; Ramberg 1976) think oligotrophic lakes can be distinguished by a lower number of phytoplankton species than eutrophic ones. The available data of our study area do not allow any conclusions of such a kind.

Following Küchler (1982) the yearly average of phytoplankton biomass increased distinctly in Lakes Stechlin and Nehmitz-N due to the rapid growth of nanoplanktonic diatoms of the genera *Stephanodiscus* and *Cyclotella* or of coccoid green algae after the bringing into operation of the nuclear power plant. But in checking his results we came to the conclusion that the proportion of nanoplanktonic algae had been overemphasized (see p. 187) and that the absolute values of biomass lie lower (Table 5.13). Nevertheless, an increase in the annual average of phytoplankton biomass of nearly 100–200 per cent must be calculated for Lakes Stechlin and Nehmitz-N. As to the biomass, considering the given different water volumina, there are no striking differences between both the lakes which are the results of the cooling water circulation.

The phytoplanktonic periodicity corresponds well with the cooling water circulation (see Figs. 5.19, p. 175; 5.20, p. 181). When the nuclear power plant is in operation the spring development of the phytoplankters starts nearly six weeks earlier than under natural conditions (see p. 184). Every interruption of the cooling water circulation is reflected by a break of continuity of the phytoplankton development (Fig. 5.19, p. 175).

The phenomena observed concerning the phytoplanktonic community after the bringing into operation of the nuclear power plant can be summed up as follows.

The cooling water circulation influenced the phytoplankton community like a violent blow caused by a collision. The natural competition between the different algal groups, for instance between nano- and microplanktonic diatoms and other algae, was disturbed. The dominance alterated from year to year, depending upon the intensity and regularity of the cooling water circulation.

The effects of eutrophication or fertilization are reversible: The conservative standing stock of the community gradually recovers from the shock, establishing a new equilibrium on a somewhat higher trophic level, stimulating higher average annual standing crops and generating intermediate changes in community structure as a consequence.

With regard to the apparently enormous potential of reactions of the Lake Stechlin-phytoplankton community-type to environmental influences, it is unwise to be hasty in generalization and to overemphasize or oversimplify the few examples studied.

There are signs that similar annual cycles of characteristic assemblages of phytoplankton species coincide with the established trophic status of the lake (Hutchinson 1967; Reynolds 1980). Lake Stechlin is by nature an oligotrophic lake due to its small phytoplankton productivity characteristically limited by phosphorus (see p. 87). But the phytoplankton productivity as well as the known species periodicity and composition represent more than the oligotrophic status. Depending on many interactions between the ranges of fluctuations in the environment and the specific responses of the species thereto, there are 'periods' or 'phases' of 'oligotrophy', 'mesotrophy', and even 'eutrophy'. Nevertheless, we ought to be able to predict the future development of the phytoplankton community in response to possible nutrient enrichment. Now Lake Stechlin is a mature oligotrophic aquatic ecosystem. It is in the position to ward off environmental dangers such as the cooling water circulation of a nuclear power plant.

On April 24th, 1926, Krieger (1927) established the oligotrophic status of Lake Stechlin on the basis of only one net-sample. Today hundreds of net and Ruttner samples are at our disposal. Therefore, we feel encouraged to give soon a sufficient answer to the question – why and how do specific populations wax and wane?

References

Ascherson, P. 1864. Flora der Provinz Brandenburg. Berlin.

Benkert, D. 1973. Floristische Neufunde aus Brandenburg und der Altmark. 1. Folge. Gleditschia 1: 51–61.

Bernatowicz, S. & Pieczynska, G. 1965. Organic matter production of macrophytes in the Lake Taltowisko (Mazurian Lakeland). Ekol. Pol. A, 13: 113–124.

Bernatowicz, S., Pieczynska, E. & Radziej, J. 1968. The biomass of macrophytes in Lake Sniardwy. Bull. Acad. Pol. Sc. Cl. II, 16: 625–629.

Bernatowicz, S. & Radziej, J. 1964. Jährliche Makrophyten-Produktion im Mauer-See-Komplex. Pol. Arch. Hydrobiol. 12: 307–348 (in Polish).

Braune, W. 1975. Studien zur Algenbesiedlung der Saale im Raum Jena, II. Vergleich der jahreszeitlichen Besiedlungsdynamik ober- und unterhalb der Stadt. Limnologica 9: 323–419.

Brettum, P. 1972. The phytoplankton of Lake Övre Heimdalsvatn, Central South Norway, 1969–1970. Norw. J. Bot. 19: 79–90.

Busse, G. 1965. Biologische Untersuchungen im Gebiet des Stechlin-Sees. Teilprogramm 1: Floristisch-ökologische Bestandsaufnahme der Protophyten. In: Abschlußbericht . . . Jena: 1–36 (unpublished).

Busse, G. 1972. Phytoplankton. In: Grundlagen zur Modellierung des Stoffhaushaltes der Gewässer am Beispiel des Stechlin-Sees. Abschlußbericht . . . Jena: 1–44 (unpublished).

Castenholz, R.W. 1960. Seasonal changes in the attached algae of freshwater and saline lakes in the Lower Grand Coulee, Washington. Limnol. Oceanogr. 5: 1–28.

Corillion, R. 1957. Les Charophycées de France et d' Europe Occidentale. Bull. Soc. Sci. Bretagne 32, hors ser. 1: 1–259; 2: 260–499.

Dor, I. 1970. Production rate of the periphyton in Lake Tiberias as measured by the glass-slide-method. Israel Bot. 19: 1–15.

Fischer, W. 1964. Flora des Ruppiner Landes. Veröff. Bez.-Mus. Potsdam 5.
Forsberg, C. 1964. Phosphorus, a maximum factor in the growth of Characeae. Nature 201: 517–518.
Freitag, H. 1962. Flora und Vegetation des Stechlin-Sees (Vorläufige Mitteilung). Mitt. flor.-soz. Arbeitsgem. 9: 261–264.
Heitmann, M.-L., Richter, D. & Schumann, D. 1969. Der Wärme- und Wasserhaushalt des Stechlin- und Nehmitzsees. Abh. Meteorol. Dienst. DDR 96. Berlin: Akademie-Verlag.
Huber-Pestalozzi, G. 1938. Das Phytoplankton des Süßwassers. Systematik und Biologie. 1. Teil. Allgemeiner Teil. Blaualgen. Bakterien, Pilze (Die Binnengewässer 16, 1). Stuttgart: E. Schweizerbart'sche Verlagsbuchhandlung.
Hübl, E. 1966. Stoffproduktion von Phragmites communis Trin. im Schilfgürtel des Neusiedlersees im Jahre 1966 (Ergebnisse nach der Erntemethode). Anz. Österr. Akad. Wiss., Math.-Naturw. Kl. 14: 271–278.
Hueck, K. 1925. Vegetationsstudien auf brandenburgischen Hochmooren. Beitr. Naturdenkmalpflege (Berlin) 10: 311–408.
Hutchinson, G.E. 1967. A treatise on limnology. Vol. 2. Introduction to lake biology and the limnoplankton. New York: Wiley & Sons.
Hutchinson, G.E. & Löffler, H. 1956. The thermal classification of lakes. Proc. Nat. Acad. Sci. Washington 42.
Ikusima, J. 1975. Primary production of aquatic macrophytes. In: Productivity of freshwater communities in Lake Biwa. Contrib. Otsu Hydrobiol. Station, University of Kyoto. Nr. 251: 28–31.
Järnefelt, H. 1956. Zur Limnologie einiger Gewässer Finnlands. XVI. Mit besonderer Berücksichtigung des Planktons. Ann. Zool. soc. 'Vanamo' 17: 1–201.
Klarer, D.M. & Hickman, M. 1975. The effect of the thermal effluent upon the standing crop of an epiphytic algal community. Int. Rev. ges. Hydrobiol. 60: 17–62.
Kloss, K. 1980. Pollenanalysen zur Vegetations-, Siedlungs- und Moorgeschichte am Südrand der ostmecklenburgisch-brandenburgischen Seenplatte (Kreis Gransee). Arch. Naturschutz Landschaftsforsch. 20: 203–212.
Koschel, R. & Mothes, G. 1976. Änderung der biologischen Produktivität in einem Seensystem durch den Kühlwasserzufluß eines Kernkraftwerkes. In: Eutrosym '76. 20–25.IX.1976 Karl-Marx-Stadt DDR. Materialien Int. Sympos. Eutroph. Sanierung Oberflächengewässern 3: 219–233.
Krausch, H.-D. 1962a. Nachrichten über Abfluß und Seespiegelveränderungen des Stechlinsee-Gebietes. Auf Grund archivalischer Quellen zusammengestellt. Limnologica 1: 211–215.
Krausch, H.-D. 1962b. Die Menzer Heide. Jb. brandenburg. Landesgesch. 13: 96–118.
Krausch, H.-D. 1964. Die Pflanzengesellschaften des Stechlinsee-Gebietes. I. Die Gesellschaften des offenen Wassers. Limnologica 2: 145–203. II. Röhrichte und Großseggengesellschaften, Phragmitetea Tx. et Prsg. 1942. Ibidem 4: 423–483.
Krausch, H.-D. 1967. Die Pflanzengesellschaften des Stechlinsee-Gebietes. III. Grünlandgesellschaften und Sandtrockenrasen. Limnologica 5: 331–366.
Krausch, H.-D. 1968a. Die Pflanzengesellschaften des Stechlinsee-Gebietes. IV. Die Moore. Limnologica 6: 321–380.
Krausch, H.-D. 1968b. Die natürliche Umwelt in Fontanes 'Stechlin'. Dichtung und Wirklichkeit. Fontane-Blätter 1: 342–353.
Krausch, H.-D. 1970. Die Pflanzengesellschaften des Stechlinsee-Gebietes. V. Wälder, Hecken und Saumgesellschaften. Limnologica 7: 357–454.
Krausch, H.-D. & Zühlke, D. 1974. Das Rheinsberg-Fürstenberger Seengebiet. Werte unserer Heimat 25, Berlin: Akademie-Verlag.
Krause, W. 1981. Characeen als Bioindikatoren für den Gewässerzustand. Limnologica 13: 399–418.
Kreeb, K.-H. 1983. Vegetationskunde. Methoden und Vegetationsformen unter Berücksichtigung ökosystemischer Aspekte. Stuttgart: 331 S.
Krieger, W. 1927. Die Gattung *Centronella* Voigt, Ber. Deutsch. Bot. Ges. 45: 281–290.
Küchler, L. 1981. Phytoplanktonuntersuchungen im Stechlinseegebiet in den Jahren 1973–1975. Limnologica 13: 83–99.

Küchler, L. 1982. Phytoplanktonuntersuchungen im Stechlin und im Nordbecken des Nehmitzsees. Limnologica 14: 231–242.
Müller-Stoll, W.R. & Krausch, H.-D. 1959. Verbreitungskarten brandenburgischer Leitpflanzen, 2. Reihe. Wiss. Z. Päd. Hochschule Potsdam, Math.-Naturwiss. R. 4: 105–150.
Nauwerck, A. 1966. Beobachtungen über das Phytoplankton klarer Hochgebirgsseen. Schweizer Z. Hydrol. 28: 4–28.
Newcombe, C.L. 1949. Attachment materials in relation to water productivity. Trans. Americ. Microscop. Soc. 68: 355–361.
Newcombe, C.L. 1950. A quantitative study of attachment materials in Sodon Lake, Michigan. Ecology 31: 204–215.
Overbeck, J. 1962. Das Nannoplankton (μ-Algen) der Rügenschen Brackgewässer als Hauptproduzent in Abhängigkeit vom Salzgehalt. Kieler Meeresforsch. 18: 157–171.
Ramberg, L. 1976. Relations between phytoplankton and environment in two Swedish forest lakes. Sc. limnol. Upsala. 426.
Raspopov, I.M. 1972. Zur Methodik der Bestimmung der Jahresproduktion der Makrophyten in den Seen der nordwestlichen UdSSR. Verh. Int. Verein. Limnol. 18: 171–175.
Retzdorff, W. 1907. Über das Vorkommen von *Aldrovandia vesiculosa* L. in der Mark Brandenburg. Verh. bot. Vereins Prov. Brandenb. 48: XXIII-XXVII.
Reynolds, C.S. 1980. Phytoplankton assemblages and their periodicity in stratifying lake systems. Holarct. Ecology 3: 141–159.
Richter, D., Neubert, W. & Klämt, A. 1979. Temperatur und Wärmehaushalt des thermisch belasteten Stechlin- und Nehmitzsees. Abh. Meteorol. Dienst. DDR Nr. 123 (16). Berlin: Akademie-Verlag.
Ruttner, F. 1929/30. Das Plankton des Lunzer Untersees; seine Verteilung in Raum und Zeit während der Jahre 1908–1913. Int. Revue ges. Hydrobiol. Hydrogr. 23: 1–138, 161–287.
Sarcander, J.F. 1865. Naturgeschichtliches Tagebuch aus Fürstenberg. Archiv Verein Naturgesch. Mecklenburgs 19: 15–49.
Scheffler, W. 1981. Quantitative Erfassung des Periphytons im Stechlinsee. Limnologica 13: 101–122.
Schmidt. D. 1981. Die Characeen – eine im Aussterben begriffene Pflanzengruppe unserer Gewässer. Gleditschia 8: 141–157.
Schwoerbel, J. 1966. Methoden der Hydrobiologie. Stuttgart: Kosmos. Franckh'sche Verlagshandlung. 207.
Sládeček, V. & Sládečková, A. 1964. Determination of the periphyton production by means of the glass slide method. Hydrobiologia 23: 125–158.
Szcepanski, A. 1969. Biomass of underground parts of the reed *Phragmites communis* Trin. Bull. Acad. Pol. Sci., Cl. II, Ser. Sci. Biol. 17: 245–246.
Sukopp, H. & Markstein, B. 1981. Veränderungen von Röhrichtbeständen und -pflanzen als Indikatoren von Gewässernutzungen, dargestellt am Beispiel der Havel in Berlin (West). Limnologica 13: 459–471.
Tscheu-Schlüter, M. 1969. Vergleichende quantitative Periphytonuntersuchungen im Litoral verschiedener Seentypen. Wiss. Z. Univ. Rostock, Math.-Naturwiss. R. 18: 783–791.
Ulbrich, E. 1929. Bericht über die Frühjahrshauptversammlung am 2. und 3. Juni 1928 in Fürstenberg i. M. und Neu-Globsow. Verh. bot. Ver. Prov. Brandenburg 71: 74–80.
Westlake, D.F. 1965. Some basis data for investigations of the productivity of aquatic macrophytes. Mem. Ist. Ital. Idrobiol. 18 (suppl.): 229–248.
Winter, H. 1870. Flora der Umgegend von Menz. Verh. bot. Ver. Prov. Brandenburg 12: 1–43.

6. Bacteria

H.-D. Babenzien & Christina Babenzien

Conceptional approaches

In recent years it has become evident that heterotrophic bacteria play a decisive role in nutrient cycling and in energy flow through the food web (Overbeck 1974; Schlegel & Jannasch 1981). The microbial mineralization processes of allochthonous and autochthonous organic matter provide the phytoplankton with nutrients that may be available in limited amounts. These metabolic activities influence substantially the physicochemical environment in lakes, e.g., oxygen consumption and hydrogen sulphide formation. The process of transformation of dissolved organic matter into bacterial biomass is one of the basic metabolic turnover patterns in a lake system. The biomass production by heterotrophic bacteria can reach the same level as the autotrophic production (Kusnetzow & Romanenko 1966; Overbeck 1972). The incidence of bacteria (number, biomass) in lakes is mostly affected by the type and amount of available energy and carbon sources.

In natural unpolluted lakes dissolved organic matter (DOM) enters the aquatic ecosystem in three ways: exudation (release of phytoplankton and macrophytes during the course of photosynthesis), autolysis (loss of DOM from decaying planktonic organisms), and excretion (loss of DOM from ingested food by zooplankton; Lampert 1978). Depending on the catchment area of the lake allochthonous organic matter, especially litter fall and atmospheric precipitation, may be quantitatively and qualitatively very significant inputs (Ohle 1982; Saunders 1980).

The pelagic zone of oligotrophic lakes as a biotope for microorganisms is characterized by spatial unhomogeneous and seasonal discontinuous conditions. In such low-nutrient environments microbial growth must be assumed to alternate with prolonged periods of starvation or exogenous dormancy (Jannasch 1979; Stevenson 1978).

Dissolved organic carbon (DOC) concentrations in lakes range generally from 1–15 mg $C \cdot l^{-1}$, but only a fraction of this amount is available to the bacteria. Only

Casper, S. Jost (ed.), Lake Stechlin. A temperate oligotrophic lake. ISBN 978-94-010-8930-2.

about 20% are low molecular weight compounds (sugars, amino acids, fatty acids) with high turnover rates (Hama & Handa 1980), and 80% have been called 'secondary substances' (Overbeck 1983) or 'humic substances'. Measured concentrations of glucose are in the order of $10\,\mu g \cdot l^{-1}$ (Stabel 1977; Gocke *et al.* 1981). However, only a small portion of the sugars and amino acids chemically determined are, in fact, utilizable in heterotrophic processes, whereas the other fraction is 'masked' (e.g. associations with polyvalent metal ions; Dawson & Gocke 1978).

As reported by Jannasch (1979), two various 'survival strategies' of procaryotes at low nutrient level environments emerged. Following this hypothesis, the survival of 'zymogenic' microorganisms (Jannasch 1955) is achieved by the occasional mass production of cells with the capacity to become dormant. In terms of population dynamics they are the so called μ_{max}-strategists with low substrate affinity. They are successful in substrate competing only in the presence of high levels of nutrients. In literature they are described as 'eutrophs' or 'saprophytes' (Kuznetsov *et al.* 1979), 'heterotrophs' (Akagi *et al.* 1977), and 'copiotrophic bacteria' (Poindexter 1981a). In contrast to the 'strategy for surviving a feast and famine existence', oligotrophic bacteria 'are conceived of as those never invited to a feast; their properties should include microbial adaptions to uninterrupted nutrient limitation' (Poindexter 1981b). These types are also designated as 'autochthonous' forms, K-strategists (Jannasch 1979), 'oligocarbophile' (Moaledj 1978), and 'oligotrophs' (Ishida & Kadota 1979; Kuznetsov *et al.* 1979). Such organisms need to possess uptake systems of relatively high affinity for substrates, large surface to volume ratios (e.g. prosthecate bacteria). Some organisms have evolved mechanisms that favour growth at interfaces. The accumulation of nutrients at solid-liquid and air-liquid interfaces enables microbial films with high bacterial density and metabolic activity (neuston-formation; Babenzien 1966; Marshall 1976). Most bacteria are able to grow over a relatively wide range of nutrient concentration with the help of physiological adaption mechanisms, but there are also indications of the existence of obligate oligotrophs. Such oligotrophic microorganisms are defined as organisms that grow in a medium containing 1 mg C per litre of organic matter (Ishida & Kadota 1981).

The conception of changing cooperation of both types is a useful heuristic model for a better understanding of mineralization processes and the fact that number, biomass production (Meyer-Reil 1977), and activity of a microflora do not run parallel in any case. As has been shown by autoradiographic studies (Hoppe 1978), estimations of the proportion of respiring bacteria (Zimmermann *et al.* 1981), and the total number (Meyer-Reil 1978), only a part of 10–50% on average are actively metabolizing bacteria in natural environments.

Very instructive measurements for this metabolic activity are given in terms of heterotrophic potential (Wright & Hobbie 1966) and the heterotrophic dark fixation of CO_2 (Overbeck 1981). These relevant methods were extensively used in our Lake Stechlin investigations and are described below.

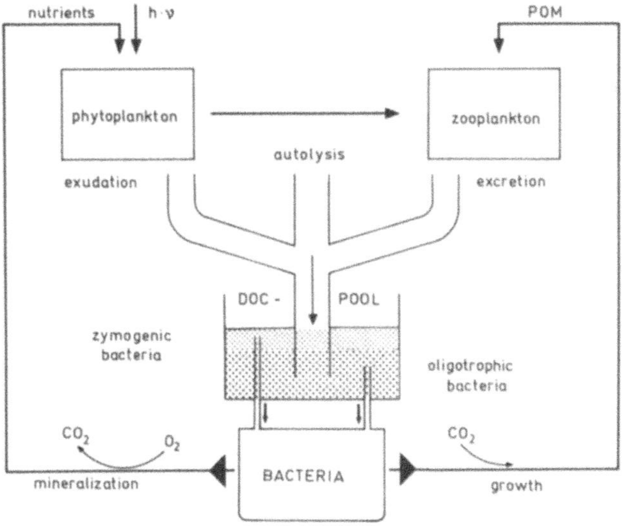

Figure 6.1 Simplified graphical model showing the dynamic equilibrium between release and uptake of dissolved organic substances and the role of bacteria in recycling process (based on Gocke 1978).

The basic relationships between the main compartments in this view are summarized in a simplified graphical model (Fig. 6.1). The detritus food chain which plays an important role in the whole ecosystem (Fenchel & Joergensen 1977) was not taken into account in this sketch.

Methods

For the purpose of this investigations we define bacteria to be oligotrophic when they form colonies (colony-forming units, $cfu \cdot ml^{-1}$) in a poor peptone agar (Wolters & Schwartz 1956) in an incubation period of 10 days at 25° C. Zymogenic bacteria were counted and isolated with standard nutrient medium (Babenzien 1964). Serial dilutions were performed with sterilized lake water. Incubation temperature of 10° C resulted only up to 60% and 37° C only up to 5% of the number at 25° C. When the counting of the colonies was carried out after 48 h (standard time in water-hygienic investigations), we got very low numbers (<10%). This difference is noticeable in view of the significance of the estimated heterotrophic bacterial numbers. The total number of bacteria were counted on membrane filters stained with erythrosine (Sorokin & Kadota 1972). Water samples were collected using the Sorokin-bottle-sampler (Sorokin & Kadota 1972).

Distribution pattern of heterotrophic bacteria

The growth and multiplication of bacteria on the one hand and dormant stages on the other are responses of microbial populations to fluctuating environmental conditions. Bacteria react more quickly to changes in their environment than other organisms. The number of bacteria in a given habitat, therefore, is a very sensitive parameter and reflects the content of easily degradable organic substances in the water. Investigations of seasonal and spatial bacterial distribution in the holomictic stratified lake, therefore, can be considered to be structural analysis of the ecosystem (Overbeck 1974).

Some principle parameters characterizing the Lake Stechlin System are presented in Table 6.1. Considering the investigations into the vertical and seasonal distribution of bacteria, which have been carried out for 20 years, first of all we can state that in Lake Stechlin fluctuations take place at a very low level. Low production of organic substances enables aerobic mineralization also in the hypolimnion; and distinct gradients in frequency of physiological groups do not occur. Table 6.2 shows the annual mean values of bacterial number from Lake Stechlin. Starting with very low numbers (120 cfu $\cdot$ ml^{-1} oligocarbophilic respectively 50 cfu $\cdot$ ml^{-1} zymogenic bacteria 1961) a slight increase has occurred since 1965. This tendency is assumed to be caused by the connection between Lake Stechlin and Lake Nehmitz in 1965. In this year the exchange of water for cooling purposes in the nuclear power plant began, first of all without heat.

Since that time the values have fluctuated at higher levels. At the present a number of 500 cfu $\cdot$ ml^{-1} (oligocarbophilic bacteria) is representative. The increase of the proportion of copiotrophic bacteria, in present time about 200 cfu $\cdot$ ml^{-1}, is remarkable. In principle, these parameters in the epilimnetic region of Lake Nehmitz are very similar. Generally, vertical distribution patterns are in the same order of horizontal patchiness. Colony counts simultaneously determined at various measuring sites of the Lake Stechlin System are presented in Table 6.3. Higher numbers obtained in the shallow east bay (near the beach) are obvious.

Table 6.1 Characteristic parameters of the Lake Stechlin System with microbiological relevance (spring overturn).

Lake	Primary production ($g\ C \cdot m^{-2} \cdot d^{-1}$)	DOC ($mg\ C \cdot l^{-1}$)	PO_4-P ($\mu g\ P \cdot l^{-1}$)	Tot-P ($\mu g\ P \cdot l^{-1}$)	NO_3-N ($\mu g \cdot l^{-1}$)	SO_4 ($mg \cdot l^{-1}$)	O_2
Stechlin	0.260	2– 6	2.5	10	100	20	Saturation
Nehmitz-N	0.383	3–10	2.0	10	100	35	Saturation
Dagow	0.794	4–12	100.0	140	300	80	Saturation

In Fig. 6.2 we demonstrate the situation during the summer stagnation period. Curves of temperature show the typical shapes of stratified lakes and the striking hypolimnetic maxima of oxygen. But no distinct maxima of bacterial numbers were obtained. However, in recent years we have observed in the epilimnion a tendency to higher differences.

Corresponding investigations from Lake Nehmitz yielded similar results. Of course, the results in the anaerobic hypolimnion are not comparable with respect to aerobic bacteria.

The metabolic coupling between bacteria and phytoplankton should result in coincident phases or temporal shifted successions of populations (Overbeck & Babenzien 1964). However, many references indicate that this coupling is not a

Table 6.2 Mean values of bacterial numbers (cfu · ml^{-1}, 10 d, 25° C) in Lake Stechlin, central basin, 1961–1983.

Year	Peptone-agar	Nutrient-agar
1961	120	50
1962	80	45
1964	60	70
1965	560	290
1966	270	60
1967	190	25
1968	180	30
1969	250	–
1976	650	10
1977	360	20
1978	230	10
1980	450	130
1981	1100	230
1982	300	100
1983	430	250

Table 6.3 Horizontal distribution patterns of colony counts of bacteria (cfu · ml^{-1}, 10 d, 25° C, peptone agar, samples taken from 2 m water depth) in the Lake Stechlin System.

Date	Lake Stechlin				Lake Nehmitz-N	Lake Gerlin
	Central basin	West basin	East basin	South basin		
Jun 1962	135	450	1890	240	–	–
Jul 1964	100	160	230	190	880	1750
Jun 1966	1300	1200	2600	–	55	1500
Sep 1966	65	130	510	110	100	250

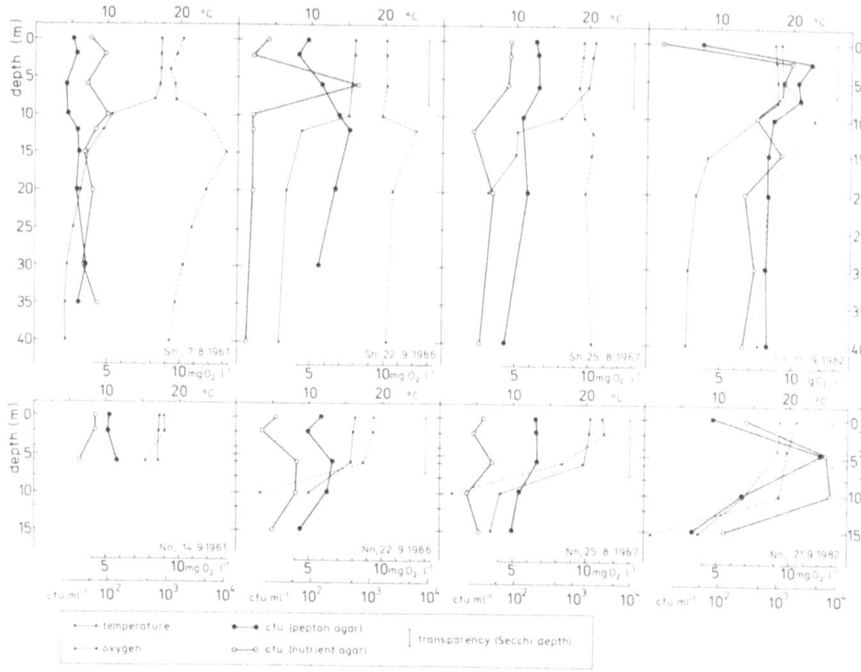

Figure 6.2 Vertical profiles of some parameters in Lakes Stechlin and Nehmitz-N at the time of stagnation.

simple one (Overbeck 1974). In our oligotrophic lake with low productivity, relations between primary production and distribution patterns of bacterial numbers are not apparent.

Haubold (1982) analysed in detail the seasonal and spatial distribution pattern of bacterial number and heterotrophic potential for a period of three years in Lake Stechlin. Some of his findings are shown in Fig. 6.3.

Normally, maximal numbers occurred in the late spring; however, there was no evidence of a distinct rhythm.

Moreover, the fluctuations obtained in the various years were quite different.

Summarizing the results mentioned above, in the pelagic zone of Lake Stechlin with respect to bacterial numbers and their distribution pattern, no distinct compartments were obvious.

Bacterial biomass

Nowadays it has become evident that not only do bacteria play an important role in nutrient cycling but they are also basic members of the food web, e.g., for filter feeding organisms like protozoa and rotifers. Within the group of secondary

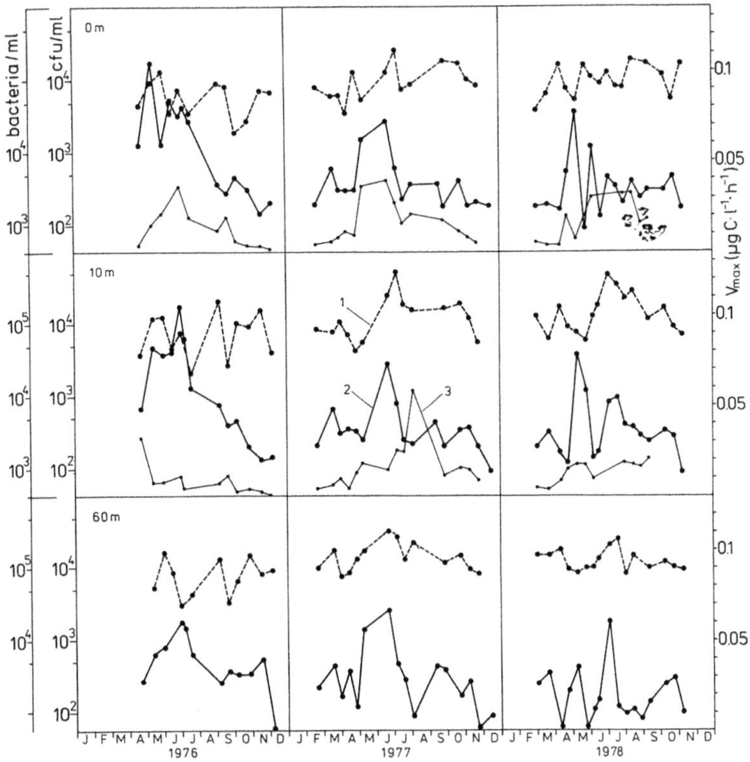

Figure 6.3 Seasonal and spatial distribution pattern of bacterial numbers and heterotrophic potential, Lake Stechlin. 1) direct counts (membrane filters, erythrosine); 2) plate counts (peptone agar); 3) heterotrophic potential (V_{max}).

producers – bacteria and zooplankton – only the former are able to transform dissolved organic substances into particulate form. By this means a part of the energy will be available in the food web once more. Improvements in determining bacterial biomass and reproduction rates led to a better understanding of these relationships.

Bacterial standing crop

Biomass values for various types of waters are summarized in Table 6.4. It seems to be easy to calculate the biomass by the values of bacterial counts, biovolume, and the factor of carbon to biomass ratio. But recent investigations using the SEM have shown that variations of bacterial size up to a factor of 1000 are possible within a water (Rheinheimer 1981). It has been shown that a distinct ratio of the small cells ('dwarfs') has not been registered yet by the membrane filter method.

The span is demonstrated by the following data for the cell volume:

Kusnetzow & Romanenko (1966)	0.5 μm^3
Straškrabová & Komarková (1979)	0.2 μm^3
Saunders (1980)	0.2–0.4 μm^3
Rheinheimer (1981)	0.005–5.0 μm^3
Krambeck *et al.* (1981)	0.05–0.06 μm^3
Riemann *et al.* (1982)	0.042 μm^3
Pedros-Alio & Brock (1982)	0.159 μm^3
Bell *et al.* (1983)	0.186 μm^3

It is difficult to determine a conversion factor for carbon to biomass, as can be seen by the following published values:

Kusnetzow & Romanenko (1966)	$7.5 \cdot 10^{-11}$ mg C $\cdot \mu m^{-3}$
Watson *et al.* (1977)	$1.21 \cdot 10^{-10}$ mg C $\cdot \mu m^{-3}$
Saunders (1980)	$1.0 \cdot 10^{-9}$ mg C $\cdot$ m^{-3}
Van Es & Meyer-Reil (1982)	1.6–0.8 $\cdot 10^{-9}$ mg C $\cdot \mu m^{-3}$

To get an impression of the bacterial biomass in Lake Stechlin we calculated the biomass on the basis of the total bacterial counts, and the assumption of a mean cell volume of 0.1 μm^3 and a carbon to biomass ratio of $1.21 \cdot 10^{-10}$ mg $\cdot$ C μm^{-3}.

Table 6.4 Bacterial biomass values for different types of waters.

Water	Biomass (mg C $\cdot$ m^{-3})	Source
Kličava Reservoir	6.4–37.4	Straškrabová & Komarková (1979)
Slapy Reservoir	8.6–10.6	Straškrabová & Komarková (1979)
Lake Frederiksborg Slotssø	50–91	Riemann *et al.* (1982)
Lake Pääjärvi	10–40	Salonen (1981)
Coastal water	5.5 ± 1.9	Ferguson & Rublee (1976)
Lake Biwa	16.6–21.3	Kadota & Tanaka (1975)

The results for 1977 and 1978 are presented in Table 6.5. The highest values were obtained for the metalimnion and for the zone near the sediment. Values between 0.4 and 10 mg C $\cdot$ m^{-3} show the very low level of the bacterial biomass in comparison to the literature data. On average the ratio of bacterial biomass to the phytoplankton biomass amounts to 8 per cent.

On this basis we also calculated the bacterial standing crop in the euphotic zone and for the whole lake (Table 6.6).

Biomass production

Arising from the previously described bacterial biomass data we tried to calculate the biomass production. Taking a mean generation time of 24 h (Saunders 1980; Rheinheimer 1981; Gorlenko *et al.* 1983) we obtained a biomass production rate of 10 g C $\cdot$ m$^{-2} \cdot$ a^{-1} (up to 17.5 m) and 35 g C $\cdot$ m$^{-2} \cdot$ a^{-1} (up to 68 m). That means for the whole lake 10–15% of the primary production.

Table 6.5 Bacterial and phytoplankton biomass, Lake Stechlin, 1977 and 1978.

Date	Bacterial biomass					Phytoplankton biomass
	(mg C · m^{-3})			(mg C · m^{-2})		(mg C · m^{-2})
	0 m	15 m	60 m	0–17.5 m	0–68 m	0–17.5 m
1977						
Feb	0.94	0.77	1.22	15	58	314
Mar	0.73	0.79	2.08	17	62	250
Mar	0.73	1.54	0.94	21	57	–
Apr	0.44	0.77	1.03	13	48	280
Apr	1.54	1.13	1.57	19	55	–
May	0.64	1.40	2.15	17	84	–
Jun	1.63	5.12	3.88	46	167	390
Jul	3.12	9.70	3.27	92	237	–
Jul	0.90	5.28	1.63	63	162	–
Aug	1.05	2.40	2.78	28	106	210
Sep	2.16	1.54	1.45	30	80	550
Oct	2.10	1.29	1.85	35	87	230
Nov	1.25	1.19	1.21	24	82	270
Nov	1.00	0.83	0.99	16	58	–
Dec	–	–	–	–	–	260
$\bar{x}$	1.3	2.4	1.9	31	96	
1978						
Mar	0.45	1.39	1.90	22	91	–
Mar	0.77	0.62	1.83	14	71	330
Apr	1.98	1.67	2.10	33	94	–
Apr	0.90	1.05	1.20	19	58	450
May	0.65	1.05	1.08	16	67	–
May	1.91	1.73	1.23	24	73	820
Jun	1.36	2.59	1.25	29	101	–
Jun	1.10	4.34	1.66	35	122	330
Jul	1.56	7.76	2.63	69	184	570
Jul	0.98	6.66	3.07	52	170	740
Aug	0.96	4.49	0.97	35	86	–
Aug	2.26	4.14	1.90	57	146	–
Sep	2.06	2.31	1.26	34	106	620
Oct	1.46	1.83	1.51	32	104	–
Oct	0.68	0.92	1.26	21	89	–
Nov	1.15	0.82	1.15	20	66	–
Dec	–	–	–	–	–	310
$\bar{x}$	1.3	2.7	1.6	32	102	

Table 6.6 Mean values of bacterial biomass and biomass production (based on 24 h generation time), Lake Stechlin 1977/78.

Bacterial standing crop	0–17.5 m	0–68 m
mg $C \cdot m^{-2}$	30	100
t C (whole lake)	0.1	0.14
Bacterial production	**0–17.5 m**	**0–68 m**
g $C \cdot m^{-2} \cdot a^{-1}$	10	35
t $C \cdot a^{-1}$ (whole lake)	35	50

A further widely used but also criticized method (Overbeck 1979; Joint & Morris 1982) for determining biomass production was developed by Kusnetzow & Romanenko (1966). This method based on the CO_2-dark assimilation (see p. 348). On the assumption that in normal bacterial growth 6% of the cell carbon is caused by CO_2-fixation the biomass production can be calculated. Because the bacterial production does not only take place in the euphotic zone (and also allochthonous material can serve as carbon source) production rates are possible that are comparable to or higher than primary production.

From our calculations with the 6% method follow biomass production rates, which are demonstrated in Table 6.7. Production rates of 50 g $C \cdot m^{-2} \cdot a^{-1}$ for the euphotic zone and 130 g $C \cdot m^{-2} \cdot a^{-1}$ up to 60 m were obtained. Values of 170 t

Table 6.7 Bacterial biomass production (6% method) and primary production, Lake Stechlin.

1982	Primary production (mg $C \cdot m^{-2} \cdot d^{-1}$)	Bacterial biomass production (mg $C \cdot m^{-2} \cdot d^{-1}$)	
	0–15 m	0–15 m	0–68 m
Jan	–	33	116
Mar	330	28	101
Apr	715	115	349
May	410	256	568
Jun	370	133	317
Jul	490	404	939
Aug	495	150	400
Sept	230	148	440
Oct	195	150	250
Nov	–	167	417
Dec	85	150	433
g $C \cdot m^{-2} \cdot a^{-1}$	110	50	130
t $C \cdot a^{-1}$ (whole lake)	360	170	185

$C \cdot a^{-1}$ (0–15 m) and 185 t $C \cdot a^{-1}$ (0–68 m) for the whole lake mean that the bacterial production may be about 50% of the primary production. The values resulting from this estimate are higher than the calculations with the generation time.

Jordan & Likens (1980) compared different methods for the determination of biomass production for the oligotrophic Lake Mirror. From the total carbon budget, the budget of dissolved organic carbon, the total annual losses of planktonic bacteria, and from the measurement of $35S\text{-}S0_4$ uptake by bacteria, they obtained bacterial production rates ranging between 3 and 8 g $C \cdot m^{-2} \cdot a^{-1}$, i.e., 8–22% of the primary production. Using the 6% method, the measured bacterial production was significantly higher (20 g $C \cdot m^{-2} \cdot a^{-1}$).

Our results were similar to those which have been published by Jordan & Likens (1980). Irrespective of which method will best reflect reality, the magnitude of the bacterial production in Lake Stechlin between 10 and 50% of the primary production obviously shows the importance of bacterial biomass in the food web.

Taxonomical and physiological features of bacterial populations

Most studies on the qualitative composition of aquatic microflora have centered on the copiotrophic group of bacteria (Straškrabová 1974; Holder-Franklin 1981). However, in recent years considerable interest has arisen concerning the role of oligotrophs whose growth is favoured by a low-nutrient media (Moaledj 1978; Witzel *et al.* 1982a).

The characterization of bacterial populations from natural habitats by using numerical methods, as described by Witzel *et al.* (1981, 1982b) and Rosswall & Persson (1982), seems to confirm the assumption of the changing ecological importance of these groups.

Our taxonomical analysis was carried out with a set of 150 strains, isolated from peptone-agar and nutrient-agar plates and examined by conventional techniques. Performance of physiological tests was described earlier (Babenzien 1964).

A high proportion of bacteria forming pigmented colonies are noticeable. Normally, yellow-orange coloured strains dominate, but there are also short-term situations with 50% blue-coloured colonies (*Chromobacterium violaceum*, Lake Große Fuchskuhle) indicating significant variations in the composition of the microbial population. Following Wolters & Schwartz (1956), a high proportion of pigmented bacteria is assumed to be typical in low-nutrient biotops. In fact, in sediment samples their number is unimportant.

Another general feature of the aquatic microflora is the high ratio of gramnegative rods (60%).

The strains isolated were classified according to Bergey's Manual (Breed *et al.* 1957; Buchanan & Gibbons 1974). The following genera were identified:

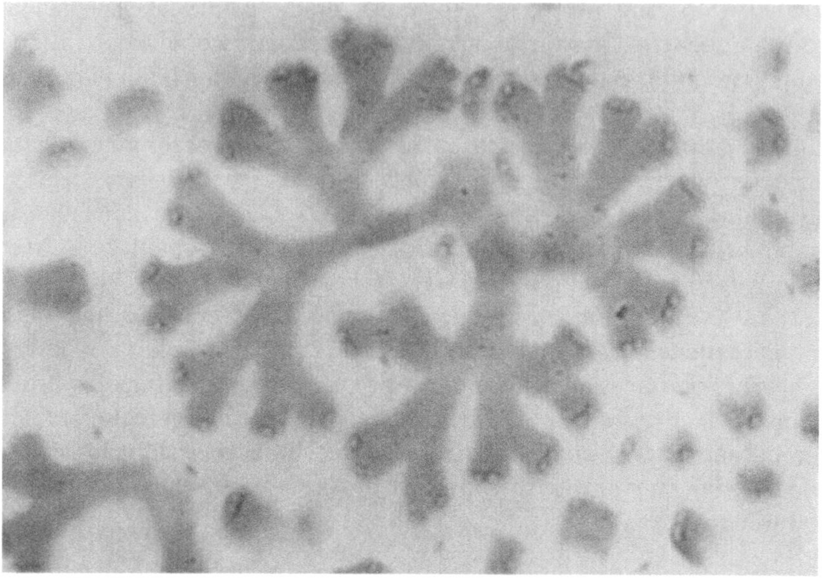

Figure 6.4 *Nevskia ramosa,* microcolony stained with toluidine blue.

Pseudomonas, Achromobacter, Micrococcus, Flavobacterium, Aerobacter, Alcaligenes, Proteus, Caulobacter, Chromobacterium, Brevibacterium, Bacillus, Protaminobacter.

In special enrichment cultures we identified the genera: *Hyphomicrobium, Beneckea, Cellvibrio, Desulfovibrio,* and *Nevskia* (Fig. 6.4). Analysing the methane cycling (Heyer & Babenzien 1984), we found the following strains to be common: *Methylocystis, Methylosinus* and *Methylomonas.*

Table 6.8 Proportion of pigmented bacteria of the total number (cfu $\cdot$ ml^{-1}) in Lake Stechlin 1961/62.

Number of colonies detected	Colonies pigmented (%)				
	Yellow	Orange	Pink	Blue	Green
2859	16.6	2.6	27.0	0.07	0.28

References

Akagi, Y., Taga, N. & Simidu, U. 1977. Isolation and distribution of oligotrophic marine bacteria. Can. J. Microbiol. 23: 981–987.

Babenzien, H.-D. 1964. Hydrobakteriologische Untersuchungen in Stechlinsee. Limnologica 2: 9–34.

Babenzien, H.-D. 1966. Untersuchungen zur Mikrobiologie des Neustons. Verh. Int. Verein. Limnol. 16: 1503–1511.

Bell, R.T., Ahlgreen, G.M. & Ahlgreen, I. 1983. Estimating bacterioplankton production by measuring 3H-thymidine incorporation in a autrophic Swedish lake. Appl. Environ. Microbiol. 45: 1709–1721.

Breed, R.S., Murray, E.G.D. & Smith, N.R. 1957. Bergey's manual of determinative bacteriology. Baltimore: Williams and Wilkins, 7. ed., 1094.

Buchanan, R.E. & Gibbons, N.E. 1974. Bergey's manual of determinative bacteriology. Baltimore: Williams and Wilkins, 8. ed., 1246.

Dawson, R. & Gocke, K. 1978. Heterotrophic activity in comparison to the free amino acid concentrations in Baltic Sea water samples. Oceanol. Acta 1: 45–54.

Fenchel, T.M. & Joergensen, B.B. 1977. Detritus food chains of aquatic ecosystems: the role of bacteria. Adv. Microb. Ecol. 1: 1–58.

Ferguson, R.L. & Rublee, P. 1976. Contribution of bacteria to standing crop of coastal plankton. Limnol. Oceanogr. 21: 141–145.

Gocke, K. 1978. Untersuchungen über den Jahreszyklus der heterotrophen Aktivität in der Kieler Förde. Arch. Hydrobiol. 82: 123–141.

Gocke, K., Dawson, R. & Liebezeit, G. 1981. Availability of dissolved free glucose to heterotrophic microorganisms. Mar. Biol. 62: 209–216.

Gorlenko, V.M., Dubinina, G.H. & Kuznetsov, S.I. 1983. The ecology of aquatic micro-organisms. (Die Binnengewässer 28). Stuttgart: E. Schweizerbarth'sche Verlagsbuchhandlung. 252.

Hama, T. & Handa, N. 1980. Molecular weight distribution and characterization of dissolved organic matter from lake waters. Arch. Hydrobiol. 90: 106–120.

Haubold, G. 1982. Untersuchungen zur Erfassung der bakteriellen Aktivität in der euphotischen Zone des Stechlinseesystems. Dresden: Thesis. 186 + XX.

Heyer, J. & Babenzien, H.-D. 1985. Untersuchungen des Methankreislaufes in einem oligotrophen See (Stechlinsee). Limnologica 16: 267–276.

Holder-Franklin, M.A. 1981. Methods of studying population shifts in aquatic bacteria in response to environmental change. Sci. Ser. 124: 1–107. Ottawa, Canada.

Hoppe, H.-G. 1978. Relations between active bacteria and heterotrophic potential in the sea. Netherlands J. Sea Res. 12: 78–98.

Ishida, Y. & Kadota, H. 1979. A new method for enumeration of oligotrophic bacteria in lake water. Arch. Hydrobiol. Beih. Ergebn. Limnol. 12: 77–85.

Ishida, Y. & Kadota, H. 1981. Growth patterns and substrate requirements of naturally occurring obligate oligotrophs. Microb. Ecol. 7: 123–130.

Jannasch, H.W. 1955. Zur Ökologie der zymogenen planktischen Bakterienflora natürlicher Gewässer. Arch. Mikrobiol. 23: 146–180.

Jannasch, H.W. 1979. Microbial ecology of aquatic low nutrient habitats. In: Shilo, M. (ed.), 1979. Strategies of microbial life in extreme environments. Berlin: Verlag Chemie. Dahlem Konferenzen: 243–260.

Joint, I.R. & Morris, R.J. 1982. The role of bacteria in the turnover of organic matter in the sea. In: Barnes, H. & Barnes, M. (eds.), 1982. Oceanogr. Mar. Biol. Ann. Rev. 20: 65–118.

Jordan, J.M. & Likens, G.E. 1980. Measurement of planktonic bacterial production in an oligotrophic lake. Limnol. Oceanogr. 25: 719–732.

Kadota, H. & Tanaka, N. 1975. Bacterial production and decomposition of organic matter. In: Mori, S. & Yamamoto, G. (eds.), 1975. Productivity of communities in Japanese inland waters. JIBP Synthesis 10: 13–15.

Krambeck, Ch., Krambeck, H.-J. & Overbeck, J. 1981. Microcomputer – assisted biomass determination of plankton bacteria on scanning electron micrographs. Appl. Environ. Microbiol. 42: 142–149.

Kusnetzow, S.I. & Romanenko, V.J. 1966. Produktion der Biomasse heterotropher Bakterien und die Geschwindigkeit ihrer Vermehrung im Rybinsker Stausee. Verh. Int. Verein. Limnol. 18: 176–187.

Kuznetsov, S.I., Dubinina, G.A. & Lapteva, N.A. 1979. Biology of oligotrophic bacteria. Ann. Rev. Microbiol. 33: 377–387.

Lampert, W. 1978. Release of dissolved organic carbon by grazing zooplankton. Limnol. Oceanogr. 23: 831–834.

Marshall, K.C. 1976. Interfaces in microbial ecology. Cambridge – Massachusetts and London – England: Harvard University Press. 156.

Meyer-Reil, L.A. 1977. Bacterial growth rates and biomass production. In: Rheinheimer, G. (ed.), 1977. Microbial ecology of a brackish water environment. Berlin-Heidelberg-New York: Springer-Verlag. 223–236.

Meyer-Reil, L.A. 1978. Autoradiography and epifluorescence microscopy combined for the determination of number and spectrum of actively metabolizing bacteria in natural waters. Appl. Environ. Microbiol. 36: 506–512.

Moaledj, K. 1978. Qualitative analysis of an oligotrophic aquatic microflora in the Plußsee. Arch. Hydrobiol. 82: 98–113.

Ohle, W. 1982. Nährstoffzufuhren des Grebiner Sees durch atmosphärische Niederschläge und Oberflächenabschwemmung des Einzugsgebietes. Arch. Hydrobiol. 95: 331–363.

Overbeck, J. 1972. Experimentelle Untersuchungen zur Bestimmung der bakteriellen Produktion im See. Verh. Int. Verein. Limnol. 18: 176–187.

Overbeck, J. 1974. Microbiology and biochemistry. Mitt. Int. Verein. Limnol. 20: 198–228.

Overbeck, J. 1979. Studies on heterotrophic functions and glucose metabolism of microplankton in Plußsee. Arch. Hydrobiol. Beih. Ergebn. Limnol. 13: 56–76.

Overbeck, J. 1981. A new approach for estimating the overall heterotrophic activity in aquatic ecosystems. Verh. Int. Verein. Limnol. 21: 1355–1358.

Overbeck, J. 1983. Bakterien und Kohlenstoffumsatz im Plußsee – Probleme und Möglichkeiten der Mikrobenökologie. Forum Mikrobiologie 6: 292–303.

Overbeck, J. & Babenzien, H.-D. 1964. Bakterien und Phytoplankton eines Kleingewässers im Jahreszyklus. Z. Allg. Mikrobiol. 4: 59–76.

Pedros-Alio, C. & Brock, Th.D. 1982. Assessing biomass and production of bacteria in eutrophic Lake Mendota, Wisconsin. Appl. Environ. Microbiol. 44: 203–218.

Poindexter, J.S. 1981a. The Caulobacters: Ubiquitous unusual bacteria. Microbiol. Rev. 45: 123–179.

Poindexter, J.S. 1981b. Oligotrophy: Fast and famine existence. Adv. Microb. Ecol. 5: 63–90.

Rheinheimer, G. 1981. Mikrobiologie der Gewässer. Jena: VEB Gustav Fischer Verlag. 251.

Riemann, B., Fuhrmann, J. & Azam, F. 1982. Bacterial secondary production in freshwater measured by 3H-thymidine incorporation method. Microb. Ecol. 8: 101–114.

Rosswall, Th. & Persson, I.-B. 1982. Functional description of bacterial populations from seven Swedish lakes. Limnologica 14: 1–16.

Salonen, K. 1981. The ecosystem of the oligotrophic Lake Pääjärvi. 2. Bacterioplankton. Verh. Int. Verein. Limnol. 21: 448–453.

Saunders, G.W. 1980. Organic matter and decomposers. In: Le Cren, E.D. & Lowe-McConnell, R.H. (eds.), 1980. The functioning of freshwater ecosystems. IBP Handbook 22: 341–392. Cambridge: University Press.

Schlegel, H.G. & Jannasch, H.W. 1981. Prokaryotes and their habitats. In: Starr, M.P., Stolp, H., Trüper, H.G., Balows, A. & Schlegel, H.G. (eds.), 1981. The Prokaryotes Vol. 1: 43–82. Berlin-Heidelberg-New York: Springer Verlag.

Sorokin, Y.I. & Kadota, H. (eds.), 1972. Techniques for the assessment of microbial production and decomposition in fresh waters. IBP Handbook 23. Oxford-London-Edinburgh-Melbourne: Blackwell Sci. Publ. 112.

Stabel, H.-H. 1977. Gebundene Kohlenhydrate als stabile Komponenten im Schöhsee und in *Scenedesmus*-Kulturen. Arch. Hydrobiol. (Monogr. Beiträge) 53: 159–254.

Stevenson, L.H. 1978. A case for bacterial dormancy in aquatic systems. Microb. Ecol. 4: 127–133.

Straškrabová, V. 1974. Seasonal occurrence of several groups of heterotrophic bacteria in two reservoirs. Int. Revue ges. Hydrobiol. 59: 9–16.

Straškrabová, V. & Komarková, J. 1979. Seasonal changes of bacterioplankton in a reservoir related to algae. I. Numbers and biomass. Int. Revue ges. Hydrobiol. 64: 285–302.

Van Es, F.B. & Meyer-Reil, L.A. 1982. Biomass and metabolic activity of heterotrophic marine bacteria. Adv. Microb. Ecol. 6: 111–170.

Watson, S.W., Novitsky, T.J., Quinby, H.L. & Valois, F.W. 1977. Determination of bacterial number and biomass in the marine environment. Appl. Environ. Microbiol. 33: 940–946.

Witzel, K.-P., Krambeck, H.J. & Overbeck, H.J. 1981. On the structure of bacterial communities in lakes and rivers – a comparison with numerical taxomony of isolates. Verh. Int. Verein. Limnol. 21: 1365–1370.

Witzel, K.-P., Moaledj, K. & Overbeck, H.J. 1982. A numerical taxonomic comparison of oligocarbophilic and saprophytic bacteria isolated from Lake Plußsee. Arch. Hydrobiol. 95: 507–520 ['Witzel *et al.* 1982a'].

Witzel, K.-P., Overbeck, H.J. & Moaledj, K. 1982. Microbial communities in Lake Plußsee – an analysis with numerical taxonomy of isolates. Arch. Hydrobiol. 94: 38–52 ['Witzel *et al.* 1982b'].

Wolters, N. & Schwartz, W. 1956. Untersuchungen über Vorkommen und Verhalten von Mikroorganismen in reinen Grundwässern. Arch. Hydrobiol. 51: 500–541.

Wright, R.T. & Hobbie, J.E. 1966. Use of glucose and acetate by bacteria and algae in aquatic ecosystems. Ecology 47: 447–464.

Zimmermann, R., Itturiaga, R. & Becker-Birk, J. 1978. Simultaneous determination of the total number of aquatic bacteria and the number thereof involved in respiration. Appl. Environ. Microbiol. 36: 926–935.

7. The invertebrate communities

D. Flößner, P. Kasprzak, G. Mothes,
D. Ronneberger & W. Schönborn

Studies of invertebrate communities have been carried out to a certain extent in the whole Lake Stechlin area, but especially with reference to Lake Stechlin. The chapter summarizes important results of detailed investigations over a long period, made by several researchers in the periods before and after the nuclear power plant came into operation.

The following sections deal with the combination of species, frequency, types of life-form, structure and dynamics of the living community of zoobenthos and zooplankton. Not dealt with or only considered in passing are Amoebina, Heliozoa, Ciliata, Turbellaria (excl. Tricladida), Nematoda, Tardigrada, Gastrotricha, and partly Oligochaeta.

The research into micro- and meiozoobenthos are limited to the years 1959–1968. Data after the bringing into operation of the nuclear power plant refer only to macrozoobenthos.

Before the operation of the nuclear power plant the planktonic Rotatoria and Crustacea were examined only qualitatively.

The first quantitative analysis of the zooplankton-community was undertaken in 1968, and only from 1978 onwards has continuous and comprehensive research information about the zooplankton in Lake Stechlin been available.

The microzoobenthos (W. Schönborn)

Protozoa

There are detailed investigations about Testacea of Lake Stechlin, but other groups of Protozoa were only studied sporadically, or not at all.

Amoebina. Astramoeba radiosa (especially its var. *granulifera*), *Dactylosphaerium vitraeum, Thecamoeba striata* and species of the *Vahlkampfia limax*-group dominate. Frequently filose amoebae (Aconchulina and Athalamia) of

Casper, S. Jost (ed.), Lake Stechlin. A temperate oligotrophic lake. ISBN 978-94-010-8930-2.

Biomyxa and *Penardia* are present. *Elaeorhanis cincta* is a regular inhabitant of Lake Stechlin too. The named species were mainly recorded as growing on sticks and stones.

Testacea. In Lake Stechlin 140 taxa of Testacea were observed. The new taxa described from the Lake are shown on Fig. 7.2 (Schönborn 1962b, 1965).

Characteristic habitats of Testacea as of other members of the microfauna are aufwuchs (growths) on *Typha, Phragmites, Chara* and *Myriophyllum,* sticks and stones, further littoral and profundal sediments, and under-water deposits of litter from beech and alder trees (see p. 406).

Random tests have shown that Testacea amount to 73% of the total microfauna (Schönborn 1962c). Maximum density of Testacea was established in the aufwuchs on stems of *Typha* (up to $265 \cdot 10^3$ $i \cdot 100\,cm^{-2}$). In summer maximum density of most species was registered.

Fig. 7.1 shows a comparison of density of different habitats. In the row aufwuchs (on *Typha→Phragmites→Potamogeton* submerged plants) sediment increases the number of species with prolongation of shell-axis, xenosomes (foreign bodies) and large shells. In the same row the number of species with appendages, spines, and flattened shells decreases. The aufwuchs on *Typha* and *Phragmites* contains most shell types.

Fig. 7.3 gives information about some dominant shell types developed by organisms growing in different habitats. In these habitats the following species dominate (Schönborn 1962b):

Aufwuchs on *Typha, Phragmites* and *Potamogeton: Arcella discoides, Centropyxis aculeata, C. discoides, Difflugia pulex, D. elegans, D. mica, Paraquadrula discoides, P. globulosa, Cochliopodium bilimbosum, Difflugiella spp., Phryganella paradoxa, Ph. acropodia, Lecythium hyalinum, Euglypha com-*

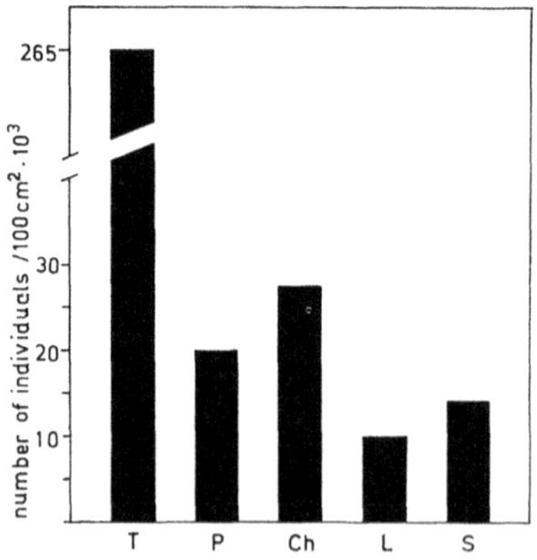

Figure 7.1 Abundance of Testacea in different habitats of Lake Stechlin (July 1960, during a density maximum). T = aufwuchs on *Typha* (height of aufwuchs 5 cm); P = aufwuchs on *Potamogeton;* Ch = aufwuchs on *Chara;* L = submerged leaves; S = sediment.

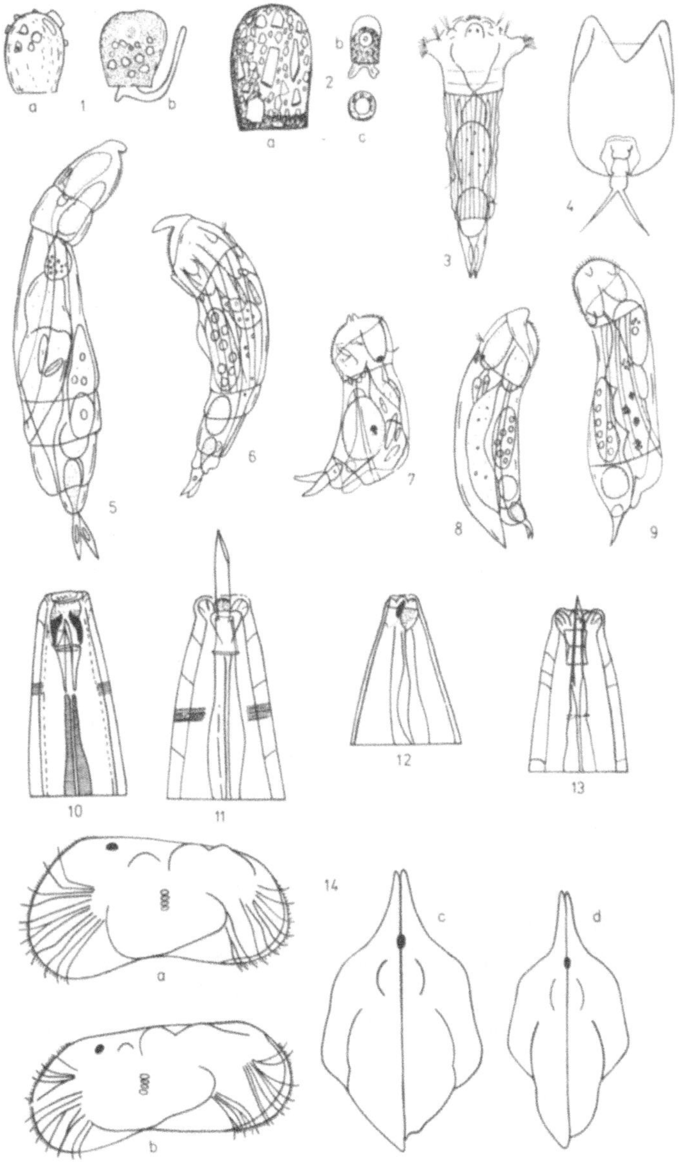

Figure 7.2 New species described from Lake Stechlin. Testacea: 1a, b. *Difflugiella horrida* Schönborn 1965; 2a, b, c, *Difflugia stechlinensis* Schönborn 1962. Rotatoria: 3. *Synchaeta asymmetrica* Koch-Althaus 1963; 4. *Lepadella pseudosimilis* Koch-Althaus 1963; 5. *Encentrum stechlinensis* Koch-Althaus 1962; 6. *E. spinosum* Koch-Althaus 1962; 7. *Cephalodella theodora* Koch-Althaus 1961; 8. *C. tenuis* Koch-Althaus 1962; 9. *C. glandulosa* Koch-Althaus 1962. Nematoda: 10. *Paractinolaimus intermedius* Altherr 1968, head; 11. *Prodorylaimus longicaudatoides* Altherr 1968, head; 12. *Dorylaimoides microdentatus* Altherr 1968, head; 13. *Labronema stechlinensis* Altherr 1968, head. Ostracoda: 14. *Limnocythere psammophila* Flößner 1965, a. ♀ left, b. ♂ left, c. ♀ dorsal, d. ♂ dorsal.

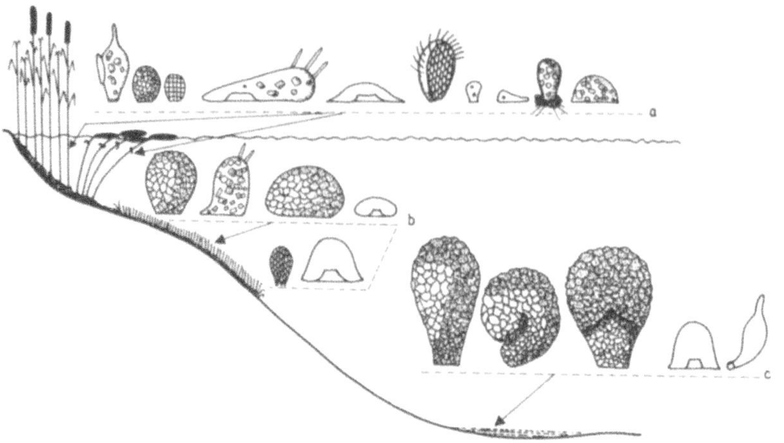

Figure 7.3 Distribution of the testacean shell types in Lake Stechlin. a. aufwuchs on *Typha* and *Potamogeton* (from left to right): *Difflugia elegans, D. mica, Paraquadrula globulosa, Centropyxis discoides, Arcella discoides, Euglypha compressa, Difflugiella sacculus, Trinema enchelys, Pseudodifflugia fascicularis, Phryganella acropodia.* b. aufwuchs on *Chara* (from left to right): *Difflugia globulosa, Centropyxis gibba, Cyclopyxis arcelloides, Arcella rotunda, Euglypha acanthophora, Arcella vulgaris.* c. sediment (from left to right): *Difflugia oblonga, Lesquereusia modesta, Pontigulasia bigibbosa, Arcella hemisphaerica,* Cyphoderia trochus.

pressa, Tracheleuglypha dentata, Trinema spp. and *Pseudodifflugia* spp.

Aufwuchs on *Chara: Arcella vulgaris, Centropyxis gibba, Cyclopyxis arcelloides, Difflugia mica, D. globulosa, Phryganella acropodia, Ph. nidulus,* and *Euglypha acanthophora.*

Sediment: *Arcella hemisphaerica, Difflugia* spp., *Pontigulasia bigibbosa, Lesquereusia modesta, Cyphoderia trochus, Campascus triqueter.*

Species which normally inhabit aufwuchs and have colonized sediment, develop characteristics typical of species normally inhabiting sediment. For instance, the species of *Centropyxis* reduce the appendages and cover their shell with little stony xenosomes. Their shell-dome is often higher than of aufwuchs-inhabiting individuals.

The populations living in the aufwuchs on sticks, stones or *Chara* show intermediate characteristics, because aufwuchs is already impregnated with sediment. The variability of these characteristics and the change in characteristics, which the aufwuchs-inhabiting individuals show, demonstrate their modificatory nature.

The common species *Centropyxis aculeata* was found in all the aufwuchs samples and about 20% in sediment samples. About 55% of individuals in sediment but only 18% in aufwuchs have shells with a high dome.

There is also a stratification in the testacean community in the aufwuchs on stems of *Typha.* Most species live in the upper part of the stems. But only *Paraquadrula discoides* occurs in a very high abundance in the basal part of the stems. Therefore, *P. discoides* is isolated from the related *Paraquadrula globulosa* in the upper part.

Centropyxis aërophila, a dominant species in terrestrial mosses and soils with typical adaptation to these habitats, migrates regularly into oligotrophic lakes. In Lake Stechlin it is observed in 80% of the *Typha* samples and in 10% of the sediment samples. The specimens from sediment are covered with little stones, that of aufwuchs are nearly membranous. Therefore, this species can be called an indicator of oligotrophy (Schönborn, Flößner & Proft 1965). Other testacean indicators almost exclusively of oligotrophic lakes, found in Lake Stechlin are *Arcella hemisphaerica, Difflugia stechlinensis, Cyclopyxis kahli* and *Trigonopyxis arcula.* The last two species have migrated like *Centropyxis aërophila* from terrestrial habitats into the lake. Other species which were seldom found in aufwuchs – such as *Arcella arenaria, A. artocrea, Difflugia bacillaria, Heleopera sylvatica, Hyalosphenia elegans* – come from terrestrial habitats (mosses, *Sphagnum*) too. They can migrate into oligotrophic waters in small numbers. *Hyalosphenia elegans,* a dominant species of bogs, also has migrated into oligotrophic Lake Stechlin.

Beech-Pine forests surround the lake near its shore. In autumn, Beech leaves fall into the littoral and form a special habitat, especially in the bays. The leaves lie in the lake water up to one year before they have decomposed (see p. 406). Testacean species of mosses and soils migrate in this habitat, becoming part of the lake fauna. They demonstrate the typical characteristics of adaptation from terrestrial habitats: globe form and (or) narrowing of pseudostoms. The most important species are: *Centropyxis aërophila* var. *sphagnicola, C. orbicularis, Cyclopyxis eurystoma, Nebela militaris, N. collaris, Quadrulella symmetrica* and *Corythion dubium. Quadrulella symmetrica* and *Nebela collaris* do not exclusively inhabit leaf deposits in Lake Stechlin.

The annual fall of leaves causes the development of a dy-like profundal sediment (C : N = 10 and more; Schönborn, Flößner & Proft 1965; Uhlmann & Albrecht 1968; but see p. 383 and 407). The water is rich in lime (see p. 381). Among Testacea indicators of dystrophy are *Arcella gibbosa, Lesquereusia spiralis, Nebela collaris, N. dentistoma* and *Hyalosphenia cuneata;* indicators of lime are the two *Paraquadrula* spp.

In the neighbouring Lakes Nehmitz and Dagow no indicators of oligotrophic or dystrophic conditions can be found. *Difflugia urceolata* and *D. amphora* occur here, known as indicators of eutrophy (Schönborn, Flößner & Proft 1965; see p. 375). *D. corona* may be an indicator of eutrophy too. It was seldom found in Lake Stechlin, only in littoral sediment (Schönborn 1967).

Ciliata. In aufwuchs-inhabiting associations of ciliated protozoa predaceous species, e. g., *Dileptus anser, Lacrymaria olor, Litonotus lamella, Stentor coeruleus, St. polymorphus* and *Trachelius ovum* dominate. Algae- and bacteria-feeders are *Aspidisca costata, Spirostomum ambiguum, Vorticella convallaria, V. monilata* and *V. nebulifera* var. *similis*. The last species also occurs as plankton attached to *Anabaena*.

Dileptus anser, Vorticella monilata and *V. nebulifera* var. *similis* are indicators of oligotrophy.

On the littoral sediment there are jelly colonies of *Ophrydium versatile*, which reach the size of a fist.

Nematoda

Collection of nematodes by Ronneberger (examined by Altherr 1968) only took place in ground water near the shore (1 m distance from open water).

The most frequent species are *Tripyla glomerans, Tobrilus steineri, T. allophysis* and *Paractinolaimus intermedius*. Most of the recorded species were found only once (see p. 509). The new species, decribed according to the habitat named above, are shown in Fig. 7.2.

Rotatoria

The following statement refers to results published by Koch-Althaus (1963).

In Lake Stechlin 159 taxa of rotifers were found. For taxa new to this see Fig. 7.2 (Koch-Althaus 1961, 1962, 1963). 134 taxa are benthonic (littoral) living; the profundal is not settled. Rotifers live in the same habitats which were mentioned for Protozoa: aufwuchs on *Typha, Phragmites, Potamogeton, Nymphaea, Myriophyllum, Chara* and in epibenthonic algae.

In epibenthonic algae lying on littoral sediment, species of *Testudinella* and *Trichocera* dominate. *Mytilina ventralis* is typical of aufwuchs on *Nymphaea;* only 7 species are associated, among them *Notommata voigti.*

Most species of rotifers live in aufwuchs on *Typha, Phragmites* and *Chara.* These habitats also show the most varying forms of rotifers. On *Typha* and *Phragmites* 85 taxa were found, on *Chara* as many as 106 taxa.

Rotifers show a maximum of their density from July to October. Minimum density was found in November and December. As early as January, density begins to increase. The most important species of aufwuchs-inhabiting associations of rotifers which occur on *Typha* as well as on *Chara* are *Lepadella patella, L. ovalis, Lecane flexilis, L. lunaris, Trichocerca* spp., *Cephalodella apocolea, C. gibba* (perennial); *Lecane luna, Lepadella triptera* (from spring to autumn); *Colurella adriatica* (winter); *Dicranophorus epicharis, D. robustus* (summer). In aufwuchs, 20–30 species were registered in each sample.

The meiozoobenthos (D. Flößner)

Introduction

In this chapter the term meiozoobenthos will be understood to mean the microcrustaceans, i. e., the orders of Cladocera, Copepoda and Ostracoda. The study of these groups of benthonic organisms was carried out in the years 1959–1968 (Flößner 1962a, 1962b, 1963, 1964, 1965, 1967, 1972). Previous studies on the microcrustacean fauna of Lake Stechlin include only Hartwig (1897) and Keilhack (1908) who reported 23 species of Cladocera and Copepoda. In addition Naidenow (1970) published a short list of microcrustaceans from the Lake Stechlin region.

The shores of Lake Stechlin are totally surrounded by deciduous and coniferous woods, whose fallen leaves are deposited on the bottom of the littoral and profundal region, giving the lake a limnological pecularity designated by the high organic carbon content of the profundal sediments (Schönborn, Flößner & Proft 1965; Uhlmann & Albrecht 1968). The sub-aqueous shores descend steeply and, therefore, the bands of vegetation in the upper littoral are small. Only in the shallow bays do they cover greater areas. On the other hand, due to the high transparency of the water (maximum of Secchi depth 10–14 m) the benthonic macrophytes reach down to a depth of nearly 20 m (Krausch 1964). Lake Stechlin has an unusually high percentage of open wind-exposed sandy and stony shore line which is free from aquatic vegetation, the so-called surf zone. The lake contains a diverse macrophyte assemblage especially rich in species of Characeae. Patches of bare muddy sediment are scattered among and along the stands of the macrophytes. Krausch (1964; p. 130) describes the horizontal zones of the littoral vegetation and distinguishes the following associations: Alnetum (0.1% of the lake area), Phragmitetum and Cladietum (2.3%), Myriophyllo-Nupharetum (<0.1%), Potametum lucentis (3.9%), *Chara* beds of the upper littoral region down to depths of 9 m (18%) and *Chara* beds and a moss community of the sublittoral region (15.6%).

Thus Lake Stechlin has a rich and diversified littoral vegetation and, therefore, there is a high diversity of benthonic microcrustacean fauna. Including the eulimnetic species, 57 species of Cladocera, 45 species of Copepoda, of it 14 species of Harpacticoida, and 20 species of Ostracoda were found in Lake Stechlin; that is more than in almost any other lake or pond in Central Europe (see the exact list on page 511). Among these species I have found some very rare members of the European limnofauna, such as *Bunops serricaudata, Alona protzi, A. weltneri, A. karelica, Moraria mrazeki, Bryocamptus vejdovskyi* and *Parastenocaris glacialis*. The ostracod *Limnocythere psammophila* is new to science (Fig. 7.2). The most abundant species are among the Cladocera: *Chydorus sphaericus, Sida crystallina, Acroperus harpae, Pleuroxus truncatus, Alona affinis, Eurycercus lamellatus* and *Alonopsis elongata;* among the Copepoda:

Megacyclops viridis, Canthocamptus staphylinus, Mesocyclops leuckarti, Attheyella dentata, Nitocra hibernica and *Attheyella trispinosa;* among the ostracods: *Cyclocypris ovum, Cypridopsis vidua, Cyclocypris serena* and *C. laevis.*

Phenology

The abundance and biomass of microcrustaceans is quite variable during the year. Almost all cladocerans can be placed in the group of bivoltine species, i.e., they show a late spring and early summer peak in abundance followed by a sharp decline in density and then increase to a second peak in autumn. *Chydorus sphaericus* and *Alonella nana* are vernal species with a maximum in spring; the former has a lower peak in autumn too. Only a few species are univoltine with one maximum in spring – *Simocephalus vetulus* – or in fall: *Camptocercus rectirostris, C. lilljeborgi, Chydorus piger* and *Pseudochydorus globosus,* whose peak of abundance may be related to the availability of food during late autumn.

The population growth of Cladocera is limited by different environmental factors at different times of the year. In spring the populations are probably limited primarily by temperature, because population growth rates are determined by birth rates during this season. On the one hand, the precipitous decline in density in late autumn is caused by habitat reduction in the course of the dying back of the higher water plants in the zone of floating leaves, and in the *Potamogeton* zone, on the other hand, by reduced food level, perhaps also in changes of spectrum and the quality of food resources. In Lake Stechlin many species (33%) winter under ice cover. The most important species in winter months are *Chydorus sphaericus, Acroperus harpae, Megacyclops viridis, M. gigas, Eurytemora lacustris, Canthocamptus staphylinus, Nitocra hibernica* and *Cyclocypris ovum.* Winter populations are to be found mainly in the littoral sediments and in the *Chara* beds which are winter-green.

Due to the reproductive cycle almost all cladocerans in Lake Stechlin are strongly monocyclic with a distinctive gamogenesis in autumn. The individual species differ in length and intensity of sexual period. Gamogenesis in approximately fourteen species is initiated before the end of September, and most of the species have become gamogenetic by the end of October. In August I have found a sexual phase in only three species: *Ceriodaphnia laticaudata, C. rotunda* and *Kurzia latissima,* which mainly inhabit smaller bodies of water. During the period from November–January, the different species completed their period of gamogenesis at various times and then died out as active populations. The percentage of sexual animals of the populations varies from species to species. Only after a period of several weeks in which both parthenogenetic and gamogenetic reproduction takes place, does complete gamogenesis occur. The perennial species, as *Acroperus harpae, Alona affinis* and *Eurycercus lamellatus* never become purely gamogenetic but winter after completing their reduced period of gamogenesis as active individuals with a low level of parthenogenesis. An unusually high percentage of species has a quite insignificant period of gamogenesis in autumn, with only very few males and ephippial females among hundreds of individuals. They are in fact acyclic and include the following species: *Chydorus sphaericus, Ch. gibbus, Alonella nana, Alona quadrangularis, A. guttata, Lathonura rectirostris, Simocephalus vetulus* and *Bosmina longirostris*. Only two species – *Camptocercus rectirostris* and *Alonopsis elongata* (Keilhack 1908) – have also a brief period of very low intensity of gamogenesis along with parthenogenesis in June. The almost complete absence of an early summer period and the great number of species with a strong tendency to a purely parthenogenetic reproduction is a very characteristic peculiarity of deep oligotrophic lakes with their stable environmental conditions.

As to temperature conditions in inland waters of temperate regions, we can distinguish three main groups. Besides the large group of eurythermal species there are in Lake Stechlin two other groups which are of special interest: cold-water-stenothermal and warm-water-stenothermal organisms, whose temperature preference is quite distinct. To the cold-water-stenothermal microcrustaceans belongs a series of species which are to be met only from October–April at water temperatures below 9°–11° C: *Cyclops kolensis, C. insignis, Megacyclops gigas, Bryocamptus vejdovskyi, Moraria mrazeki* and *Candona candida. Eurytemora lacustris* and *Canthocamptus staphylinus* occur throughout the year, but with a population peak in winter periods with water temperatures below 8°–11° C. Due to the poor heat absorbtion of the lake's water in summer time, only four warm-water-stenothermal species – *Ectocyclops phaleratus, Cryptocyclops bicolor, Microcyclops varicans* and *Notodromas monacha* – are to be found in small numbers. The species *Mesocyclops leuckarti, Thermocyclops oithonoides* and *Eucyclops macrurus,* designated as warm-water-stenothermal by Rylov (1963), are eurythermal perennial organisms in Lake Stechlin.

Habitat analysis

The littoral region is a highly complicated environment with a complex structure and a mosaic pattern of habitat arrangement. Many microcrustaceans are closely adapted to the different structures of substrate, mostly plant surfaces, in the benthonic habitats (Fryer 1968, 1974).

A general ecological division of the microcrustacean fauna of Lake Stechlin shows the following groups (Flößner 1964):

1. Littoral plankton; free-swimming species, which inhabit the open-water gaps in emerged and submerged vegetations and the water layers just above the bottom: *Ceriodaphnia* species, *Polyphemus pediculus, Mesocyclops leuckarti.* Some true limnetic forms of the genera *Daphnia, Bosmina, Eurytemora* and *Heterocope* are also to be found on open shores.

2. Phytophilic forms such as *Camptocercus lilljeborgi, Graptoleberis testudinaria, Pleuroxus aduncus* and other cladocerans, *Cryptocyclops bicolor, Ectocyclops phaleratus* and *Attheyella dentata.*
3. Occuring in dense vegetation as well as on plantless bottoms, but not ubiquitous are: *Sida crystallina, Alonopsis elongata, Alona affinis, Eurycercus lamellatus, Acroperus harpae, Alonella excisa* and others; *Megacyclops viridis,* the genera *Eucyclops* and *Macrocyclops, Microcyclops rubellus, Canthocamptus staphylinus, Attheyella trispinosa* and others; *Cypridopsis vidua, Cyclocypris ovum* and others.
4. Bottom-dwellers which inhabit sand, periphyton on stones and layers of leaf-litter in the surf zone: *Chydorus piger, Nitocra hibernica, Attheyella crassa, Bryocamptus vejdovskyi, Potamocypris villosa, Limnocythere psammophila.*
5. Iliophilic forms: *Latona setifera, Disparalona rostrata, Alona quadrangularis, Macrothrix rosea, Pleuroxus trigonellus, Herpetocypris reptans.*
6. Iliobiontic forms: *Ilyocryptus sordidus, Drepanothrix dentata, Pleuroxus uncinatus, Cypria lacustris, Cytherissa lacustris,* the *Candona* species except of *C. candida* and *C. albicans.*
7. Psammophilic forms: *Chydorus gibbus, Monospilus dispar, Paracyclops fimbriatus.*
8. Psammobiontic forms: *Rhynchotalona falcata, Darwinula stevensoni.*
9. Stygobiontic species: *Parastenocaris* species.
10. Ubiquitous with a slight preference for vegetational habitats, ecological generalists: *Chydorus sphaericus* and *Alonella nana.*

As to methods of feeding, we can distinguish between the main categories: herbivores, detritivores, predators, scavengers and parasites.

Herbivores are the most important group in the littoral region, where they are special inhabitants of the various water plant beds. They are scrapers and filter-feeders, which collect their food from firm subtrates, mostly plant surfaces. We designate them as substrate-filtrators (Flößner 1964) and add to this group, among others, *Alonopsis elongata, Pleuroxus truncatus* and *Alona affinis.* Another sub-category of herbivores are the seston-filtrators. They feed like filter-feeding zooplankters on fine detritus and algae gained from the water around the macrophytes or other solid substrates in the plantless littoral. A typical seston-filtrator in the littoral region of Lake Stechlin is *Sida crystallina,* which has an intermediate character between the littoral and limnetic forms, often attached by the gland at the back of the head to broad-leaved plants (*Nymphaea, Potamogeton*) or to dead leaves, stones and so on. Specialized seston-filtrators are *Scapholeberis mucronata* and *Notodromas monacha,* which feed on the neuston algae and bacteria.

The detritivores, as bad swimmers, inhabit mostly bottom deposits, rich in organic substances, or the periphyton on stones, less on reed plants. Most of them don't have a filter apparatus, but collect food particles separately with trunk limbs or peristome limbs. Such detritivores are the harpacticoids and almost all ostracods, but of the cladocerans there are only a few highly specialized bottom-dwellers such as *Leydigia, Rhynchotalona, Monospilus, Chydorus piger, Drepanothrix* and *Macrothrix rosea,* whereas *Ilyocryptus, Disparalona* and *Latona* are bottom-dwelling substrate-filtrators.

Few species of Cladocera are known to have distinct ecological niches (Fryer 1968, 1974). The scavenger *Pseudochydorus globosus* and the ectoparasite *Anchistropus emarginatus* occur in Lake Stechlin only in small numbers. The chydorid *Graptoleberis* is a snailscraper living on leaves (Fryer 1968). *Scapholeberis* and *Notodromas,* specialized for inverted creeping beneath the surface film of water, need calm water ranges protected from wave action and are, therefore, only sporadically distributed in Lake Stechlin. The same applies to true mud-dwellers such as *Ilyocryptus, Disparalona, Latona, Macrothrix rosea* which burrow into flocculent surface sediments which in the littoral region of Lake Stechlin are of relatively restricted occurrence.

Some species of cladocerans, especially mud-dwellers exhibit extremely clumped patterns of spatial distribution within littoral habitats. In Lake Stechlin the following species show a high degree of clumping: *Drepanothrix, Macrothrix rosea, Ilyocryptus, Alona rectangula, Disparalona,* but also *Sida crystallina* and *Polyphemus pediculus.* The latter may build up 'plankton swarms' in open-water gaps of the littoral vegetation or on local places of open shores. Patchiness of cladoceran distribution may be created and maintained by heterogeneity of habitat, concentration of decomposing organic material and water movements which accumulate the free-layed resting eggs of *Sida* and *Polyphemus.*

True bottom-dwellers have evolved convergent adaptations in body structure or behavioural pattern to similar habitats, such as the presence of long screening setae at the ventral margins of the carapace (*Ilyocryptus, Latona*), retention of moulted carapaces at *Alonopsis, Ilyocryptus* and *Monospilus* or the collection of camouflaging burden of detritus on carapace at *Ilyocryptus, Pleuroxus uncinatus,* and *Chydorus gibbus. Limnocythere psammophila* has lateral bulges on each side of the carapace (Fig. 7.2) as do many other bottom-dwelling ostracods.

A division of the numerous phytophilic inhabitants of the various littoral macrophyte beds into distinct microhabitats is not generally practicable, but some species show a preference for certain plant communities or structural peculiarities of water plants. *Camptocercus lilljeborgi, Macrocyclops fuscus, Ectocyclops* and *Microcyclops rubellus* prevail among floating-leaved plants (*Nymphaea, Potamogeton natans*). Most abundant in *Chara* beds are *Eurycercus, Nitocra* and *Cyclocypris laevis.* Quade (1969) reports the same grouping of species in Minnesota lakes with respect to cladoceran fauna. *Alonopsis* and *Sida* are largely confined to broad-leaved submerged plants (*Potamogeton, Stratiotes*), *Graptoleberis, Alonella exigua* and *Alona rectangula* to fine-leaved submerged plants (*Myriophyllum, Ceratophyllum*). In medium-hard-water lakes like Lake Stechlin, the preference for fine-leaved submerged plants by *Graptoleberis,* a distinctive species with a 'molluscan-like' (Fryer 1968) way of life, may be due to

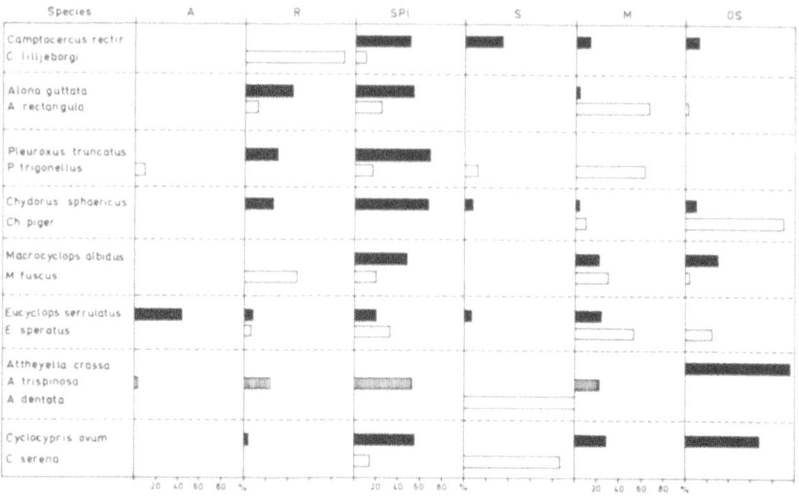

Figure 7.4 Habitat preference (percentage values of numbers of individuals) of closely related species in Lake Stechlin. A = Alder carrs, R = reed beds and floating plants, SP1 = submerged plants, S = sublittoral, M = littoral mud, OS = shores on the windward side.

difficulty in obtaining food from the marlencrusted substrate of the broad-leaved *Potamogeton* species (Whiteside *et al.* 1978).

Several closely related species show a competitive exclusion, i.e., they have their optimal population density in different habitats (Fig. 7.4). Moreover, the two *Camptocercus* and *Chydorus* species listed in this figure are also slightly temporal separated. *Camptocercus lilljeborgi* has its maximum abundance in September/October, whereas *C. rectirostris* is most abundant two months later in November/December. The population peaks of *Ch. sphaericus* and *Ch. piger* are from March to early June and in October/November, respectively. A very distinct ecological classification of the littoral microcrustacean fauna follows the subdivision of the littoral region according to the degree of in-filling of the lake basin. In Lake Stechlin two main types of stands may be discerned: lacustrine habitats and semilacustrine habitats. Lacustrine habitats are characterized as sandy-stony substrate type in erosive littoral or mud substrate type of sparse vegetation stands on open, lakeward sites. The typical plant-associations at lacustrine habitats are Potametum lucentis and diverse *Chara*-communities which are poor in periphyton. Semilacustrine habitats are of the mud-vegetation substrate type in the protected shore zone. The vegetation is characterized by broad dense reedbeds in the shoreward zone rich in periphyton and detritus and by the zone of floating-leaved macrophytes (Myriophyllo-Nupharetum), both with abundant growth of emerged and submerged aquatic plants which function ecologically in the local reduction of wave action. The lacustrine group of microcrustaceans includes in Lake Stechlin the psammophilic and psammobiontic forms, the bottom-dwellers on sand, in periphyton on stones and on layers of fallen leaves, mentioned on p. 222). From the phytophilic forms typical lacustrine species are *Alonopsis, Sida, Pleuroxus truncatus, Megacyclops viridis* and *Cyclocypris laevis*. Lacustrine shore stands, with their abundance of the free-swimming *Polyphemus* and the stationary seston-filtrator *Sida*, form a link to the open water zone of the lake.

The semilacustrine group of microcrustaceans contains the iliophilic and iliobiontic forms (without the typical species of the profundal region, see p. 222) and such phytophilic forms as *Camptocercus lilljeborgi, Alonella exigua, Alona affinis, A. rectangula, Pleuroxus aduncus, Ceriodaphnia pulchella* and *C. quadrangula, Macrocyclops fuscus* and *Ectocyclops phaleratus*. Also *Scapholeberis* and *Notodromas* are typical semilacustrine species. Free-swimming species of semilacustrine shore stands are *Bosmina longirostris* and the ceriodaphnids which are typical members of the pond zooplankton. Owing to the steepness of the shores in Lake Stechlin, the semilacustrine microcrustacean fauna is only poorly developed, and many semilacustrine species occur in low densities.

A distinctive difference exists among microcrustaceans, especially cladoceran fauna in submerged *Myriophyllum* either of semilacustrine or of lacustrine stands, which proves the low level or absence of affinities of Cladocera for particular plants but shows the preference of definite plant associations with specific environmental conditions. Thus *Scapholeberis, Ceriodaphnia reticulata, C. megops, Pleuroxus aduncus, Camptocercus lilljeborgi* and *Microcyclops rubellus* inhabit exclusively *Myriophyllum* at semilacustrine stands but are absent in *Myriophyllum* at lacustrine stands. The ecological valence of microcrustaceans mentioned seems to be narrower than that of members of the plant community.

Our studies demonstrate the great importance of habitat structure in determining the environmental preference of the benthonic microcrustaceans in Lake Stechlin. On the other hand, the correlation between distribution and the physical and chemical conditions of the environment are often likely to be indirect and reflect the chemical needs of aquatic plants which are inhabited by microcrustaceans (the presence of adequate supply of food). Reference must be made to the enrichment of humic substances in the profundal sediments of Lake Stechlin, which limit the densities of almost all microcrustacean species owing to the low nutritional value of the sediment.

Lake Stechlin proves to be a typical oligotrophic lake, not only by its physical and chemical conditions (high transparency, high oxygen supply of the hypolimnion) and the low primary production, but also by the occurrence of several indicator organisms among the benthonic microcrustaceans. To these indicators of oligotrophy belong the following species: *Alonopsis elongata, Drepanothrix dentata, Latona setifera, Chydorus piger, Ch. gibbus, Rhynchotalona falcata, Eurytemora lacustris, Heterocope appendiculata, Bryocamptus vejdovskyi, Moraria mrazeki, Limnocythere psammophila* and the *Parastenocaris* species, especially *P. glacialis*. Added to this is the limnetic species *Bythotrephes longimanus*. Further, the high abundance of *Sida crystallina* is typical for clearwaters.

All these species can also be found in mesotrophic or even weakly eutrophic lakes as *Ch. gibbus* (Frenzel 1982), but the whole assemblage of all these indicators in partly high densities is almost absent in mesotrophic waters. The abundant occurrence of *Alonopsis* alone designates Lake Stechlin as oligotrophic.

Zoogeographical aspects

A characteristic feature of the microcrustacean fauna of Lake Stechlin is the high rate of species with boreal, respectively, boreo-alpine distribution which is favoured by the oligotrophic state of the lake. These northern species are not restricted to Lake Stechlin, but occur in numerous other mesotrophic lakes in the surrounding area. Moreover, the numerous bog lakes of this district contain additional remarkable boreal species (Flößner 1963, 1972). The Lake Stechlin district is shown by this to be one of the most important centres of accumulation of boreal and boreo-alpine species in the northern part of Central Europe. In this respect it can be compared with the heath and bog districts in the northern parts of the Federal Republic of Germany and the Lusatia (Herr 1917; Hollwedel 1978; Klie 1937; Lundbeck 1951). Many of these species occur frequently throughout Lake Stechlin, such as *Alonopsis, Chydorus piger,* and *Drepanothrix;* others are very infrequent, such as *Alona karelica, Moraria mrazeki, Parastenocaris glacialis* or *Diacyclops nanus.* Several are more widespread in the lowlands of Central Europe; others have a disjunct distribution in the southern part of its range and Lake Stechlin is an isolated, very forward outpost of the main area, such as for *A. karelica, Parastenocaris glacialis* and *Moraria mrazeki* (Enckell 1969; Flößner 1972; Lang 1948). Together with *Drepanothrix, Chydorus piger,* and *Bryocamptus vejdovskyi,* they are relict species from the early post-glacial times. Almost all boreal species prefer cold water situations with population peaks in Lake Stechlin in winter times. Cold-stenotherm are *Bryocamptus vejdovskyi, Moraria mrazeki* and *Parastenocaris glacialis.* In Lake Stechlin most of these northern species are confined to plantless, open wind-exposed sandy and stony shores or to vegetation beds of lacustrine littoral stands. Moreover, they are indicators of meso- to oligotrophy.

Microcrustacean communities

A first insight into the synecological structure of the benthonic microcrustacean fauna of a lake should describe and classify the various assemblages of species into distinct communities. We have described it by character-species which are largely confined to a certain type of habitat or to a definite association of a taxonomic group of animals (taxocoenosis). In addition predominant associated species are used for describing communities. Thus similarities and differences between the communities associated with different habitats become apparent.

Lake Stechlin contains eight communities of benthic microcrustaceans which are described in the following.

The *community of Alder-swamps* is of little importance because the habitat occupies only very small areas of the littoral region. Owing to this and the inconstancy of the environmental conditions (high oscillations of water level and water temperature with long period of ice cover), the alder-swamps of Lake Stechlin are similar to small water bodies. In total 26 species of cladocerans, 16 of copepods and 3 of ostracods were found. All typical species – *Daphnia curvirostris, Kurzia latissima, Cyclops insignis* and *C. furcifer* – are known as characteristic inhabitants of shallow waters.

The *community of the reed beds and the floating-leaved plants* inhabits the Phragmitetum and partially the Myriophyllo-Nupharetum. In Lake Stechlin it is poorly developed and occurs with a moderate number of species only in a somewhat restricted number of places. Diagnostic species are *Ceriodaphnia pulchella, C. quadrangula, C. reticulata, Scapholeberis, Camptocercus lilljeborgi, Ectocyclops* and *Notodromas*. The taxocoenosis is subdivided into two sub-communities. One occurs between the stalks of the plants; the other inhabits the periphyton on the stalks of the reed plants. The first mentioned consists of plankton organisms such as ceriodaphnids, *Bosmina longirostris, Mesocyclops* and also true limnetic elements such as *Eurytemora, Heterocope* and *Diaphanosoma*. The periphyton on *Phragmites* and *Typha* are, in contrast, dominated by benthonic cyclopids, harpacticoids and chydorids with the predominant species *Alona affinis, A. costata* and *Alonopsis*.

The *submerged vegetation* which consists mainly of *Potamogeton* species, *Myriophyllum, Stratiotes, Ceratophyllum* and the widely enlarged meadows of Characeae is one of the habitats in Lake Stechlin with the greatest diversity of the fauna. 82 species – 46 cladocerans, 26 copepods, 10 ostracods – were found, but only three of them are characteristic ones with a high fidelity to the microcrustacean community: *Pleuroxus aduncus, Graptoleberis* and *Lathonura*.

Some subdivisions into sub-communities are necessary. The broad-leaved *Potamogeton* species of the Potametum lucentis are dominated by lacustrine species such as *Sida, Alonopsis elongata, Polyphemus* and *Megacyclops viridis*, which are filter-feeders or predators and often swim in the open water between the plants. The sub-community on *Myriophyllum* of the Nupharetum differs by the occurrence of *Pleuroxus aduncus* and the high proportion of other semi-lacustrine species such as *Acroperus, Alonella exigua* and *Eucyclops macruroides*. A quite different structure occurs in the *Chara* beds which have the highest rate of submerged vegetation, resulting in a distinctive sub-community of submerged vegetation. It is characterized by the abundant occurrence of *Eurycercus, Chydorus sphaericus, Acroperus, Simocephalus, Canthocamptus* and *Cyclocypris* species. On the other hand, the share of *Pleuroxus aduncus, Graptoleberis, Sida, Alona costata, Pleuroxus truncatus, Macrocyclops fuscus* and *Eucyclops mac-*

rurus in the sub-community of *Chara* beds is far less than that for other submerged plants. *Pleuroxus trigonellus, Nitocra* and some other bottom-dwelling species inhabit the lower muddy layers of Characeae, thus explaining its presence in *Chara*.

The *sublittoral community* is found in the dense vegetation of *Nitella flexilis, Vaucheria dichotoma* and aquatic mosses which cover the deeper littoral region from 9 to nearly 20 m, especially on steeply descending parts of the sub-aqueous shore. A characteristic feature of the microcrustacean fauna is the abundant occurrence of *Attheyella dentata* and *Cyclocypris serena*. They are detritivores and confined to substrates. To the associated species belong *Camptocercus rectirostris, Graptoleberis, Megacyclops viridis* and *Eucyclops serrulatus*. With increasing depth beyond the eulittoral zone, the number of microcrustaceans diminishes strikingly. The sublittoral community contains only 41 species (23 cladocerans, 14 copepods, 4 ostracods) in contrast to the neighbouring and similar structured *Chara* beds. Frequent inhabitants of the upper *Chara* beds, which are absent in sublittoral vegetation, are, among others, *Nitocra hibernica, Acroperus, Alonella excisa, Cypridopsis vidua, Eucyclops* and *Macrocyclops* species.

The *community of littoral mud* is characterized by a series of substratum-specific true bottom-dwellers such as *Pleuroxus uncinatus, Alona quadrangularis, Ilyocryptus, Drepanothrix, Macrothrix rosea* and the ostracods *Candona hyalina, C. acuminata* and *Herpetocypris reptans*. Actually, the community is very rich in species (41 cladocerans, 30 copepods, 10 ostracods), but it can hardly be designated as typical for Lake Stechlin, owing to the reduction of wind protected, semilacustrine parts of the shores which are optimal places for the community.

The littoral mud community can be subdivided into two ecological groups: the mud-layer sub-sommunity and the water-mud-contact layer sub-community. The cladocerans and ostracods of the mud-layer microhabitat are true benthonic species, burrowing into the surface layer of flocculent deposits and presenting many striking adaptations to this environment (Fryer 1968, 1974). The harpacticoids creep through the interstitial gaps of the loose mud layers. Specific inhabitants of this microhabitat are *Ilyocryptus, Drepanothrix, Macrothrix rosea, Disparalona,* some harpacticoids and ostracods. The sub-community in the water-mud-contact layer consists of epibenthonic species directly dependent upon the mud layer for sustenance. They do not creep through the mud layer or burrow into it, but rest upon it and can swim about immediately and rapidly searching for a new resting place. To these epibenthonic mud-dwellers belong *Latona, Alona rectangula, Pleuroxus uncinatus* and many cyclopids, especially the bigger ones such as *Megacyclops* and *Macrocyclops*.

The most typical *microcrustacean community* of Lake Stechlin occurs on *sandy and stony deposits of open wind-exposed shores* which cover great areas. The erosive littoral is characterized by a great number of species (38 cladocerans, 25 copepods, 12 ostracods). Many of them are ± restricted to the community: *Chydorus piger, Ch. gibbus, Rhynchotalona, Monospilus, Paracyclops fim-*

briatus, Nitocra, Attheyella crassa, Bryocamptus vejdovskyi, Moraria mrazeki, Potamocypris villosa, Darwinula and the new species *Limnocythere psammophila.* High densities of *Alonopsis, Acroperus* and *Alona affinis* also occur. Sandy deposits overlain by a thin layer of organic detritus are inhabited by a special sub-community with the iliophilic species *Latona setifera, Alona quadrangularis, Pleuroxus trigonellus* and *Leydigia leydigi.* Another sub-community is found in the periphyton on stones preferred by the lasiophilic (Meuche 1939) species *Alonopsis elongata, Alonella excisa* and *Alona costata,* whereas *Monospilus, Chydorus gibbus, Rhynchotalona* and *Darwinula* are true sand-dwellers avoiding the periphyton. These forms together with the already mentioned iliophilic cladocerans have also developed striking adaptations to the way of life on sandy bottoms. The postabdomens are usually broad and powerfully developed for use in creeping. The eye has a tendency to become small (*Leydigia*) and may even disappear (*Monospilus*). Old carapaces are retained at each moult and the burdened animals are thus protected from being washed away by water movements (*Monospilus, Alonopsis*). *Latona* is provided with specialized setae on the ventral margins of carapace, preventing the clogging of the filter chamber with detritus particles. *Rhynchotalona* possesses a short, thick postabdomen with a strong claw and four powerful denticles, a very efficient organ for pushing away from firm substrates. *Paracyclops fimbriatus* is a true creeper with short antennae and a vermiform extended body like harpacticoids.

The *interstitial ground-water* in the sand layers of the shores, hardly investigated in European lakes, is inhabited by a little diversified microcrustacean fauna containing three character-species of the genus *Parastenocaris* (*P. brevipes, P. phyllura, P. glacialis*) which are typical stygobionts. *Candona albicans* can be designated as a stygophilic organism. This assemblage may be typical for oligotrophic lakes, because the microcaverns of the sandy sediments are rich in oxygen and not filled up with decaying detritus particles. In the ground-water of the sandy shores 33 species in low densities are to be found; most of them are not indigenous but stygoxenic freshwater organisms washed in by wave action.

The *bottom deposits in the profundal zone* are very poorly inhabited by microcrustaceans, both in species and individuals. Characteristic species are without exception ostracods: *Candona neglecta, Cytherissa lacustris* and *Cypria lacustris.* The cold-water-stenothermal *Megacyclops gigas* occurs also with high constancy, but is not confined to the profundal zone. Only 10 species altogether were found there: 4 copepods and 6 ostracods, but no cladoceran species. The strikingly low population density in the profundal zone of Lake Stechlin is caused by the nutritional impoverishment of the sediment, which is rich in humic substances (Möller 1935), especially in fulvic acids and organic carbon and has a C: N-index of about 10 (Schönborn, Flößner & Proft 1965; Uhlmann & Albrecht 1968), owing to the allochthonous import of leaf-litter of the surrounding trees (see p. 219).

The macrozoobenthos (G. Mothes)

Introduction

The analyses of the macrozoobenthos were carried out from 1960–1970 and concentrated on Lake Stechlin, Lake Gerlin and the two basins of Lake Nehmitz. They resulted from a question relating to the impact of the nuclear power station on the lakes.

Changes expected in the qualitative and quantitative composition of the fauna following the bringing into operation of the nuclear power plant were to be analysed both in the entire lake and in structural parts of it. For this purpose the macrozoobenthos of the waters according to the habitat structure and species composition was subdivided into communities and sub-communities as typical ecological units.

Abundance and stock of species in Lakes Stechlin and Nehmitz

Table 7.1 contains data for Lake Stechlin and Lake Nehmitz. The analysis of other lakes showed similar results. Unfortunately, there is no appropriate methodology for abundance analysis of the reed and swamp communities; there are only qualitative collections. The abundance analysis shows maximum values in the community of the submerged plants. The sublittoral and profundal zones have a considerably lower abundance than the community of the submerged plants.

The community of the submerged plants contains the largest range of species. The swamp community shows a comparatively similar number of species. The

Table 7.1 Medium abundance and number of species in Lake Stechlin (S) and the two basins of Lake Nehmitz (N; NN = north basin, NS = south basin) before the bringing into operation of the nuclear power plant. Number of species in per cent of the known species in the whole body of water (S = 408 species, N = 452 species).

Community	Reed zone	Swamp zone	Surf zone	Zone of submerged and floating plants	Sublittoral zone	Profundal zone
Abundance (individuals · m^{-2})						
S	?	?	4,807	17,101	4,854	457
NN	?	?	15,912	14,225	4,930	2,106
NS	?	?	7,205	7,769	306	649
Number of species (%)						
S	26	41	28	66	24	6
N	23	46	29	57	10	1

other communities are clearly below those numbers. All the analysed groups of species occur, with several representatives each, in the community of the submerged plants. In the other communities certain groups of species are absent. The relatively high number of species in the zone of marshy meadows and alder-swamps is due to an high species-count of the groups Hydracarina, Odonata, Heteroptera and Coleoptera. Because of its richness of species and individuals, the zone of the submerged plants in the upper littoral must be considered the optimum community of the lacustrine fauna development. The other communities show restricted conditions of existence. Of such restrictions, only light (Phragmitetum, sublittoral zone, profundal zone), current (surf zone), and low summer temperatures (profundal zone) should be mentioned.

Analyses focussed on Lake Stechlin in the period from August 1960 to the spring of 1965, when the outer cooling-water circulation of the nuclear power plant was brought into trial operation. Lake Nehmitz was examined on a smaller scale. Nevertheless, only 408 species were found in the larger and more divided Lake Stechlin, in contrast to over 452 in Lake Nehmitz. The paucity of nutrients in Lake Stechlin did not offer acceptable conditions of existence for many species. A change took place soon, with nutrients being supplied from Lake Nehmitz through the water exchange between Lake Gerlin, Lake Nehmitz (north basin) and Lake Stechlin after the outer cooling-water circulation of the nuclear power plant had become operational. While the stock of species in Lake Nehmitz increased from 452 to only 462 in the course of further analyses, there was an increase by 94 species to a total number of 502 species in Lake Stechlin. This strong increase affected, above all, the communities of the submerged plants. The species maximum of a standing body of water during the transitional stage between oligotrophy and eutrophy is, accordingly, to be found in the sphere of mesotrophy.

The biggest number of species in Lakes Stechlin and Nehmitz is made up of the group of Chironomidae (178 resp. 171), followed by Hydracarina (78 resp. 103), Mollusca (35 resp. 33), Odonata (31 resp. 20), Coleoptera (31 resp. 24), and Trichoptera (47 resp. 42).

In contrast to the high number of species of Chironomidae, one individual species, *Dreissena polymorpha,* reaches the highest density of individuals in many communities and sometimes exceeds 50 per cent of total abundance. Thus very high abundance values are also found in the surf zone of Lake Nehmitz (Table 7.1).

The macrozoobenthonic communities before the nuclear power plant was brought into operation

Combinations of species are especially suited to define the individual characteristic of a body of water if vicariant species are available as biological indicators for different types of waters. The group of Chironomidae which is particularly rich in species offers very favourable conditions in this respect. The theory of the lake types based on analyses of the benthonic fauna draws on the research work done by Thienemann and Lundbeck and has been generalized by Brundin. In this work I shall also draw on these results which have been put forward by other authors, too (see Thienemann 1954; Brundin 1956).

The results are based on quantitative bottom catches which were completed by qualitative samples and visual observations in the water (Mothes 1965a, 1966c, 1967d). It has been possible to corroborate statistically the subdivision of the benthal of Lake Stechlin into zones settled by different communities and sub-communities (Mothes & Blume 1969).

The analysed waters contain up to six benthonic communities: 1) zone of dense reed stand, 2) swamp zone, 3) surf zone, 4) zone of submerged and floating plants in the epilimnetic zone, 5) sublittoral zone, 6) profundal zone.

Zone of dense reed. The zone of dense reed was qualitatively analysed only in Lake Stechlin and Lake Nehmitz. Due to strong overshadowing by emersed shoots, there is a very poor growth of algae. Therefore, algophagous organisms (Gastropoda, Orthocladiinae, etc.) are almost totally absent. The dense reed functions as sedimentation area for the seston of penetrating surface water. We found *Chironomus macani* and *Arctopelopia griseipennis* of the rarer species and *Camptochironomus pallidivittatus* and *Hydrozetes parisiensis* of the more common ones, which live preferably in the reed-beds. The paucity of light and the low exchange of water can cause anoxis (Lake Nehmitz). In that case, there is no macrobenthonic fauna.

Towards the open lake, the reed-beds border on the zone of the submerged plants. A marginal sub-community with an extremely great number of species and individuals develops in the light-transmissive border area on the ground and on the submerged stalks. It consists of filter feeders (e.g. sponges, *Anodonta cygnea* in Lake Gerlin and Lake Nehmitz) and rich periphyton.

Swamp zone. The swamp zone occupying only very small areas was also analysed only qualitatively. There are mostly well lighted shallow water biotopes, where the changing water level strongly influences the living organisms. Such habitats are marshy meadows in the terminal bays of Lake Nehmitz and Alder carrs of Lakes Stechlin and Nehmitz. The nutrient content for Carnivora is so high that the large hydric predators from among the invertebrates are to be found here almost exclusively. Fish as competitors of the invertebrates are largely absent.

Big predators are *Aeshna* spp. (Mothes 1965c). From among the Dytiscidae, various species of *Dytiscus* live in the marshy meadows, together with the Hemiptera *Nepa cinerea* and *Ilyocoris cimicoides*. Other predators are *Argyroneta aquatica* and a great number of water mites (Mothes 1965b, 1967b, 1970).

In contrast to other habitats, the marshy meadows contain large snails, such as

Lymnaea stagnalis, Planorbis corneus and *Planorbis planorbis*. Large Trichoptera of the genera *Limnephilus* and *Phryganea,* too, prefer these calm, productive marginal biotopes (Mothes 1964c, 1967a). Alder carrs which are rare in this region have been classified under the habitats of the swamp zone because of the stock of macrobenthonic species. Herbivora are largely absent and foliage decomposers dominate.

Surf zone. The surf zone has a distinct characteristic only in the larger lakes. The premise for a typical development is the rise and fall of the water caused by the waves. In the lakes we analysed, its effect is limited to the shore region up to a depth of about a metre. There is mostly a sandy or stony soil, often without macrophytes. In shore regions which are less exposed to the wind, sparse reed-beds may have developed. There are no extensive organogenous permanent sediments. Larger stones often bear periphyton coatings of diatoms or filamentous tufts of algae. Charetum asperae develops on sandy bottom. The community of the surf zone is marked by the following characteristics:

- Organisms moving freely on the soil without being able to stick to the substrate are mostly absent. They inhabit this zone, coming from adjacent ones, only for a short time during periods with a low wind intensity (Gammaridae, Trichoptera, etc.).
- Organisms living on the bottom have passive (e.g., stone weights of the conches of *Goera pilosa* and *Molanna angustata*) or active mechanisms so as not be washed away (e.g., *Theodoxus fluviatilis* and *Dreissena polymorpha* stick to stones; *Helobdella stagnalis* has suction cups).
- Most organisms live in the soil and measure only a few millimetres or possess special fossorial devices (e.g., *Ephemera vulgata*).

Other typical forms of this zone are *Paranais litoralis, Pisidium amnicum, Unio pictorum, Tinodes waeneri, Oulimnius troglodytes,* Gyrinidae larvae, *Arctopelopia barbitarsis,* species of the Chironomidae genera *Thienemannimyia, Ablabesmyia, Stictochironomus, Cryptochironomus* and *Demicryptochironomus,* subgenus *Tripodura* of *Polypedilum, Cladotanytarsus* (Mothes 1966a).

There is a relatively high abundance of Carnivora (over 70 per cent) at places exposed to the current, where the predators carry off plankton Crustacea (observations of *Arctopelopia barbitarsis*). The energy supply in the form of organic substance from the pelagic zone obviously plays an important part for the existence of this community.

The extreme factor 'wave action' exerts its chief effect on the surf zone. It is strongest in the open parts of Lake Stechlin. Its effects are reduced in the sheltered terminal bays and in sparse *Phragmites* stands. Accordingly, three sub-communities were distinguished (Table 7.2) which extend to a depth of about 1.2 m.

In the sub-community at open shores from among Mollusca only *Dreissena polymorpha* and *Theodoxus fluviatilis* are of importance. Small Oligochaeta live on the detritus washed into the sand; 35 per cent of them are *Paranais litoralis*. Fifty per cent of the Hirudinea are *Helobdella stagnalis*. Some 60 per cent of Chironomini belong to *Stictochironomus pictulus*, 30 per cent to *Polypedilum bicrenatum* and *P. pullum*, 10 per cent to *Paratendipes albimanus*. Sixty per cent of Tanytarsini are *Cladotanytarsus* species, mostly *Cl. dispersopilosus*. This paucity of species indicates the extreme character of the sub-community.

The sparse *Phragmites* stands consolidate with their rhizomes the bottom of the lake, keep back coarse detritus and offer shelter to larger organisms such as *Gammarus* species and *Asellus*. This is the main distribution area of species considered rheophilous: *Ephemera vulgata, Goera pilosa, Molanna angustata* and *Nemoura cinerea*. From among Heleidae the genus *Dasyhelea* is typical. Although this sub-community shows the lowest abundance, it possesses a larger range of species than the former one and harbours larger animals which do not like the bottom of the lake made choppy by the waves.

The number of species on the sandy soil of the terminal bays also increases as against the open region of the lake. There are remarkable quantities of *Pisidium* (typical *P. amnicum* beside the more widespread *P. casertanum*) and *Stylaria lacustris*. From among Hydracarina there are mainly species moving on the

Table 7.2 Medium abundance (individuals · m^{-2}) of animal groups in the surf zone. Sub-communities: a = surf shore in the open lake region; b = sparse Phragmitetum on open shores; c = sandy beach in the terminal bays.

Sub-communities	a	b	c
Number of samplings	11	4	11
Total abundance	5,262	3,429	4,854
Hydrozoa	46	13	9
Turbellaria	9	–	14
Nematoda	–	–	5
Mollusca	848	651	1,030
Oligochaeta	1,022	191	339
Hirudinea	70	–	60
Hydracarina	14	64	236
Amphipoda/Isopoda	–	191	660
Ephemeroptera/Plecoptera	413	700	478
Odonata	–	–	14
Hemiptera	42	102	–
Coleoptera	186	115	102
Trichoptera	79	140	112
Heleidae	144	153	139
Tanypodinae	653	407	473
Orthocladiinae	60	13	46
Chironomini	1,155	485	794
Tanytarsini	502	204	320
Diptera (remainder)	19	–	23

bottom (25 per cent genus *Atractides*). The tanypodine *Arctopelopia barbitarsis*, dominating in the open region of the lake, is replaced by *Thienemannimyia vitellina* and others, above all *Procladius* species. This is the main range of *Paratendipes albimanus*. *Cladotanytarsus* is no longer of importance; other species of the big genus *Tanytarsus* live here.

Zone of submerged and floating plants in the epilimnetic zone. The community of the zone of submerged macrophytes in the epilimnetic depth layer occupies the well-lighted sector. It gives a different impression due to the multiple species or the absence of macrophytes communities (floating plants, pondweeds, *Ceratophyllum*, *Myriophyllum*, *Utricularia*, *Stratiotes* stocks, moss and Characeae lawns, algae tufts, etc.; see p. 131). The result was the distinction of a multitude of macrobenthonic sub-communities distinguished by their predominance of herbivora.

The zone of the submerged macrophytes in the epilimnetic depth zone is settled by the community which has most species and individuals in an open, standing body of water (Table 7.3). There all types of nutrition exist. This community is

Table 7.3 Medium abundance (individuals · m^{-2}) of animal groups of the zone of submerged and floating plants in the epilimnetic depth zone. Sub-communities: a = Charetum filiformis in the open lake region; b = Nitellopsidetum obtusae in the open lake region; c = Characeae associations in the terminal bays; d = sandy soil; e = muddy deposits; f = foliage deposits.

Sub-communities	a	b	c	d	e	f
Number of samplings	13	6	17	2	7	2
Total abundance	33,396	9,890	13,654	1,659	5,245	19,051
Hydrozoa	138	153	30	–	22	51
Turbellaria	138	17	63	–	–	51
Nematoda	–	–	–	–	7	–
Mollusca	24,200	8,020	6,618	204	815	10,820
Oligochaeta	738	196	600	255	262	280
Hirudinea	78	51	36	–	36	–
Hydracarina	317	119	294	26	153	434
Amphipoda/Isopoda	949	170	558	77	124	5,150
Ephemeroptera/Plecoptera	1,260	144	1,192	230	1,087	938
Odonata	27	24	36	–	–	26
Hemiptera	8	–	–	51	7	–
Coleoptera	232	51	84	26	29	102
Trichoptera	1,020	153	588	102	109	153
Heleidae	67	9	129	51	36	51
Tanypodinae	830	136	438	255	850	689
Orthocladiinae	1,460	289	1,164	–	44	51
Chironomini	1,420	314	1,218	331	1,430	77
Tanytarsini	490	17	603	51	219	–
Diptera (remainder)	20	17	3	–	15	178

marked especially by snails: *Valvata cristata, V. piscinalis, Bithynia tentaculata, Physa fontinalis, Lymnaea peregra, Gyraulus albus* and *G. riparius, Acroloxus lacustris,* moreover sponges and Bryozoa, the bivalve *Dreissena polymorpha,* the Oligochaeta *Stylaria lacustris,* the water mites *Hygrobates longipalpis, Midea orbiculata* and *Mideopsis orbicularis,* the mayfly genera *Centroptilum, Leptophlebia* and *Caenis,* the caddis flies *Oxyethira flavicornis, Cyrnus flavidus* and *C. trimaculatus,* the Orthocladiinae genera *Cricotopus, Parakiefferiella* and *Psectrocladius,* the Chironomini genera *Dicrotendipes, Glyptotendipes, Polypedilum* and *Pseudochironomus* as well as from the Tanytarsini groups of species of the genera *Tanytarsus, Paratanytarsus,* and *Corynocera.*

The richness of species of this community, too, strongly depends on the productivity of the water. Before the nuclear power plant began to work, Lake Stechlin had only the Zygoptera *Enallagma cyathigerum* from among frequent large predators. It may largely live on microcrustaceans; the larvae were fixed on the tips of the macrophytes. After the power plant started work, *Coenagrion puella* and *C. pulchellum* restricted the distribution area of *E. cyathigerum, Cricotopus* species and the genus *Psectrocladius.* Proof of the better nutritional conditions in the sediment are species of the Chironomidae genera *Chironomus* and *Einfeldia* as well as the doubling of the *Tanytarsus* share and the simultaneous halving of the *Paratanytarsus* share compared to the first sub-community (Table 7.3a).

Another sub-community is to be found on sandy soil (Table 7.3d) on steep sided shore areas. It sometimes directly borders on the surf zone. *Polypedilum scalaenum, Stictochironomus pictulus* and *Apatania fimbriata* also live here. However, species which characterize the surf zone and are adapted to the current are absent. Among the Hydracarina, *Atractides pavesii* dominates here.

On plant-free gyttja (Table 7.3e) the fifth sub-community settles, above all in the terminal bays. *Caenis horaria, Dreissena polymorpha, Procladius* sp., and *Pagastiella orophila* make up over 50 per cent of the total population. It differs conspicuously from the third sub-community due to the low occurrence of the genus *Chironomus* and the relatively large frequency of *Microtendipes pedellus.*

The sixth sub-community (Table 7.3f) settles the extensive foliage deposits (with an area of up to several hundred square metres and a thickness of over 1 metre, for instance Neuglobsow bay, north bay) which shelter a special grouping of the genera *Valvata, Bithynia, Gammarus, Asellus, Leptophlebia* and *Caenis. Nemoura cinerea* achieves its maximum abundance. Often 'stenotopic bog-dwellers' are to be found here, such as *Vejdovskyella comata* and *Monopelopia tenuicalcar.*

The absence of *Chironomus plumosus* is evidence of the low productive capacity of the waters. *Chironomus nuditarsis* exists in the lakes in the place of *Ch. plumosus.* In the plant stands *Dreissena polymorpha* accounts for over 50 per cent of the total population with slow growing (Lake Stechlin) or medium growing (Lake Nehmitz) specimens (with a length of 6 mm in 3 summers and 1 summer, respectively). As a whole, the Mollusca can exceed 75 per cent of the total population.

The zone of the submerged and floating plants in the epilimnetic range extends from a depth of 1.20–8.50 m and was subdivided into six sub-communities, of which only the first three carry a dense plant cover (Table 7.3). Both pondweeds and floating plants (*Nymphaea, Nuphar,* etc.) grow between the submerged Characeae carpets.

The first sub-community (Table 7.3a) comprises mainly the plant association of Charetum filiformis in the open region of the lake at a depth of 1.20–5 m (see p. 132). It has the highest abundance, with 60 per cent of a very slowly growing population being made up of *Dreissena polymorpha* (Mothes 1964c). From among the frequently occurring Trichoptera there is a maximum abundance of *Oxyethira flavicornis, Cyrnus trimaculatus* and, to a smaller extent, *Erotesis baltica.* From among Orthocladiinae there are mainly species of the *Cricotopus sylvestris* group and minor *Psectrocladius* species.

The second sub-community (Table 7.3b) is also situated on the open shore region of the lake and borders on the first one with increasing depth. The only plant association to grow at a depth of 5–8.5 m is Nitellopsidetum obtusae. The very densely growing and large plants probably impede the exchange of water up to the black, often smelly, organogenous soil. Most animals live on Characeae. Compared with the adjacent sub-communities, there are less than 35 per cent of individuals. There is only a maximum quantity of *Physa fontinalis* and *Arrenurus perforatus.* The particular role of gathering food from the pelagic zone is underlined by *Dreissena polymorpha,* with a total abundance of over 75 per cent.

The third sub-community (Table 7.3c) includes the stock of plants in the terminal bays at a depth range of 1.2–8.5 m. As a whole, it differs from the first one by a greater diversity of species, a decrease of filter feeders and an increase in sediment eaters and predators.

Here *Caenis horaria* reaches its maximum abundance. The Odonata *Enallagma cyathigerum* are complemented by *Erythromma naias* and *Pyrrhosoma nymphula.* Among Trichoptera there are often major forms such as species of the genera *Mystacides* and *Limnephilus.* Typical species are *Cyrnus flavidus, Ablabesmyia monilis* and the Orthocladiinae *Pogonocladius consobrinus* which is rare in the northern Central European region and which reduces the minor *Cricotopus* species and the genus *Psectrocladius.*

Sublittoral zone. The macrozoobenthonic community of the sublittoral zone lies in the transitional area between epilimnion and hypolimnion, or in the sector with decreasing light intensity, which no longer permits a positive assimilation balance of benthonic plants in depths of between 8.5–20 metres. In Lake Stechlin macroscopically visible plants grow in the whole zone, in contrast to Lake Nehmitz. Typical of the sublittoral zone is a relatively high amount of seston from the pelagic zone and a supply of fresh organic substance from the littoral communities situated above.

Despite the lush submerged plant carpets Lake Stechlin lacks the density of species of submerged plants in the epilimnetic depth zone (Table 7.1). In Lake Stechlin, where the number of species decreases at the transition from the epilimnetic to the metalimnetic zone to about a third, the sublittoral zone contains the following important species. From among planarians there are *Dugesia polychroa* and *Planaria torva;* from among the many snails of the epilimnetic region mention need only be made of *Physa fontinalis* and *Lymnaea peregra* in the sublittoral zone. From among Naidida, *Stylaria lacustris* reaches its maximum abundance; *Chaetogaster diaphanus* as well as the Hirudinea *Piscicola geometra* are worth mentioning. From among Trichoptera only *Oecetis ochracea* are to be found to a greater extent; Odonata are totally absent; Hemiptera and Coleoptera are negligible. From among Chironomidae five genera are worth mentioning: *Procladius, Microtendipes* (*chloris*), *Stictochironomus* (*pictulus*), *Polypedilum* and *Tanytarsus.*

From among particle catchers and filter feeders which mainly feed on the pelagic zone, there are in Lake Stechlin in certain places large numbers of *Cristatella mucedo,* at steep-sided places of Lake Nehmitz, banks of *Dreissena* as well as large numbers of *Anodonta. Microtendipes chloris,* as a fresh detritus eater, achieves a high population density. *Procladius nigriventris* sensu Brundin living there eats mainly large bottom diatoms.

Several sub-communities are to be distinguished (Table 7.4), of which only the most important species will be mentioned here.

The first sub-community (Table 7.4a) is marked by *Nitella flexilis* stocks in the open region of Lake Stechlin. It is the main range of distribution of Trichoptera *Oecetis ochracea* and Oligochaeta *Stylaria lacustris* (apart from periphyton covers in shallow water). *Ablabesmyia phatta* replaces *A. monilis* from the community of submerged macrophytes. Orthocladiinae are represented, above all, by large *Psectrocladius species.* The low abundance of Chironomini (mainly *Microtendipes chloris* occurring) is compensated by *Tanytarsus* species (among them *T. norvegicus*). In the upper region there is a great number of the Bryozoa *Cristatella mucedo.*

The second sub-community (Table 7.4b) is marked by poor growth of *Vaucheria* thalli (part of the Nitello-Vaucherietum association). Deposits of detritus are typical. *Polypedilum pullum* and *Asellus aquaticus* are widespread.

The moss stands are settled by the third sub-community (Table 7.4c). Here *Dreissena polymorpha* again accounts for a considerable share of 30 per cent. *Pisids* are, however, most conspicuous (e.g. *P. conventus*). Few species of Chironomini are found. Among them, *Stictochironomus pictulus* makes up 40 per cent. Tanytarsini and Hydracarina, however, are represented by a great number of species.

The *Nitella* stocks in the east bay of Lake Stechlin are inhabited mainly by *Chironomus* and *Einfeldia* species in contrast to those of the open region of the lake. Species depending on oxygen-rich water are absent (Table 7.4d).

The fifth sub-community (Table 7.4e) inhabits the sandy soil and, as the community of the surf zone, mainly harbours *Polypedilum pullum* from among Chironomini as well as *Microtendipes chloris*, both eurytopic sublittoral species, as fresh detritus eaters. Both account for a share of 90 per cent of Chironomini, i.e. over 25 per cent of the total stock. Of Mollusca, *Dreissena polymorpha* represents 60 per cent and the genus *Pisidium* 35 per cent, above all *P. casertanum*.

On gyttja rich in nutrients in the east bay of Lake Stechlin a population of small abundance lives mainly, made up of *Microtendipes chloris*. During spring no macrobenthonic fauna is to be found (Table 7.4f).

The borders of the above mentioned littoral communities correspond to those which were found in the analysis of the macrophytobenthos (see Krausch, p. 130–140). This points to a general validity of associations. On the other hand, it is impossible to assign zoocoenoses to every phytosociological unit. There are several examples where different associations harbour only one zoocoenosis. In the case of the Nitello-Vaucherietum, however, there are two sub-communities. Both components occur clearly separated in Lake Stechlin. The phytosociological

Table 7.4 Medium abundance (individuals · m^{-2}) of animal groups in the sublittoral zone. Sub-communities: a = Nitella populations in the open lake region; b = Vaucheria meadows in the open lake region; c = moss stands; d = Nitella populations in the east bay; e = sandy soil; f = detritus rich soil, east bay.

Sub-communities	a	b	c	d	e	f
Number of samplings	3	7	8	5	5	5
Total abundance	4,758	3,274	11,477	1,776	3,539	920
Hydrozoa	–	7	38	–	–	–
Turbellaria	221	22	6	–	10	–
Nematoda	–	–	6	31	–	–
Mollusca	1,020	276	8,620	592	662	51
Oligochaeta	1,835	611	547	122	327	51
Hirudinea	34	–	6	–	10	10
Hydracarina	272	73	38	10	113	31
Amphipoda/Isopoda	136	510	389	82	480	10
Ephemeroptera/Plecoptera	–	–	19	–	10	73
Odonata	–	–	–	–	–	–
Hemiptera	–	–	–	–	–	–
Coleoptera	–	7	–	–	–	–
Trichoptera	136	73	198	143	31	20
Heleidae	–	80	6	–	20	–
Tanypodinae	102	350	388	51	318	163
Orthocladiinae	51	29	19	31	31	10
Chironomini	170	850	707	490	1,120	459
Tanytarsini	781	364	484	347	285	31
Diptera (remainder)	–	22	6	–	10	–

and the zooecological subdivision systems are not identical, but reflect ecologically effective gradients in the ecosystem (Mothes & Blume 1969).

Profundal zone. The analysis of the profundal zone in Lake Stechlin results in the distinction of three ring-shaped sub-communities not clearly separated (Table 7.5).

The sub-community of the upper profundal, despite a modest total abundance, shows species that point to a nutrient supply from the sublittoral zone, such as *Microtendipes chloris* and *Stictochironomus pictulus.* It directly joins the sublittoral zone and extends to a depth of 35 m.

At a depth of 35–60 m the zones have but a very poor fauna, consisting of Nematoda (25 per cent), Tubificidae (25 per cent) and Chironomidae (barely 50 per cent). In this sub-community, which is the fundamentally characteristic one of Lake Stechlin's profundal zone, Nematoda and Tubificidae increase in absolute terms.

Table 7.5 Abundance (individuals · m^{-2}) of the profundal zone of Lake Stechlin. Depth zones of the sub-communities in metres: a = 20–35, b = 35–60, c = 60–68.

	Before			After		
	the bringing into operation of the outer cooling water circulation of the nuclear power plant					
Sub-communities	a	b	c	a	b	c
Number of samplings	17	26	6	19	27	4
Total abundance	582	256	975	2,733	2,013	1,364
Turbellaria	12	6	9	54	39	13
Nematoda	21	61	230	1,626	1,224	600
Dreissena polymorpha	–	–	–	3	2	–
Pisidium conventus	96	6	–	317	143	102
Tubificidae	51	67	440	228	224	357
Hirudinea	–	2	–	–	–	–
Hydracarina	13	2	–	13	6	–
Chaoborus flavicans	–	–	–	11	6	–
Caenis	–	–	–	3	–	–
Procladius	38	12	60	67	47	64
Psectrocladius	–	2	–	–	–	13
Cricotopus	–	–	–	–	4	38
Endochironomus	–	–	–	5	–	–
Microtendipes chloris	24	4	–	24	6	–
Stictochironomus pictulus	12	4	–	3	–	–
Polypedilum nubeculosum	–	–	–	–	2	–
Sergentia coracina	21	16	77	113	102	102
Lauterbornia coracina	30	25	17	21	37	25
Micropsectra lindebergi	210	49	142	21	2	–
Tanytarsus	–	–	–	221	150	51

A third sub-community is restricted to the largest depression of the lake which is 60–68 m deep. Here relatively much detritus is washed together. That is why the highest abundance of the profundal zone is to be found here, especially Nematoda (45 per cent) and Tubificidae (25 per cent). Regarding Chironomidae, there is an increase in *Sergentia coracina* typical of mesotrophic lakes.

The macrozoobenthonic communities after the nuclear power plant was brought into operation

The arrangement and make-up of the macrozoobenthonic communities totally changed after the outer cooling water circulation of the nuclear power station had become operational. The nutrient content of the profundal zone normally decreases with growing depth and distance from the littoral zone and again increases in the deepest depressions. Due to the cooling water circulation of the nuclear power station, nutrient is transported from Lake Nehmitz into Lake Stechlin and deposited mainly in the west bay into which the cooling water canal flows. A totally new population structure has developed; the former sub-communities are no longer recognizable.

Table 7.5 treats the same areas with regard to the changes after the starting of the cooling water circulation. The medium abundance is higher than during the initial condition; the increase is mainly due to the Nematoda. Beside Nematoda which have not been analysed in their species, *Pisidium conventus* and *Sergentia coracina* have increased most perceptibly.

With the change in the productivity of the ecosystem (see Koschel, p. 423), a change of species takes place in the community. Some examples need to be mentioned.

Before the nuclear power plant started operating, we collected only larvae of the *pictulus*-group of *Stictochironomus* in the surf zone of Lake Stechlin. After the nuclear power station began to work we found *St. rosenschöldi* living in this community of Lake Nehmitz as well as a small population of *St. histrio,* which is typical of eutrophic lakes. There are similar displacements as regards the genus *Cladotanytarsus.*

In Lake Stechlin Bryozoa were mostly limited to the metalimnetic zone, the sublittoral zone or the reed-beds. Only after the start of the nuclear power plant did they wander from the sublittoral zone into the zone of submerged macrophytes.

Before the nuclear power plant began to work Lake Stechlin had only the Zygoptera *Enallagma cyathigerum* from among frequent large predators. It may largely live on microcrustaceans; the larvae were fixed on the tips of the macrophytes. After the power plant started work *Coenagrion puella* and *C. pulchellum* restricted the distribution area of *E. cyathigerum.*

In the profundal zone individual species occur which normally do not live there and were transported there by the current (*Psectrocladius, Cricotopus*) or came from Lake Nehmitz (*Chaoborus*).

The community of the profundal zone of Lake Stechlin as a whole has not changed by the operation of the nuclear power plant. Due to the dominating exogenous nutrient influence the sub-communities have changed. Parts of the sublittoral zone of Lake Stechlin moved into the profundal zone owing to the decreasing transparency of the water when the outer cooling water circulation of the nuclear power plant was brought into operation; the sublittoral zone moved to a shallower part of the lake.

Such clear water was flowing through Lake Gerlin that the profundal zone became the littoral zone and the lake became a pond.

In the north basin of Lake Nehmitz the littoral zone increased, too, and the profundal zone was populated by *Chironomus* species, along with the *Chaoborus* larvae formerly found there.

Summarizing, we can establish that in the different lakes similar but contrary changes took place due to the inflow of the cooling water circuit. The areas and the species quantities of the communities have shifted in their structures, however, and the communities themselves remained preserved as units.

The biological indicators

According to Brundin (1956), Lake Stechlin is an oligotrophic lake and a *Tanytarsus* lake. The species which are of importance as biological indicators are Chironomidae *Lauterbornia coracina, Micropsectra lindebergi, Tanytarsus niger, Tanytarsus norvegicus* and Mollusca *Pisidium conventus* (Meier-Brook & Mothes 1966; Mothes 1966b; 1968; 1971a, b). They all demand a hypolimnion rich in oxygen and are absent in the other lakes of our study area. The low abundance of *Sergentia coracina* and the absence of *Chironomus anthracinus* point to the low productive capacity of that lake. Other less typical inhabitants of the profundal zone in Lake Stechlin are *Procladius choreus, P. fuscus, Microtendipes chloris* and *Stictochironomus pictulus*. Characteristic forms of eutrophic lakes have not been found in the profundal zone of Lake Stechlin.

Lake Nehmitz, as an example of a less deep lake, is of the eutrophic type due to its *Chironomus* population in and above the chemocline. However, this lake is, likewise, a poor producer, for it only harbours the characteristic species *Chironomus anthracinus* and not *Ch. plumosus*. Moreover, in the lakes there are the Chironomidae *Tanytarsus miriforceps, T. sylvaticus* and *Stictochironomus rosenschöldi*. In the deep profundal zone only *Chaoborus flavicans* is to be found. Rarer species are *Corynocera ambigua* and *Propsilocerus lacustris*. Therefore, the term mesotrophic would better suit the type of Lake Nehmitz in regard to its productivity (see also p. 465). After the supply of nutrient-poor (oligotrophic) water from Lake Stechlin into the north basin of Lake Nehmitz after the bringing into operation of the nuclear power plant, a small population of *S. coracina* developed there.

Although 'stenotopus bog-dwellers' have been found in the littoral zone of Lake Stechlin (p. 239, p. 510), there are no characteristic species of dystrophic lakes (e.g. *Chironomus tenuistylus*) in the sediments under the open-water areas. The dystrophic type or a widely spread dy sediment (Schönborn *et al.* 1965) cannot be established by biological indicators of the macrobenthic fauna (Mothes 1965d) in the study area.

The zooplanktonic population (D. Ronneberger & P. Kasprzak)

Introduction

In the past the zooplankton of Lake Stechlin was only occasionally the object of investigations. Existing papers deal mainly with taxonomic – structural aspects (Hartwig 1897; Schönborn 1962; Koch-Althaus 1962; Flößner 1972). For the determination of matter- and energyflux in a nutritional web, quantitative datas are necessary. This demand was considered the first time by Grüneberg & Witte (1972). However, intensive investigations about zooplankton distribution in time and space started at the end of the seventies (Kasprzak & Ronneberger 1982). The interest was directed to Cladocera, Copepoda and Rotatoria, because they represent the main part of the zooplankton community. With a certain regularity *Difflugia limnetica* (Testacea; Schönborn 1962; cf. Fig. 7.5) and some ciliates (Grüneberg & Witte 1972) are found in the pelagic zone of Lake Stechlin.

Methods

The zooplankton samples were taken at the selected measuring points on Lake Stechlin, using closing nets with a mesh size of 50 μm and 110 μm (from knot to knot) for rotifers and crustaceans, respectively. The filtration efficiency of the rotifer net was only 12% (calibration with a current meter). Hence a correction factor was introduced. The fixation of animals was carried out using a formalin-sugar-mixture (Haney & Hall 1973). In the laboratory part of the original sample was counted in the microscope. This subsample contained about 100 animals of the numerically dominating species. Biomass values are given as carbon equivalents. For this purpose literature datas (Latja & Salonen 1978; Dumont *et al.* 1975; Bottrell *et al.* 1976) as well as our own results (Kasprzak 1983a) were used. The conversion from dry weight to carbon was made on the assumption $DW \cdot 0.5 \triangleq$ carbon (Winberg 1971).

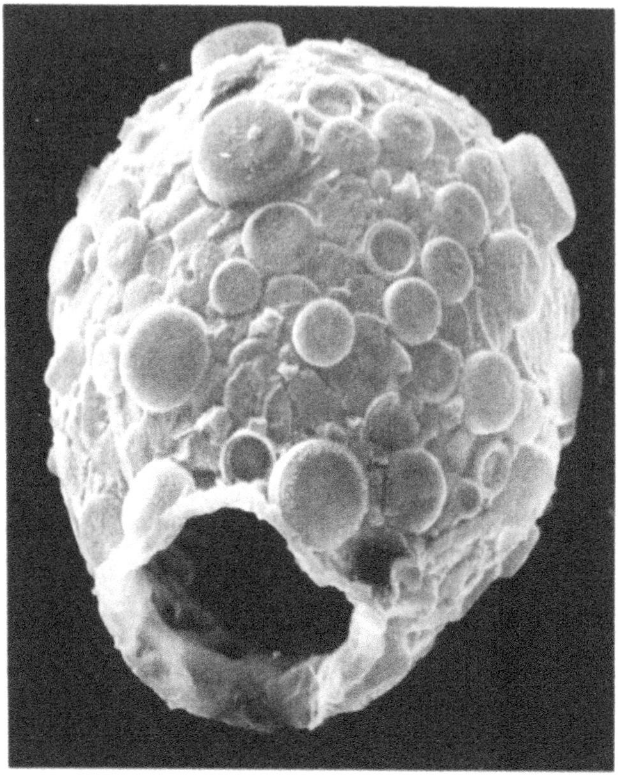

Figure 7.5 *Difflugia limnetica*, REM, × 1,000. Photo: Th. Peschke.

Annual distribution of individual numbers and biomass of crustaceans and rotifers

The number of zooplankton species living in Lake Stechlin is included in the species list (see p. 500). Eight species of cladocerans, 3 species of calanoid and 4 species of cyclopoid copepods, as well as 30 species of rotifers were found.

The growth of crustaceans starts in April (Fig. 7.6), reaching maximal quantities of individuals and biomass in summer and decreases in November/December. With regard to biomass the spectrum of dominance is rather clear (Fig. 7.7). In spring and autumn *Eudiaptomus gracilis* and/or *Eurytemora lacustris* are dominant and, in addition, during summer *Daphnia cucullata*.

The individual numbers and biomass of rotifers show a similar pattern with a maximum in summer (monthly mean from 1978–1982). The number of individuals varies from $5 \cdot 10^4$ to $60 \cdot 10^4$ individuals $\cdot$ m^{-3} or $5 \cdot 10^5$–$10 \cdot 10^6$ individuals $\cdot$ m^{-2}, respectively. Expressed as biomass, it amounts to 1–8 mg C $\cdot$ m^{-3} or 10–270 mg C $\cdot$ m^{-2} (Figs. 7.8, 7.9). Comparing certain groups, the dominance structure is similar from year to year. In March and April the species of *Synchaeta* (*S.*

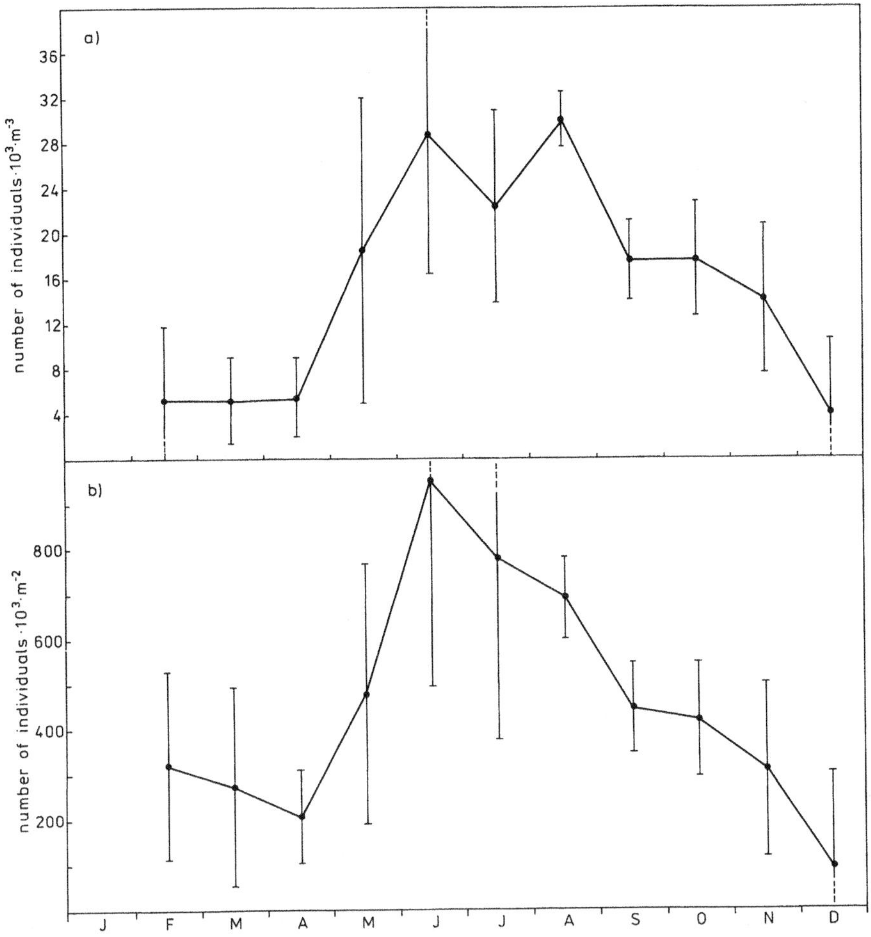

Figure 7.6 Annual distribution of crustaceans. Number of individuals per m^3 and m^2, monthly means 1978–1980; 95% confidence limits.

pectinata, S. oblonga, and *S. tremula*) are important, and, in addition, in some years species of the genus *Ascomorpha* (*A. ecaudis* and *A. saltans*) and *Pompholyx sulcata.* In May the species composition changes. Now *Polyarthra dolichoptera, Conochilus unicornis* and *Filinia terminalis* become dominant.

During summer the perennial forms *Keratella cochlearis* and *Kellicottia longispina* reach their greatest abundance. In autumn the growth of rotifer zooplankton decreases and species like *Filinia terminalis, F. longiseta, Polyarthra dolichoptera* and *P. vulgaris* are the main constituents of the rotifer community (Kasprzak & Ronneberger 1982).

The mentioned dominance structure coincides with the generalization of Morgan *et al.* (1980) for Baltic lakes, and in accordance with the classification scheme

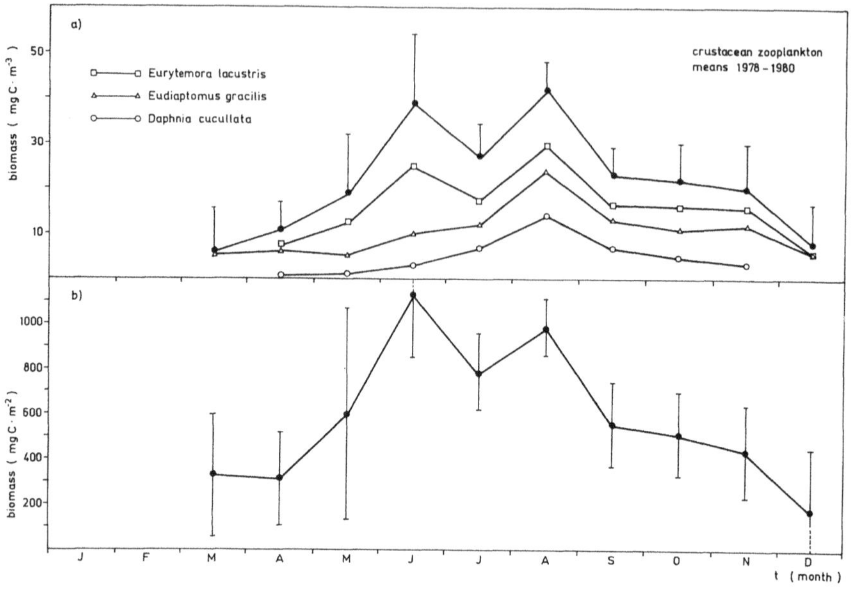

Figure 7.7 Annual distribution of crustacean biomass mg C per m^3 and m^2, monthly means 1978–1980; 95% confidence limits.

of Geller & Müller (1981) the crustacean community of Lake Stechlin reflects the transition stage from oligotrophy to mesotrophy. In spring and autumn calanoid copepods dominate, and, in addition, a fine filter-feeder (*Daphnia cucullata*) appears during summer. Similar structures of dominance were also found in more productive lakes of the area, but *Eurytemora lacustris* disappeared in eutrophic lakes, so that *Eudiaptomus gracilis* alone represents the group of calanoid copepods. Moreover, in spring and autumn there is a large population of *Cyclops kolensis* (Kasprzak 1983b) in these lakes. Such relatively simple structures are of great advantage for determining matter- and energy-flux between primary and secondary producers, because only a few species are the subject of investigation.

Temperature and available food play an important role in zooplankton growth. According to Richter *et al.* (1979, see also p. 52) the temperature in Lake Stechlin shows a sharp increase from March to June. The monthly means are 1.2, 4.9, 11.2 and 17.5° C. Thus egg duration and food assimilation are stimulated more and more. Moreover, the primary production of phytoplankton is high and the biomass frequently reaches a first maximum (Koschel 1981), so that the growth conditions for animals are relatively good (see Table 11.6, p. 339).

Not much is known about the coexistence of the two calanoid copepods under limiting food conditions such as in Lake Stechlin (see pp. 324 ff.). According to Pesta (1928) and Kiefer (1960), both species are common in the crustacean plankton of the larger lakes in the Baltic area. Nauwerck (1963) assumes similar environmental demands for both species. *Eurytemora affinis* is a slightly smaller

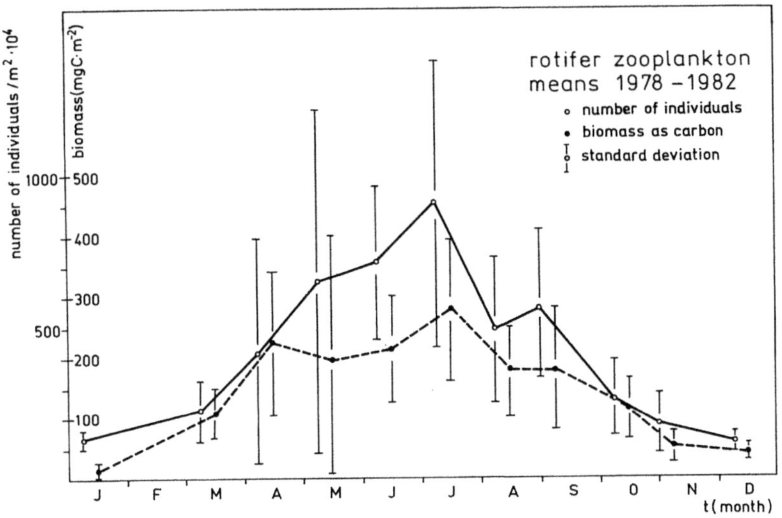

Figure 7.8 Annual distribution of rotifers. Number of individuals and biomass per m², monthly means 1978–1982; standard deviations.

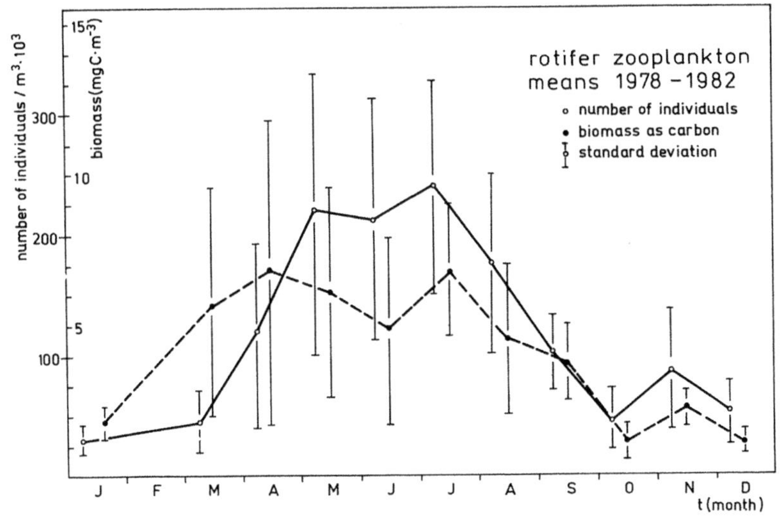

Figure 7.9 Annual distribution of rotifers. Number of individuals and biomass per m³, monthly means 1978–1982; standard deviations.

form from North American lakes (Friedman 1980). The niche separation in nutrition biology against other calanoid crustaceans is effected by filter-feeding of greater particles (Richman *et al.* 1980). *Daphnia cucullata* is regarded as a filtrator of small particles compared with calanoid copepods (Morgan *et al.* 1980; Geller & Müller 1981). For this reason the coexistence of the three species in discussion may be possible, assuming that *Daphnia cucullata* feeds on small particles, *Eudiaptomus gracilis* on average and *Eurytemora lacustris* on larger ones.

Rotifers such as herbivorous filter-feeders utilize mainly bacteria, small detritus and nano- or ultraplanktonic algae smaller than 15 μm (Gliwicz 1974). Only species of the genus *Polyarthra, Synchaeta* and *Gastropus* also feed on larger particles. Thus nanoplanktonic diatoms, flagellates and other ultraplanktonic algae dominant in Lake Stechlin (13–50 $\mu g\ C \cdot l^{-1}$; see p. 185) can cause the growth of rotifers. Even in other months, e.g. June, a close connection between phytoplankton and rotifers development is evident (Kasprzak & Ronneberger 1982). This is possible because of the high rate of eggs produced per time unit and the short duration time. Halbach & Halbach-Keup (1974) found in *Brachionus calyciflorus* a hatching time of two days (related to the food conditions of Lake Stechlin). Rotifers can thus quickly react to varying food conditions.

The 95% confidence limit and the standard deviations in Figs. 7.7–7.9 indicate a wide scatter of biomass production.

One reason is the difference in primary production. In 1972 it amounted to 378 mg $C \cdot m^{-2} \cdot d^{-1}$ and in 1975 only 186 mg $C \cdot m^{-2} \cdot d^{-1}$ (Koschel 1981). In Fig. 7.10 the zooplankton biomass of some lakes is related to their primary production. Though these results can show no more than a trend, the positive influence of biomass production by algae on secondary production is obvious.

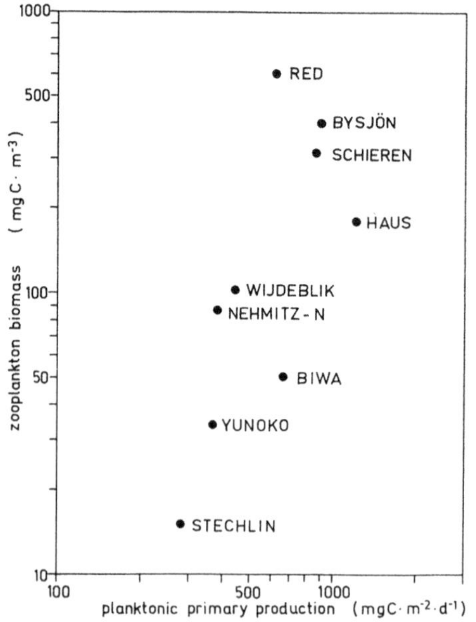

Figure 7.10 Mean crustacean biomass of some selected lakes as a function of their planktonic primary productivity. Lake Stechlin, Lake Nehmitz, Lake Haussee: Crustacean biomass after Kasprzak (1983b), primary productivity after Koschel *et al.* (1981); Red Lake (Andronikowa 1972), Lake Bysjön (Coveney *et al.* 1977), Lake Schieren (Nowak 1975), Lake Wijde Blik (Gulati 1972), Lake Biwa (Syuiti 1975), Lake Yunoko (Tadashi 1975).

Vertical distribution and diurnal migration

With respect to vertical distribution of zooplankton, one can distinguish between longterm and shortterm patterns. Longterm patterns, in our sense, will be the vertical distribution of biomass according to the regular sampling occasions. Shortterm patterns are established first of all by diurnal migration.

Table 7.6 shows some characteristical constellations of crustacean depth distribution. The animals have a tendency to stratification in the upper region and/or in the metalimnion throughout almost the whole year. Only during circulation in spring and autumn are deeper layers also more inhabited. This pattern is superimposed in a specific way by the diurnal migration (Fig. 7.11). In the morning (9:00 a.m.) the animals prefer the depth around 15 m, but shortly after midnight the surface layer. Because the results are based on dry weight estimates of the total samples, it is impossible to show to what degree the single species takes part in the daily migration. But according to former investigations by Grüneberg & Witte (1972) *Eudiaptomus gracilis, Eurytemora lacustris, Daphnia cucullata, D. hyalina* and *Bosmina coregoni* are the main migrating forms.

The vertical distribution of rotifers is summarized in Fig. 7.12. The main occurrence is situated in the layer from 0–25 m. During circulation an enrichment is evident in upper regions from 0–15 m, whereas in stratification the animals prefer deeper strata from 15–20 m. The distribution of single species is rather different. In spring the dominant *Synchaeta pectinata* as well as *Filinia terminalis* change their main centres with the start of stratification to deeper layers.

Diel vertical migration in rotifers is not so clearly expressed as in the case of crustaceans and the amplitude is rather small (up to 5 m; George & Fernando 1970; Pivoda 1977). Grüneberg & Witte (1972) confirmed these results. They

Table 7.6 Selected depth distribution of crustacean zooplankton in Lake Stechlin 1980/81 (E = epilimnion, M = metalimnion, H = hypolimnion).

Date	Biomass ($mg\ C \cdot m^{-3}$)			(%)		
	Depth (m)			Depth (m)		
	0–20	20–40	40–60	0–20	20–40	40–60
25. 3.80	3.95	1.75	2.63	46	21	33
	E	M	H	E	M	H
21. 7.81	23.53	64.58	18.96	25	58	17
10. 9.81	17.29	48.38	8.54	23	65	17
27.10.81	43.60	–	5.68	85	–	15
10.11.81	25.55	–	44.95	36	–	64

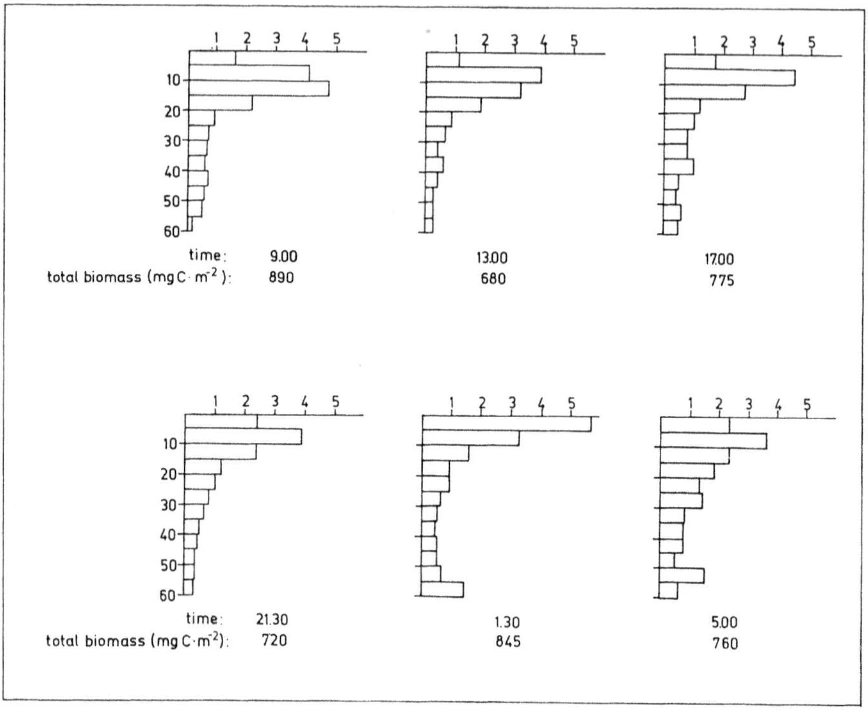

Figure 7.11 Diurnal migration of crustacean biomass, June 1980.

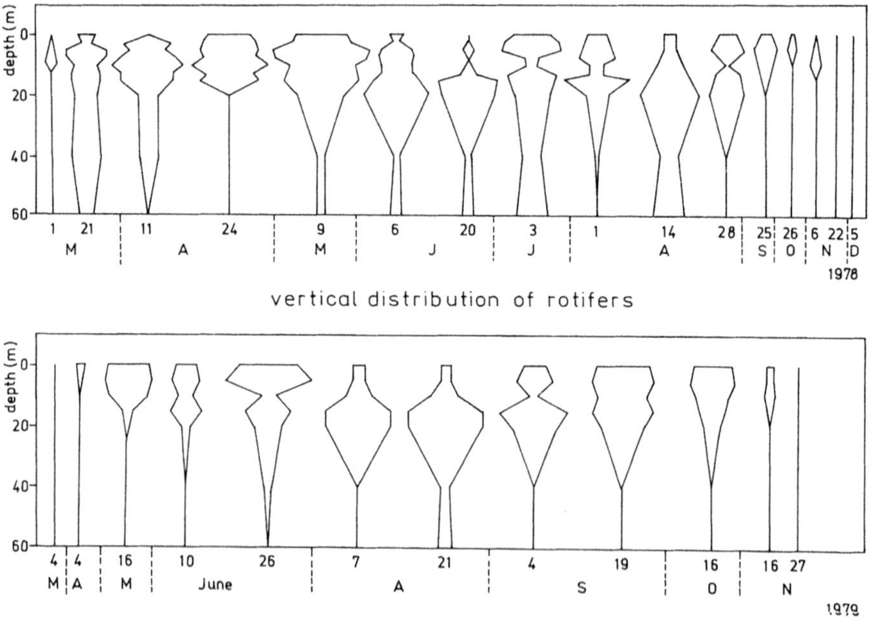

Figure 7.12 Vertical distribution of rotifer plankton in Lake Stechlin, 1978 and 1979.

found no regular migration pattern in *Keratella cochlearis, Kellicottia longispina, Polyarthra dolichoptera* and *P. vulgaris* living in Lake Stechlin.

Much work was done on the adaptive significance of different depth distributions on the level of population, as well as single individuals. The predation-hypothesis (Zaret & Suffern 1976; Wright *et al.* 1980), light response (Nauwerck 1963; Siebeck 1960), quality and quantity of food (Parre 1979; Bohrer 1980) and the rhythmic nature of some physiological performances (Haney & Hall 1975; Duval & Geen 1976; Gyllenberg 1981) were of special interest. Bohrer (1980) showed in vitro an intensive migration of zooplankton to the stratum with the highest particle concentration. According to Küchler (1981) and Koschel (1981), it can be assumed that the enrichment of phytoplankton is most pronounced in the 0–20 m layer. The biomass of bacteria is distributed more homogeneously over the whole water column (Babenzien, pers. comm.). The same seems to be valid for detritus (Proft, pers. comm.), which means that the crustaceans prefer the layer of high temperature where a part of their food is produced.

Comparing respiration, assimilation, ingestion and filtration, there is another interesting aspect (Table 7.7). During migration the animals cross a temperature gradient and consequently their metabolism is submitted to a diurnal rhythm. Concerning temperature respiration decreases faster than assimilation. That means, the growth of the animals becomes smaller, because the energy and matter input is also a function of temperature, but the faster decrease of respiration losses may be interpreted as a compensation mechanism (McLaren 1963).

The shift of main centres of distribution of certain rotifer species in deeper layers of water can be understood as a separation of food niches between crustaceans and rotifers, where the temperature has a controlling influence. Ruttner-Kolisko (1980) found the maximum population size in *Filinia terminalis* at 12° C. Similar findings in Lake Stechlin confirm this result (Kasprzak & Ronneberger 1982). However, the priority between temperature and food distribution (nanoplanktonic algae) on determination of the rotifer abundance is obscure. According to Küchler (1981) sometimes in summer there are maxima of nanoplanktonic diatoms at a depth of around 10–15 m. Koschel (1981) has shown an occasional increasing primary production in the same depth during summer.

Table 7.7 Temperature dependency of respiration, assimilation, ingestion and filtration in *Daphnia* (relative values). Respiration: Goss & Bunting (1980), *Daphnia pulex,* egg-carrying females, no notice about size. Assimilation: Lampert 1977, *D. pulex,* females, size 2 mm. Ingestion and filtration: Geller (1975) *D. pulex,* females, size 2 mm.

Temperature (° C)	Respiration (%)	Assimilation (%)	Ingestion (%)	Filtration (%)
20	100	100	100	100
15	62	85	73	70
10	41	43	48	60

Another possibility for abundance maxima of *Keratella cochlearis* in deeper water is the so-called 'plankton rain' concentrated in the intervening depth between epi- and metalimnion which contains detritus and living particles. It may be that this detritus mixture causes the enrichment of the omnivorous *Keratella cochlearis*. Gilbert & Starkweather (1981) have found no selection of food by this species. *Polyarthra dolichoptera* and *P. vulgaris* are selective filter-feeders, preferring flagellated crypto- and chrysomonades (Gilbert & Starkweather 1981), which appear in Lake Stechlin in spring and autumn at low temperatures (Küchler 1981). The growth of *Polyarthra* species takes place at the same time.

Horizontal distribution

Vertical differences in biomass concentration as well as horizontal gradients are of great importance for the intensity of metabolism in the lake. Investigations carried out in June 1980, October 1981 and May 1982 should give an indication to such inhomogeneities. The measuring points are to be seen in Fig. 7.13. The cooling water circulation of the nuclear power plant was inoperative on both occasions (1980, 1981). In the case of crustaceans, down to a depth of around 30 m as a first approximation, the biomass seems to be a function of depth. In other regions deeper than 30 m the values vary in a similar range. Not only the absolute values but also the specific biomass content per cubic metre is different in various lake areas at identical depths (Fig. 7.14; Table 7.8). Contrary to this, the individual numbers and biomass distributions of rotifers are more homogeneous throughout the whole lake (Table 7.9), excepting station 1.

There are a number of reasons for horizontal changes in biomass standing crops. Mullin & Brooks (1976) found a positive correlation between phytoplankton and zooplankton biomass in marine biotops. Stavn (1971) deals with the influence of Langmuir-spirals on zooplankton distribution. Nauwerck (1978), researching a Swedish lake, found several independent growth centres for *Bosmina obtusirostris*. In Lake Stechlin the decreasing biomass of crustaceans in the littoral region is quite obvious. This finding accords well with the data of Siebeck (1980) about 'Uferflucht' (avoidance of the shore), but the values in the pelagic zone cannot be so explained. The small horizontal differences in primary production of phytoplankton cannot be the cause of such a scatter (Koschel, pers. comm.). Ward (1981) also found a good correlation between biomass distribution of *Daphnia pulicaria* and water depth. Patchy distributions were exceptional.

The 'Uferflucht' in the sense of Siebeck (1968) seems to be unimportant for rotifers at a distance of more than 40 m from the shore (Pressler 1977), whereas wind causes drifting and Langmuir-spirals have a strong influence (Margalef 1974; Miracle 1977).

Summarizing the results of vertical and horizontal distribution, there is a tendency for the crustacean biomass to decrease in the direction of shore and

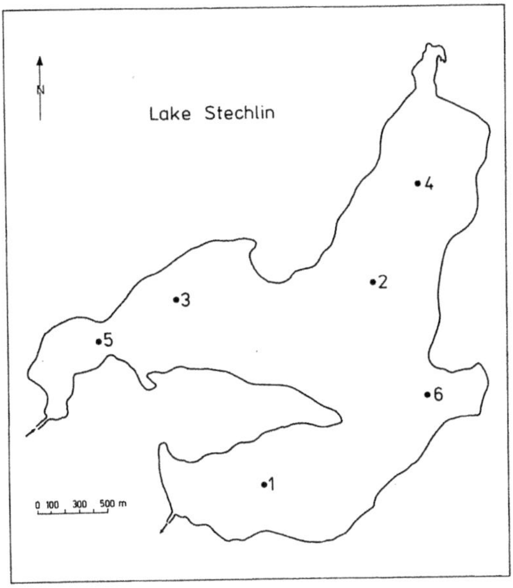

Figure 7.13 Sampling stations for the estimation of horizontal distribution of zooplankton in Lake Stechlin.

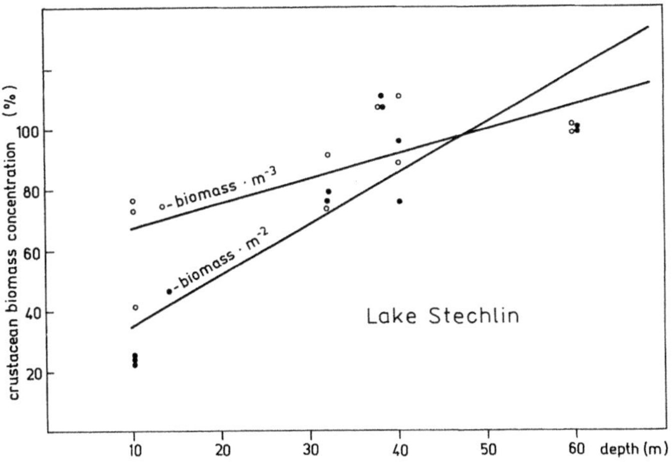

Figure 7.14 Crustacean biomass in Lake Stechlin as a function of depth; results above 68 m (station 2) are defined as 100%.

bottom, and the vertical distribution is modified in a characteristic way by diurnal migration. The rotifer horizontal distribution is more constant. The vertical distribution has a characteristic summer maximum at 15–20 m and no diurnal migration.

Table 7.8 Crustacean biomass at various locations of Lake Stechlin.

Date	Location	Depth (m)	Biomass	
			(mg C · m^{-3})	(mg C · m^{-2})
June 1980	1	0–10	–	
		10–20	16.50	
		20–38	10.60	
	2	0–10	20.50	
		10–20	24.25	
		20–40	5.35	
		40–60	3.95	638.0
	3	0–10	18.75	
		10–20	20.70	
		20–32	7.35	482.5
	4	0–10	22.85	
		10–20	17.95	
		20–40	3.75	483.5
	5	0–10	15.75	158.0
	6	0–10	15.10	151.0
October 1981	1	0–10	33.50	
		10–20	19.75	
		20–38	2.10	570.5
	2	0–10	31.35	
		10–20	13.25	
		20–40	1.40	
		40–60	1.95	512.5
	3	0–10	23.2	
		10–20	15.05	
		20–32	2.20	409.0
	4	0–10	27.75	
		10–20	16.85	
		20–40	2.30	492.0
	5	0–10	12.85	128.5

Longterm variations

The ecosystem of Lake Stechlin is characterized by increasing nutrient loading from different sources such as the nuclear power plant or the water discharge from Lake Dagow (p. 411 ff.). The process of nutrient enrichment is also accompanied by changes at the level of zooplankton (Ravera 1977). In the case of Lake Stechlin it is difficult to exclude or determine a tendency in the development of crustacean community because quantitative data are scarce. But compared with the results of Hartwig (1897) and such of the sixties (Flössner, pers. comm.) no obvious change took place. Both indicated *Daphnia cucullata, Eudiaptomus gracilis* and *Eurytemora lacustris* as the most important species, as is the case at present.

Also the rotifer community remained constant from a quantitative point of

Table 7.9 Horizontal distribution of rotifer biomass May 1982, Lake Stechlin.

Location	Depth (m)	Biomass	
		(mg C · m^{-3})	(mg C · m^{-2})
1	0–10	14.59	
	10–20	17.37	
	20–40	8.87	497.0
2	0–10	33.53	
	10–20	16.81	
	20–40	4.48	
	40–60	1.97	632.4
3	0–10	27.75	
	10–20	22.73	
	20–30	9.98	604.6
4	0–10	27.93	
	10–20	12.9	
	20–40	3.78	483.9
5	0–10	27.65	276.5
6	0–10	28.99	289.9

view. The dominant species (see p. 218, Koch-Althaus 1964; Kasprzak & Ronneberger 1982) were the same before and after the nuclear power plant started operation. But some changes in species composition may indicate a beginning process of eutrophication at the level of rotifer population. *Notholca acuminata, N. foliaceae* and *Sinantherina socialis* disappeared, also the regular and frequently occuring *Collotheca libera* and *Synchaeta grandis.* Other species, more common in eutrophic lakes – like *Anuraeopsis fissa, Brachionus angularis, Conochilus natans, Pompholyx complanata, P. sulcata* and *Trichocerca capucina* – appeared after the power plant began to operate.

References

Altherr, E. 1968. Nématodes de la nappe phréatique du réseau fluvial de la Saale (Thuringe) et psammiques du Lac Stechlin (Brandenbourg du nord). Limnologica 6: 247–320.

Andronikowa, I.N. 1972. Biological productivity of the main communities of the Red Lake. In: Productivity problems of fresh waters. Eds. Z. Kajak & A. Hillbricht-Ilkowska. Warszawa-Krakow: Polish Science Publishers. 57–71.

Bohrer, R.N. 1980. Experimental studies on the diel vertical migration. In: Evolution and ecology of zooplankton communities. Ed. W.C. Kerfoot. New England: Univ. Press. 111–121.

Bottrell, H.H., Duncan, A., Gliwicz, Z.M., Grygierek, E., Herzig, A., Hillbricht-Ilkowska, A., Kurasawa, H., Larsson, P. & Weglenska, T. 1976. A review of some problems in zooplankton production studies. Norw. J. Zool. 24: 419–456.

Brundin, L. 1956. Die bodenfaunistischen Seetypen und ihre Anwendbarkeit auf die Südhalbkugel. Inst. Freshw. Res. Drottningholm, Report 37: 186–235.

Coveney, M.F., Cronberg, G., Enell, M., Larsson, N.K. & Olofsson, L. 1977. Phytoplankton, zooplankton and bacteria standing crop and production relationships in a eutrophic lake. Oikos 29: 5–21.

Dumont, H.H., Van de Velde, J. & Dumont, S. 1975. The dry weight estimate of biomass in a selection of cladocera, copepoda and rotifera from the plankton, periphyton and benthos of continental waters. Oecologia (Berl.) 19: 75–97.

Duval, W.S. & Geen, G.H., 1976. Diel feeding and respiration rhythm in zooplankton. Limnol. Oceanogr. 21: 823–829.

Enckell, P.H. 1969. Distribution and dispersal of Parastenocarididae (Copepoda) in northern Europe. Oikos 20: 493–507.

Flößner, D. 1962a. Zur Cladocerenfauna des Stechlinseegebietes. I. Über Morphologie und Variabilität einiger Formen und über Funde seltener Arten. Limnologica 1: 217–229.

Flößner, D. 1962b. Zur Variabilität und Synonymie von *Chydorus piger* G. O. Sars. Zool. Anz. 169: 397–401.

Flößner, D. 1963. *Scapholeberis microcephala* (Lilljeborg MS) G. O. Sars 1890 und *Alona karelica* Stenroos 1897 (Crustacea, Cladocera) in Norddeutschland gefunden. Zool. Anz. 170: 253–254.

Flößner, D. 1964. Zur Cladocerenfauna des Stechlinseegebietes. II. Ökologische Untersuchungen über die litoralen Arten. Limnologica 2: 35–103.

Flößner, D. 1965. *Limnocythere psammophila* n. sp., ein neuer Muschelkrebs (Crustacea, Ostracoda) der deutschen Fauna. Zool. Anz. 175: 466–470.

Flößner, D. 1967. *Alona rustica* La Scott und *Alona costata* Sars (Cladocera, Chydoridae). Zur Kenntnis der Morphologie und Ökologie einer Zwillingsart. Limnologica 5: 417–434.

Flößner, D. 1972. Krebstiere, Crustacea. Kiemen- und Blattfüßer, Branchiopoda. Fischläuse, Branchiura (Die Tierwelt Deutschlands 60). Jena: VEB Gustav Fischer. 1–501.

Frenzel, P. 1982. Die Coenosen der litoralen Cladocera (Crustacea) des Bodensees. Arch. Hydrobiol. Suppl. 62: 291–331.

Friedmann, M.M. 1980. Comparative morphology and functional significance of copepod receptors and oral structures. In: Evolution and ecology of zooplankton community. Ed. W. C. Kerfoot. New England: Univ. Press. 185–197.

Fryer, G. 1968. Evolution and adaptive radiation in the Chydoridae (Crustacea: Cladocera): a study in comparative functional morphology and ecology. Phil. Trans. Roy. Soc. London, Ser. B. Biol. Sci. 254 (795): 221–385.

Fryer, G. 1974. Evolution and adaptive radiation in the Macrothricidae (Crustacea: Cladocera): a study in comparative functional morphology and ecology. Phil. Trans. Roy. Soc. London, Ser. B. Biol. Sci. 269 (898): 137–274.

Geller, W. & Müller, H. 1981. The filtration apparatus of cladocera: filter mesh-sizes and their implications on food selectivity. Oecologia (Berlin) 49: 316–321.

George, M.G. & Fernando, C.H. 1970. Diurnal migration in three species of rotifers in Sunfish Lake, Ontario. Limnol. Oceanogr. 15: 218–223.

Gilbert, J.J. & Bogdan, K.G., 1981.Selectivity of *Polyarthra* and *Keratella* for flagellate and aflagellate cells. Verh. Int. Verein. Limnol. 21: 1515–1521.

Gliwicz, Z.M. 1964. Trophic status of freshwater zooplankton species. Wiadom. Ecol. 20: 197–206.

Grüneberg, G. & Witte, W. 1972. Erste Beiträge zur Populationsdynamik des Zooplanktons im Großen Stechlinsee. Diplom-Arbeit, Pädagogische Hochschule Potsdam. 1–100 (manuscript).

Gulati, R.D. 1972. Limnological studies on some lakes in the Netherlands. Freshw. Biol. 2: 37–54.

Gyllenberg, G. 1981. *Eudiaptomus gracilis* (Copepoda, Calanoida): diurnal migration in the field and diel oxygen consumption rhythm in the laboratory. Ann. Zool. Fennici 18: 229–232.

Halbach, U. & Halbach-Keup, G. 1974. Quantitative Beziehungen zwischen Phytoplankton und der Populationsdynamik des Rotators *Brachionus calyciflorus*. Arch. Hydrobiol. 73: 273–309.

Haney, F.J. & Hall, D.J. 1973. Sugar-coated *Daphnia*, a preservation technique for Cladocera. Limnol. Oceanogr. 18: 331–333.

Haney, J.F. & Hall, D.J. 1975. Diel vertical migration and filter feeding activities of *Daphnia*. Arch. Hydrobiol. 75: 413–441.

Hartwig, W. 1897. Zur Verbreitung der niederen Crustaceen in der Provinz Brandenburg. Forschungsber. Biol. Stat. Plön 5: 115–149.

Herr, O. 1917. Die Phyllopodenfauna der preußischen Oberlausitz und der benachbarten Gebiete. Abh. naturforsch. Ges. Görlitz 28: 1–162.

Hollwedel, W. 1978. Zur Cladoceren-Fauna des Sager Meeres. Teil II: Litorale Cladoceren und eine Mitteilung über Ostracoden-Funde. Oldenburger Jb. 75/76: 145–182.

Kasprzak, P. 1983a. Bestimmung des Körperkohlenstoffs von Planktoncrustaceen. Limnologica 15: 191–194.

Kasprzak, P. 1983b. Struktur, Populationsdynamik und Assimilationsleistung des Crustaceenplanktons im oligotrophen Stechlinsee. Diss. A Techn. Univ. Dresden (manuscript).

Kasprzak, P. & Ronneberger, D. 1982. Vergleichende Untersuchungen zur Struktur und Dynamik des Zooplanktons im Stechlinsee, Nehmitzsee und Haussee 1978/79. Limnologica 14: 263–295.

Keilhack, L. 1908. Zur Cladocerenfauna der Mark Brandenburg. Mitt. Zool. Mus. Berlin 3: 435–488.

Kiefer, F. 1960. Ruderfußkrebse (Copepoden). Stuttgart: Kosmos Verlag. Frankh. 1–97.

Klie, W. 1937. Die Entomostrakenfauna kalkarmer Seen Norddeutschlands mit vergleichsweiser Berücksichtigung normal kalkhaltiger Seen des gleichen Gebietes. Arch. Hydrobiol. 31: 85–100.

Koch-Althaus, B. 1961. *Cephalodella theodora* n. sp., ein neues Rotator aus dem Stechlinsee. Zool. Anz. 166: 219–221.

Koch-Althaus, B. 1962. Weitere neue und bemerkenswerte Rotatorien aus dem Stechlinsee. Limnologica 1: 65–82.

Koch-Althaus, B. 1963. Systematische und ökologische Studien an Rotatorien des Stechlinsees. Limnologica 1: 375–456.

Koschel, R. 1981. Bilanz der Primärproduktion des Phytoplanktons in den Gewässern des Stechlinseegebietes. Limnologica 13: 55–81.

Krausch, H.-D. 1964. Die Pflanzengesellschaften des Stechlinseegebietes. I. Die Gesellschaften des offenen Wassers. Limnologica 2: 145–203.

Küchler, L. 1981. Phytoplanktonuntersuchungen im Stechlinseegebiet in den Jahren 1973–1975. Limnologica 13: 55–81.

Lang, K. 1948. Monographie der Harpacticiden. II. Lund: Håkan Ohlssons Boktrykeri.

Latja, R. & Salonen, K. 1978. Carbon analysis for the determination of individual biomass of planktonic animals. Verh. Int. Verein. Limnol. 20: 2556–2560.

Lundbeck, J. 1951. Zur Kenntnis der Lebensverhältnisse in sauren Binnenseen. Arch. Hydrobiol. Suppl. 20: 18–117.

Margalef, R. 1974. Ecologiá. Barcelona: Omega. 1–951.

McLaren, J.A. 1963. Effects of temperature on growth of zooplankton and the adaptive value of vertical migration. J. Fish. Res. Board Can. 20: 685–727.

Meier-Brook, C. & Mothes, G. 1966. Die Mollusken des Stechlinsees. Berichtigungen und Ergänzungen. Limnologica 4: 483–487.

Meuche, A. 1939. Die Fauna im Algenbewuchs. Nach Untersuchungen im Litoral ostholsteinischer Seen. Arch. Hydrobiol. 34: 349–520.

Miracle, M.R. 1977. Migration, patchiness and distribution in time and space of planktonic rotifers. Arch. Hydrobiol. Beih. Ergebn. Limnol. 8: 19–37.

Möller, L. 1935. Stechlinsee und Sakrower See. Arch. Hydrobiol. 29: 137–156.

Morgan, N.C., Bachiel, T., Bretschko, G., Duncan, A., Hillbricht-Ilkowska, A., Kajak, Z., Kitchell, J.F., Larsson, P., Leveque, C., Nauwerck, A., Siemer, F. & Thorpe, J.E. 1980. Secondary production. In: The function of freshwater ecosystems. Eds. E. D. Le Cren & R. H. Lowe-McConnel. 247–340.

Mothes, G. 1964. Die Mollusken des Stechlinsees. Limnologica 2: 411–421.

Mothes, G. 1965a. Der Wert des autonomen Tauchens bei limnologischen Arbeiten. Limnologica 3: 11–16.

Mothes, G. 1965b. Die Hydracarinen kleiner seichter Randbiotope des Stechlinsees. Limnologica 3: 23–28.

Mothes, G. 1965c. Die Odonaten des Stechlinsees. Limnologica 3: 389–397.

Mothes, G. 1965d. Ein Beitrag zur Sediment-Charakterisierung des Stechlinsees und zur biologischen Typisierung von Seen. Limnologica 3: 383–388.

Mothes, G. 1966a. Die Tanypodinen (Diptera, Chironomidae) des Stechlinsees. Limnologica 4: 13–25.

Mothes, G. 1966b. Ein Beitrag zur Kenntnis der Chironomiden des Stechlinsees. Gewässer Abwässer 41/42: 85–93.

Mothes, G. 1967a. Die Trichopteren des Stechlinsees. Limnologica 5: 1–10.

Mothes, G. 1967b. Einige Tiergruppen mit geringer Artendichte innerhalb der makroskopischen Bodenfauna des Stechlinsees. Limnologica 5: 11–21.

Mothes, G. 1968. Einige ökologisch interessante Chironomiden aus dem Stechlinseegebiet. Ann. Zool. Fenn. 5: 92–96.

Mothes, G. 1970. Die Hydracarinen des Nehmitz- und Gerlinsees. Qualitative Untersuchungen. Limnologica 7: 295–307.

Mothes, G. 1971a. Ökologische Einheiten bei Chironomiden. Limnologica 8: 143–150.

Mothes, G. 1971b. Probleme der Ernährung sympatrischer Chironomiden-Gruppen. Limnologica 8: 193–199.

Mothes, G. & Blume, R. 1969. Versuch einer biozönotischen Gliederung des Stechlinsees anhand der ökologischen Analyse des Makrobenthos in Verbindung mit statistischen Auswerteverfahren. Z. wiss. Zool. 180: 67–147.

Mullin, M.M. & Brooks, E.R. 1976. Some consequences of distributional heterogeneity of phytoplankton and zooplankton. Limnol. Oceanogr. 21: 784–796.

Naidenow, W. 1970. Über die Cladoceren und Copepoden einiger Gewässer der norddeutschen Seenplatte. Izv. Zool. Inst. Muz. Sofia 32: 329–342.

Nauwerck, A. 1963. Die Beziehungen zwischen Zooplankton und Phytoplankton im See Erken. Symb. Bot. Upsala 17: 1–163.

Nauwerck, A. 1978. *Bosmina obtusirostris* Sars im Latnjajaure. Arch. Hydrobiol. 82: 387–418.

Nowak, K.E. 1975. Die Bedeutung des Zooplanktons für den Stoffhaushalt des Schierensees. Arch. Hydrobiol. 75: 149–224.

Parre, S. 1979. On the adaptive significance of vertical migration. Limnol. Oceanogr. 24: 781–782.

Pesta, O. 1928. Krebstiere, Crustacea, I: Ruderfüßer, Copepoda, Calanoida, Cyclopoida. Jena: Gustav Fischer Verlag. 1–135.

Pivoda, B. 1977. Migration of planktonic rotifers in Lunzer Obersee (Austria). Arch. Hydrobiol. Beih. Ergebn. Limnol. 8: 50–52.

Preissler, K. 1977. Horizontal distribution and 'avoidance of shore' by rotifers. Arch. Hydrobiol. Beih. Ergebn. Limnol. 8: 43–46.

Quade, H.W. 1969. Cladoceran faunas associated with aquatic macrophytes in some lakes in northwestern Minnesota. Ecology 50 (2): 170–179.

Ravera, O. 1977. Effects of eutrophication on the zooplankton of a subalpine lake: Lake Lugano. In: Lake pollution – eutrophication control, proc. seminar, Killarney, Ireland, May 1977. National Science Council.

Richman, S., Bohon, S.A. & Robbins, S.E. 1980. Grazing interactions among freshwater calanoid copepods. In: Evolution and ecology of zooplankton communities. Ed. W.L. Kerfoot. New England: Univ. Press. 219–233.

Richter, D., Neubert, W. & Klämt, A. 1979. Temperatur und Wärmehaushalt des thermisch belasteten Stechlinsees. Abh. meteorol. Dienst. DDR 16: 1–40.

Rylov, V.M. 1963. Fauna of U.S.S.R. Crustacea. Freshwater Cyclopoida. Jerusalem: Oldbourne Press.

Schönborn, W. 1962a. Über Planktismus und Cyclomorphose bei *Difflugia limnetica* (Levander) Penard. Limnologica 1: 21–34.

Schönborn, W. 1962b. Neue Testaceen aus dem Großen Stechlinsee und dessen Umgebung. Limnologica 1: 83–91.

Schönborn, W. 1962c. Die Ökologie der Testaceen im oligotrophen See, dargestellt am Beispiel des Großen Stechlinsees. Limnologica 1: 111–182.

Schönborn, W. 1965. Studien über die Gattung *Difflugiella* Cash (Rhizopoda, Testacea). Limnologica 3: 315–328.

Schönborn, W. 1967. Taxozönotik der beschalten Süßwasser-Rhizopoden. Eine raumstrukturanalytische Untersuchung über die Lebensraumerweiterung und Evolution bei der Mikrofauna. Limnologica 5: 159–207.

Schönborn, W., Flößner, D. & Proft, G. 1965. Die limnologische Charakterisierung des Profundals einiger norddeutscher Seen mit Hilfe von Testaceen-Gemeinschaften. Limnologica 3: 371–380.

Siebeck, O. 1960. Untersuchungen über die Vertikalwanderungen planktischer Crustaceen unter Berücksichtigung der Strahlungsverhältnisse. Int. Revue ges. Hydrobiol. 45: 381–454.

Siebeck, O. 1980. Optical orientation of pelagic crustaceans and its consequence in the pelagic and littoral zones. In: Evolution and ecology of zooplankton communities. Ed. W.C. Kerfoot. New England: Univ. Press. 28–38.

Stavn, R.H. 1971. The horizontal-vertical distribution hypothesis: Langmuir-circulations and *Daphnia* distribution. Limnol. Oceanogr. 16: 453–466.

Syuiti, M. 1975. Productivity of freshwater communities in Lake Biwa. In: Productivity of communities in Japanese inland waters. Eds. S. Mori & G. Yamamoto. Tokyo: Univ. Press. 1–45.

Tadashi, A. 1975. Productivity of the community in Lake Yunoko. In: Productivity of communities in Japanese inland waters. Eds. S. Mori & G. Yamamoto. Tokyo: Univ. Press. 47–104.

Thienemann, A. 1954. *Chironomus* (Die Binnengewässer 20). Stuttgart: Schweizerbart'sche Verlagshandlung. 1–834.

Uhlmann, D. & Albrecht, E. 1968. Biogeochemische Faktoren der Eutrophierung von Trinkwasser-Talsperren. Limnologica 6: 225–245.

Ward, J.F. 1981. Annual variations in the abundance of *Daphnia pulicaria* Forbes in West Blue Lake, Manitoba. Verh. Int. Verein. Limnol. 21: 648–651.

Whiteside, M.C., Williams, J.B. & White, C.P. 1978. Seasonal abundance and pattern of Chydorid, Cladocera in mud and vegetative habitats. Ecology 59: 1177–1188.

Winberg, G.G. 1971. Symbols, units and conversion factors in studies of freshwater productivity. London: IBP Control Office. 1–23.

Wright, D., Obrion, J. & Vinyard, G. 1980. Adaptive value of vertical migration: a simulation model argument for the predation hypothesis. In: Evolution and ecology of zooplankton communities. Ed. W.L. Kerfoot. New England: Univ. Press. 138–148.

Zaret, T.M. & Suttern, J.S. 1976. Vertical migration in zooplankton as a predater avoidance mechanism. Limnol. Oceanogr. 21: 804–813.

8. The vertebrate communities

D. Flößner, D. Ronneberger & W. Scheffler

The complex research programme of Lake Stechlin did not include special studies of the vertebrate communities, but some observations on the species composition of fish and birds and especially of waterfowl using the lake as a resting and wintering site are made below.

Fish (D. Ronneberger)

No special study of the population structure or feeding ecology of the fish stock in the Lake Stechlin area exists. This section describes the species composition, and some aspects of occurrence, the yield of fish, and the beginning of breeding experiments of European Cisco, the main and indicator fish species of Lake Stechlin (Fig. 8.1).

The dominant fish species and fishing

In the Lake Stechlin area there 24 fish species are present (p. 525). In the Lakes Stechlin, Nehmitz, and Dagow the species composition is nearly identical. Data on the occurrence and relative frequency of some individual species are given in Table 8.1. Carp and Rainbow Trout have been introduced into Lake Stechlin and Silver Carp into Lake Dagow by fishermen.

Fishing is practised with drag net (European Cisco, Roach, Perch; Fig. 8.2), gill net (European Cisco), bow net (Eel, Pike), and by electro fishing. In Lake Stechlin the average yield amounts to 7 $kg \cdot ha^{-1} \cdot yr^{-1}$, in Lake Nehmitz to 16*, and in Lake Dagow to 31, e.g., with the increasing trophic state the fish yields increase.

In Lake Stechlin the European Cisco is the dominant fish species, with an

* In the following sections all numbers stand for $kg \cdot ha^{-1} \cdot yr^{-1}$.

Casper, S. Jost (ed.), Lake Stechlin. A temperate oligotrophic lake. ISBN 978-94-010-8930-2.

Figure 8.1 The European Cisco (*Coregonus albula*). Photo: W. Scheffler.

Figure 8.2 Drag-net fishing at Lake Stechlin. Photo: W. Scheffler.

Table 8.1 Average annual fish yield (kg · ha^{-1})[a] of Lakes Stechlin (S), Nehmitz (N)[b], and Dagow (D).

Species	Stechlin	Nehmitz	Dagow
Abramis brama (Bream)	0.8	3.0	–
Anguilla anguilla (Eel)	0.5	1.3	2.6
Coregonus albula (European Cisco)	2.4	0.3	–
Cyprinus carpio (Carp)	0.3	0.06	0.44
Esox luceus (Pike)	1.3	4.8	6.3
Rutilus rutilus (Roach)	0.6	2.1	–
Lucioperca lucioperca (Pike-Perch)	–	–	1.3
Perca fluviatilis (Perch)	0.2	0.4	0.04
Tinca tinca (Green Tench)	0.5	1.3	0.7
Food fish	0.9	2.6	19.8
Total	7.5	15.86	31.18

[a] Evaluation periods: S 1970–1982, N 1974–1982, D 1975–1982; area of fishing: S 412 ha, N 167 ha, D 23 ha.
[b] Including Lakes Breutzen and Gerlin.

average yield between 1970 and 1982 of 2.4. The yield of Pike amounts to 1.3, of Eel to 0.5, and of Bream to 0.8. The other fish species (Table 8.1) do not exceed 0.5. The proportion of small individuals of Perch, Roach, Red-Eye, and Bleak are relatively high (total sum ~1).

In Lake Nehmitz Pike (4.8), and Bream (3) are dominant. The catch of Roach (2.1), Eel (1.3), and Green Tench (1.1) is more profitable than in Lake Stechlin.

In Lake Dagow, the lake with the highest yield, Pike (6.3), Eel (2.6), and Pike-Perch (1.3) are the dominant species. Roach and Green Tench are of little importance.

Lake Stechlin – a Coregonus-lake

Lake Stechlin is one of the few typical *Coregonus*-lakes in the GDR (Bauch 1956). The rich underwater flora of the lake, consisting mainly of extensive *Chara*-meadows (Krausch 1964; see p. 130), offers optimal spawn conditions for the European Cisco.

In the Baltic region the species can be found in large, deep, oxygen-rich, and cool lakes. It is divided into many subspecies and local forms (Nicolski 1957; Ladiges & Vogt 1979). It is a plankton feeder and, according to the particular food conditions in the colonized lakes, it has correspondingly different body sizes.

In Lake Stechlin in 1982 the European Cisco grew to an average length of 17 to 20 cm. At four years and a body length of 16 cm the reproductive phase is reached. Throughout the year the species lives in the pelagic area, and only during

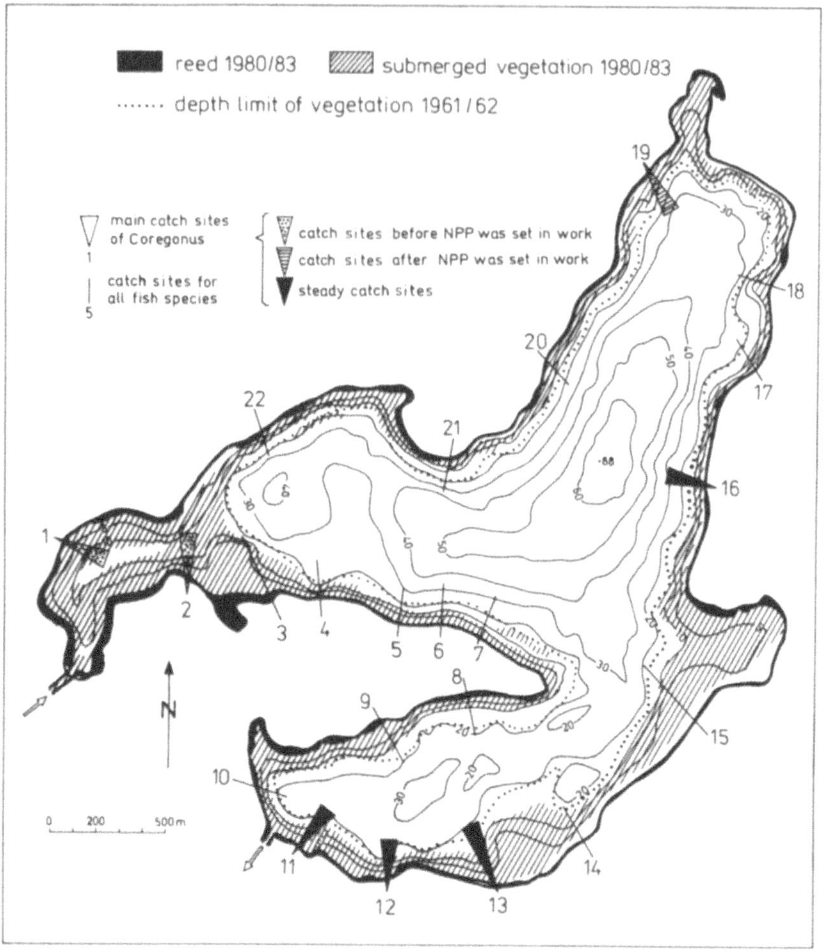

Figure 8.3 Map of fishing-sites from Lake Stechlin.

spawning does it migrate in fairly large shoals into the littoral zones. The spawning state is correlated to the water temperature. Spawning and breeding of the juvenile fish larvae occurs at low temperature (5–8° C) and in the presence of oxygen (Nicolski 1957). In Lake Stechlin homoeothermy starts at the end of November or at the beginning of December (Richter, see p. 53). Then the spawn-trip starts. This is a suitable time to catch the European Cisco with drag nets. Fig. 8.3 shows the main hauls of fish from Lake Stechlin. During the last forty years the yield of European Cisco has decreased from 8 to 1 kg $ha^{-1} \cdot yr^{-1}$ (1947–1982) (Fig. 8.4), but the size of individuals increased (Böttcher, pers. comm.). The cause is unknown; the results of special investigations made by Bauch in the fifties are not available. Hartmann (1977) has calculated a mean fish yield for oligotrophic lakes at 7 kg $ha^{-1} \cdot yr^{-1}$.

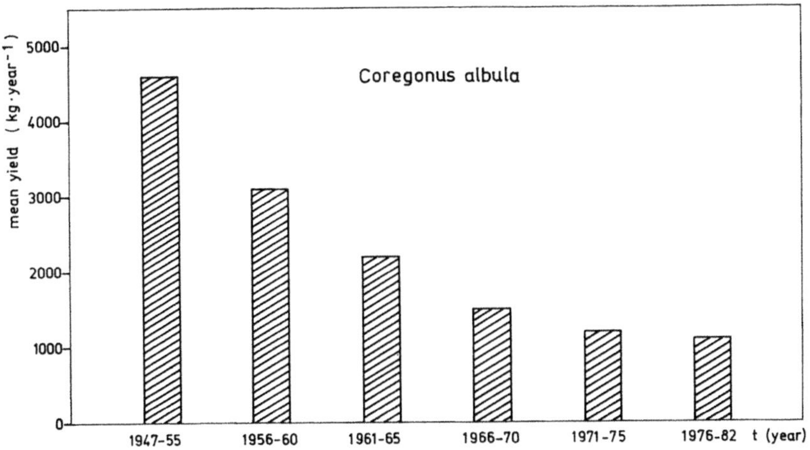

Figure 8.4 Mean yield of European Cisco at Lake Stechlin.

The special features of the development and abundance of European Cisco populations can be understood only by a complete insight into the whole lake ecosystem. The high complexity of Lake Stechlin and the absence of any research in this field do not allow the formation of a definitive judgement on the factors responsible for the decrease of the yield in Lake Stechlin. Two aspects may be discussed.

When the nuclear power plant was brought into operation the lake water transparency declined (p. 81) and the lower limit of the *Chara*-meadows was raised from 20 to 12–14 m (p. 422). These changes resulted generally in a loss of spawning sites. They were especially strengthened by the cooling water circuit impairing the essential spawning sites 1 and 2 (Fig. 8.3) in the west bay of Lake Stechlin. The yield there is extremely low now. Together with the influx of cooling water, living and more or less damaged plankton, as well as great masses of detritus, enter the lake. This additional plankton and detritus 'rain' prevents the eggs of European Cisco from developing.

The newly observed increase of lake water transparency and the extension of the *Chara*-meadows (p. 144) did not affect the important spawning areas in the west bay. After 1970 a new spawn site was found (catch site 19, Fig. 8.3) not impaired by the cooling water circuit.

Overfishing of the *Coregonus* stock in Lake Stechlin may also have contributed to the decrease in yield of European Cisco. Overfishing is a well-known phenomenon. Bekmann (1751) mentioned high yields from Lake Stechlin and considerable 'exports' to Neuruppin, but Bratring (1799) announced significant decrease in yield.

As early as 1781 references to overfishing in some lakes of the Lake Stechlin area are made in some archives of the Rheinsberg area, and also in the nineteenth century when European Cisco was transported to Berlin. Overfishing of European Cisco may happen repeatedly: the catch sites are identical with the spawning places. Loss of spawning places and overfishing result in the squandering (Verbuttung) of the fish stock. The young *Coregonus* larvae are overcome by the great predation pressure from small Perch in the littoral area.

The breeding of juvenile Coregonus in Lake Stechlin

To increase the fish stock breeding experiments started in 1981. Hitherto significant results are not available. The experiments will be continued.

Artificially fertilized eggs are put into 'Zugergläser' (Fig. 8.5) at the end of November. Lake water of low temperature (5–8° C) flows continuously through the eggs. 100 days later, in March and April respectively, the fish larvae slip out and swim free after one week. The young larvae are bred to a length of 15 to 20 mm and then are released into the lake pelagic zone. The larvae so bred have a better chance of survival.

The remains of food for breeding are retained in the breeding glasses to avoid any input of nutrients into the lake. The growth of the young *Coregonus* larvae is impaired by the very low natural standing stock biomass of the relevant food organisms. To enrich the zooplankton density phototaxis was used. The *Coregonus* larvae were put into a net (volume: 8 m^3, mesh size: 0.3 mm; Fig. 8.6) equipped with two 100 volt lamps for nightly illumination. The mesh size retains the larvae in the net, but the zooplankton can immigrate. One test made between 7:30 p.m. and 7:00 a.m. showed that the number of whole zooplankton individuals increased by 60 per cent. The number of copepods increased by 700 per cent, of nauplii by 400 per cent, and of rotifers by 85 per cent. Daily control of the plankton development during breeding time (Fig. 8.7, Fig. 8.8) shows constant or increasing values. Examination of the gut contents of larvae and juvenile fishes shows that all types of zooplankters had been consumed.

Firstly the larvae up to a body length of 10 mm favour rotifers. Larger young fish feed on crustaceans especially copepods, nauplii, and small cladocerans (*Bosmina* spp.). The food concentration allowed a growth rate of average 0.3 mm per day. Nevertheless the measured food concentration of average 400 to 500 zooplankton individuals per litre (or 50 mg carbon per litre) limits the growth of young fish. Vitro experiments, without food limitation, resulted in maximal growth rates of 0.5 mm per day. A body length of 20 mm is reached 14 days earlier.

After 45 days the larvae had grown to length of 20 mm. Then they were released into the lake.

Figure 8.5 'Zugergläser' used for breeding of *Coregonus albula*-eggs. Photo: W. Scheffler.

Figure 8.6 Net-container used for breeding of fish larvae at Lake Dagow. Photo: W. Scheffler.

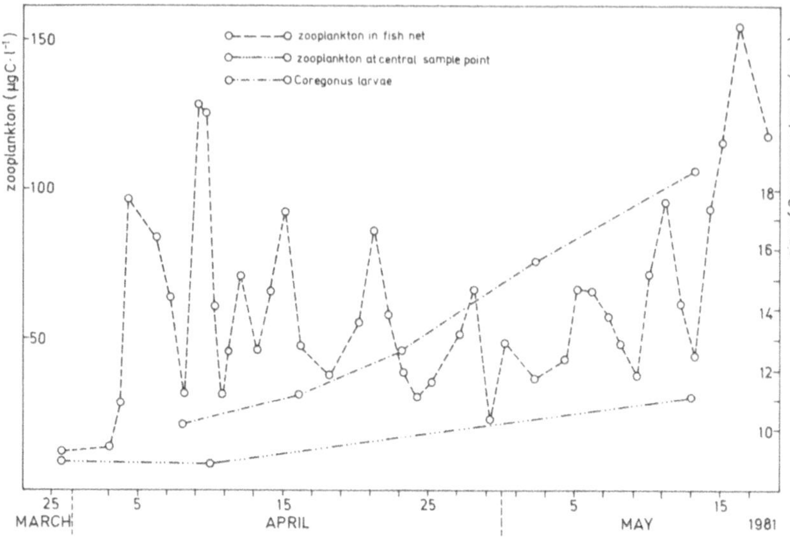

Figure 8.7 Zooplankton biomass enriched by nightly illumination in the net-container.

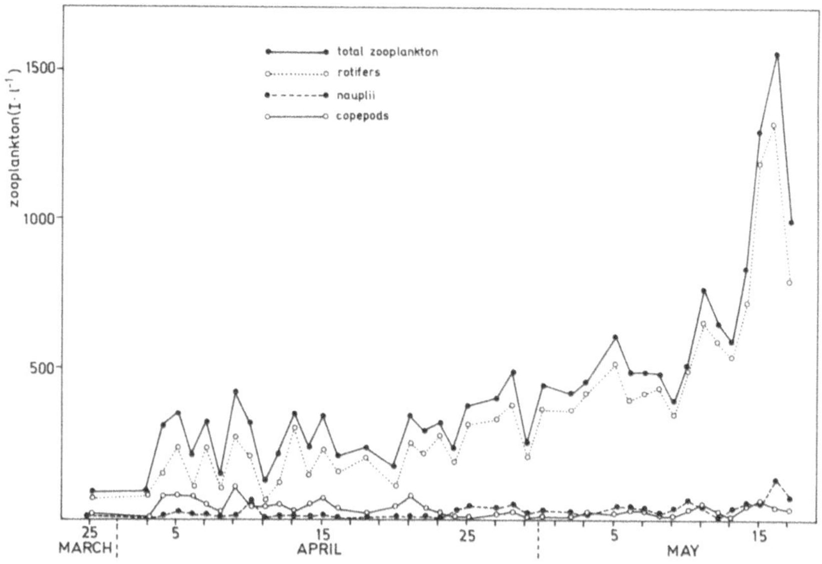

Fig. 8.8 Numbers of individuals of different groups of zooplankton in the net-containers.

Birds (D. Flößner & W. Scheffler)

Lake Stechlin as breeding site

The birds of Lake Stechlin have been observed regularly since 1959 (Flößner 1962, 1967, 1971, 1974). Altogether, only ten species were found as breeding birds with very low densities (Table 8.2). The low number of breeding birds is caused by the high proportion of open wind-exposed shores having a narrow, steeply descending bank of shore without higher aquatic vegetation (see p. 29). Thus both the nesting sites and the food supply are very restricted (Flößner 1971).

Characteristic inhabitants of Lake Stechlin are Goldeneye (*Bucephala clangula*) and Goosander (*Mergus merganser*) which live in open water and feed on fish and small invertebrates respectively (Flößner 1962). They breed in tree-hollows of the surrounding woods. Goldeneye and Goosander are indicator species of oligotrophic lakes. Typical associated species of the bird community of Lake Stechlin are the Great Crested Grebe (*Podiceps cristatus*) and the Mallard (*Anas platyrhynchos*). The Goosander (Fig. 8.10) has disappeared as a breeding bird from Lake Stechlin since 1967, probably due to increased disturbances by boats and anglers. On the other hand, the appearance of the Coot (*Fulica atra*) and the Mute Swan (*Cygnus olor*) and increasing numbers of breeding Great Crested Grebe in the late sixties and seventies (Tables 8.2 and 8.3) are indications of the eutrophication of Lake Stechlin from the influx of heated cooling water from the nuclear power plant and of nutrient loaded water from neighbouring Lake Dagow. Of the birds of the reed belts only the Sedge Warbler (*Acrocephalus schoenobaenus*) and the Reed Bunting (*Emberiza schoeniclus*) breed regularly. Savi's Warbler (*Locustella luscinioides*) and Reed Warbler (*Acrocephalus scirpaceus*) are recent immigrants in the seventies or breed only spo-

Table 8.2 Number and density (breeding pairs/10 ha) of breeding birds at Lake Stechlin 1959–1975 (values for the Goosander only to 1966).

Species	Number	Density
Great Crested Grebe	1–15	0.02–0.35
Mute Swan	1–2	0.02–0.04
Mallard	2–7	0.04–0.17
Goldeneye	1–6	0.02–0.14
Goosander	1–5	0.02–0.12
Coot	1–7	0.02–0.17
Savi's Warbler	1	1.0
Sedge Warbler	2–6	2.1–6.3
Reed Warbler	3	3.1
Great Reed Warbler	1–2	1.5
Reed Bunting	2–4	2.1–4.2

Figure 8.9 An Osprey nesting on an artificial 'tree'. Photo: W. Scheffler.

Figure 8.10 Goosanders. Few pairs bred at Lake Stechlin to 1966. Photo: W. Scheffler.

Figure 8.11 Wintering Coots on the only hole at the frozen over Lake Stechlin. Photo: W. Scheffler.

Table 8.3 Annual changes in number of breeding Great Crested Grebe at Lake Stechlin 1959–1983.

	1959–1972	1975	1978	1980	1983
Number of pairs breeding (PB)	4 (average)	12–15	12–15	20	32
PB/10 ha	0.09 (average)	0.35	0.35	0.47	0.75

radically, as does the Great Reed Warbler (*Acrocephalus arundinaceus*). All these species are widespread on inland waters, including deep oligotrophic lakes situated in extended woodlands. Birds typical for eutrophic lakes such as Rallidae, Bittern, Little Bittern, Marsh Harrier and all dabbling ducks with the exception of the Mallard are absent from Lake Stechlin. Birds of prey have as top carnivores an important role in the food-web of aquatic ecosystems, especially the Osprey (*Pandion haliaetus*) feeding exclusively on fish. In the Lake Stechlin area the numbers of breeding pairs (Fig. 8.9) have increased in the last decade, but the breeding success changed periodically (Table 8.4).

Lake Stechlin as resting and wintering site

Lake Stechlin fulfils an important function as a resting place for migratory waterfowl, particularly in winter (Flößner 1962, 1974). Significant environmental factors regulating the occurrence of waterfowl at Lake Stechlin are firstly the trophic state determining the food resources, secondly the morphometry of the lake basin affecting the ability to obtain the food and the ice-coverage and thirdly the climate, which determines the ice-coverage and the site of the wintering areas.

Altogether twenty species of waterfowl were found. The predominant role is played by the Coot (Fig. 8.11) with a peak of about 8000 birds. Large groups of Mallards and Pochards (*Aythya ferina*) have also been noted. They have maximum numbers of 550 and 850 birds, respectively. The Great Crested Grebe, the Tufted Duck (*Aythya fuligula*) and the Goldeneye appear regularly, but only in

Table 8.4 Breeding success of Ospreys in the Lake Stechlin area.

Year	Number of pairs breeding	Number of young fledged	Juveniles per pairs breeding
1971	4	3	0.8
1975	7	9	1.3
1980	7	6	0.9
1983	7	11	1.6

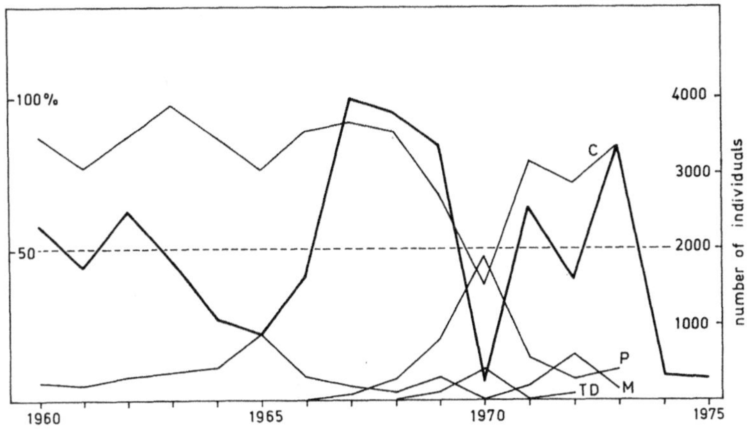

Figure 8.12 Mid-winter numbers (average values of January and February) of waterfowl at Lake Stechlin and percentage values of dominant species 1959–1975. C = Coot; P = Pochard; M = Mallard; TD = Tufted Duck.

smaller groups of less than 150 individuals. *Mergus* species, sea ducks and divers such as the Black-throated Diver and Horned Grebe, and particularly dabbling ducks occur only sporadically and in very low numbers. All these species with the exception of the Great Crested Grebe have their population maximum at Lake Stechlin in mid-winter. The highest count of waterfowl recorded for Lake Stechlin between 1959 and 1975 was 8328 birds on 14 January 1968, i.e., approximately 1300 individuals per 10 ha shore area. Table 8.5 shows the average maximum numbers of dominant species during the present study, Fig. 8.12 the average mid-winter numbers of all waterfowl and the percentage values of dominant species in the various seasons. Both show Lake Stechlin is an important wintering area for Coots and diving ducks.

The phenology of the waterfowl of Lake Stechlin is determined by its function as a resting place of birds evading winter conditions. The Great Crested Grebe migrates in low numbers of 15 to 45 birds from August into late October. The first wintering birds appear in early December and then peak in January (Fig. 8.13). Migrating in spring begins in mild winters already in early- to mid-February, but

Table 8.5 Average maximum numbers of dominant waterfowl species at Lake Stechlin, 1959–1975.

Species	Number	Percentage value	Birds/10 ha shore area
Coot	2,893	84.6	526
Mallard	261	7.6	–
Pochard	188	5.5	30
Goldeneye	43	1.3	6.8
Tufted Duck	26	0.6	4.1
Great Crested Grebe	14	0.4	2.2

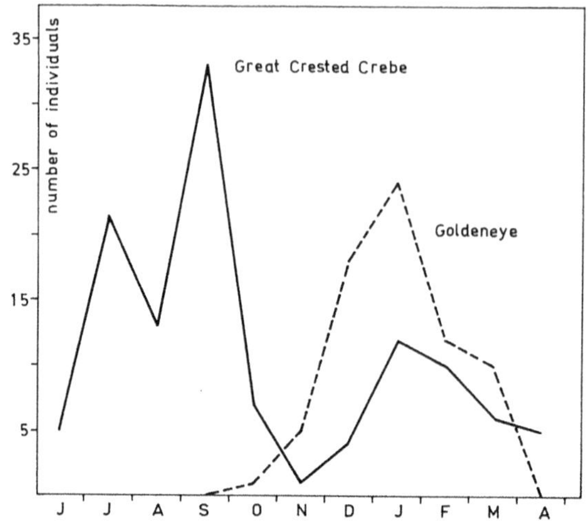

Figure 8.13 Average monthly numbers of Great Crested Grebe and Goldeneye at Lake Stechlin 1959–1975.

usually from the end of March to mid-April. During the moulting season in some years there are accumulations of more than a hundred birds on the open lake. Goldeneyes and Pochards show, like other waterfowl, a marked mid-winter maximum (Figs. 8.13, 8.14). Wintering Pochards have increased their numbers more than fivefold since 1966 (Fig. 8.14).

The phenology of the Coot, the predominant species at Lake Stechlin, is the characteristic one for a waterfowl evading winter conditions (Fig. 8.14). The numbers increase suddenly at the end of December, shortly after ice has covered the surrounding shallow lakes and then peak in mid-winter. The population maximum usually reaches 2000 to 3000 birds, exceptionally 8000 (14 January 1968), but in mild winters only about 1000 birds are found. In some years the Whitetailed Eagle (*Haliaeetus albicilla*) is to be seen hunting the Coot flocks. The Coots remain until the lake is frozen over totally (Fig. 8.11). According to field observations the most important diet for Coots are Characeae, especially the species *Chara filiformis, Ch. fragilis, Nitellopsis obtusa* and *Nitella flexilis* which are so widely distributed in Lake Stechlin that food shortage does not occur during a resting period of one to six months. *Dreissena polymorpha* was not preferred, as it is reported from many authors for eutrophic lakes in Central Europe (Feiler & Paepke 1964; Geroudet 1966; Stichmann, Prünte & Raus 1969; Suter 1982). The restricting factor for the length of the wintering stay of Coots at Lake Stechlin is the climate, i.e., the beginning of ice cover. Total winterings are, therefore, dependent on the weather. This occurred eight times in sixteen winters (1959 to 1975).

The suitability of Lake Stechlin as a wintering area for Coots and other waterfowl is dependent on two factors: the late and light ice cover of the lake, due

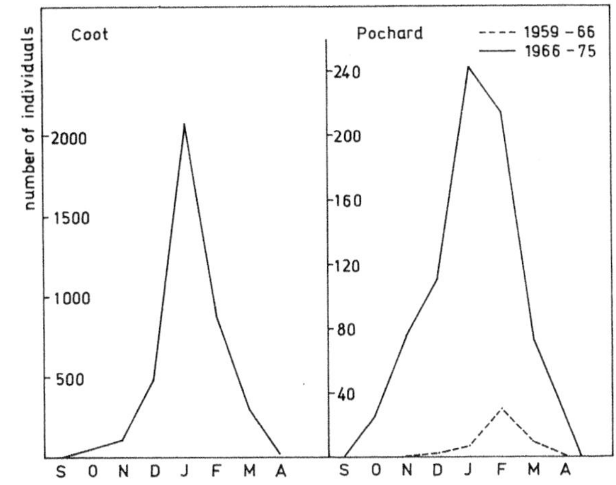

Figure 8.14 Average monthly numbers of Coot and Pochard at Lake Stechlin 1959–1975.

to the beginning of homoeothermy not before mid-December, and the existence of a shallow shore area extending approximately 55 ha with enlarged dense *Chara* beds as the main food of wintering Coots. The *Chara* beds of the shallow littoral are sufficient for feeding of 2000 to 3000 Coots for a period of 80 to 100 days.

References

Bauch, G. 1956. Chemismus und Fischnährtierproduktion in den von der Kleinmaräne (*Coregonus albula* L.) bewohnten norddeutschen Gewässern. Abh. Ber. Naturk. Vorgesch. Magdeburg 10: 15–36.

Bekmann, J.C. 1751. Historische Beschreibung der Chur und Mark Brandenburg. Ergänzt und herausgegeben von Bernhard Ludwig Bekmann. 1. Theil. Berlin.

Bratring, F.W.A. 1799. Die Grafschaft Ruppin in historischer, statistischer und geographischer Hinsicht. Ein Beitrag zur Kunde der Mark Brandenburg. Berlin: G. Hayn.

Feiler, M. & Paepke, H.-J. 1964. Über die Biologie und die Verluste des Bleßhuhns auf den Potsdamer Havelseen währand des Winters 1962/63. Falke 11: 119–126.

Flößner, D. 1962. Ornithologische Beobachtungen im Gebiet des Stechlinsees. Falke 9: 343–344.

Flößner, D. 1967. Weitere ornithologische Beobachtungen im Gebiet des Stechlinsees. Beitr. Tierw. Mark 4: 131–137.

Flößner, D. 1971. Die Brutvögel des Naturschutzgebietes Stechlin. Brandenburg. Naturschutzgeb. 13: 1–13.

Flößner, D. 1974. Die Vogelwelt des Naturschutzgebietes Stechlin. Naturschutzarb. Berlin Brandenburg 10: 59–61.

Geroudet, P. 1966. Premières conséquences ornithologique de l'introduction de la 'moule zébrée' *Dreissena polymorpha* dans le lac Léman. Nos Oiseaux 28: 301–307.

Hartmann, J. 1977. Fischereiliche Veränderungen in kulturbedingt eutrophierenden Seen. Schweiz. Z. Hydrol. 39: 243–254.

Krausch, H.-D. 1964. Die Pflanzengesellschaften des Stechlinsee-Gebietes. I. Die Gesellschaften des offenen Wassers. Limnologica 2: 145–203.

Ladiges, W. & Vogt, D. 1979. Die Süßwasserfische Europas. Hamburg, Berlin: Parey.
Nicolski, G.W. 1957. Spezielle Fischkunde. Berlin: VEB Deutscher Landwirtschaftsverlag.
Stichmann, W., Prünte, W. & Raus, T. 1969. Beiträge zur Avifauna des Möhnesees. Quantitative und phänologische Studie. Anthus 6: 45–148.
Suter, W. 1982. Die Bedeutung von Untersee-Ende/Hochrhein (Bodensee) als wichtiges Überwinterungsgewässer für Tauchenten (*Aythya, Bucephala*) und Bläßhuhn (*Fulica atra*). Ornith. Beob. 79: 73–96.

9. Flora and fauna of the Lake Stechlin area

*Short comments upon the lists of taxa (appendix)**

S.J. Casper & W. Schönborn

Lake Stechlin and its surroundings are a typical post-glacial area (p. 9) like most of the lake districts of the northern hemisphere. Lake Stechlin itself is not older than 12 000 years. This represents a very brief period for the post-glacial flora and fauna to have produced new species, and our lakes under investigation generally show a very low level of species differentiation. The present flora and fauna are mainly the result of present environmental factors. Nevertheless, some characteristics show that a specific regional past history has influenced the composition of flora and fauna on the species and infra-species level.

Of the lakes mentioned in our lists (see appendix p. 485–527), different intensive studies were made with regard to their species composition. Phytoplankton, macrophytobenthos, and zoobenthos of Lakes Stechlin and Nehmitz-N were thoroughly investigated; only random samples were taken from Lake Dagow; the bog-lakes and the other lakes in the woodland were sporadically examined.

During our studies we observed more than 1900 taxa of plants and animals, 1825 of which are listed (p. 485–527). In Lake Stechlin itself, nearly 1425 taxa (~78 per cent of the whole stock of taxa of the study area) were determined; 1050 (~73%) of them account for animals.

Flora

More than 550 taxa were observed in our study area: 112 aquatic and semi-aquatic macrophytes (incl. 10 taxa of Charophyceae), 19 fungi, and 410 algae. More than 50 per cent of them occur regularly and may be said to be common. Eighty-one taxa of aquatic macrophytes, 14 of fungi, and 278 of algae live in Lake Stechlin itself. In phytoplankton samples, 240 taxa were determined, more than 30 per cent of which are littoral- or tychoplankters. The diatoms, with 164 taxa (~43 per cent), respresent the main proportion; only 23 taxa of them are euplanktonic.

* See p. 485.

Casper, S. Jost (ed.), Lake Stechlin. A temperate oligotrophic lake. ISBN 978-94-010-8930-2.

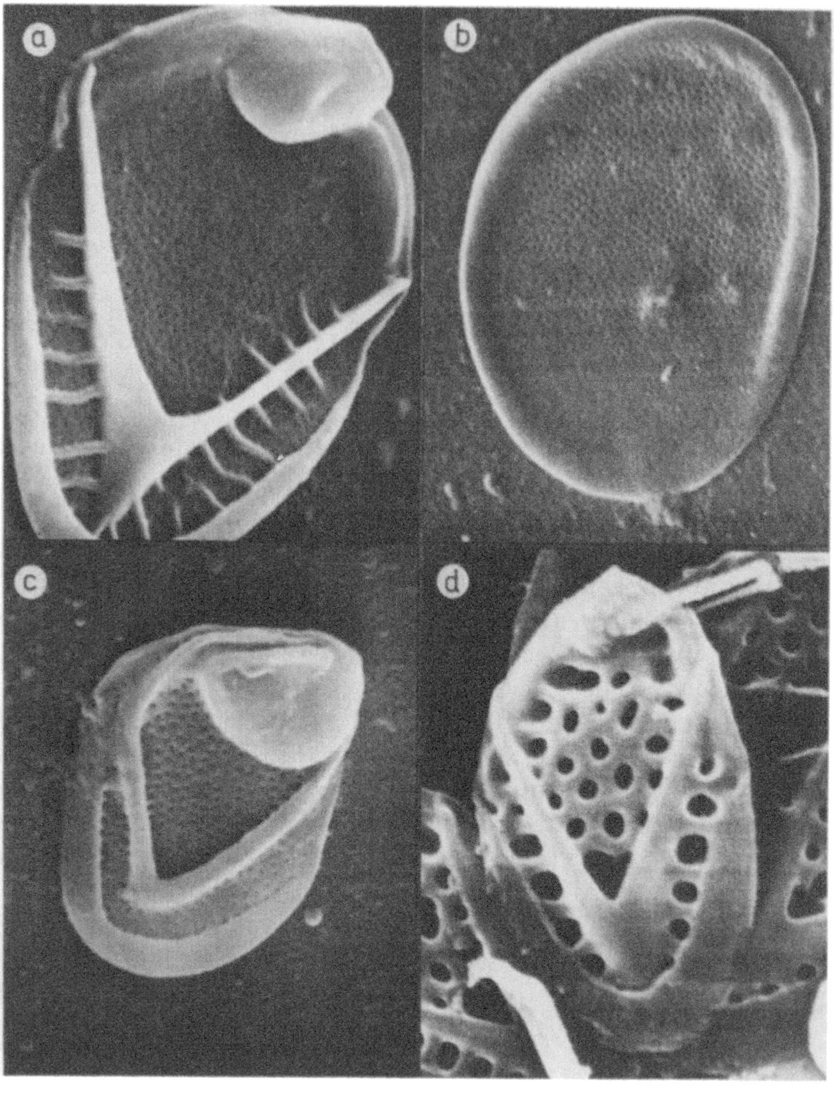

Figure 9.1 REM-photos of scales of *Mallomonas* species. a *Mallomonas acaroides* var. *striatula;* b *Mallomonas fastigata;* c *Mallomonas tonsurata;* d *Mallomonas crassisquama*. Photos K. Augsten.

The numbers mentioned above differ greatly from those of Busse (1965, 1972) and Küchler (1981, 1982). Busse (1965) determined 95 phytoplankton taxa in Lake Stechlin, 94 in Lake Nehmitz, and 31 in Lake Dagow. But it must be considered that she only listed taxa which she had determined exactly; nearly 50 taxa (mostly nanoplanktonic) observed remained disregarded. Küchler (1982) pointed out 144 phytoplankton taxa in Lake Stechlin, 128 in Lake Nehmitz-N, and 73 in Lake Dagow.

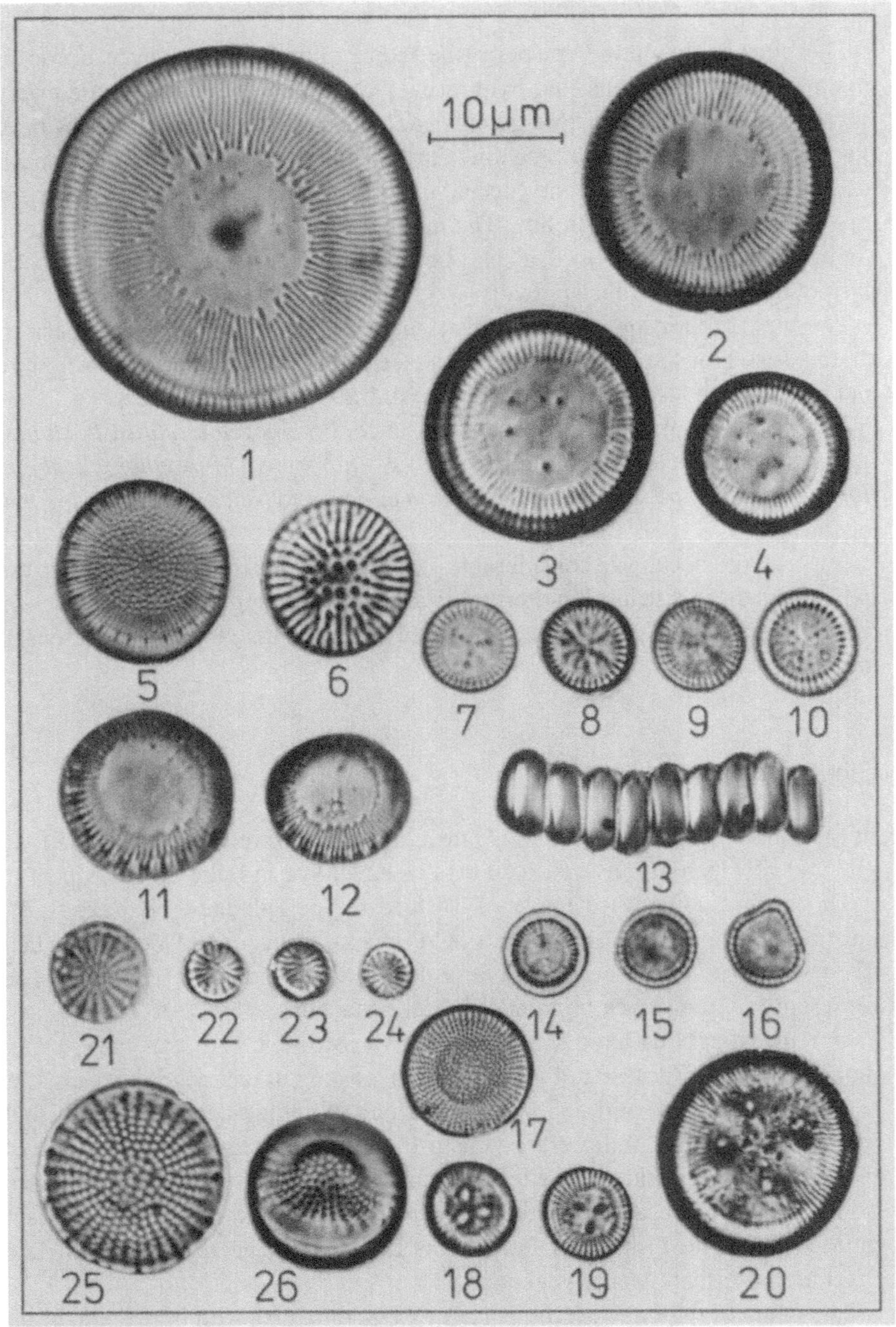

Figure 9.2 Planktonic diatoms of Lake Stechlin. 1, 2 *Cyclotella kützingiana;* 3, 4 *Cyclotella kützingiana* var. *planetophora;* 5, 6 *Cyclotella comta;* 7, 8, 9, 10 *Cyclotella comta* var. *oligactis;* 11, 12 *Cyclotella operculata;* 13, 14, 15, 16 *Cyclotella glomerata;* 18, 19, 20 *Cyclotella ocellata;* 21 *Stephanodiscus hantzschii;* 22, 23, 24 *Stephanodiscus spec. hantzschii-group;* 17, 25, 26 *Stephanodiscus rotula* var. *minutulus*. Photos W. Scheffler.

Thorough studies have been performed concerning Bacillariophyceae, coccoid green algae, Dinobryonaceae, Synuraceae, Charophyceae, and aquatic macrophytes. Together with the blue-green algae, they represent the main stock of the plant communities in Lake Stechlin. Further investigations have to pay special attention to the monadoid and coccoid nanoplanktonic and ultraplanktonic species and the microphytobenthos. The number of algal taxa will then amount to about 400. The so-called 'μ-alga', mentioned in chapter 5 (pp. 185–186; Fig. 5.15), has not been listed in our appendix.

The high proportion of boreal distributed macrophytic and algal species is characteristic for the Lake Stechlin area (see p. 13). Most of them grow in mires and bogs, which are cold stands, but also in the oligotrophic waters.

Boreal aquatic macrophytes are, for instance, *Potamogeton rutilus, P. alpinus, P. nitens, P. praelongus, P. gramineus, P. filiformis, Myriophyllum alterniflorum, Sparganium minimum, Utricularia minor,* and *Najas marina* subsp. *intermedia.*

Busse (1965) declared a considerable part of the observed phytoplanktonic taxa to be of northern distribution. Perhaps *Dictyosphaerium sphagnale, Scenedesmus costato-granulatus, Synura sphagnicola* and *Uroglena eustylis* may respresent the 'northern' type.

Fauna

In the waters of our study area 1275 taxa of animals have been determined, 466 (~37%) of which are insects. 1050 taxa (~82%) live in Lake Stechlin itself.

Thorough studies were made of Testacea-, Spongillidae-, Hydrozoa-, Rotatoria-, Bryozoa-, Mollusca-, Acari-, Crustacea-, Insecta-, and Vertebrata taxa, 537 (~51%) of which occur regularly and can be said to be common. They represent the main stock of animals in Lake Stechlin.

Our investigations have shown that an oligotrophic lake can have a great number of taxa. Because not all of the taxa existing in fact were detected – 200 ciliates, 40 amoebae and heliozoan protozoa, 80 nematodes, 50 micro-turbellaries, 20 annelides, 20 dipteres, 8 gastrotriches and tardigrades hitherto unobserved ought to be added – the total number of animal taxa in Lake Stechlin can be estimated to be about 1500. In recent years aquatic Mammalia were not pointed out. The Fish-otter, *Lutra lutra* L., had last been observed in Lake Stechlin more than twenty years ago.

The fauna of the Lake Stechlin area possess some taxa with boreal and boreoalpine distribution, respectively. Their rate is very high in the microcrustaceans (nearly 15 taxa, inclusive such taxa occuring in the neighbouring bog lakes; see p. 226). But other groups contain taxa of this northern type of distribution too.

Of boreal distribution are, for instance, the mite *Thyas pachystoma* (Lake Nehmitz), the beetles *Haliplus apicalis, Gyrinus natator,* and *Lacornis oblongus,*

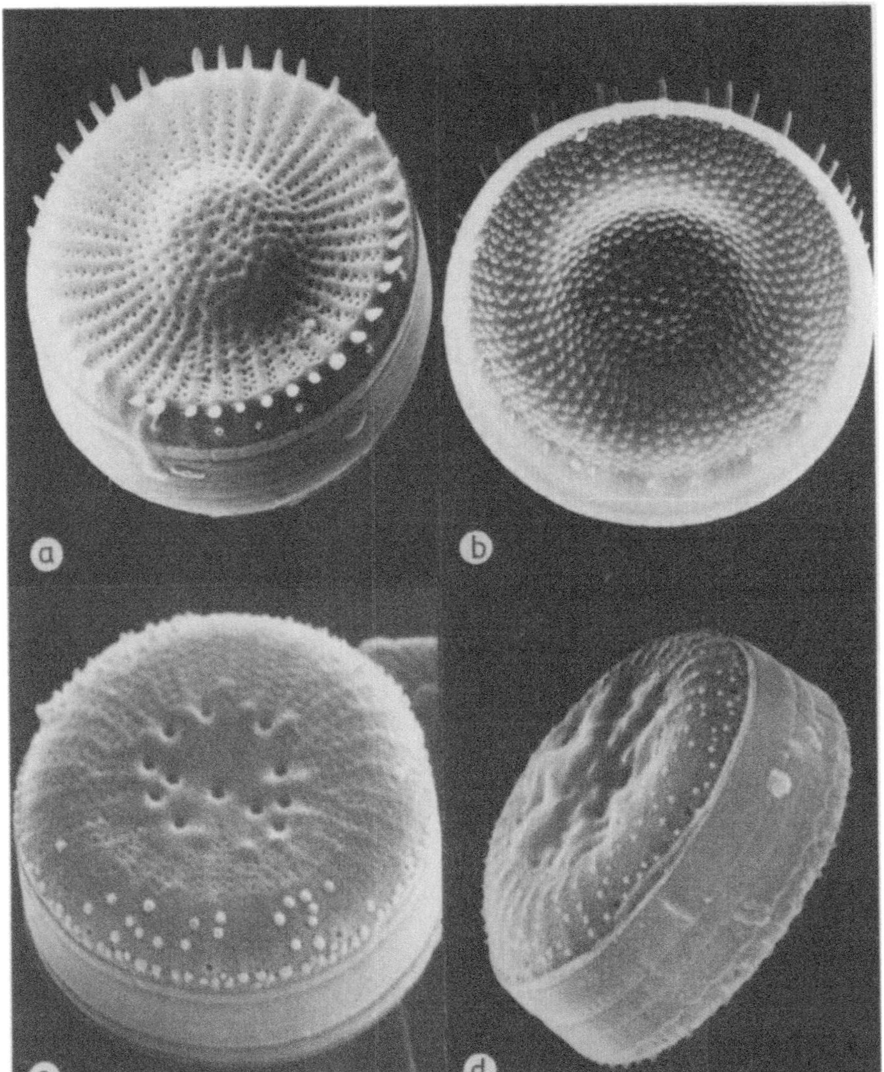

Figure 9.3 REM-photos of a, b *Stephanodiscus rotula* var. *minutulus* (Photos K. Augsten); c, d *Cyclotella comta* var. *oligactis*. (Photos Raidt).

and the trichoptera *Limnephilus elegans*.

Of typical boreo-alpine distribution are the testate amoeba *Difflugia globulosa*, the mussels *Pisidium lilljeborgi* and *P. hibernicum*, the chironomides *Lauterbornia coracina* and *Tanytarsus lugens*, moreover, the dragon-fly *Leucorrhinia dubia* (in bog-lakes), and the beetle *Coelambus marklini* (in Lake Nehmitz).

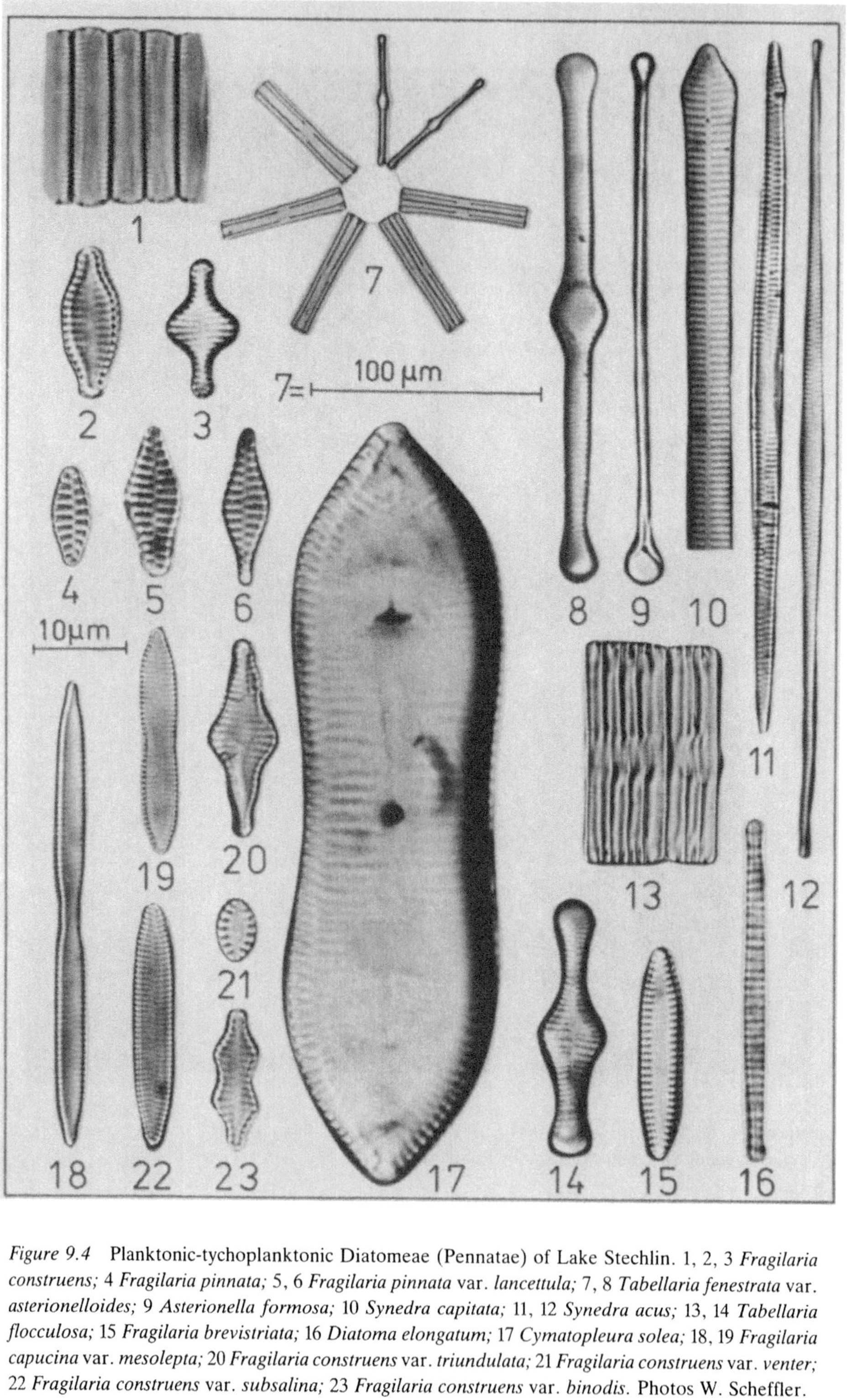

Figure 9.4 Planktonic-tychoplanktonic Diatomeae (Pennatae) of Lake Stechlin. 1, 2, 3 *Fragilaria construens;* 4 *Fragilaria pinnata;* 5, 6 *Fragilaria pinnata* var. *lancettula;* 7, 8 *Tabellaria fenestrata* var. *asterionelloides;* 9 *Asterionella formosa;* 10 *Synedra capitata;* 11, 12 *Synedra acus;* 13, 14 *Tabellaria flocculosa;* 15 *Fragilaria brevistriata;* 16 *Diatoma elongatum;* 17 *Cymatopleura solea;* 18, 19 *Fragilaria capucina* var. *mesolepta;* 20 *Fragilaria construens* var. *triundulata;* 21 *Fragilaria construens* var. *venter;* 22 *Fragilaria construens* var. *subsalina;* 23 *Fragilaria construens* var. *binodis*. Photos W. Scheffler.

Conclusions

The Lake Stechlin area – Lake Stechlin itself and expecially its neighbouring bog-lakes – is one of the most important centres of boreal and boreo-alpine taxa in the northern part of Central Europe. This distribution – once having come in to existence during post-glacial periods – is now favoured by the oligotrophic clear-cold-water status of Lake Stechlin.

References

Busse, G. 1965. Biologische Untersuchungen im Gebiet des Stechlin-Sees. Teilprogramm 1: Floristisch-ökologische Bestandsaufnahme der Protophyten. In: Abschlußbericht ... Jena: 1–36 (unpublished).

Busse, G. 1972. Phytoplankton. In: Grundlagen zur Modellierung des Stoffhaushaltes der Gewässer am Beispiel des Stechlin-Sees. Abschlußbericht ... Jena: 1–44 (unpublished).

Küchler, L. 1981. Phytoplanktonuntersuchungen im Stechlinseegebiet in den Jahren 1973–1975. Limnologica 13: 83–99.

Küchler, L. 1982. Phytoplanktonuntersuchungen im Stechlin und im Nordbecken des Nehmitzsees. Limnologica 14: 231–242.

PART FOUR

The metabolism of the Lake Stechlin area

10. The primary production

R. Koschel & W. Scheffler

The primary production of the phytoplankton (R. Koschel)

Introduction

The primary production of the phytoplankton is one of the most important sources of the energy input in freshwater ecosystems (Moss 1980). A broad knowledge of endogenous and exogenous control mechanisms of phytoplanktonic primary production is necessary to obtain basic information on the energy flow in the ecosystem of Lake Stechlin. Pretreatment conditions of the phytoplankton population as well as factors concerning the regulation of the overall carbohydrate metabolism of the algae and the integration of the photosynthetic products into development and growth of phytoplankton allow a better understanding of the lake metabolism as a whole (Harris 1978).

In Lake Stechlin phytoplanktonic primary production was measured first in 1967 (Javornicky 1974). To obtain more detailed information on the energy flow and the carbon transfer in space and time, long-term research program was started in 1970. The vertical, horizontal and annual distribution of the phytoplanktonic primary production rates have been analysed regularly using the ^{14}C-technique (Koschel 1974a, b; 1976; 1981a; Koschel & Kasprzak 1984). Data of specific endogenous control mechanisms of phytoplankton growth (e.g. phosphate metabolism) and of varying biotic and abiotic factors (e.g., phytoplankton and zooplankton biomass, different nutrients, temperature, radiation) were collected (Koschel 1976, 1980, 1981b; cf. pp. 178–182, 87–118, 47–56, 72–85).

My own investigation on phytoplanktonic primary production is based on a system-analytic outlook. Fig. 10.1 demonstrates the close correlation between experimental registration and measurement of primary production and theoretical generalization of growth and loss of phytoplankton, which lead to an ecological model of phytoplankton growth (pp. 438 ff.).

The results are described and discussed in the following sections, if need be complemented by similar observations in neighbouring lakes. The various data

Casper, S. Jost (ed.), Lake Stechlin. A temperate oligotrophic lake. ISBN 978-94-010-8930-2.

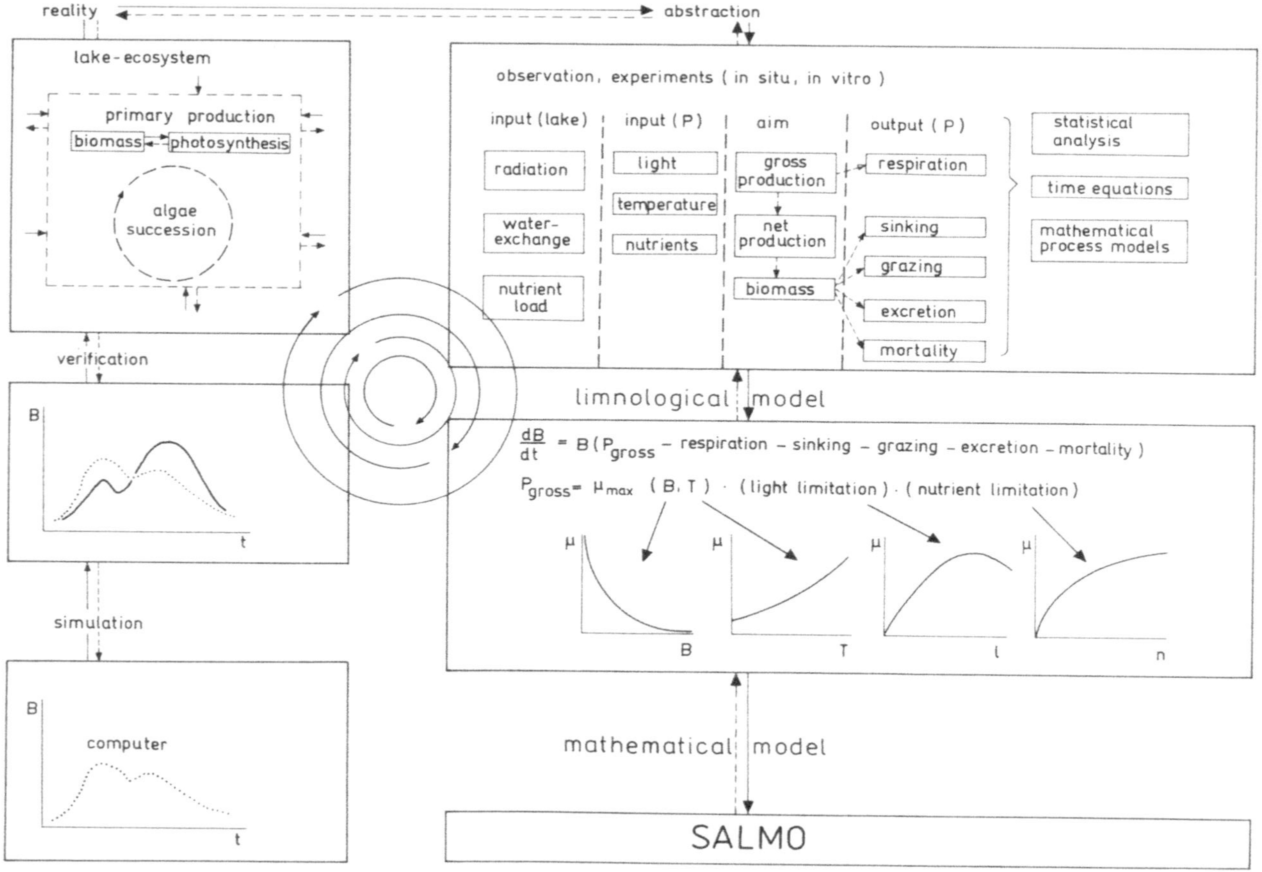

Figure 10.1 System-analytical scheme of the analysis of primary production of the phytoplankton.

have been brought together to portray the role and place of phytoplanktonic primary production in the metabolism of Lake Stechlin and lake ecosystems in general.

Questions will be answered such as: Which variety of phytoplanktonic primary production and effectivity of primary production were observed in Lake Stechlin? How is the nanoplanktonic primary production correlated with the total primary production? What role does phosphorus play in the primary production processes and which relationships exist between phosphorus, primary production and eutrophication in Lake Stechlin and neighbouring lakes?

Seasonal and vertical distribution of primary production of phytoplankton in Lake Stechlin

Fig. 10.2 summarizes seasonal variations in daily carbon assimilation rates per square metre. Fig. 10.3 shows records of depth profiles of primary production per unit volume of water of Lake Stechlin (deepest point) from 1970–1982.

As early as February and March remarkable primary production rates were measured in years without ice-cover and with input of cooling water from the nuclear power plant. Maximum primary production rate (a_{opt}) varied between 25–50 mg $C \cdot m^{-3} \cdot d^{-1}$ and the daily primary production between 80–400 mg $C \cdot m^{-2} \cdot d^{-1}$. A first peak is normally obvious in March or April (50–70 mg $C \cdot m^{-3} \cdot d^{-1}$ in a_{opt} and 250–400 mg $C \cdot m^{-2} \cdot d^{-1}$). a_{opt} was found in depths between 2.5 m–5 m (Fig. 10.3, 10.4). After a well-developed decreasing phase (March/April) the primary production increases again rapidly. Maximum values were found in June or July. The highest recorded primary production rate was about 100 mg $C \cdot m^{-3} \cdot d^{-1}$ in a_{opt} and 1 g $C \cdot m^{-2} \cdot d^{-1}$ (Table 10.1). The daily primary production averaged 60 mg $C \cdot m^{-3} \cdot d^{-1}$ in a_{opt} and 500 mg $C \cdot m^{-2} \cdot d^{-1}$ in the period from June–August. Vertical distribution is characterized by maximum values near the surface (Fig. 10.4). Nevertheless, more profiles were observed with a maximum in greater depth (2.5–7.5 m) in 1975 and in the following years too. By late August primary production decreased to low values. During the autumn primary production varied between 40–60 mg $C \cdot m^{-3} \cdot d^{-1}$ in a_{opt} and 200–300 mg $C \cdot m^{-2} \cdot d^{-1}$. The minimum values occurred from November–January (10–20 mg $C \cdot m^{-3} \cdot d^{-1}$ in a_{opt} and 20–40 mg $C \cdot m^{-2} \cdot d^{-1}$).

An other picture of the seasonal variation of primary production was found in years with a long ice cover (1970) or a long phase without the circulation of thermally loaded cooling water (1975). In 1970 the February/March peak of primary production was not observed. After ice-thaw (March) primary production increased quickly with the increase of irradiance levels. After that primary production shows two seasonal peaks, one in early summer and another in early autumn. The two seasonal peaks are separated by an increasing-phase in July/August (Fig. 10.2). In 1975 primary production showed, compared with other

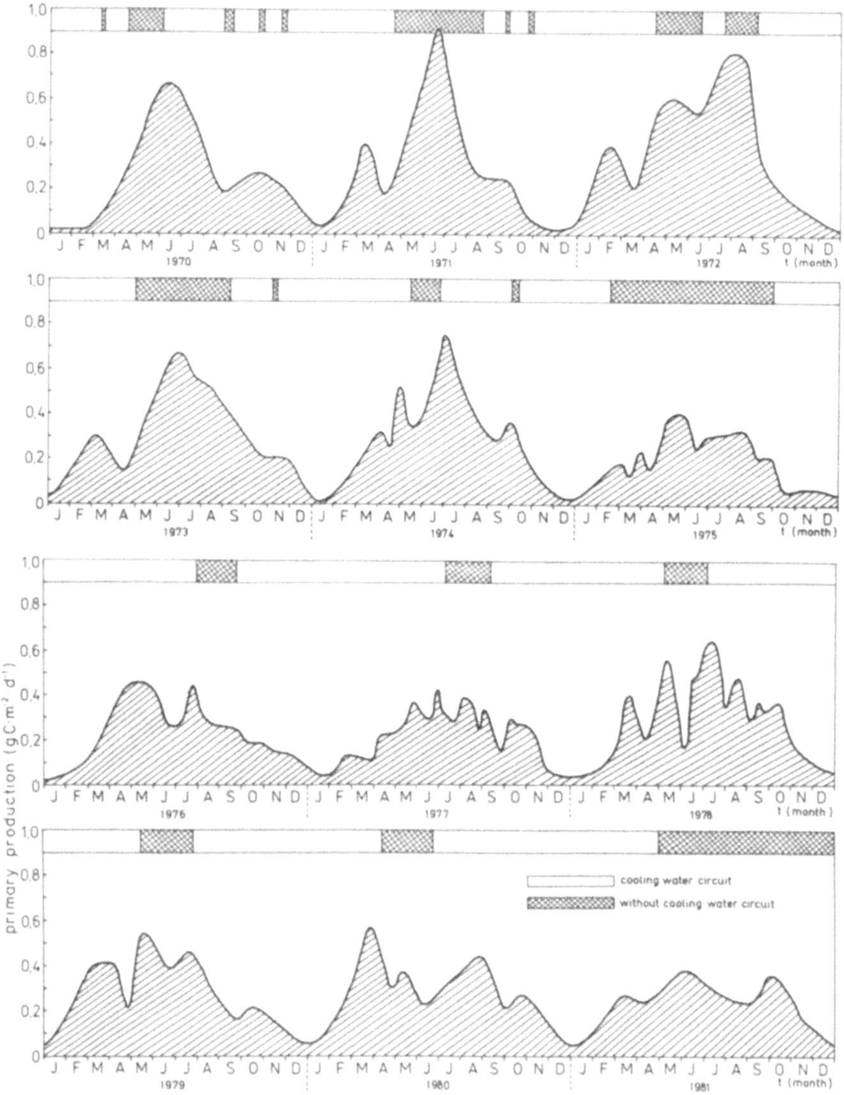

Figure 10.2 Seasonal periodicity of primary production of phytoplankton (and cooling water input) in Lake Stechlin from 1970 to 1982.

years, an identical seasonal periodicity in winter and early spring, but by May primary production rates decreased rapidly. Values of primary production in summer and autumn were lower than in the other years (Table 10.1). In 1975 annual primary production was estimated at 60% of annual average (1970–1982).

Altogether the annual primary production averaged 97 g $C \cdot m^{-2}$ from 1970–1982 (Table 10.2).

Table 10.1 Maximum of primary production of phytoplankton during winter/spring and summer from 1970–1982 in Lake Stechlin.

Year	Winter/Spring-maximum			Summer-maximum		
	a_{opt} (mg C · $m^{-3} \cdot d^{-1}$)	Daily production (mg C · $m^{-2} \cdot d^{-1}$)	Month	a_{opt} (mg C · $m^{-3} \cdot d^{-1}$)	Daily production (mg C · $m^{-2} \cdot d^{-1}$)	Month
1970	not observed			80	610	Jun
1971	50	400	Mar	100	950	Jul
1972	55	350	Feb	110	800	Jul/Aug
1973	50	300	Mar	75	640	Jul
1974	40	300	Apr	120	740	Jul
1975	35	260	Mar	80	400	Jun
1976	55	420	May	55	450	Jul
1977	30	180	Feb	50	490	Jun
1978	30	420	Mar	75	640	Jul
1979	45	410	Apr	80	470	Jul
1980	45	570	Mar	55	440	Aug
1981	30	265	Mar	65	380	Jun
1982	65	715	Apr	55	540	Jul
$\bar{x}$	45	385	Mar	75	580	Jul

Table 10.2 Maximum daily production, mean daily primary production and annual primary production of phytoplankton from 1970–1982 in Lake Stechlin.

Year	Maximum daily primary production (mg C · $m^{-2} \cdot d^{-1}$)	Mean daily primary production (mg C · $m^{-2} \cdot d^{-1}$)	Annual primary production (g C · $m^{-2} \cdot a^{-1}$)
1970	610	277	101
1971	950	277	101
1972	800	378	138
1973	640	315	115
1974	740	266	97
1975	400	186	68
1976	450	216	79
1977	490	203	74
1978	640	277	101
1979	470	244	89
1980	570	274	100
1981	380	238	87
1982	715	299	109
$\bar{x}$	604	265	97

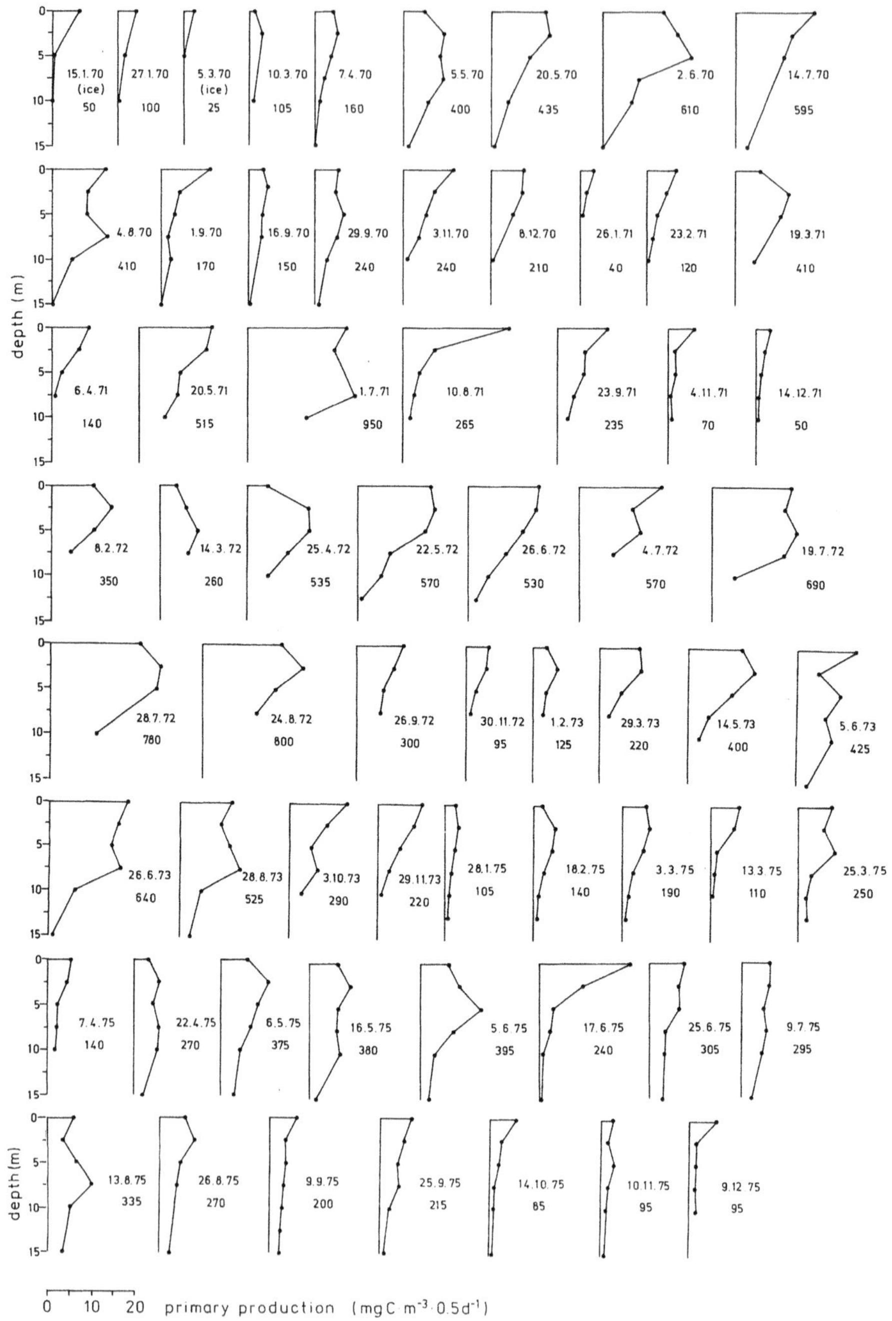

a

Figure 10.3 Vertical distribution of primary prouction of phytoplankton in Lake Stechlin from 1970 to 1982 (values under the dates are daily primary production rates, in mg $C \cdot m^{-2} \cdot d^{-1}$).

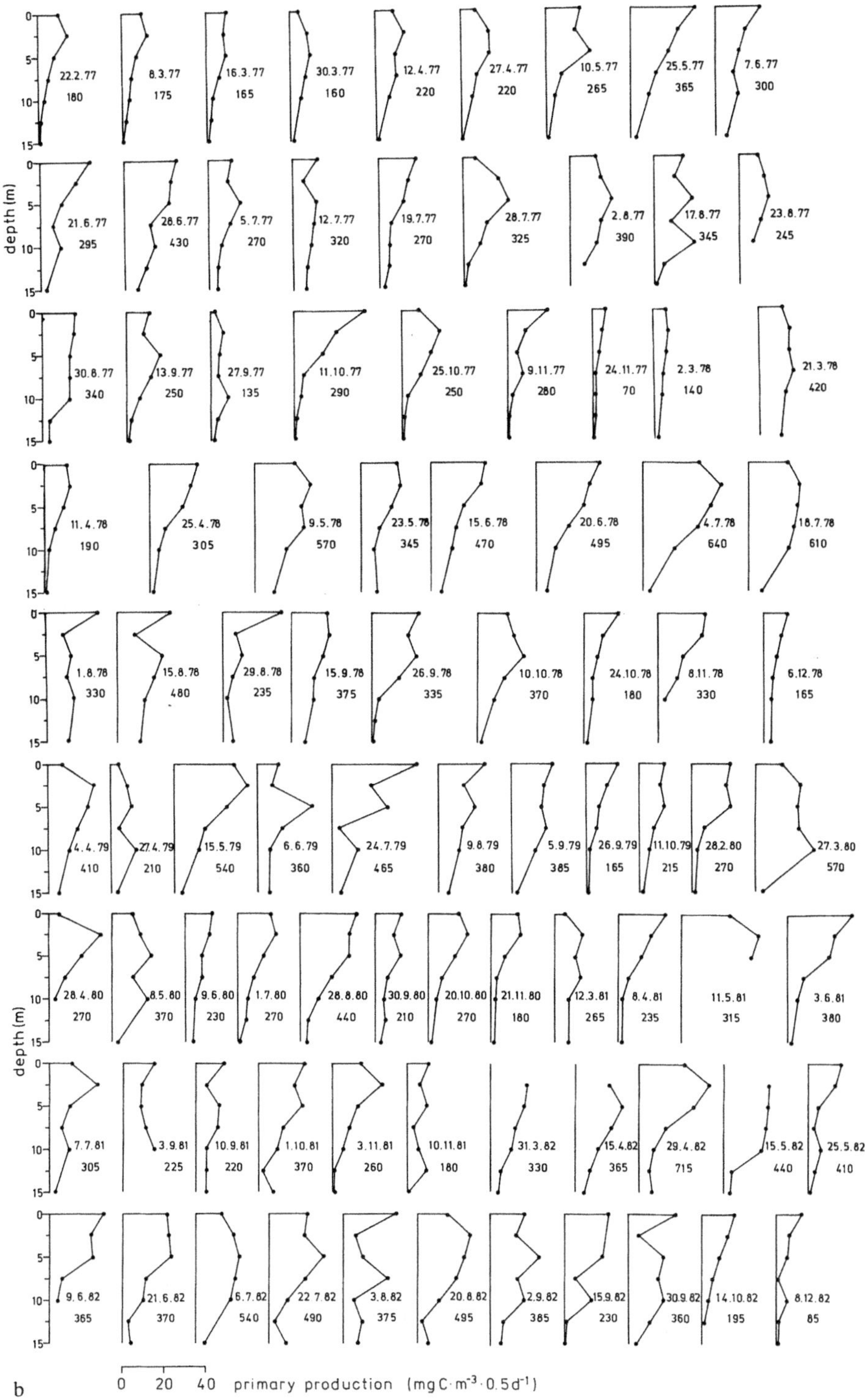
depth (m)
0
5
10
15
22.2.77
180
8.3.77
175
16.3.77
165
30.3.77
160
12.4.77
220
27.4.77
220
10.5.77
265
25.5.77
365
7.6.77
300
21.6.77
295
28.6.77
430
5.7.77
270
12.7.77
320
19.7.77
270
28.7.77
325
2.8.77
390
17.8.77
345
23.8.77
245
30.8.77
340
13.9.77
250
27.9.77
135
11.10.77
290
25.10.77
250
9.11.77
280
24.11.77
70
2.3.78
140
21.3.78
420
11.4.78
190
25.4.78
305
9.5.78
570
23.5.78
345
15.6.78
470
20.6.78
495
4.7.78
640
18.7.78
610
1.8.78
330
15.8.78
480
29.8.78
235
15.9.78
375
26.9.78
335
10.10.78
370
24.10.78
180
8.11.78
330
6.12.78
165
4.4.79
410
27.4.79
210
15.5.79
540
6.6.79
360
24.7.79
465
9.8.79
380
5.9.79
385
26.9.79
165
11.10.79
215
28.2.80
270
27.3.80
570
28.4.80
270
8.5.80
370
9.6.80
230
1.7.80
270
28.8.80
440
30.9.80
210
20.10.80
270
21.11.80
180
12.3.81
265
8.4.81
235
11.5.81
315
3.6.81
380
7.7.81
305
3.9.81
225
10.9.81
220
1.10.81
370
3.11.81
260
10.11.81
180
31.3.82
330
15.4.82
365
29.4.82
715
15.5.82
440
25.5.82
410
9.6.82
365
21.6.82
370
6.7.82
540
22.7.82
490
3.8.82
375
20.8.82
495
2.9.82
385
15.9.82
230
30.9.82
360
14.10.82
195
8.12.82
85
0 20 40 primary production (mgC·m⁻³·0.5d⁻¹)

b

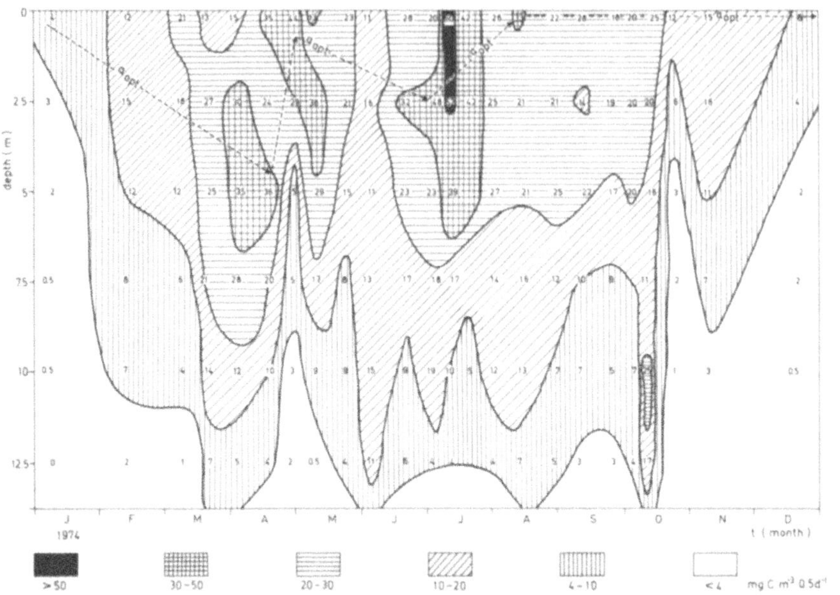

Figure 10.4 Spatial and temporal distribution of primary production of phytoplankton in Lake Stechlin in 1974.

Seasonal and vertical distribution of primary production of phytoplankton in Lakes Nehmitz and Dagow

Fig. 10.5 and Fig. 10.6 represent seasonal and vertical distribution of primary production in Lake Nehmitz.

The primary production of Lake Nehmitz (north basin) resembles that of Lake Stechlin. As early as March/April, relatively high primary production rates can be observed. In 1974 primary production rates of 50–90 mg $C \cdot m^{-3} \cdot d^{-1}$ in a_{opt} and 350–575 mg $C \cdot m^{-2} \cdot d^{-1}$ were measured. A short decline in primary production (May or June) followed. Maximum values of primary production were found in July and August (100–140 mg $C \cdot m^{-3} \cdot d^{-1}$ in a_{opt} and 750–850 mg $C \cdot m^{-2} \cdot d^{-1}$. a_{opt} was determined between surface and depth of 5 m (Fig. 10.6). A second a_{opt}-value occurred in metalimnetic layer in August and September (Fig. 10.6). Especially between July and September (and also in March and April) the primary production is higher in Lake Nehmitz than in Lake Stechlin (Fig. 10.5). In other months primary production per square metre is about the same in both lakes. Annual primary production of phytoplankton was estimated at 140 g $C \cdot m^{-2}$ in 1974.

In the south basin of Lake Nehmitz maximum primary production rates of 150–210 mg $C \cdot m^{-3} \cdot d^{-1}$ in a_{opt} and 850–1050 mg $C \cdot m^{-2} \cdot d^{-1}$ were observed (Fig. 10.5). a_{opt} was determined on an average of 2.5 m. Primary production showed two seasonal peaks, one in late spring (June), the other in late summer (August/September) (Fig. 10.5).

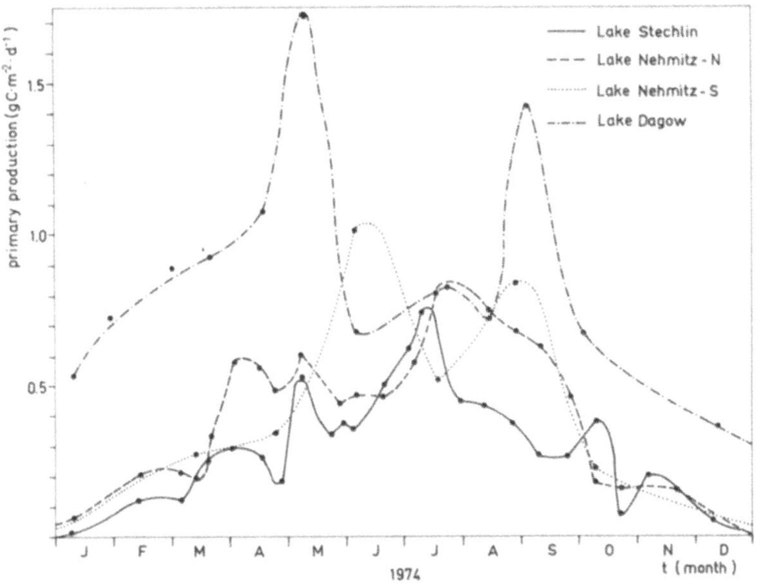

Figure 10.5 Seasonal periodicity of primary production of phytoplankton in lakes of the Lake Stechlin area.

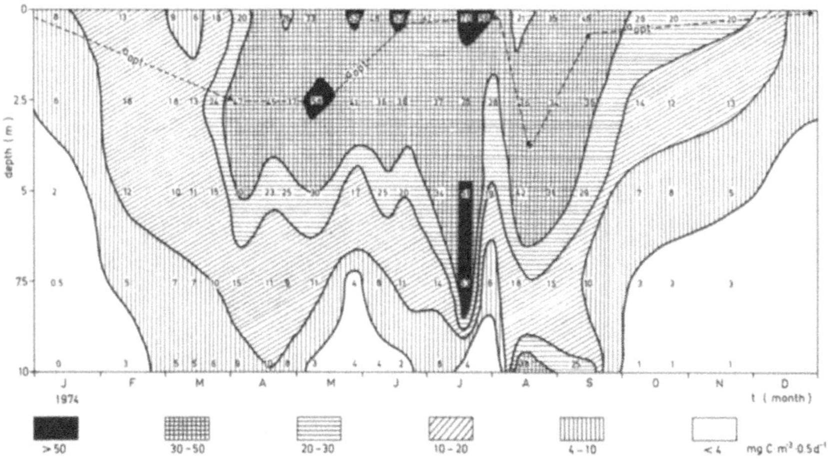

Figure 10.6 Spatial and temporal distribution of primary production in Lake Nehmitz-N in 1974.

Annual primary production was 145 g C · m^{-2} in 1974.

In contrast to primary production of phytoplankton in Lake Stechlin and Lake Nehmitz, markedly higher values were measured in Lake Dagow (Figs. 10.5, 10.7). In Lake Dagow primary production rates ranged from 15 mg C · m^{-2} · d^{-1} (under ice- and snow-cover) to 4250 mg C · m^{-2} · d^{-1} (June 1971). a_{opt} was observed between surface and depth of 1 m. Two major peaks of primary produc-

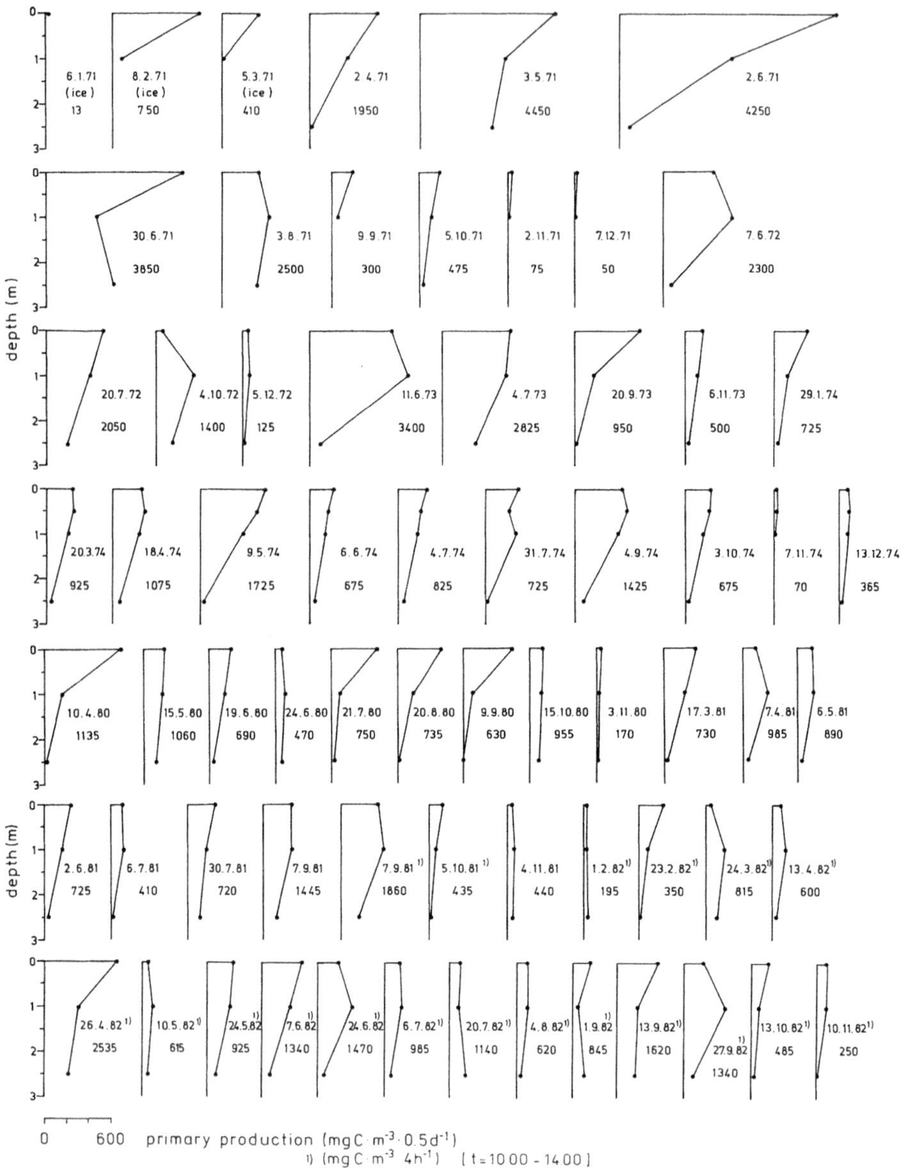

Figure 10.7 Vertical distribution of primary production of phytoplankton in Lake Dagow from 1974 to 1982 (cf. Figure 10.3).

tion were found in annual cycle, a first (greatest) in June and a second (smaller) in September. During the summer months, July and August, a well-developed depression-phase of primary production was observed (Fig. 10.5).

Annual primary production decreased from 290 g $C \cdot m^{-2} \cdot a^{-1}$ in 1971–1974 to 230 g $C \cdot m^{-2} \cdot a^{-1}$ in 1978 and reached approximately identical values in the following years with 233 g $C \cdot m^{-2} \cdot a^{-1}$ in 1980 and 228 g $C \cdot m^{-2} \cdot a^{-1}$ in 1981. In 1982 annual primary production increased again to 268 g $C \cdot m^{-2} \cdot a^{-1}$.

Light utilization efficiency of primary production

Efficiencies of phytoplanktonic primary production represent a physiological response by the phytoplankton population to light environment and other ecological factors. Efficiency of primary production in converting solar energy into potential energy is defined as: Eff = PSR/E (Eff = efficiency, PSR = photosynthetically stored radiation (in our cases PSR = P, P = gross primary production), E = light energy; cf. Dubinsky 1980).

The light utilization efficiency of phytoplanktonic primary production (Eff_{tot} or Eff_{PAR}) and per unit biomass ($Eff_{tot/B}$ or $Eff_{PAR/B}$) are calculated on the basis of energy equivalents of gross primary production (0.1 g C = 4.17 kJ) phytoplanktonic biomass and the global radiation (E_{tot}) or the photosynthetically active radiation (PAR, PAR = 46% of E_{tot}; see p. 74).

Mean annual light utilization efficiency by gross primary production of phytoplankton per square metre amounted to 2.6‰ (Eff_{PAR}), respectively 1.2‰ (Eff_{tot}) from 1970–1982 in Lake Stechlin (Table 10.3). Eff_{PAR} ranged from 1.7–3.8‰ in different years (Table 10.3). Eff_{PAR} shows high values in the period between

Table 10.3 Annual values of global radiation (E_{tot}), photosynthetically active radiation (PAR), primary production of phytoplankton (P), light utilization efficiency of primary production (Eff_{tot}, Eff_{PAR}) and light utilization efficiency of primary production per unit biomass ($Eff_{PAR/B}$).

Year	E_{tot} ($kJ \cdot cm^{-2} \cdot a^{-1}$)	PAR ($kJ \cdot cm^{-2} \cdot a^{-1}$)	P ($kJ \cdot m^{-2} \cdot a^{-1}$)	Eff_{tot} (‰)	Eff_{PAR} (‰)	$Eff_{PAR/B}$* (‰ $[g\ B_{FM}]^{-1}$)
1970	354.2	162.9	4,212	1.2	2.6	0.6
1971	351.0	161.5	4,212	1.2	2.6	1.9
1972	321.6	147.7	5,755	1.8	3.8	0.6
1973	345.4	158.3	4,796	1.4	3.1	(0.7)
1974	324.3	149.2	4,045	1.3	2.8	(0.6)
1975	366.7	168.7	2,836	0.8	1.7	(0.2)
1976	360.4	165.8	3,294	0.9	1.9	0.2
1977	314.8	144.8	3,086	1.0	2.1	0.7
1978	333.7	153.5	4,212	1.3	2.8	0.5
1979	328.2	150.9	3,711	1.1	2.4	–
1980	333.5	153.4	4,170	1.2	2.7	–
1981	311.9	143.5	3,628	1.2	2.1	–
1982	377.4	173.6	4,545	1.2	2.6	–
$\bar{x}$	340.0	156.4	4,045	1.2	2.6	0.7

* phytoplankton biomass (cf. p. 178); B_{FM} = fresh weight of phytoplankton biomass.

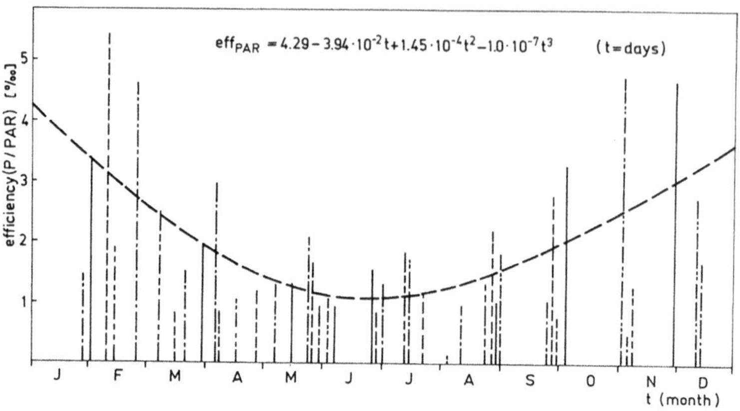

Figure 10.8 Seasonal periodicity of light utilization efficiency of primary production of phytoplankton (Eff_{PAR}) in Lake Stechlin.

autumn and early spring (2.0–5.5‰) and low values in late spring and summer (0.2–2.0‰) (Fig. 10.8).

Generally higher efficiencies and also other seasonal changes of efficiencies were observed in eutrophic and polytrophic lakes of the Lakes Stechlin and Breiter Luzin areas (Fig. 10.9). Mean Eff_{PAR} was 2.9–3.7‰ in eutrophic Lake Dagow and 5.1‰ in polytrophic Lake Haussee.

Reduced light utilization efficiencies with the increase of light intensities were observed in seasonal periodicity as well as in depth profiles (Fig. 10.8–10.10). Eff_{PAR} is generally influenced to a high degree by short-term mechanisms (stochasticity of global radiation, light inhibition – all caused by high light intensities) and by long-term mechanisms (adapted phytoplankton algae in time and space).

Efficiencies do not increase or decrease uniformly. Efficiencies can be understood more easily as an oscillating process in different levels of light utilization efficiency or light adaptation. Fig. 10.11 shows such oscillation of Eff_{PAR} at 0 m and 5 m depths of Lake Stechlin. The phytoplankton algae are thus exposed to two opposite disadvantages: light inhibition near the lake surface during strong illumination and reduced utilization of light during low light intensities as a consequence of light adaptation (Tilzer 1973). These rhythms are overlapped by the long-term succession of various phytoplankton species with different light adaptation or by vertical migration of flagellated phytoplankton species.

A deeper understanding of light utilization efficiencies of phytoplankton needs, therefore, analysing the depth-variation of vertical profiles of photosynthetic activity concerning not only the vertical attenuation of light energy but also the vertical distribution of phytoplankton structure and biomass. Profiles of depth-variation of utilization efficiencies of primary production can be of especially varied shape in stratified lakes.

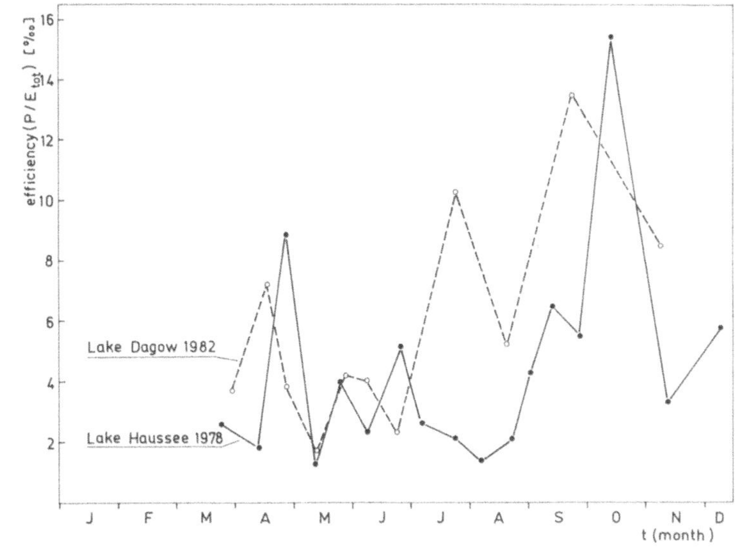

Figure 10.9 Seasonal periodicity of light utilization efficiency of primary production of phytoplankton (Eff_{tot}) in the Lakes Haussee and Dagow.

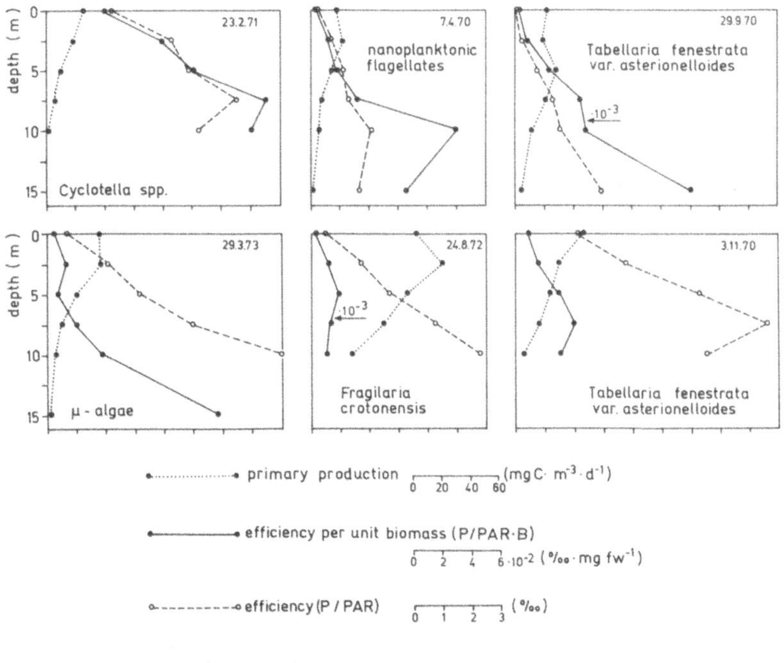

Figure 10.10 Vertical distribution of primary production of phytoplankton and light utilization efficiencies (Eff_{PAR}, $Eff_{PAR/B}$) in Lake Stechlin.

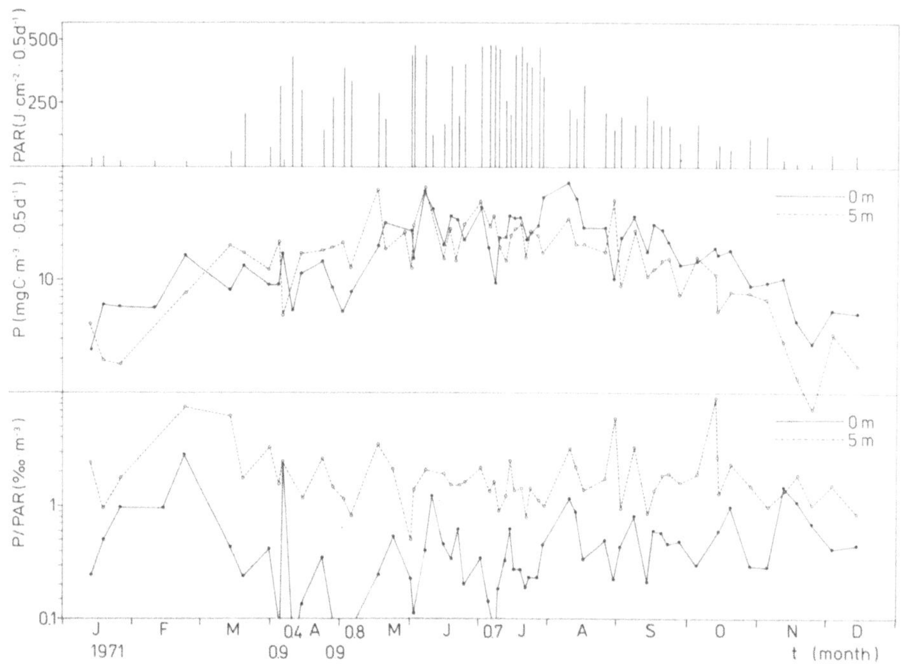

Figure 10.11 Photosynthetically active radiation (PAR), primary production of phytoplankton (P) and light utilization efficiency of primary production (Eff_{PAR}) at 0 m and 5 m levels in Lake Stechlin in 1974.

Differences of algae biomass and composition can exist in various layers as a result of active migration, sinking rates, differential growth-and mortality rates or thermal conditions, and influence light utilization efficiency (pp. 301–303).

Light utilization efficiency per unit phytoplankton biomass ($Eff_{PAR/B}$) characterizes the light adaptation of algae better than Eff_{PAR} alone. In Lake Stechlin $Eff_{PAR/B}$ came to a maximum similar to Eff_{PAR} during low light intensities in winter and early spring. At the time, low phytoplankton biomass but high specific activities were observed. During summer months, $Eff_{PAR/B}$ is greatly decreased (Koschel 1981a). Altogether, $Eff_{PAR/B}$ shows considerable variability in different years (Table 10.3).

Vertical profiles of $Eff_{PAR/B}$ show maxima as a rule at a depth of 10 m (Fig. 10.10). Successive zones of light inhibition, light saturation and light limitation can be distinguished, particularly if the activity coefficients of phytoplankton (P/B) are plotted against inputs of light energy from surface to the depth (Fig. 10.12). Plots of Fig. 10.12 demonstrate similar differences in the photosynthetic capacity of algae-species.

Phytoplankton populations exist with relatively low values of biomass but high growth efficiencies. These algae can assimilate the same carbon quantities as other phytoplankton populations with high values of biomass but low growth efficiencies (Table 10.3; Figs. 10.10, 10.12).

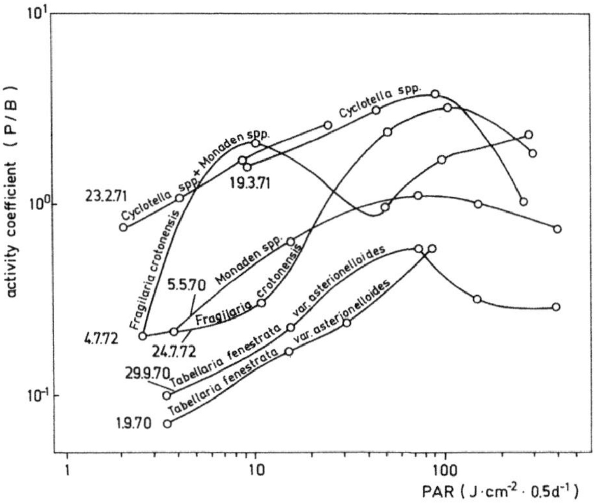

Figure 10.12 Relationship between activity coefficient of phytoplankton (P/B) and photosynthetically active radiation (PAR) for various phytoplankton populations in Lake Stechlin (P/B = mg assimilated C · 0.5 d^{-1} : mg C of phytoplankton biomass).

Photosynthesis – light relations

The quantitative description (generalization) of photosynthetic light curves is one of the most advanced limnological submodels in phytoplankton growth models. Talling (1957) developed such mathematical submodel for photosynthesis – light relations in various types of lakes by a general pattern of the response of pelagic photosynthesis to vertical attenuation of light.

By Talling's equation

$$P = a_{opt} \cdot z_{0.5\ I_k} \qquad (1)$$

calculated values of primary production of phytoplankton were compared with observed primary production (Fig. 10.13). Comparison detected that this modification can be used for an approximation of primary production in cases where the light penetration of visible spectral components in waters and the primary production rate of light saturation are known. But the comparison also underlined that many conditions are given, where calculated and observed values differ from each other. Differences of more than 100% were found. a_{opt} lies near the surface of the lake and cannot be explained on the basis of PAR and their distribution alone. Disproportionate distribution of phytoplankton, nutrients and other factors influences the light utilization efficiency of primary production. These influences must be considered also in other modifications of photosynthesis – light submodels, which were developed from Talling's fundamental work (e.g., Steel 1962; Vollenweider 1965; Straškrabá 1976a).

These model-modifications allow at any rate a better description of photo-

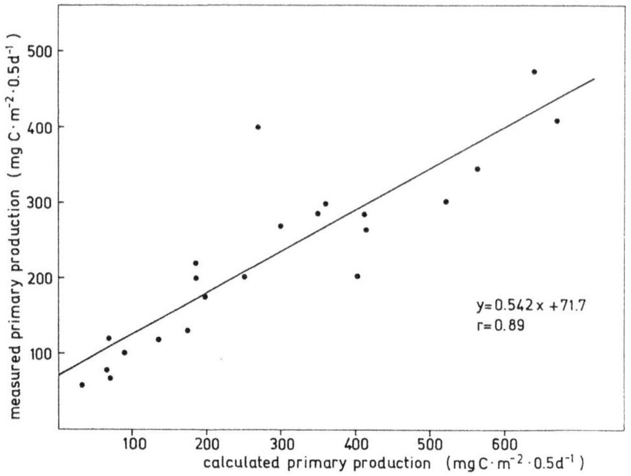

Figure 10.13 Relationship between measured and calculated daily primary production of phytoplankton. Daily primary production calculated on basis of Talling's equation (cf. p. 301).

synthetic light sensivity of different phytoplankton populations. The photosynthetic light sensitivity of algae can be described by the I_k-parameter. I_k varies in different algae groups and shows seasonal variations. I_k-values of between 11.2–128.2 $J\,cm^{-2} \cdot d^{-1}$ in Lake Stechlin were found (Fig. 10.14). They were calculated on the basis of an equation from Steel (in Straškrabá 1976a), where P/B_{max} to P/B relations and PAR in different depths were considered:

$$(P/B)_{max} = \frac{2\, I/2\, I_k}{1 + (I/2\, I_k)^2} \qquad (2)$$

Comparison between calculated and observed P/B-values (Fig. 10.15) showed especially during summer stratification that various I_k-values can also exist in vertical profiles, a relatively high I_k-value for the population near the lake surface and a lower value for the deep population. Exact differentiations of I_k from in situ-observations effect many difficulties. Heterogeneity of phytoplankton, motile and immotile forms or different mixing rates influence measurements. Only on individual data or on defined conditions is an exact separation and measurement of light and dark adapted algae practicable.

Some I_k-values for definite phytoplankton algae of our lake district are separated in Table 10.4. It indicates a high variability of I_k, not only by different algae species and different temperature but also, if possible, by the modification of the same species or by various nutrient conditions. Highest I_k-values were generally found for summer-algae by high water temperatures (Table 10.4). How far these variations of I_k are directly dependent on temperature cannot be answered from our observations. Benndorf (1979) explained the variability of I_k as an adaptation mechanism.

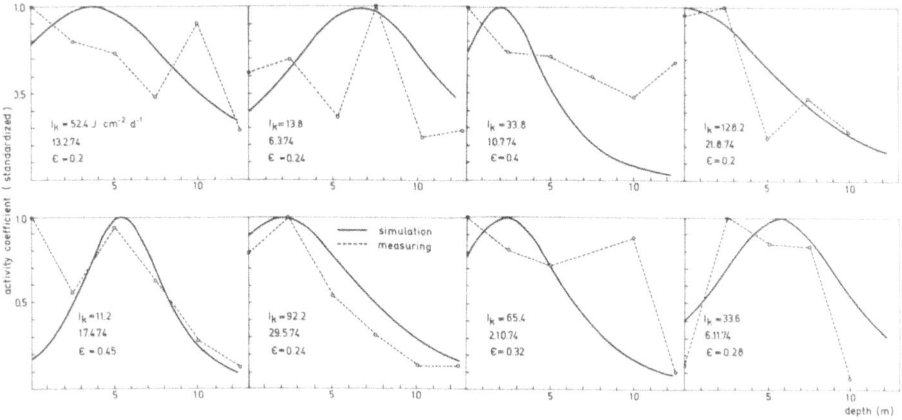

Figure 10.14 Measured and calculated depth profile of activity coefficient of phytoplankton in Lake Stechlin. I_k-values calculated on basis of Steel's equation (cf. p. 302).

Table 10.4 I_k–values of spring- and summer algal species.[a]

Algal species	Date	Origin	Temp (°C)	I_k-value (PAR) ($J \cdot cm^{-2} \cdot h^{-1}$)
Stephanodiscus hantzschii[b]	19.4.79	Lake Breiter Luzin (mesotrophic)	8.3	5.8
Stephanodiscus hantzschii[b]	19.4.79	Lake Haussee (polytrophic)	8.3	8.3
Stephanodiscus hantzschii[b]	19.4.79	Lake Dagow (eutrophic)	9.4	9.4
Fragilaria crotonensis	15.7.72	Lake Stechlin (oligotrophic)	20.7	22.2
Fragilaria crotonensis	3.7.80	Lake Roofen (mesotrophic)	18.0	22.7
Tabellaria fenestrata var. *asterionelloides*	29.9.70	Lake Stechlin (oligotrophic)	–	20.0
μ-alga[c]	March/April 1980–82	Lake Fuchskuhle (acid bog lake)	5.0–7.0	6.0
μ-alga[c]	July 1980	Lake Barsch (acid bog lake)	20.0	11.2
	November 1981	Lake Barsch (acid bog lake)	8.0	8.9

[a] In situ observations
[b] *St. hantzschii* complex (Lund & Reynolds 1982)
[c] Undetermined chlorophycean algae (cf. p. 280)

Primary production and phytoplankton population

Understanding of the factors determining phytoplankton productivity requires a detailed knowledge of the interactions between primary production, biomass (pigments, enzymes etc.) and the composition of phytoplankton. That is not only one of the most interesting fields of phytoplankton ecology but also one of the most difficult problems in experimental work, especially in field ecology. Many problems arise, e.g., in the determination of phytoplankton composition and biomass, in the differentation of active or inactive phytoplankton species, in the different photosynthetic behaviour of individual species or in the discrepancies between gross and net carbon assimilation.

In addition to the primary production of phytoplankton of Lake Stechlin and of other lakes in the study area, many data of phytoplankton composition and biomass within the same sample were measured (pp. 158–159). This extensive data accumulation makes it possible to examine some relationships prevailing between primary production of phytoplankton and biomass of whole population or fractions (or taxonomic units) of phytoplankton community.

A great number of research publications deal with photosynthesis-biomass relations of phytoplankton in freshwater and in marine environments (e.g., Rodhe *et al.* 1958; Findenegg 1971; Elster 1965; Goldman *et al.* 1968; Javornický & Komárková 1973; Malone 1980). Data on differences in phytosynthetic behaviour between individual species of phytoplankton populations have been provided, as a rule, by autoradiography (Gutelmacher 1975; Gutelmacher & Petrova 1982; Desertová 1976; Höhne & Nixdorf 1982). But there also exist some indirect possibilities of differentiation of phytoplankton productivity. Indirect evidence has been found by comparison of the phytosynthetic rates or the specific activity of phytoplankton populations dominated by one species or separated by defined size fractions (Findenegg 1971; Tilzer 1972; Koschel 1974a, 1976, 1981a; Munawar *et al.* 1978; Rai 1982).

In Lake Stechlin the relationship between net- and nano-planktonic primary production of phytoplankton indicated that nanoplanktonic proportion contributes more to total primary production of phytoplankton than netplanktonic proportion (Table 10.5). Nanoplanktonic primary production contributes much more to total primary production than its biomass indicated (Koschel 1974b; 1981; pp 187–192). From data sampled in Lake Stechlin it was concluded that nanoplankton shared about 30%–100% of total daily phytoplankton production. From 1971–1978 the quota of yearly nanoplanktonic primary production is greater than 60% (60%–78%), on average 70% (Table 10.5). The nanoplanktonic fraction '<20μm' within the whole nanoplanktonic fraction '<50μm' is the main bulk of primary production. I analysed the primary production of the '<20μm' fraction in Lake Stechlin from 1979–1982. The quota of yearly primary production of the '<20μm' fraction on total yearly primary production averages 68% (Table 10.5).

The importance of nanoplanktonic primary production was also found in other lakes of the Lakes Stechlin and Breiter Luzin areas (Table 10.5). As a rule a quota of >50% of yearly nanoplanktonic primary production of total primary production of phytoplankton was established in oligotrophic as well as in polytrophic lakes. Rai (1982) investigated the primary production of Lake Schöhsee, a typical moderately eutrophic Baltic Lake, and found that the phytoplankton fraction with a size <10μm dominated the total primary production of phytoplankton. But he found also that not only the percentage quota for carbon assimilation of nanoplankton is higher but also the absolute biomass of nanoplankton could be high. Nanoplanktonic cell biomass occupied 64–90% of the total phytoplankton biomass in Lake Schöhsee. These observations would lead to the negation of our investigations and of the general hypothesis that nanoplankton often accounts for greater a proportion of primary production than their biomass would indicate. In Lake Stechlin our investigations mostly showed a higher activity coefficient for nanoplanktonic algae than for netplanktonic algae (Fig. 10.15). The activity coefficient of the nanoplanktonic fraction was on an average about ten-fold higher than the activity coefficient of net planktonic fraction of phytoplankton. In addition, the specific activity of phytoplankton compared with the phytoplankton biomass showed a negative parabolic correlation (Fig. 10.16).

Similar findings have been made by Benndorf & Stelzer (1973) and Javornický (1980). But this general relation does not consider that in any case the specific activity is dependent upon the size of the algae. A large number of other factors, particularly the given structure of cells, pigment concentrations or other inherent species characteristics, could be affecting the activity coefficient.

Measurements of photosynthetic behaviour on ultraplanktonic algae in Lake Stechlin and other lakes showed a lower activity coefficient than expected on the

Table 10.5 Quota of nanoplanktonic primary production of total primary production of phytoplankton in different lakes.

Lake	Oservation period	Yearly primary production		Quota (%)
		Total ($g\ C \cdot m^{-2}$)	Nanoplanktonic ($g\ C \cdot m^{-2}$)	
Stechlin	1971–1978	91	63[a]	70
Stechlin	1980–1982	99	67[b]	68
Nehmitz-N	1974	140	105[a]	75
Nehmitz-S	1974	145	88[a]	60
Dagow	1974	290	130[a]	45
Dagow	1979–1982	248	139[b]	56
Breiter Luzin	1982	210	135[b]	64
Haussee	1978	420	240[b]	57

[a] Nanoplanktonic fraction cell size <50 μm
[b] Nanoplanktonic fraction cell size <20 μm

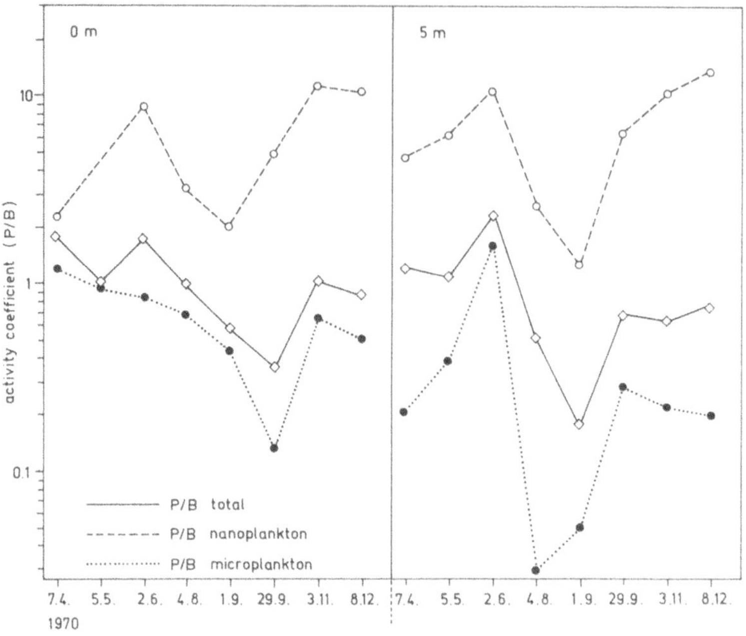

Figure 10.15 Seasonal periodicity of activity coefficient (P/B) and quota of nanoplanktonic and microplanktonic activity coefficients at 0 m and 5 m in Lake Stechlin.

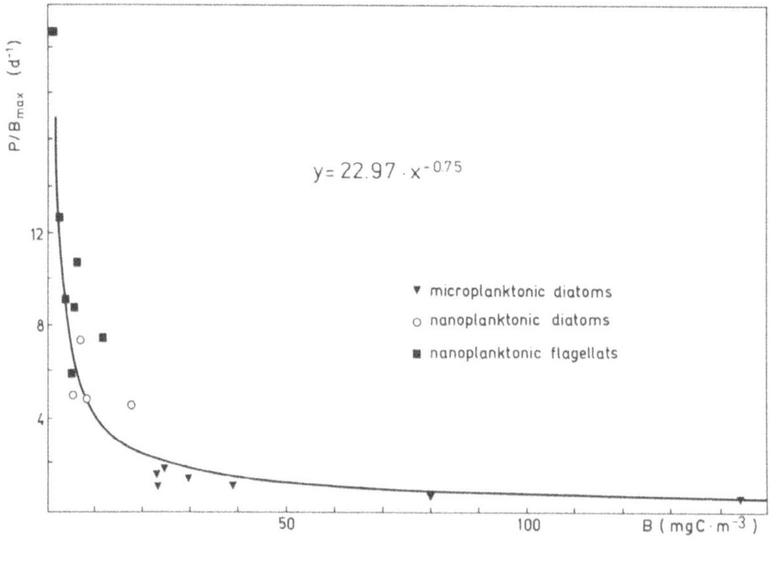

Figure 10.16 Relationship between activity coefficient (P/B_{max}) and biomass of phytoplankton in Lake Stechlin.

basis of cell size. During ultraplanktonic mass developments in early spring 1973 and 1974 (p. 185 ff.), we measured only very low specific activities of ultraplanktonic algae in Lake Stechlin. Low specific activities of ultraplanktonic green algae (2–3 μm cell size) were established also in an acid bog-lake (Lake Große Fuchskuhle).

To sum up: the variation of specific activity within a defined cell size can be relatively high. It appears that the specific activity of a defined biomass is not only dependent on the cell size but also on other factors. Photosynthetic behaviour of phytoplankton population in relation to biomass must be due to: (1) the density of phytoplankton species, (2) the cell size, (3) the photosynthetic characteristics of individual species (e.g., I_k-, T_{opt}- or K_s-values), (4) the heterotrophic carbon assimilation of individual species and (5) the age and history of population.

Primary production and phosphorus uptake of phytoplankton

Phosphorus uptake kinetic. Phosphorus is the primary nutrient limiting the biomass production in Lake Stechlin and most neighbouring lakes (Koschel 1974a, 1976; pp. 87–92). In this connection phosphorus uptake – algal growth relationships are especially interesting for estimation of phosphorus-limited primary production. Using carrier-free $^{32}PO_4$-P we analysed the phosphorus uptake of pelagic microorganisms in samples of water from various lakes of our study area (Koschel 1974a, 1980, 1981b). The phosphorus influx can be approximated with a kinetic of first order (Fig. 10.17). The kinetic velocity constant (k_1) and the turnover time (T_t) for phosphorus uptake can be relatively easily determined (Fig. 10.17). T_t and K_1 are data which propose as gross-values of phosphorus demand of pelagic planktonic microorganisms.

Phosphate uptake depends on the intracellular and extracellular phosphorus content (Button 1978). The assimilation of phosphorus by planktonic microorganisms may also be used to characterize or to identify phosphorus starvation or the phosphorus status of planktonic populations.

Periodicity of phosphorus uptake. Seasonal distribution of planktonic phosphate uptake shows a characteristic rhythm in all the lakes studied (Fig. 10.18). In Lake Stechlin minimum values of phosphate uptake were found as early as late autumn and winter (k_1<0.001 min^{-1}, T_t>100 h). After January phosphate uptake increased quickly. Extremely high values were measured from April – September. Its intensity is maximum in May –July (k_1 = 0.3–0.5 min^{-1}, T_t = 0.03–0.05 h). Phosphurus uptake decreased after September. Maximum values in spring and summer are about 10^4 fold higher than minimum values in winter and autumn. High values of phosphate uptake were only found in the euphotic zone (Fig. 10.19). During the stratification period the phosphorus uptake at 20 m reaches only a low percentage of the phosphorus uptake near the surface, at 40 m only few

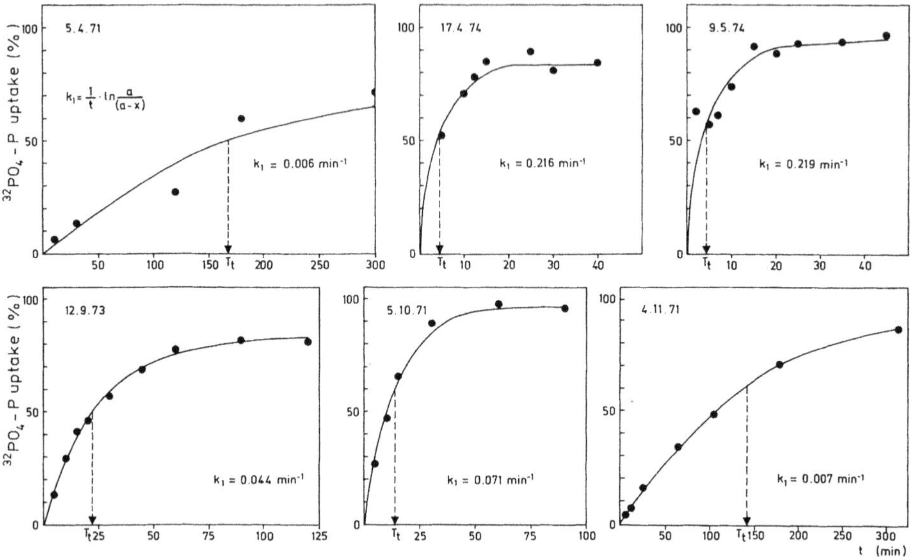

Figure 10.17 Phosphate uptake ($^{32}PO_4$-P) in surface water of Lake Stechlin (k_1 = velocity constant, T_t = turnover time).

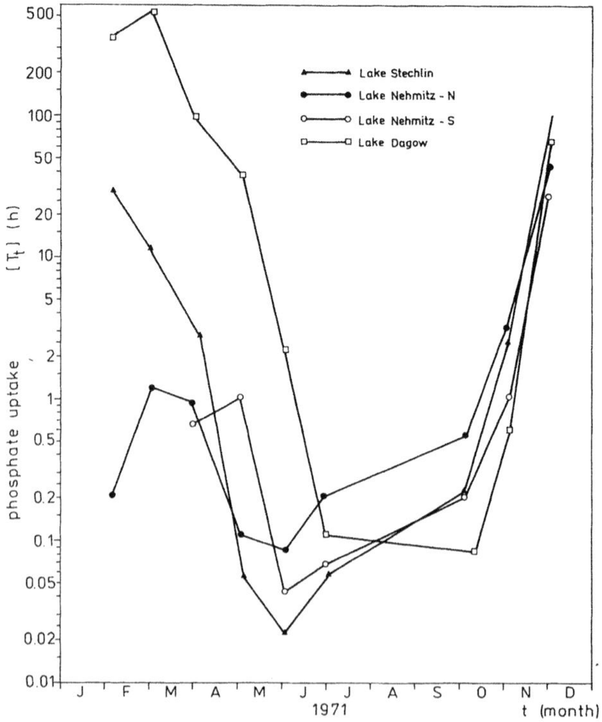

Figure 10.18 Seasonal periodicity of phosphate uptake (T_t) in pelagic surface water of Lakes Stechlin, Nehmitz and Dagow.

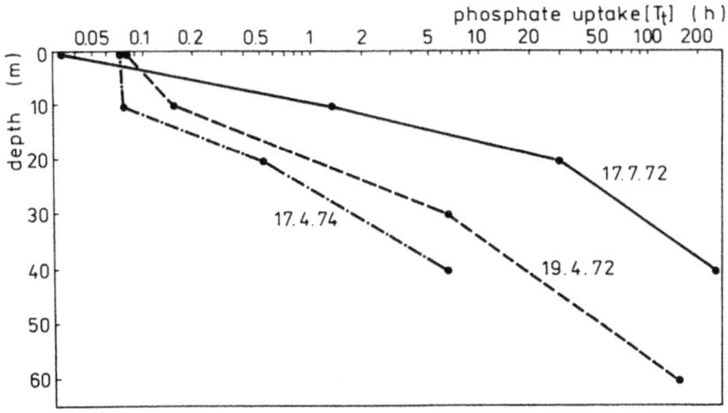

Figure 10.19 Vertical distribution of phosphate uptake (T_t) in Lake Stechlin.

pars pro mille. Further, the phosphate uptake is not uniform in the whole water column during circulation phases. A sharp drop of k_1 or T_t were measured in deep water samples (Fig. 10.19).

Phosphate uptake in Lake Nehmitz resembles the phosphate uptake in Lake Stechlin (Fig. 10.18), but greater differences exist in winter and early spring. Under the ice cover phosphate uptake velocities are higher in Lake Nehmitz-N than in Lake Stechlin. But the absolutely highest phosphate uptake velocities were measured in Lake Stechlin, followed by Lake Nehmitz-N, Lake Nehmitz-S and Lake Dagow (Fig. 10.18).

In eutrophic Lake Dagow phosphate uptake also shows relatively high values. The period with maximum phosphate uptake velocities is substantially shorter and only in a shallow water column (0 m–2.5 m) than in the other three lakes (Fig. 10.18). Comparisons between phosphate uptake and extracellular orthophosphate concentration indicated that phosphate uptake increased with decrease of orthophosphate concentration (Koschel 1974a, 1981b).

Relation between phosphate uptake and primary production. High initial phosphate uptake velocities are reflected, as a rule, in limited primary production rates of phytoplankton (Koschel 1981b). The connection will be clearer, if phosphate uptake (k_1 or T_t) is contrasted with light utilization efficiency per unit biomass and the saturation quotient (soluble PO_4-P/phytoplankton biomass). During periods with a high saturation quotient, small phosphate uptake velocities and light utilization efficiencies were measured (Fig. 10.20). Phosphate uptake increases with the decrease of the saturation quotient. Light utilization efficiency per unit biomass decreased at the same time. The relation between exogenous orthophosphate concentration and primary production of phytoplankton becomes more and more disadvantageous, and it leads to maximum phosphate uptake velocities in June and July, with greatly reduced light utilization efficiencies per unit biomass.

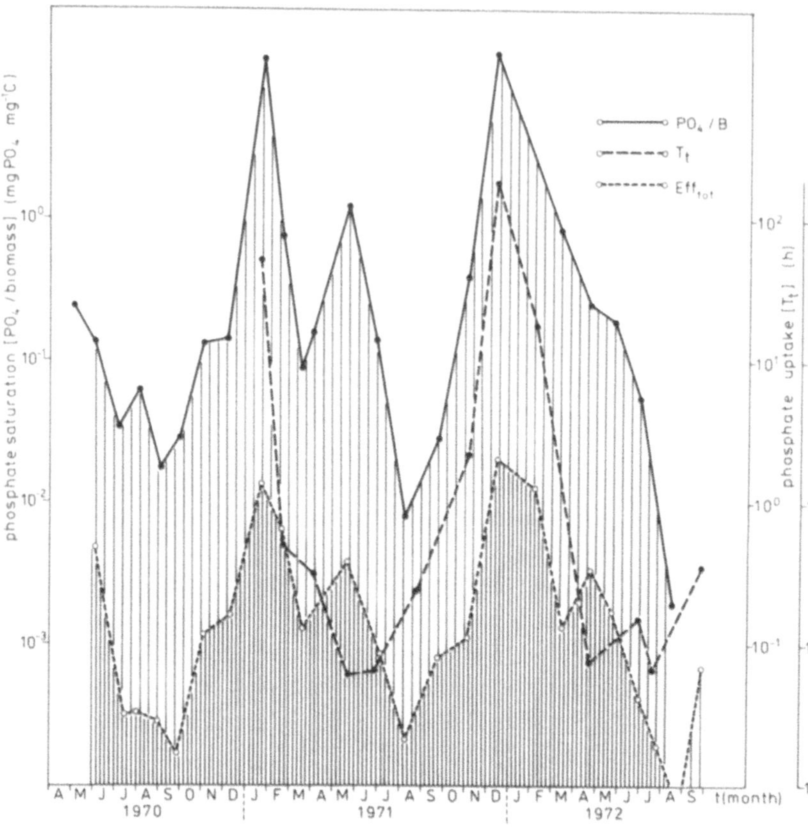

Figure 10.20. Phosphate uptake (T_t) light utilization efficiency of primary production per unit phytoplankton biomass ($Eff_{tot/B}$) and phosphate saturation coefficient (PO_4/B) in Lake Stechlin.

Phosphate uptake velocity is a gross value expressing initial phosphate uptake. Growth activity and phosphate starvation dominated initial phosphate uptake. A rapid phosphate uptake indicated phosphorus starved planktonic microorganisms. Phosphorus pools of cells are unsaturated. A slow phosphate uptake speaks for saturated phosphorus pools and an adequate phosphorus supply of plankton. Corresponding primary production (or efficiencies of primary production) are mostly smaller by high phosphate uptake velocities (P-starvation) than by comparatively slow phosphate uptake velocities (P-saturation). Measurement of phosphate uptake parameters permits a preliminary approximation of quantitative scale of P-limitation of planktonic productivity (Fig. 10.21). K_1-constants of phosphate uptake $>0.02\,min^{-1}$ indicate P-limited production processes.

The phosphate uptake of different phytoplanktonic fractions. The percentage quota of different planktonic fractions (>10 μm; 10–1.5 μm; 1.5–0.6 μm) of total

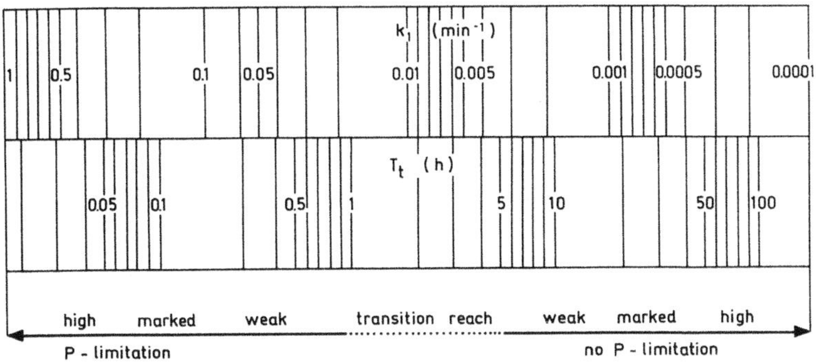

Figure 10.21 Relation between phosphate uptake (k_1, T_t) and P-limitation of primary production of phytoplankton.

phosphate uptake velocities is shown in Table 10.6. The fraction >10 μm in size is represented by an average of 7% on total phosphate uptake, the fraction 10–1.5 μm an average of 56% and the fraction 1.5–0.6 μm an avarage of 37%.

One trend is especially interesting. Large fractions are more inclined to phosphate uptake as phosphate uptake velocity decreases, and, vice versa, increases of phosphate uptake bring about an important decrease of phosphate uptake velocities in smaller fractions.

Table 10.6 Percentage quota of different planktonic fractions of total planktonic phosphate uptake in various lakes[a].

Lake	Date	Phosphate uptake T_t (h)	Perentage quota of fractions >10 μm (%)	10–1.5 μm (%)	1.5–0.6 μm (%)
Lake Dagow	07.07.72	0.11	6	65	29
Lake Werbellin	17.08.72	0.43	24	36	40
Lake Müritz-G[b]	01.08.72	1.10	2	43	55
Lake Müritz-S[c]	01.08.72	2.60	9	40	51
Lake Nehmitz-N	10.08.72	0.15	2	68	30
Lake Nehmitz-S	10.08.72	0.14	4	63	33
Lake Parstein	14.08.72	0.50	5	65	30
Lake Stechlin	17.07.72	0.03	5	67	28

[a] Percentage quota of phosphate uptake was measured at beginning of steady state between extra- and intracellular radiophosphorus (see Fig. 10.17)

[b] G = Außenmüritz

[c] S = Binnenmüritz

Primary production and trophic status

Since Naumann (e.g. 1932) and Thienemann (e.g. 1928) introduced the terms 'oligotrophic', 'mesotrophic', 'eutrophic' and 'dystrophic' into limnology, many authors have described the term 'trophy' and the process 'eutrophication' (e.g. Vollenweider 1968; Eutrophication 1969; Uhlmann & Hrbaček 1967; Landner 1976).

Correspondingly, it was established that the trophic state must be determined on the basis of many facts (pp. 456–463) and that the process of eutrophication is an extremely complex one. But there is a basic relationship between the trophic or nutrient state of a lake and its biological productivity. Primary production of phytoplankton, the biomass or chlorophyll content of phytoplankton and the nutrient concentration (especially phosphorus) may be indicators for trophic state of lakes. The most sensitive index is based solely on changes in the rate and vertical distribution of primary production of phytoplankton (e.g., Koschel 1981a). However, the level proposed to indicate 'eutrophy' varies widely from author to author (e.g., Findenegg 1964; Hübel 1966; Rodhé 1969; Vollenweider 1968; Landner 1976; Koschel 1981a, p. 461; Zdanowski 1982). The levels are influenced through different physiographical and other factors (Uhlmann 1982). Geochemical basis conditions play a major role in regulation of trophic levels. The regulation of trophic state is obviously different for hard-water lakes on the one hand and for soft-water lakes on the other hand.

For the hard-water lakes of the Lake Stechlin area and neighbouring lake districts, we must set the upper limits between oligotrophic and eutrophic conditions at the level of primary production of phytoplankton of about 120 g $C \cdot m^{-2} \cdot a^{-1}$ (p. 461). This high level is determined through the primary production of phytoplankton in Lake Stechlin. Lake Stechlin belongs to those oligotrophic lakes with the highest annual primary production of phytoplankton known (Vollenweider 1968; Le Cren & Lowe-McConnel 1980).

There is an exponential connection between the annual primary production and the total phosphorus level (spring-maximum) for the hard-water lakes of the Lake Stechlin area and neighbouring lake districts (Fig. 10.22; Koschel *et al.* 1981). The curve shows that the primary production of phytoplankton is especially sensitive to changes in the phosphorus content in the oligotrophic to mesotrophic states. In the eutrophic state the sensivity of the primary production to such changes declines. A similar dependence has been established by various authors for the connection between the chlorophyll content of the phytoplankton and the concentration of phosphorus (Sakomoto 1966; Dillon & Rigler 1974; Straškrabá 1976b; 1978; Schindler 1978; Zdanoswki 1982). The phytoplankton must, however, be otherwise evaluated than the primary production. Phytoplankton can change its composition as well as its mass. With a constant mass but varying composition very differing rates of primary production are possible. The importance which is to be attached to this functional relation becomes especially clear

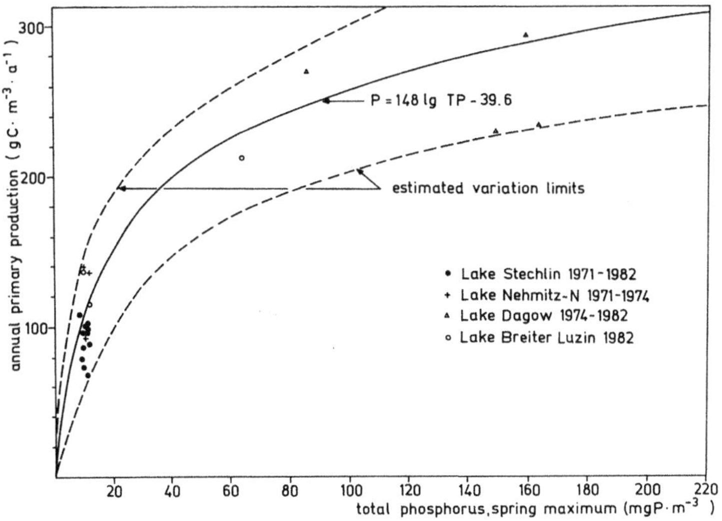

Figure 10.22 Relationship between annual primary production of phytoplankton and total phosphorus (spring maximum) in various lakes of the Mecklenburg Lake District.

from our observations in Lake Stechlin (p. 178–185, 289–294). Whereas the annual primary production of phytoplankton in Lake Stechlin in comparable years (1970–1972) only varied by a small percentage, the middle phytoplankton biomass changed by more than 100 per cent (cf. pp. 178–185).

The connection between phytoplankton biomass or its chlorophyll content and the level of phosphorus concentration is best made clear by means of a sigmoid function (Straškrabá 1976b; 1978; Fig. 10.23). This link shows three phases. The first phase (the lag-phase) indicates the defensive capability of the phytoplankton towards the increased phosphorus concentration. In oligotrophic to mesotrophic lakes the raising of the phosphorus concentration leads first not to a comparable raising of the phytoplankton biomass but mostly to a change in the specific activity of the phytoplankton community (for example changes in the composition, changes in the nano- and microplankton biomass). In the second phase (exponential growth phase) the increased supply of phosphorus can no longer be absorbed by such changes alone. The biomass now also grows exponentially with increasing phosphorus. In the third phase (saturation phase) the limitation of the phytoplankton by the phosphorus is increasingly reduced timewise, or completely stopped. Changes in the phosphorus level lead to only comparatively small changes in the phytoplankton biomass. The light (reduced transmission, high self-shading effect) and other nutrients (nitrogen) now limit mainly the primary production.

The lakes examined in our research can be included in respect to the relations between the chlorophyll content and the total phosphorus content in this sigmoid

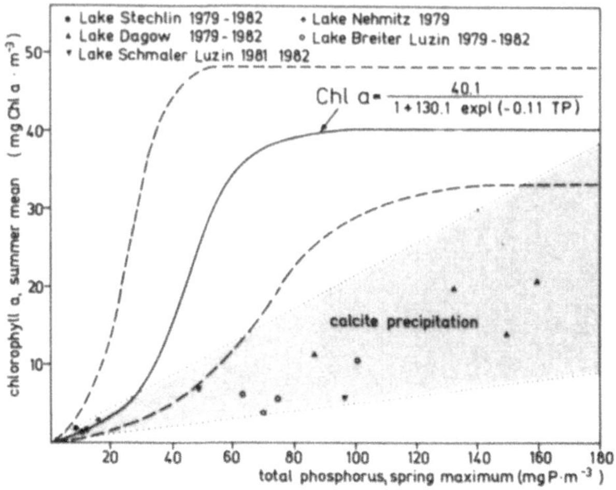

Figure 10.23 Relationship between chlorophyll content of phytoplankton (mean of epilimnion) and total phosphorus (spring maximum) in various lakes of the Mecklenburg Lake District.

function (Fig. 10.23). The lakes are, however, exclusively in the lower reaches of the toleration limits put forward by Straškrabá (1976b, 1978). In the case of Lake Stechlin and Nehmitz this deviation is not so clear since both lakes are in the 'lag-phase'. In the meso- to slightly eutrophic lakes, however, the strong depression of the phytoplankton becomes clear. In all the lakes dealt with, the water is markedly hard with autochthonous pelagic calcite precipitation phases (pp. 102–106, 412–419). Due to the calcite precipitation the phytoplankton biomass is reduced during the summer stagnation period. Calcite precipitation slows up an 'accelerated' eutrophication (Koschel *et al.* 1983; pp. 417–419) and diminishes the exponential increase in the phytoplankton biomass in relation to the phosphorus supply (Fig. 10.23).

The primary production of the microphytobenthos (W. Scheffler)

The primary production of the periphyton was studied using the C^{14}-method (Scheffler 1981).

The annual course of the primary production in the littoral of Lake Stechlin

The primary production in the littoral of Lake Stechlin was continuously measured from April 1974–September 1975 at measuring point Sl 1 (Table 10.7; Fig. 10.24). After a relatively high production rate with 422 mg C · m^{-2} · d^{-1} in September 1974, the primary production strongly decreases to the winter and reaches the

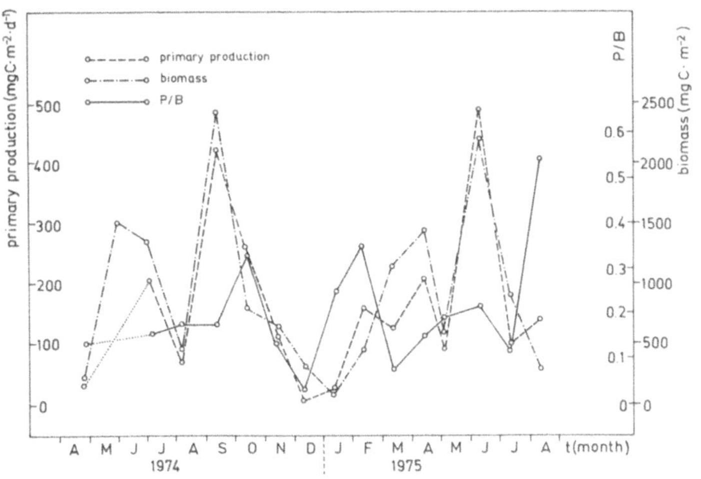

Figure 10.24 Primary production, biomass, and P/B-quotient for the annual periodicity (April 1974–August 1975; measuring point Sl 1; depth 0.05 m).

measured minimum in December, with about 6 mg $C \cdot m^{-2} \cdot d^{-1}$. From February–May 1975 the rates vary between 96 and 215 mg $C \cdot m^{-2} \cdot d^{-1}$. In June the curve reaches its climax with a rate of 476 mg $C \cdot m^{-2} \cdot d^{-1}$. With 68 mg $C \cdot m^{-2} \cdot d^{-1}$ in August 1974 and with 98 mg $C \cdot m^{-2} \cdot d^{-1}$ in July 1975, we find relatively low rates for summer. Fig. 10.24 shows that the curves of the biomass and the primary production usually exhibit a synchronous tendency.

With an increasing biomass the primary production rates rise, and decrease of the biomass rates keeps pace with falling production rates. There are exceptions to this rule. In March 1975 a larger biomass brings a lower primary production than in the months before, and in August 1975 the primary production rate ascends in spite of the biomass getting smaller. Information on the physiological efficiency of periphyton gives the activity coefficient primary production: biomass (P/B). The course of the curves of the specific activity only occasionally corresponds to that of the biomass (for instance from October–December 1974, January–February 1975 and March–April 1975). Here it corresponds to the appropriate ascent or descent of the rates of biomass. In other periods both curves develop in contrary trends. So from July–August 1974, from December 1974–January 1975 and from July–August 1975 the biomass rates decreased but the specific activity of the biomass increased. In August 1975 it was especially high, with a rate of 0.54.

Table 10.7 Primary production, biomass, and P/B-quotient of the periphyton in the littoral of Lake Stechlin (April 1974–September 1975; measuring point Sl 1; depth 0.05 m; the primary productivity and the biomass were measured on the last day of exposure).

Exposure	Biomass ($mg\ C \cdot m^{-2}$)	Primary production ($mg\ C \cdot m^{-2} \cdot d^{-1}$)	P/B
1974			
28.03–23.04	201.0	26.3	0.13
24.04–30.05	1514.9	–	–
31.05– 3.07	1340.4	204.4	0.15
4.07– 8.08	390.5	67.9	0.17
9.08–13.09	2408.3	421.8	0.17
14.09–11.10	785.2	262.4	0.33
12.10–14.11	633.5	109.3	0.17
15.11–11.12	260.2	6.4	0.03
1975			
12.12–14.01	104.1	25.6	0.25
15.01–12.02	476.2	165.0	0.35
13.02–12.03	1161.7	124.6	0.11
13.03–14.04	1431.4	214.6	0.15
15.04– 5.05	559.8	96.1	0.17
6.05–11.06	2224.2	475.7	0.21
12.06–11.07	860.4	97.8	0.11
12.07–11.08	264.7	142.4	0.54

Annual course of the primary production in the artificially warmed part of Lake Stechlin

In Fig. 10.25 all data obtained under continuous thermal loading at measuring point Sk are brought together. In November and December all rates are very low. From January–February the primary production rates, biomass and specific activity increase. The standstill of the power plant prevented continuous measuring. But the high rates from April of the preceding year (1974) show that the increasing development can extend into the spring. Both rates of August and September 1974 are contrary. In August a small biomass has a high specific activity and production rate. In September biomass, primary production rate and P/B-quotient are likewise low.

Depth distribution of primary production

The rates obtained at seven levels from 0.05 m up to 10 m of water depth are summarized in Table 10.8. The primary production rate at the surface is relatively low; it is highest at a depth of 1 m. Then it continuously declines down to a depth

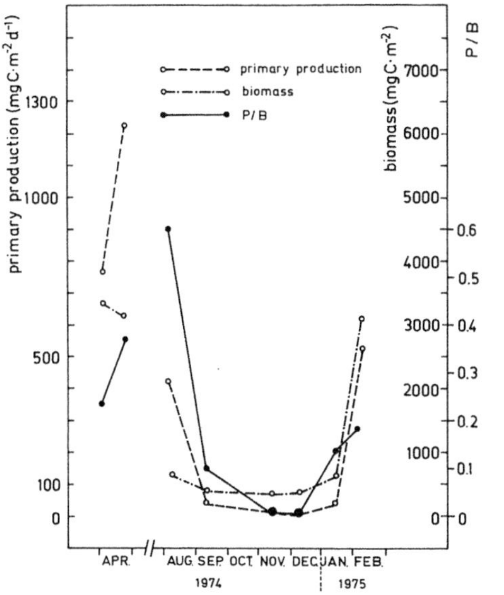

Figure 10.25 Primary production, biomass, and P/B-quotient under thermal loading 1974–1975; measuring point Sk; depth 1.0 m).

of 10 m, where still 5.3 per cent of the maximum rate was measured. The compensation point is about 15 m in depth.

The specific activity of the periphyton (P/B) is likewise very small at the surface. At a depth of 2 m it is highest and then declines constantly down to 5.5 m. At 7.5 m it is higher again. At a depth of 10 m the P/B-quotient is hardly less than at a depth of 5.5 m. This gradient speaks for a different utilization of light at different depths.

The low surface rate indicates an escape of light. The optimal production horizon (max at 2 m) is situated at 1 m and 4 m depth. The increasing P/B-quotient

Table 10.8 Primary production, biomass, and P/B-quotient of the periphyton at different depths (measuring point St; exposure time 11.8.–22.9.1976; day of measurement 22.9.1976).

Depth (m)	Primary production ($mg\ C \cdot m^{-2} \cdot d^{-1}$)	Biomass ($mg\ C \cdot m^{-2}$)	P/B
0.05	167.7	2004.0	0.08
1.0	585.8	1998.5	0.29
2.0	501.0	1110.0	0.50
3.5	267.5	888.7	0.30
5.5	142.7	875.7	0.16
7.5	72.4	278.4	0.26
10.0	31.2	217.7	0.14

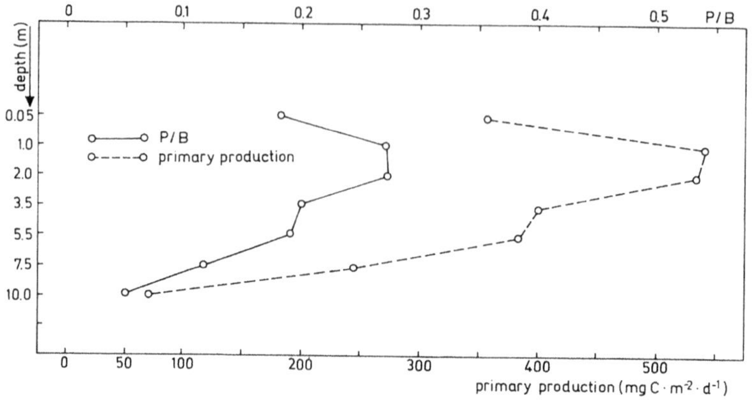

Figure 10.26 Primary production and P/B-quotient, established of biomass taken at a depth of 1.0 m and transferred to various depths (22.9.1976; measuring point St). The biomass of periphyton after the exposure time 11.8.–22.9.1976 amounted to 1998.5 mg C · m^{-2}.

between 5.5 m and 7.5 m might indicate that a biocoenosis adapted to low light conditions has established itself. The biomass rates correspond with those of the primary production with the exception of the 0.05 m range near the surface. Fig. 10.26 gives a reference to the possible role of the supply of light, too. Here are shown the results of an experiment where the primary production of biomass produced at a depth of 1 m was measured at various depths. The primary production rates are highest in the range between 1 m and 2 m, and they continuously decrease with greater depth. The surface rates (0.05 m) are lower. They indicate an activity reduction due to strong surface radiation.

The size of primary production

Table 10.9 shows a balance of the primary production of periphyton in Lake Stechlin. The bases of computations were the results of the year from the 31st of May, 1974 up to the 31st of May, 1975, obtained at the littoral measuring point Sl 1 at 0.05 m depth (see Table 10.7).

For this balance the area of the lake was observed down to a depth of 11.25 m. This corresponds to about 1 205 000 m^{-2} or 28.1 per cent of the whole area of the lake (Heitmann *et al.* 1969). The basis of this definition and the partition of seven areas between 0 m and 11.25 m deep were the result of a series of measurements, where primary production, biomass and specific activity had been measured at seven depths (Table 10.8). In a first approximation it was supposed that light and primary production of periphyton are connected linearly. Therefore, the primary dates (Table 10.7) were corrected with the factor 0.6. It is calculated by the relation of the medium supply of light in the year under study to the supply of light

Table 10.9 Calculated carbon-content ofthe periphyton in the littoral of Lake Stechlin (31.5.1974–31.5.1975).

Depth (m)	Lake area (m^2)	Calculated carbon-concent (kg C_{org})
0–0.5	90 000	3 484.8
0.5–1.5	135 000	18 243.8
1.5–2.75	145 000	17 067.9
2.75–4.5	185 000	11 461.2
4.5–6.5	185 000	6 088.7
6.5–8.75	210 000	3 496.4
8.75–11.25	255 000	1 876.0
0–11.25	1 205 000	61 718.0

on the day of measuring. This factor was disregarded by Scheffler (1981).

Accordingly, the gross primary production of periphyton in Lake Stechlin amounts to about 62 t C in the period between the 31st of May, 1974 and the 31st of May, 1975.

The respective calculated gross primary production of phytoplankton in the same period was 336 t C (Koschel, personal communication).

The primary production of periphyton was thus about 9.1 per cent of that of phytoplankton.

The value of this calculation is reduced by its low quantitative basis. The division of the primary production between the several depths of the littoral undertaken for the year being studied is based on only one series of measurements (Table 10.8). Nevertheless, the calculation shows that the portion of periphyton in the total primary production may be considerable in a stratified oligotrophic lake.

References

Benndorf, J., 1979. Kausalanalyse, theoretische Synthese und Simulation des Eutrophierungsprozesses in stehenden und gestauten Gewässern. Dissertation B, Techn. Univ. Dresden, Fakultät Bau-, Wasser- und Forstwesen. 1–165.

Benndorf, J. & Stelzner, W., 1973. Untersuchungen über das Regulationsverhalten hydrischer Ökosysteme und seine Bedeutung für die biogene Phosphatelimination. Int. Revue Hydrobiol. 58: 599–616.

Button, D.K., 1978. On the theory of microbial growth kinetics by limiting nutrient concentrations. Deep-Sea Res. 25: 1163–1177.

Desertová, B., 1976. Productivity of individual algal species in natural phytoplankton assemblages determined by means of autoradiography. Arch. Hydrobiol., Suppl. 49 (Algol. Stud. 17): 415–449.

Dillon, P.J. & Rigler, F.H., 1974. The phosphorus-chlorophyll relationship in lakes. Limnol. Oceanogr. 19: 767–773.

Dubinsky, Z., 1980. Light utilization efficiency in natural phytoplankton communities. In: Falkowski,

P.G. (Ed.): Primary productivity in the sea. New York & London: Plenum Press. 83–98.
Elster, H.-J., 1965. Absolute and relative assimilation rates in relation to phytoplankton populations. Mem. Ist. Ital. Idrobiol. Suppl 18: 77–103.
Eutrophication, 1969. Causes, consequences, correctives. Proceedings of a Symposium. Washington, D.C.: National Academy of Sciences. VIII + 661.
Findenegg, I., 1964. Types of planktic primary production in the lakes of the Eastern Alps as found by the radioactive carbon method. Verh. Int. Verein. Limnol. 15: 352–359.
Findenegg, I., 1971. Die Produktionsleistungen einiger planktischer Algenarten in ihrem natürlichen Milieu. Arch. Hydrobiol. 69: 273–293.
Goldman, C.R., Gerletti, M., Javornický, P., Melchiorri-Santolini, U. & De Amezaga, E., 1968. Primary productivity, bacteria, phytoplankton and zooplankton in Lake Maggiore: correlation and relationship with ecological factors. Mem. Ist. Ital.Idrobiol. 23: 49–127.
Gutelmacher, B.L., 1975. Relative significance of some species of algae in plankton primary production. Arch. Hydrobiol. 75: 318–328.
Gutelmacher, B.L. & Petrova, N.A., 1982. Production of individual species of algae and its role in the productivity of phytoplankton in Ladoga Lake. Int. Revue Hydrobiol. 67: 613–624.
Harris, G.P., 1978. Photosynthesis, productivity and growth: The physiological ecology of phytoplankton. Arch. Hydrobiol. Beih. Ergebn. Limnol. 10: 1–171.
Heitmann, M.-L., Richter, D. & Schumann, D., 1969. Der Wärme- und Wasserhaushalt des Stechlin- und Nehmitzsees. Abh. Meteorol. Dienst. DDR 96. Berlin: Akademie-Verlag.
Höhne, E, Nixdorf, B. & Krauss, M., 1982. Zur Beurteilung der Bioaktivität des Phytoplanktons der Talsperre Saidenbach unter Anwendung der Mikroautoradiographie. Limnologica 14: 221–229.
Hübel, H., 1966. Die ^{14}C-Methode zur Bestimmung der Primärproduktion des Phytoplanktons. Limnologica 4: 267–280.
Javornický, P., 1974. The relationship between productivity and biomass of phytoplankton in some oligotrophic water-bodies of the German Democratic Republic. Limnologica 9: 181–196.
Javornický, P., 1980. Density dependent effects. In: Le Cren, E.D. & Lowe-McConnel, R.H. (Eds.): The functioning of freshwater ecosystems. Cambridge: Univ. Press. 170–173.
Javonický, P. & Komárková, J., 1973. The changes in several parameters of plankton primary productivity in Slapy Reservoir 1960–1967, their annual corelations and correlation with the main ecological factors. Hydrobiol. Studies 2: 155–211.
Koschel, R., 1974a. Einfluß der physikalischen und chemischen Umweltfaktoren auf die Primärproduktion des Phytoplanktons im Stechlinsee. Diss. A, Techn. Univ. Dresden, Fakultät Bau-, Wasser- und Forstwesen. 1–72.
Koschel, R., 1974b. Primärproduktionsuntersuchungen im Stechlinsee. Limnologica 9: 143–156.
Koschel, R., 1976. Der Einfluß des Phosphorangebotes auf die Primärproduktion des Phytoplanktons in einem geschichteten Klarwassersee (Stechlinsee, DDR). Limnologica 10: 325–346.
Koschel, R., 1980. Untersuchungen zur Phosphataffinität des Planktons in der euphotischen Zone von Seen. Limnologica 12: 14–145.
Koschel, R., 1981a. Bilanz der Primärproduktion des Phytoplanktons in den Gewässern des Stechlinseesystems. Limnologica 13: 55–81.
Koschel, R., 1981 b. Die Phosphataffinität von Mikroorganismengesellschaften im Stechlinseesystem und in anderen Seen. Limnologica 13: 129–145.
Koschel, R., Haubold, G., Kasprzak, P., Küchler, L., Proft, G. & Ronneberger, D., 1981. Eine limnologische Zustandsanalyse des Feldberger Haussees. Acta Hydrochim. Hydrobiol. 9: 255–279.
Koschel, R., Benndorf, J., Proft, G. & Recknagel, F., 1983. Calcite precipitation as a natural control mechanism of eutrophication. Arch. Hydrobiol. 98: 380–408.
Koschel, R. & Kasprzak, P., 1984. Model use and verification of ecological parameters for an oligotrophic lake (Lake Stechlin, G.D.R.). Ecol. Modelling 26: 97–101.
Landner, L., 1976. Eutrophication of lakes. WHO, Regional Office for Europe, Geneva ICP/CEP, No. 210: 1–78.

Le Cren, E.D. & Lowe-McConnel, R.H. (Eds.), 1980. The functioning of freshwater ecosystems. Cambridge: Univ. Press. XXIX + 588.

Lund, J.W.G. & Reynolds, C.S., 1982. The development and operation of large limnetic enclosures in Blelham Tarn, English Lake District, and their contribution to phytoplankton ecology. In: Round, F.E. & Chapman, V.J. (Eds.): Progress in phycological research, vol. 1: Amsterdam: Elsevier Biomedical Press B.V. 1–65.

Malone, T.C., 1980. Size-fractionated primary productivity of marine phytoplankton. In: Falkowski, P.G. (Ed.): Primary productivity in the sea. New York & London: Plenum Press. 301–320.

Moss, B., 1980. The relative importance of different producing communities. In: Le Cren, E.D. & Lowe-McConnel, R.H. (Eds.): The functioning of freshwater ecosystems. Cambridge: Univ. Press. 214–215.

Munawar, M., Munawar, I.F. & Dupuis, G., 1978. Relative importance of nanoplankton in Lake Superior, phytoplankton biomass, and community metabolism. J. Great Lakes Res. Int. Assoc. Great Lakes Res. 4: 415–442.

Naumann, E., 1932. Grundzüge der regionalen Limnologie. In: Thienemann, A. (Hrsg.): Die Binnengewässer. Bd. XI. Stuttgart: Schweizerbart 'sche Verlagsbuchhandlung. XIV + 176.

Rai, H., 1982. Primary production of various size fractions of natural phytoplankton communities in a North German lake. Arch. Hydrobiol. 95: 395–412.

Rodhe, W., 1969. Crystallization of eutrophication concepts in Northern Europe. In: Eutrophication. Causes, consequences, correctives. Proceedings of a Symposium. Washington, D.C.: National Academy of Sciences. 50–64.

Rodhe, W., Vollenweider, R.A. & Nauwerck, A., 1958. The primary production and standing crop of phytoplankton. In: Buzzati-Traverso, A.A. (Ed.): Perspectives in marine biology. Berkeley & Los Angeles: Univ. of California Press. 299–322.

Sakamoto, M., 1966. Primary production by phytoplankton community in some Japanese lakes and its dependence in lake depth. Arch. Hydrobiol. 62: 1–28.

Scheffler, W., 1981. Quantitative Erfassung des Periphytons im Stechlinsee. Limnologica 13: 101–122.

Schindler, D.W., 1978. Factors regulating phytoplankton production and standing crop in the world's freshwaters. Limnol. Oceanogr. 23: 478–486.

Steel, J.A., 1962. Environmental control of photosynthesis in the sea. Limnol. Oceanogr 7: 137–150.

Straškrabá, M., 1976a. Development of an analytical phytoplankton model with parameters empirically related to dominant controlling variables. Abh. Akad. Wiss. DDR: 33–65.

Straškrabá, M., 1976b. Empirical and analytical models of eutrophications. Proc. EUTROSYM, 20.–25.9.1976. Karl-Marx-Stadt, DDR. 3: 352–371.

Straškrabá, M., 1978. Theoretical considerations on eutrophication. Verh. Int. Verein. Limnol. 20: 2714–2720.

Talling, J.F., 1957. Photosynthetic characteristics of some freshwater plankton diatoms in relation to underwater radiation. New Phytol. 56: 29–50.

Thienemann, A., 1928. Der Sauerstoff im eutrophen und oligotrophen See. Ein Beitrag zur Seetypenlehre. In: Thienemann, A. (Hrsg.): Die Binnengewässer. Bd. IV. Stuttgart: Schweizerbart'sche Verlagsbuchhandlung. 175.

Tilzer, M., 1972. Dynamik und Produktivität von Phytoplankton und pelagischen Bakterien in einem Hochgebirgssee (Vorderer Finstertaler See, Österreich). Arch. Hydrobiol. Suppl. 40: 201–273.

Tilzer, M., 1973. Diurnal periodicity in the phytoplankton assemblage of a high mountain lake. Limnol. Oceanogr. 18: 15–30.

Uhlmann, D., 1982. Evaluation of strategies for controlling eutrophication of lakes and reservoirs. Int. Revue Hydrobiol. 67: 821–835.

Uhlmann, D. & Hrbáček, J., 1976. Kriterien der Eutrophie stehender Gewässer. Limnologica 10: 245–253.

Vollenweider, R.A., 1965. Calculation models of photosynthesis-depth curves and some implications regarding day rate estimates in primary production measurements. Mem. Ist. Ital. Idrobiol. Suppl. 18: 425–457.

Vollenweider, R.A., 1968. Scientific fundamentals of the eutrophication of lakes and flowing waters, with particulars reference to nitrogen and phosphorus as factors in eutrophication. OECD, Directorate for Scientific Affairs, Paris, DAS/CSI/68.27: 1–159.
Zdanowski, B., 1982. Variability of nitrogen and phosphorus contents and lake eutrophication. Pol. Arch. Hydrobiol. 29: 541–597.

11. The secondary production

P. Kasprzak & D. Ronneberger

Introduction

The investigations of Koschel (1974, 1976) dealing with primary production of phytoplankton raised the question of the role of zooplankton in the pelagic carbon balance of Lake Stechlin. It derived from the difference between primary production and standing crop of phytoplankton, from the so-called 'excess assimilation'. The aim of the investigations was to show what part of primary production was used by zooplankton (Crustacea, Rotatoria). This demand could be established quantitatively using the dynamic ecological model SALMO (Benndorf 1979) for the analysis of the ecosystem Lake Stechlin. It became clear that a better understanding of the problems of 'primary productivity – standing crop of phytoplankton – losses' and the 'biomass production of zooplankton' is impossible without a better knowledge of its nutritional biology (Benndorf *et al.* 1981).

For that reason the carbon assimilation of some dominant crustaceans and the grazing of dominant rotifers were measured. Temperature, food concentration, type of food and animal size were selected as the most important influencing factors.

Methods and methodological problems

Preparation of the food, determination of its carbon content

Eight types of food were tested, and Table 11.1 gives information about their origin.

The determination of the food carbon content depended on the particle size. Particles larger than 5–6 μm were enriched on glassfibre filters and subsequently oxydized in concentrated sulphuric acid with 4.8 mg $\cdot$ l^{-1} $K_2Cr_2O_7$ at 120° C in a heat chamber for at least 4 h. The loss of potassium dichromate was measured photometrically. The carbon content of particles smaller than 5–6 μm was mea-

Casper, S. Jost (ed.), Lake Stechlin. A temperate oligotrophic lake. ISBN 978-94-010-8930-2.

Table 11.1 Origin of the used kinds of food and the feeders.

Food	Origin	Feeders*
Kirchneriella lunaris	Batch culture	C, R
Scenedesmus quadricauda	Batch culture	C
Chlorella spec.	Batch culture	R
μ-alga	Batch culture/Lake Fuchskuhle	C
Mixed seston	Sewage pond	C
Dinobryon pediforme	Lake Fuchskuhle	C
Detritus	Sediment	C
Oscillatoria utermöhliana	Batch culture	C

* C = crustaceans; R = rotifers

sured by conductometric determination of CO_2 after oxydation with potassium peroxodisulfate. For more details see Proft (1980).

Experimental animals, determination of the body carbon of selected planktonic crustacea

As demonstrated in the chapter on the zooplankton population (pp. 243 ff.), *Eudiaptomus gracilis, Daphnia cucullata* and/or *Eurytemora lacustris* are the dominant forms in the standing crop. This is why they were selected for assimilation experiments. As a reference to the absorbed carbon, the body carbon content was used. For *Daphnia cucullata* and *Eudiaptomus gracilis* a relation between body size and body carbon was developed. As a measure of body carbon the waste of potassium dichromate was used, as described above. For more details see Kasprzak (1983a).

The equations are as follows:

D. cucullata	$C = 1.94\ L^{1.8}$	$r = 0.85$	(1)
E. gracilis	$C = 4.37\ L^{1.75}$	$r = 0.85$	(2)

C = carbon content (μg C), L = animal size (mm)

Feeding experiments

The method used in feeding experiments was very similar to that described by Lampert (1977a) and is presented in detail in Kasprzak (1983b). Note that the animals were fed at a certain temperature and concentration of labelled food for 1 or 3 h. After this feeding period the animals were separated from the food suspension, killed in hot water (60–70° C), rinsed carefully with filtered lake water, coloured with Lugols iodine, selected according to their body size and prepared for the measurement of radioactiviy. Holtby & Knoechel (1981) refer to a loss of radioactive material during the manipulation with Lugols solutions, but within a period of 10 min it amounts to not more than 10%.

Measurement of radioactivity

To measure the radioactivity of crustaceans first a Geiger-Müller-counter (GM) was used (automatic sample changer TESLA NZ 617-A, analyzer RFT VA-V-120, counter RFT VA-G-120). The animals were mounted on paper filters, dried in an exsiccator and subsequently measured.

Using ^{14}C, a remarkable absorption must be taken into consideration (Simon & Rauschenbach 1974). That is why a method developed by Ward *et al.* (1970) for *Daphnia pulex* was adapted to estimate the self-absorption coefficient for *Daphnia cucullata* and *Eudiaptomus gracilis.*

For *D. cucullata* the equation is as follows:

$$AK = 1.608\ L^{0.759} \tag{3}$$

AK = self-absorption coefficient, L = body length (mm).
The results for *E. gracilis* are doubtful. That is why, as a first approximation, the equation obtained for *D. cucullata* was used for *E. gracilis* too.

Sometimes the liquid scintillation procedure (LSC) was used. The animals were digested with Protosol (New England Nuclear, 55° C) and subsequently 5 ml of cellutol scintillator (Romigger & Pollman 1972) was added. To prevent chemoluminescence, the scintillation vials were cooled overnight at 5° C. Quenching correction was made by an external standard.

Calculation of assimilation

Due to the determination of radioactivity it was possible to calculate the specific radioactivity of the food, expressed as counts per time and carbon unit. After the correction of self-absorption (not necessary for the LSC-technique) and respiration losses (Lampert 1977a), the alteration in the measured counts of the animals in a period between 1 and 3 h feeding time is a measure of the assimilated carbon. A variance analysis (P = 5%) gave no significant differences between GM and LSC-method. Fig. 11.1 gives the general outline of the assimilation measurements.

Temperature adaption of crustaceans

The use of animals not adapted to the temperature can lead to inaccurate

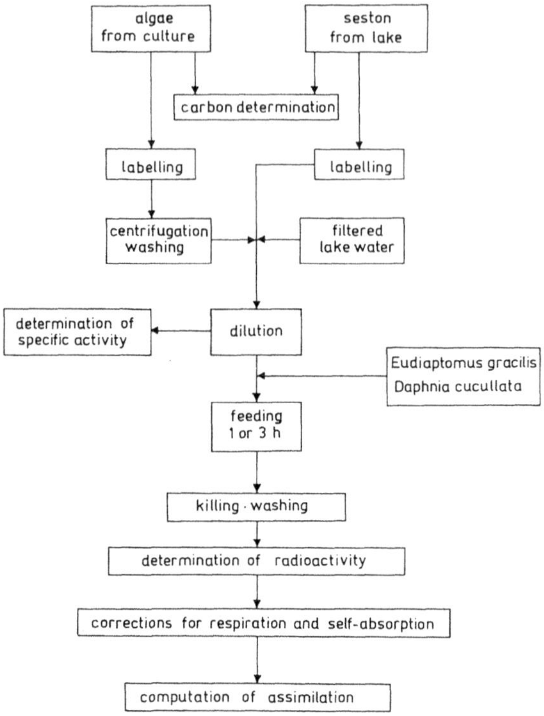

Figure 11.1 General outline of the assimilation experiments.

assimilation rates (cf. Geller 1975, Bevan *et al.* 1978). Lampert (1977b) used only such individuals of *Daphnia pulex*, grown at the same temperature that was later used in the experiments.

Because of problems in the culture of *Eudiaptomus gracilis* and *Daphnia cucullata* (Hopp 1979) the animals used were freshly isolated from the Lake Stechlin and Lake Fuchskuhle (bog lake) respectively. The temperature difference between experiment and natural location was not more than 10° C.

Assimilation – reasons for the choice of method

According to Rigler (1971) 'assimilation' in terms of animal nutrition is the absorption of digested organic material from the gut into the tissue of the animal. The term is used in this paper exclusively in this sense. Johannes & Satomi (1967), Conover & Francis (1973) and Lampert (1977a) discussed the method of assimilation measurement with radioactive labelled food and have shown many possible mistakes. For the following reasons the procedure suggested by Rigler (1971) and further developed by Lampert (1977a) has been selected: Geller (1975) showed a strong connection between ingestion and nutritional adaption of animals, but a prefeeding time of 30 min was sufficient for adaption. Before the first measurement the animals have enough time to adapt to the new milieu, including type of food.

For this type of experiment Lampert (1977a) had developed equations to correct the respiration losses of $^{14}CO_2$. They were obtained from the filter-feeder *Daphnia pulex*. Therefore, the use for *Daphnia cucullata* and *Eudiaptomus gracilis* seems to be possible. Johannes & Satomi (1967) pointed out a loss of soluble labelled organic material during the exchange of gut content. Lampert (1978) could give evidence of this phenomena on *Daphnia pulex* too.

Problems with handling stress are eliminated. There is no danger of incomplete exchange of the gut content. Finally, the method was used because of the problems in continuous culture of euplanktonic crustacea (Hopp 1979).

Methods for in-situ grazing

The method for the estimation of the 'in-situ grazing rate' of zooplankton and selected species of rotifers is very similar to that described by Gliwicz (1968) and Haney (1973). A 5 l Friedinger sampler was used, equipped with a container for radioactive food. This container opened itself just at the moment when the sampler was closed. Within 5 min the distribution of the food in the sampler was homogeneous. The food concentration was between 0.5 and 1 mg $C \cdot l^{-1}$. The feeding time of the natural zooplankton population amounted to 15 min. Starkweather & Gilbert (1977) have shown that the radioactivity of the animals increases in a linear form, and respiration losses can be neglected in short term experiments. After this feeding time the animals were separated from the food suspension and fixed with Lugols solution. A subsample was sucked into a glassfibre filter and used for the determination of the grazing of the zooplankton population. From the other part 50 individuals of rotifers, especially of *Kellicottia longispina* and *Keratella cochlearis*, were selected to determine their specific grazing rate. The experiments were carried out in 0; 2.5; 5; 10 and 20 m depth. The LSC-technique was used exclusively for the measurement of radioactivity. The handling of samples was carried out as described above for crustaceans. Instead of cellutol a simple toluene cocktail was used.

For *Keratella cochlearis, Kellicottia longispina* and *Polyarthra vulgaris* the results of direct estimation of body carbon brought by Latja & Salonen (1978) were used. Body carbon of other rotifers was calculated according to Dumont *et al.* (1975) and Bottrell *et al.* (1976). For in situ investigations the temperature adaption is not necessary. For in vitro investigations a temperature adaption time of two hours was used. The losses of radioactive material caused by fixation with Lugols solution were corrected according to Holtby & Knoechel (1981).

Assimilation of Daphnia cucullata and Eudiaptomus gracilis

Assimilation as a function of food supply and type of food

To indicate the possible spectrum of the results some of them dealing with assimilation as a function of food supply and type of food are given in Fig. 11.2. For the mathematical description two equations were used, both of saturation type. The use has been theoretically established (above a certain food concentration the limit of all mechanical and enzymatic systems of the animal is reached) and practically verified in a number of investigations (e.g., Ryther 1954; Richmann 1958; Mc Mahon 1965; Hayward & Gallup 1976; Lampert 1977b; Horn 1979). The two equations are as follows:

$$A = P + \frac{B}{F} \tag{4}$$

$$\frac{1}{A} = P + \frac{B}{F} \tag{5}$$

A = Assimilation (μg C · animal $^{-1}$ · h^{-1}), F = food concentration (mg C · l^{-1}), P, B = factors, numeric values summarized in Tables 11.2 and 11.3.

There is no biological criterion for the choice between the two functions. Both are only regression equations. Nevertheless, for further considerations equation (4) is preferred because of the realistic plateau reflected in all cases. The course of the function is similar to that found by other authors (Lampert 1977b; Bevan *et al.* 1978). After a big increase at low levels of food concentration, assimilation reaches a plateau. The level of the plateau calculated as 'mean maximum assim-

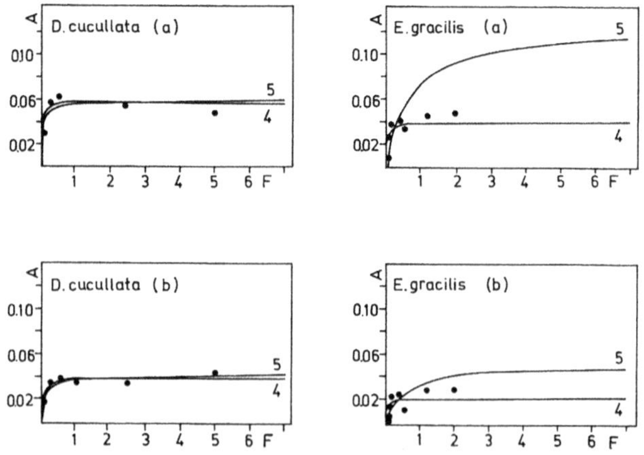

Figure 11.2 Assimilation of *Daphnia cucullata* (left; food *Oscillatoria utermöhliana)* and *Eudiaptomus gracilis* (right; food *Scenedesmus quadricauda)*. a = uncorrected (net) b = corrected for respiration losses (gross), A = assimilation (μg C · animal^{-1} · h^{-1}), F = food concentration (mg C · l^{-1}), temperature 20° C, animal size 1.1 mm. For more details see text and Tables 11.2 and 11.3

ilation rate' (average of all visually selected plateau values) varies according to the individual type of food. The differences seem to be significant (P = 5%). Similar results were achieved by other investigators (cf. Ryther 1968; Lampert 1977b). However, the reasons for the differences presented in Tables 11.2 and 11.3 remain unclear, because no information was gained about the possible influencing factors. The assimilation rate of *Eudiaptomus gracilis* fed with *Oscillatoria utermöhliana* was surprisingly high (cf. Table 11.3; Fig. 11.3). In the past blue-greens were regarded as bad food for planktonic crustaceans (Arnold 1971). More detailed investigations have shown, however, that one must distinguish between 'assimilability' and 'nutrient value' (Schindler 1971; Bernardi *et al.* 1981; Lampert 1981; 1982). Lampert (1977b) showed very high turnover rates, especially for small specimens of *Daphnia pulex* fed with *Synechococcus elongatus*. But in long-term growth experiments the increase in the population density was very low, and

Table 11.2 *Daphnia cucullata*: factors P and B for the equations (4) and (5) (see p. 328)*.

Food		Net		Gross	
		(4)	(5)	(4)	(5)
Scenedesmus	P	0.02302	19.25	0.03901	8.01
	B	– 0.00050	9.95	– 0.00055	6.19
Kirchneriella	P	0.03170	32.91	0.05457	18.41
	B	– 0.00053	0.86	– 0.00109	0.66
Oscillatoria	P	0.03863	24.41	0.05582	17.28
	B	– 0.00147	2.29	– 0.00174	1.16
Mixed seston	P	0.04268	8.75	0.06974	13.79
	B	– 0.00058	13.86	– 0.00096	1.55

* Temperature 20° C, animal size 1.1 mm, net – uncorrected, gross – corrected for respiration losses

Table 11.3 *Eudiaptomus gracilis*: factors P and B for the equations (4) and (5).

Food		Net		Gross	
		(4)	(5)	(4)	(5)
Kirchneriella	P	0.02270	46.46	0.03450	32.27
	B	– 0.00015	– 0.03	0.00040	0.83
μ-alga	P	0.02030	42.33	0.03150	26.63
	B	0.00380	42.76	– 0.00595	27.19
Oscillatoria	P	0.06260	20.40	0.10001	10.83
	B	0.00021	0.28	0.00063	– 0.09
Detritus	P	0.02270	179.97	0.04531	37.61
	B	– 0.00014	10.94	– 0.00270	4.07
Mixed seston	P	0.02850	43.58	0.03890	29.70
	B	– 0.00026	2.90	– 0.00034	1.79

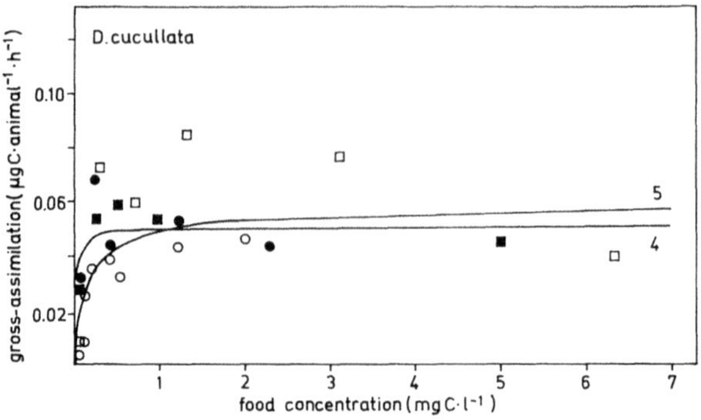

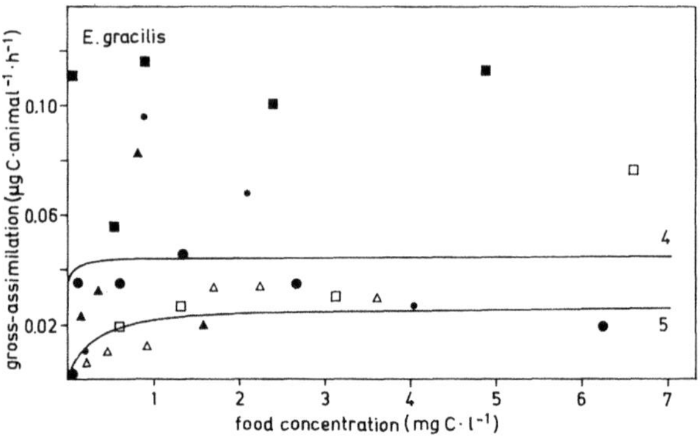

Figure 11.3 Gross assimilation of *Daphnia cucullata* and *Eudiaptomus gracilis* as a function of food concentration and type of food. Temperature 20° C, animal size 1.1 mm. Mean equations of all values:

Daphnia cucullata (4) $A = 0.0562 - \frac{0.00086}{F}$

(5) $\frac{1}{A} = 16.5600 + \frac{3.02}{F}$

Eudiaptomus gracilis (4) $A = 0.0471 - \frac{0.00033}{F}$

(5) $\frac{1}{A} = 35.1500 + \frac{7.80}{F}$

△ = μ-algae; ▲ = *Dinobryon pediforme*; ● = *Kirchneriella lunaris*; ■ = *Oscillatoria utermöhliana;* • = detritus; □ = mixed seston; ○ = *Scenedesmus quadricauda.*

that indicates a bad nutritional value. However, the variance analysis of the 'mean maximum assimilation rate' has indicated significant differences for *Daphnia cucullata* as well as for *Eudiaptomus gracilis* with respect to the individual types of food. Nevertheless the attempt should be made to 'reduce' all measured values to a mean relation (Fig. 11.3). The figure shows a wide ranging difference in the increasing assimilation. But Fig. 11.6 gives an explanation for such results. They are caused by remarkable differences in the food ingestion and assimilation between the single individuals. The turnover time of body carbon and growth rates obtained, using the average equation (4), are presented in Table 11.4. It becomes clear from Tables 11.2 and 11.3 that *Eudiaptomus gracilis* is assimilating nearly the same amount of carbon as *Daphnia cucullata* in spite of its higher biomass. Accordingly, longer turnover and lower growth rates are obtained. The result is surprising because the absolute assimilation normally increases with the biomass of the animals, whereas the specific values decrease. But it must be taken into consideration that we are dealing with two different species. This is why various assimilation and turnover rates can occur. Schindler (1971) obtained similar findings concerning *Daphnia longispina* and *Diaptomus gracilis*. Ponyi-Zankai & Ponyi (1974) and Ponyi-Zankai (1975) had given references to assimilation as a function of reserve matter in the animal. The authors suppose a negative feed-back mechanism between fat content and assimilation intensity ('appetite'). *Eudiaptomus gracilis* used in our experiments had grown in the mentioned Lake Fuchskuhle with a higher seston content than Lake Stechlin, where *Daphnia cucullata* was sampled. Hence, a higher content of reserve matter could be assumed for *Eudiaptomus gracilis*.

Comparing the growth rates with the results obtained by other authors, the values are similar. Hall (1964) found a net growth rate in *Daphnia galeata mendotae* of about $0.5 \cdot d^{-1}$. Uhlmann (1971) and Unger (1979) have shown maximal net growth rates in *Daphnia pulex* and *Daphnia magna* of $0.3 \cdot d^{-1}$. Hopp (1979) pointed out results for *Daphnia hyalina* between 0.2 and $0.3 \cdot d^{-1}$. Metz (1973) calculated net growth rates of $0.45 \cdot d^{-1}$ for *Daphnia pulex*. According to Lampert (1977b) the maximal growth rate depends strongly on the biomass of the animals. Specimens of *Daphnia pulex* from 1.5 mm size reach a maximum gross growth rate of $1.1 \cdot d^{-1}$. At 3.0 mm size this value has decreased to $0.6 \cdot d^{-1}$.

Table 11.4 The turnover time of body carbon and growth rate.*

	Daphnia cucullata		*Eudiaptomus gracilis*	
	T_t	μ	T_t	μ
Gross	1.81	0.55	4.61	0.22
Net	2.28	0.44	7.22	0.144

* Animal size 1.1 mm; corresponding carbon content ($\mu g\ C \cdot animal^{-1}$): *D. cucullata* 2.3; *E. gracilis* 5.2; T_t = turnover time (d), μ = growth rate (d^{-1}), temperature 20° C

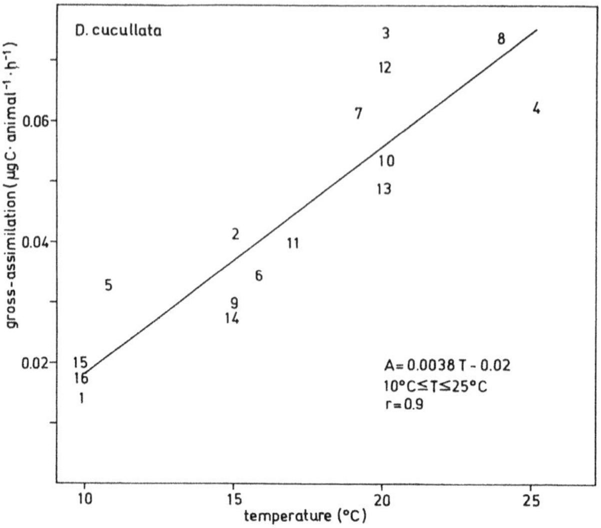

Fig.11.4 Gross assimilation of *Daphnia cucullata* as a function of temperature (food concentration 0.5 mg C · l^{-1}, animal size 1.1 mm), 1–8, 13: *Kirchneriella lunaris,* 11, 15, 16, 19: *Scenedesmus quadricauda,* 10: *Oscillatoria utermöhliana,* 12: mixed seston, 14: μ-alga.

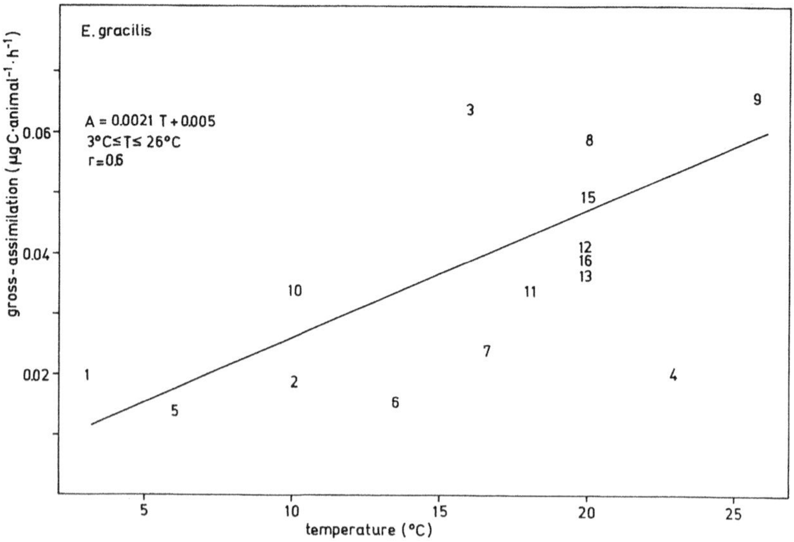

Figure 11.5 Gross assimilation of *Eudiaptomus gracilis* as a function of temperature (food concentration and animal size see Figure 11.4). 1–4, 11: μ-alga, 5–9, 13: *Kirchneriella lunaris,* 12: *Dinobryon pediforme,* 14: *Oscillatoria utermöhliana,* 15: detritus, 16: mixed seston.

Assimilation as a function of temperature

At Fig. 11.4 and 11.5 the assimilation of *Daphnia cucullata* and *Eudiaptomus gracilis* is presented as a function of temperature, tested with various types of food. The description by means of straight lines is only a first approximation. Many other authors have measured performances of planktonic crustaceans related to temperature. As a rule they found optimum curves (Hasler 1937; Lampert 1977b; Kersting 1978; Horn 1979).

Assimilation as a function of animal size

Assimilation as a function of animal size was tested at 20°C with a food concentration above the incipient limiting level using *Kirchneriella lunaris* (1.1 mg $C \cdot l^{-1}$) for *Daphnia cucullata* and the μ-alga (4.8 mg $C \cdot l^{-1}$) for *Eudiaptomus gracilis*. Some of the results are presented in Fig. 11.6 (GM-technique) after 1 and 3 h feeding time, respectively. The difference between the two functions is a measure for the size dependent assimilation. After a lag-phase a strong increase of the count numbers becomes obvious. Therefore, assimilation is increasing in an exponential way (cf. Fig. 11.7).

For the mathematical description two equations were tested:

$$A = P \cdot L^{B} \tag{6}$$
$$A = P \cdot e^{B \cdot L} \tag{7}$$

A = assimilation (μg C $\cdot$ animal$^{-1} \cdot$ h^{-1}); L = animal size (mm); P, B = factors.

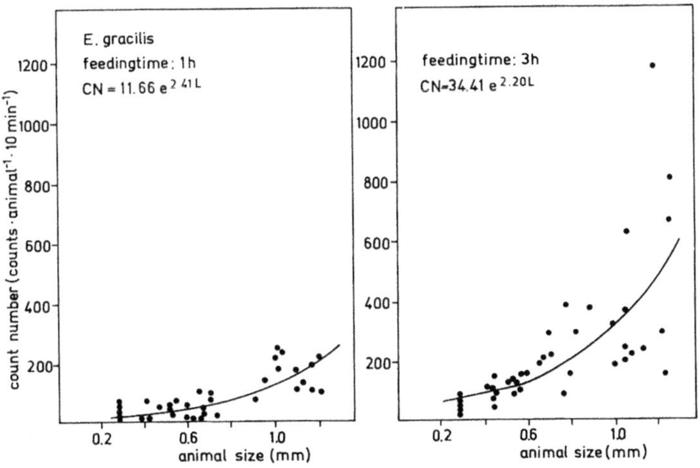

Figure 11.6 Count number of *Eudiaptomus gracilis* as a function of animal size (uncorrected for respiration losses, food: μ-alga 4.8 mg $C \cdot l^{-1}$, CN = count number, L = animal size).

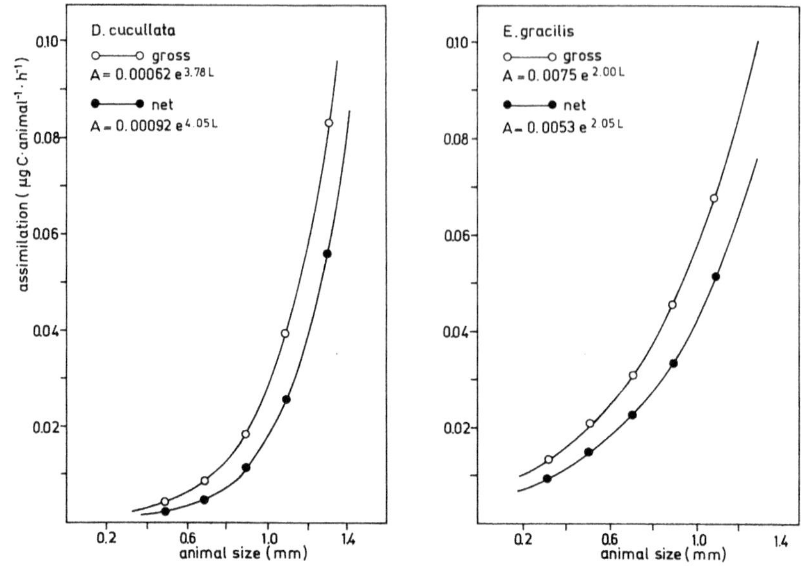

Figure 11.7 Assimilation of *Daphnia cucullata* (food: *Kirchneriella lunaris* 1.1 mg C · l^{-1}) and *Eudiaptomus gracilis* (food: μ-alga 4.8 mg C · l^{-1}) as a function of animal size (temperature 20° C).

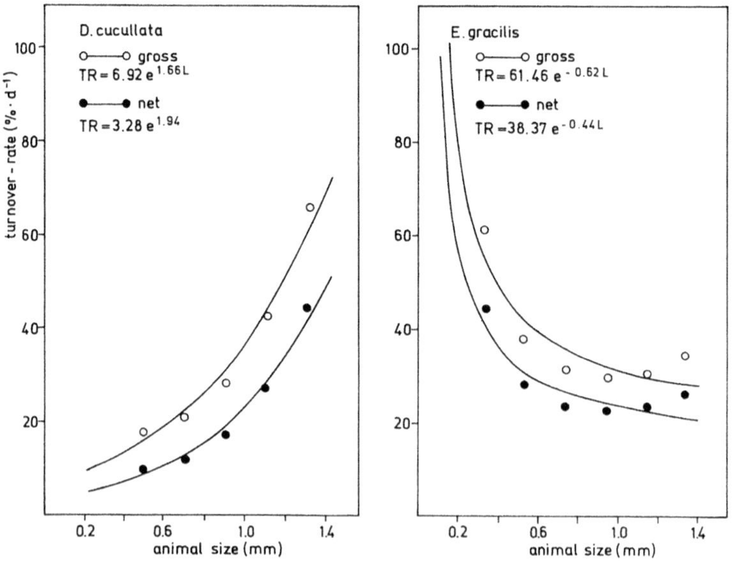

Figure 11.8 Turnover rate of *Daphnia cucullata* (food: *Kirchneriella lunaris* 1.1 mg C · l^{-1}) and *Eudiaptomus gracilis* (food: μ-alga 4.8 mg C · l^{-1}) as a function of animal size (temperature 20° C).

Equation (7) was used for further considerations, because for equation (6) an underestimation of the measured values became visible. Fig. 11.8 shows the turnover rates as a function of animal size. The results are contrary and the behaviour of *Daphnia cucullata* is not normal, because it is clear from other investigations that specific grazing or assimilation rates decrease with increasing size (Geller 1975; Lampert 1977a). It may be that the extreme scatter of count numbers (see Fig. 11.6) in larger animals is one of the reasons which caused the results in *D. cucullata.* In his regression analyses Lampert (1977b) has selected animals above a certain length.

In situ grazing of zooplankton and dominant rotifers

The results dealing with rotifer grazing are presented in Table 11.5. At food concentration above the 'incipient limiting level' the ingestion expressed as carbon turnover of the single rotifers shows great differences, ranging from 120–450% per day. These variations can be explained by different food preferences. Pourriot (1965, 1977) has found *Filinia longiseta, Kellicottia longispina* and *Polyarthra dolichoptera* as algae feeders, whereas *Brachionus angularis* and *Keratella cochlearis* prefer detritus. Erman (1962), working on *Brachionus calyciflorus*, also found differences in food ingestion according to the type of food used. For further investigations it will be necessary to have more knowledge about quality and quantity of seston in Lake Stechlin as a food source of rotifers.

The results on in situ grazing investigations for zooplankton are given in Fig. 11.9. The maximum of grazing was found between 0–5 m depth. The mean potential grazing amounts 7.6 μg C $\cdot$ h^{-1} $\cdot$ l^{-1} from May to September.

Table 11.5 Maximum grazing of dominant rotifers of Lake Stechlin (food concentration 2–4 mg C $\cdot$ l^{-1}).

Species	Grazing (ng C $\cdot$ animal^{-1} $\cdot$ d^{-1})	Body carbon (ng C $\cdot$ animal^{-1})	Grazing as % of body carbon
Filinia terminalis	202.0	45	450
Asplanchna priodonta	360.0	300	120
Keratella cochlearis	29.0	24	121
Keratella quadrata	43.2	32	135
Kellicottia longispina	46.2	15	312
Polyarthra dolichoptera	192.0	44	436
Gastropus stylifer	72.0	50	240
Brachionus angularis	33.6	18	186

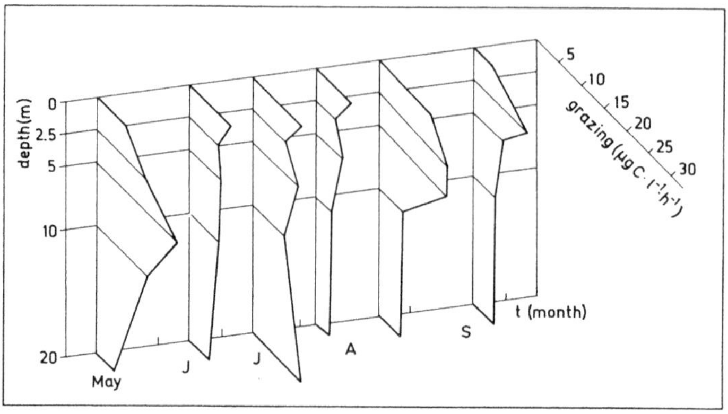

Figure 11.9 Potential grazing of zooplankton in Lake Stechlin (food: *Chlorella* spec. 0.5–1 mg · l^{-1}; grazing time: 15 min).

Grazing as a function of temperature

In the two dominant rotifer species of Lake Stechlin *Keratella cochlearis* and *Kellicottia longispina* the temperature dependency of grazing was investigated in situ and in vitro. Between 5 and 22° C, representative for temperate lakes, it can be described as a straight line (Fig. 11.10). The equations are as follows:

$$G = 1.08\,T + 2.12 \qquad \textit{K. cochlearis} \tag{8}$$
$$G = 1.44\,T + 10.60 \qquad \textit{K. longispina} \tag{9}$$

G = grazing (ng C · $\text{animal}^{-1} \cdot h^{-1}$), T = temperature (° C)

Starkweather & Gilbert (1977) have also found a linear increase of ingestion up to 28° C in *Brachionus calyciflorus*. Erman (1956) has described an optimum curve for the same species with maximum at 22° C.

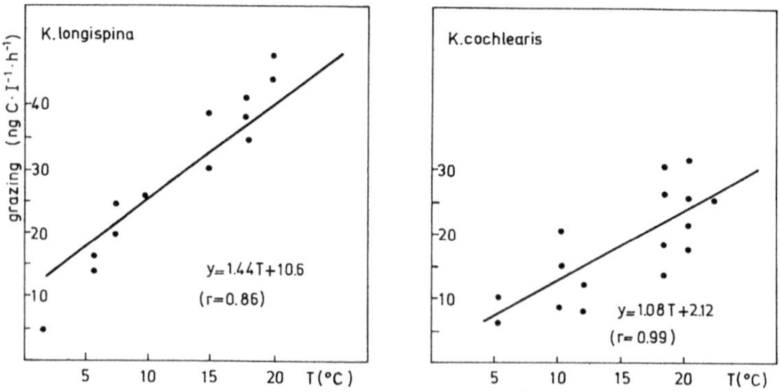

Figure 11.10 Grazing of *Kellicottia longispina* and *Keratella cochlearis* as a function of temperature (food: *Chlorella* spec. 0.5–1 mg C · l^{-1}).

Grazing as a function of food supply

Grazing as a function of food supply was examined for the zooplankton community and *Keratella cochlearis*. As an example some of the results are given in Figs. 11.11 and 11.12. For the mathematical description of the grazing, straight lines were used. The equations are as follows:

Increase $G = 0.0530\ F - 3.30$	zooplankton	(10)
Increase $G = 0.0286\ F - 5.44$	*K. cochlearis*	(11)
Plateau $G = 2.85 \cdot 10^{-5}\ F + 45.7$	*K. cochlearis*	(12)

G = grazing, zooplankton (ng $C \cdot d^{-1} \cdot \mu g\ C^{-1}$), *K. cochlearis* (ng $C \cdot d^{-1} \cdot$ animal^{-1}), F = food concentration ($\mu g\ C \cdot l^{-1}$).

The incipient limiting concentration found in *Keratella cochlearis* is high (1800 $\mu g\ C \cdot l^{-1}$) compared with crustaceans. The plateau grazing rate at food concentration above the incipient limiting level amounts to 46.8 ng $C \cdot d^{-1} \cdot$ animal^{-1}. In Lake Stechlin the mean concentration of particulate organic carbon (size fraction smaller than 50 μm; see p. 109) reaches 300 $\mu g\ C \cdot l^{-1}$, and this value is situated at the very beginning of the increase phase. That means an average grazing rate of *Keratella cochlearis* under the food conditions in the lake of about 3.1 ng $C \cdot d^{-1} \cdot$ animal^{-1}, and this equals only 6.6% of the potential maximum grazing rate. Ross & Munawar (1981) have investigated the feeding rate of *Kellicottia longispina* with natural phytoplankton divided into three size classes. By conversion of their results to carbon, the ingestion rate in the smallest fraction (1–5 μm) equals 20 ng $C \cdot d^{-1}$ animal^{-1} and in the following (5–20 μm) 10 ng $C \cdot d^{-1} \cdot$ animal^{-1}, respectively. Bogdan *et al.* (1980) found a grazing rate of 43 ng $C \cdot d^{-1} \cdot$ animal^{-1} in *Keratella cochlearis* fed with *Chlamydomonas* at food concentration of about 300 $\mu g\ C \cdot l^{-1}$ (all values converted to carbon). Although the food

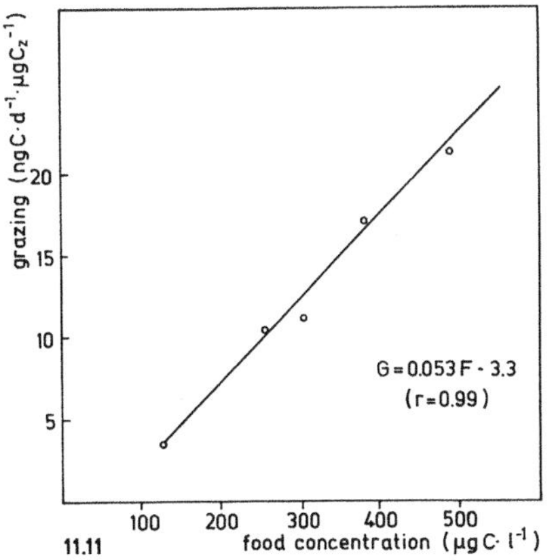

11.11

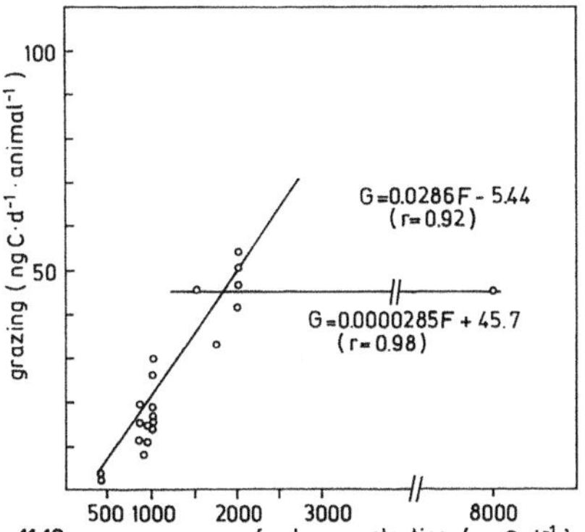

11.12

Figure 11.11 Grazing of zooplankton as a function of food concentration.
Figure 11.12 Grazing of *Keratella cochlearis* as a function of food concentration.

concentration found in Lake Stechlin is similar to that used by the authors mentioned, the grazing of *Keratella cochlearis* in this investigation is less.

Significance of structure, population dynamics and assimilation of the crustacean and rotifer plankton for the carbon budget in the pelagic zone

Structure and population dynamics

The results presented in the chapter on the zooplankton population (pp. 243 ff.) characterize the zooplankton as a dynamic part of the ecosystem of Lake Stechlin concerning time and space. This becomes clear from the distribution throughout the course of the year, as well as from the diurnal migration and horizontal dishomogeneity. It is the aim of the following part of this paper to make a balance of assimilation of selected crustaceans and rotifers (regarded as herbivores) and to compare it with phytoplankton standing crop, planktonic primary production and detritus content of the water. For this only the standing crop of selected crustaceans and rotifers in the epilimnion is considered. But is should be remembered that the samples were taken between 8 and 12 o'clock a.m. The results for diurnal migration show that this method gives only a spotlight of the distribution which is submitted, in reality, to continuous alteration throughout the day.

The rhythm of distribution and the related temperature gradient cause a diurnal rhythm in assimilation. Therefore, the following carbon balance is a minimum value concerning the influencing factors standing crop and temperature, because the animals are at the lowest point of their diurnal migration at the time mentioned (Haney & Hall 1975). The horizontal differences in the pelagic zone are obvious but not so strong.

Assimilation

Model. The dependence of assimilation on food supply, temperature and the size of the crustaceans was tested by varying one of these factors. The two others were left as constants. Therefore, a similar assimilation rate ought to be obtained at food concentrations above the incipient limiting level (around $0.2\,\mathrm{mg\ C \cdot l^{-1}}$), at 20° C and at an animal size of 1.1 mm using three different equations. The results are summarized in Tabel 11.6. The approximation is good for *Daphnia cucullata* concerning temperature and food concentration. The values of *Eudiaptomus gracilis* are nearly identical. The dependence on animal size gives under- (*D. cucullata*) and overestimations (*E. gracilis*), respectively.

In the following the equations should be united. Because of the great number of measures and by comparison with other authors (see p. 331), it is assumed that the results of assimilation as a function of food supply are quantitatively and

Table 11.6 Comparison of gross assimilation of *Daphnia cucullata* and *Eudiaptomus gracilis* calculated on the basis of the established dependencies for food concentration, temperature and animal size*.

Species Influence factor	Equation	Assimilation
D. cucullata		
Food concentration	$A = 0.5250 - \frac{0.00086}{F}$	0.052
Temperature	$A = 0.0038 \cdot T - 0.02$	0.056
Animal size	$A = 0.0006\ e^{3.78 \cdot L}$	0.0384
E. gracilis		
Food concentration	$A = 0.0471 - \frac{0.0003}{F}$	0.047
Temperature	$A = 0.0021 \cdot T + 0.005$	0.047
Animal size	$A = 0.0075 \cdot e^{2.01 \cdot L}$	0.067

* A = assimilation ($\mu g\ C \cdot animal^{-1} \cdot h^{-1}$), F = food concentration ($1.0\,mg\ C \cdot l^{-1}$), T = temperature (20° C), L = animal length (1.1 mm).

qualitatively correct (saturation curve, incipient limiting concentration). This is why the equation is used without change in the model.

As shown before the results concerning temperature and animal size are only approximations of reality. Therefore, the equations are changed to relative ones, and in this way they act only as correction terms. The results at 20° C and 1.1 mm animal size are 1.

The equations are as follows:

Gross assimilation, *Daphnia cucullata*

$$A = (0.0525 - \frac{0.00086}{F}) \cdot (0.0678\ T - 0.03572) \cdot 0.0155\ e^{3.78\ L}. \tag{13}$$

Gross assimilation, *Eudiaptomus gracilis*

$$A = (0.0472 - \frac{0.0003}{F}) \cdot (0.0447\ T + 0.1064) \cdot 0.1104\ e^{2.0\ L}. \tag{14}$$

A = assimilation ($\mu g\ C \cdot animal^{-1} \cdot h^{-1}$), T = temperature (° C), L = animal size (mm), F = food concentration ($mg\ C \cdot l^{-1}$).

Nutritional situation of the zooplankton and the relation between planktonic primary and secondary production

There are a number of investigations characterizing the nutritional situation of the zooplankton in Lake Stechlin. Küchler (1981) obtained results about the standing crop of phytoplankton, 10–180 $\mu g\ C \cdot l^{-1}$ (monthly means 1973–1975, converted from wet weight to carbon according to Winberg 1971). Proft (unpubl.) has carried out investigations about particulate organic carbon, 180–380 $\mu g\ C \cdot l^{-1}$ (monthly means 1980, 0–40 m depth). Finally, Koschel (1981) has measured the phytoplanktonic primary production, 4.2–42.1 $\mu g\ C \cdot l^{-1} \cdot d^{-1}$ (monthly means 1970–1975). The phytoplankton concentration is too low to cause a maximum assimilation (incipient limiting concentration about 0.2 mg $C \cdot l^{-1}$). An exception can be found in May (180 $\mu g\ C \cdot l^{-1}$), as the incipient limiting concentration for crustaceans is nearly reached. But to a certain extent it is a simplification to regard all algae as ingestable. As a rule the share of phytoplankton useful as food is lower than the whole standing crop. On the other hand, according to Lampert (1977c) the food concentration where the animals can equalize respiratory losses by assimilation is very low (0.05 mg $C \cdot l^{-1}$). The POC content is in the range of the incipient limiting level.

Primary production is only an indirect measure of nutritional situation. According to Koschel (unpubl.) the seston (without bacteria) in Lake Stechlin consists of detritus 66%, phytoplankton 25%, rotifers and crustaceans 9%. Such a high detritus content is also known from other lakes (Wright 1959). His nutritional value, however, is obscure. All standpoints from 'good' to 'bad' are to be found in the literature (Naumann 1921; Uhlmann 1954; Nauwerck 1963; Joergensen 1966; Heinle *et al.* 1977; Chervin 1978; Herbert 1978). Otzuki *et al.* (1969) characterize *Scenedesmus* spec. as an adequate food after bacterial decomposition for several weeks. Bell & Ward (1970) indicate low assimilability for detritus. Odum (1980) interprets the growth of detritus feeding animals by digestion of the bacteria around the detritus particles. Poulet & Marsot (1978) have shown a selective grazing of food enriched microcapsules, whereas empty ones were left. A direct digestion of structure carbohydrates seems to be impossible, because planktonic herbivores do not possess the enzymes necessary (Dehn 1930; Hasler 1935, 1937; Mayzaud & Poulet 1978; Herbert 1978). Saunders (1969) has investigated the assimilation of *Daphnia* with respect to phytoplankton, detritus and bacteria. The relative assimilation efficiencies were as follows (ratio of assimilated to ingested food): phytoplankton 54%, detritus 9% and bacteria 24%. Notwithstanding the low assimilation efficiency, detritus has a big significance for secondary production because the seston of the lake in question was ten times richer in detritus than in phytoplankton.

Some species of rotifers are known as detritus feeders; others prefer algae. According to Pourriot (1965, 1977) *Keratella cochlearis, Conochilus unicornis* and *Filinia longiseta*, abundant in Lake Stechlin, are detritus feeders. Bogdan *et al.* (1980) have pointed out that bacteria (*Aerobacter)* and yeast (*Rhodotorula)* are useful as food only for detritus feeding rotifers. Algae feeders like *Kellicottia longispina* avoid such organisms. After Gilbert & Starkweather (1978) rotifer feeding is only a function of particle size.

All in all the problem of detritus nutrition seems to be a question of the decomposition of the substrate (Schiller *et al.* 1977). Lake Stechlin must be regarded as detritus rich. This explains the existence of a pelagic detritus feeding chain in the sense of Odum (1980).

In Table 11.7 the nutritional situation of zooplankton is summarized. As a first approximation, it is assumed that the POC-values found by Proft (p. 340) are valid for other years too and that 25% of seston is phytoplankton. This postulate is a simplification because any dynamic is not considered. Expecially in the period from July to September the total seston content decreases more rapidly than the phytoplankton, and thus the ratio is changing in favour of the phytoplankton (Koschel, pers. comm.). For the desired carbon balance this problem has a significance because the calculated phytoplankton losses can only give the order of magnitude.

To make a quantitative connection between filter feeding zooplankton and the phytoplankton-detritus-mixture, it is necessary to make some additional assumptions. It should not be a selective but a proportional ingestion and assimilation of the phytoplankton as well as the detritus, so that the POC-results can serve as input for the calculation of assimilation. The assimilation of detritus by *Daphnia cucullata* ought to be comparable to that found for *Eudiaptomus gracilis*. Finally, *Eurytemora lacustris, Daphnia hyalina* and the group of nauplii should be integrated in the balance because of the similar feeding behaviour. As a first approximation, *Keratella cochlearis* can be taken as a model for the rotifer population. If

Table 11.7 Particulate organic carbon (POC), [a] phytoplankton, [b] detritus, [b] mean phytoplanktonic gross primary production (GPP) [c] and gross secondary production of crustaceans (GSP_C) [d] and rotifers (GSP_R).[d]

Month	J	J	A	S	O	N
POC	270	360	340	210	180	315
Phytoplankton	67	90	85	52	45	78
Detritus	203	270	255	158	135	237
GPP	29.7	42.1	27.0	15.1	18.0	10.2
GSP_C	–	7.3	4.7	6.3	5.6	1.5
GSP_R	4.0	2.5	0.9	0.3	0.1	0.5
GSP_{C+R}	–	9.8	5.6	6.6	5.7	2.0
% of GPP_C	–	17	17	42	31	15
% of GPP_R	13	6	3	2	0.5	5
% of GPP_{C+R}	–	23	21	44	32	20
Grazing of zooplankton[c]	10.2	15.6	14.5	7.7		

[a] Fraction <50 μm, numbers stand for $\mu g\ C \cdot l^{-1}$, according to Proft (p. 340).
[b] Cf. p. 340, numbers stand for $\mu g\ C \cdot l^{-1}$.
[c] Numbers stand for $\mu g\ C \cdot l^{-1} \cdot d^{-1}$, GPP according to Koschel (1981), monthly averages from 1970–1975.
[d] Magnitude of GSP crustaceans (C) and rotifers (R) in % of primary production.

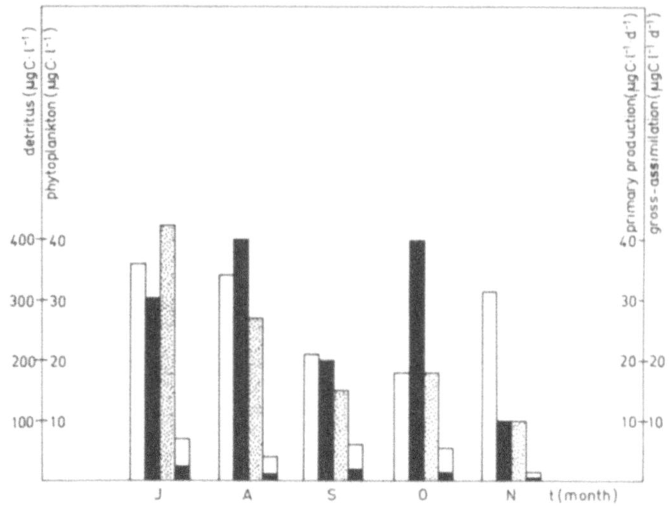

Figure 11.13 Gross assimilation of selected crustaceans (p. 341) (White-black column; white: assimilated detritus, black = assimilated phytoplankton), detritus content (white column; mean of 0–40 m, cf. Proft p. 340), phytoplankton standing crop biomass (black column; mean of 0–15 m, Küchler 1981) and phytoplanktonic primary production (dotted column; mean of 1970–1975, Koschel 1981).

the gross assimilation of selected crustaceans and the rotifer plankton is now regarded as (gross) secondary production, it is to be seen that it equals 20–44% of the planktonic gross primary production (cf. Tab. 11.7; Fig. 11.13). The results obtained by Nowak (1975) for Lake Schieren are somewhat lower than this; 50–60% of the primary production in autumn were used by the total zooplankton. However, it remains uncertain, whether the seston mixture in the pelagic region of Lake Stechlin is as easily assimilable as the sediment detritus given to *Eudiaptomus gracilis*. The same is valid for *Keratella cochlearis* fed only with freeze-dried *Chlorella* spec. Hallegraeff (1978) has found in a Dutch lake 59% ash in the total seston, and this may restrict the assimilation performance of the animals. It must be the aim of further investigations to characterize the seston of Lake Stechlin quantitatively and qualitatively to show more clearly its suitability as food for zooplankton.

References

Arnold, D.E. 1971. Ingestion, assimilation, survival and reproduction by *Daphnia pulex* fed by seven species of blue-green algae. Limnol. Oceanogr. 16: 906–920.

Benndorf, J. 1979. Kausalanalyse, theoretische Synthese und Simulation des Eutrophierungsprozesses in stehenden und gestauten Gewässern. Diss. B., Techn. Univ. Dresden.

Benndorf, J., Koschel, R. & Recknagel, F. 1981. Erste Ergebnisse der Ökosystemanalyse des Pelagials des Stechlinsees mit der Modellmethode. Limnologica 13: 195–200.

Bell, R.K. & Ward, F.J. 1970. Incorporation of organic carbon by *Daphnia pulex*. Limnol. Oceanogr. 15: 713–726.

Bernardi, E., Giussani, G. & Petretti, E.L. 1981. The significance of blue-green algae as food for filter feeding zooplankton: experimental studies on *Daphnia* spp. fed by *Microcystis aeruginosa*. Verh. Int. Verein. Limnol. 21: 477–483.

Bevan, L., Winner, J.M. & Wallen, D.G. 1978. Incorporation rates of *Simocephalus vetulus* Schoedler as influenced by food size. Verh. Int. Verein. Limnol. 20: 2474–2478.

Bogdan, K.G., Gilbert, J.J. & Starkweather, P.L. 1980. In situ clearance rates of planktonic rotifers. Hydrobiologia 73: 73–77.

Bottrell, H.H., Duncan, A., Gliwicz, Z.M., Grygierek, E., Herzig, A., Hillbricht-Ilkowska, A., Kurasawa, H., Larsson, P. & Weglénska, T. 1976. A review of some problems in zooplankton production studies. Norw. J. Zool. 24: 419–456.

Chervin, M.B. 1978. Assimilation of particulate organic carbon by estuarine and coastal copepods. Mar. Biol. 49: 265–275.

Conover, M.F. & Francis, V. 1973. The use of radioactive isotopes to measure the transfer of materials in aquatic food chains. Mar. Biol. 18: 272–283.

Dehn, M. 1930. Untersuchungen über die Verdauung bei Daphnien. Z. vergl. Physiol. 13: 334–358.

Dumont, H.J., Van de Velde, J. & Dumont, S. 1975. The dry weight estimate of biomass in a selection of cladocera, copepoda and rotifera from the plankton, periphyton and benthos of continental waters. Oecologia (Berlin) 19: 75–97.

Erman, L.A. 1956. Quantitative investigations of rotifer feeding. Zool. Ž. 35: 965–971 (in Russian).

Erman, L.A. 1962. Quantitative aspects of feeding and selection of food in the planktonic rotifer *Brachionus calyciflorus* PALL. Zool. Ž. 41: 34–48 (in Russian).

Geller, W. 1975. Die Nahrungsaufnahme von *Daphnia pulex* in Abhängigkeit von der Futterkonzentration, der Temperatur, der Körpergröße und dem Hungerzustand der Tiere. Arch. Hydrobiol., Suppl. 48: 47–107.

Gilberg, J.J. & Starkweather, P.L. 1978. Feeding in the rotifer *Brachionus calyciflorus*. III. Direct observations on the effect of food type, food density, change of food type, and starvation on the incidence of pseudotrocheal screening. Verh. Int. Verein. Limnol. 20: 2382–2388.

Gliwicz, Z.M., 1968. The use of anaesthetizing substance in studies on the food habits of zooplankton communities. Ecol. Pol. A 26: 279–295.

Hall, D.J. 1964. An experimental approach to the dynamics of a natural population of *Daphnia galeata mendotae*. Ecology 45: 94–110.

Hallegraeff, G.M. 1978. Caloric content and elementary composition of seston of three Dutch freshwater lakes. Arch. Hydrobiol. 38: 80–89.

Haney, J.F. 1973. An in situ examination of the grazing activities of natural zooplankton communities. Arch. Hydrobiol. 72: 87–132.

Haney, J.F. & Hall, D.J. 1975. Diel vertical migration and filter feeding activities of *Daphnia*. Arch. Hydrobiol. 75: 413–441.

Hasler, A.D. 1935. The physiology of digestion in plankton crustacea. I. Some digestive enzymes of *Daphnia*. Biol. Bull. mar. biol. Lab. Woods Hole 68: 207–214.

Hasler, A.D. 1937. The physiology of digestion in plankton crustacea. II. Further studies on the digestive enzymes of (A) *Polyphemus*, (B) *Diaptomus* and *Calanus*. Biol. Bull. mar. biol. Lab. Woods Hole 72: 290–298.

Hayward, R.S. & Gallup, D.N. 1976. Feeding, filtering and assimilation in *Daphnia schoedleri* SARS as effected by environmental conditions. Arch. Hydrobiol. 77: 139–163.

Heinle, D.R., Harris, R.P., Ustach, J.F. & Flemmer, D.A., 1977. Detritus as food for eastuarine copepods. Mar. Biol. 40: 341–353.

Herbert, P.D.N. 1978. The population biology of *Daphnia* (Crustacea, Daphnidae). Biol. Rev. 53: 378–426.

Holtby, L.B. & Knoechel, R. 1981. Zooplankton filtering rates: Error due to loss of radioisotopic

label in chemically preserved samples. Limnol. Oceanogr. 26: 774–780.
Hopp, I. 1979. Die Abhängigkeit der spezifischen Leistung planktischer Crustaceenpopulationen von deren Alters- und Größenzusammensetzung. Dipl.-Arb., Techn. Univ. Dresden.
Horn, W. 1979. Der Anteil des Crustaceenplanktons an der Verringerung des Phytoplanktongehaltes in der Talsperre Saidenbach. Diss. A, Techn. Univ. Dresden.
Joergensen, G.B. 1966. The biology of suspension feeding. London: Pergamon Press.
Johannes, R.E. & Satomi, M. 1967. Measuring organic matter retained by aquatic invertebrates. J. Fish. Res. Board Can. 24: 2467–2471.
Kasprzak, P. 1983a. Bestimmung des Körperkohlenstoffs von Planktoncrustaceen. Limnologica 15: 191–194.
Kasprzak, P. 1983b. Struktur, Populationsdynamik und Asssimilationsleistung des Crustaceenplanktons im oligotrophen Stechlinsee. Diss. A, Techn. Univ. Dresden.
Kersting, K. 1978. Some features of feeding, respiration and energy conversion of *Daphnia magna*. Hydrobiologia 59: 113–117.
Koschel, R. 1974. Einfluß der physikalischen und chemischen Umweltfaktoren auf die Primärproduktion des Phytoplanktons im Stechlinsee. Diss. A, Techn. Univ. Dresden.
Koschel, R. 1976. Der Einfluß des Phosphatangebotes auf die Primärproduktion des Phytoplanktons in einem geschichteten Klarwassersee (Stechlinsee, DDR). Limnologica 10: 325–346.
Koschel, R. 1981. Bilanz der Primärproduktion des Phytoplanktons in den Gewässern des Stechlinseesystems. Limnologica 13: 55–81.
Küchler, L. 1981. Phytoplanktonuntersuchungen im Stechlinseegebiet in den Jahren 1973–1975. Limnologica 13: 83–99.
Lampert, W. 1977a. Studies on the carbon balance of *Daphnia pulex* DE GEER as related to environmental conditions. I. Methodological problems of the use of ^{14}C for the measurement of carbon assimilation. Arch. Hydrobiol., Suppl. 48: 287–309.
Lampert, W. 1977b. Studies on the carbon balance of *Daphnia pulex* DE GEER as related to environmental conditions. II. The dependence of carbon assimilation on animal size, temperature, food concentration and diet species. Arch. Hydrobiol., Suppl. 48: 310–335.
Lampert, W. 1977c. Studies on the carbon balance of *Daphnia pulex* DE GEER as related to environmental conditions. IV: Determination of the 'threshold' concentration as a factor controlling the abundance of zooplankton species. Arch. Hydrobiol., Suppl. 48: 361–368.
Lampert, W. 1978. Release of dissolved carbon by grazing zooplankton. Limnol. Oceanogr. 23: 831–834.
Lampert, W. 1981. Inhibitory and toxic effects of blue-green algae on *Daphnia*. Int. Rev. ges. Hydrobiol. 66: 285–298.
Lampert, W. 1982. Further studies on the inhibitory effect of the toxic blue-green *Microcystis aeruginosa* on the filtering rate of zooplankton. Arch. Hydrobiol. 95: 207–220.
Latja, R. & Salonen, K. 1978. Carbon analysis for the determination of individual biomass of planktonic animals. Verh. Int. Verein. Limnol. 20: 2556–2560.
McMahon, J.W. 1965. Feeding rate of *Daphnia magna* STRAUS in different foods labelled with radioactive phosphorus. Limnol. Oceanogr. 10: 105–113.
Mayzaud, P. & Poulet. S.A. 1978. The importance of the time factor in the response of zooplankton to varying concentrations of naturally occurring particulate matter. Limnol. Oceanogr. 23: 1144–1154.
Metz, H. 1973. Kohlenstoffinhalt, Aufnahme und Abgabe des Kohlenstoffs bei *Daphnia pulex*. Arch. Hydrobiol., Suppl. 42: 232–252.
Naumann, E. 1921. Spezielle Untersuchungen über die Ernährungsbiologie des tierischen Limnoplanktons. Lunds Univ. Arsskrift, N.F. Avd. 2, Bd. 17, Nr. 4, 1–27.
Nauwerck, E. 1963. Die Beziehungen zwischen Zooplankton und Phytoplankton im See Erken. Symb. Bot. Uppsal. 17: 1–163.
Nowak, K.E. 1975. Die Bedeutung des Zooplanktons für den Stoffhaushalt des Schierensees. Arch. Hydrobiol. 75: 149–224.

Odum, E.P. 1980. Grundlagen des Ökologie. Stuttgart: Thieme.
Otzuki, A., Hanya, T. & Yamagishi, H. 1969. Residue from bacterial decomposition of green algal cells as food for *Daphnia*. Nature 122: 1182.
Ponyi-Zankai, N. & Ponyi, J.E. 1974. On the seasonal fluctuation in the food incorporation of *Eudiaptomus gracilis*. Ann. Biol. (Tihany) 41: 357–362.
Ponyi-Zankai, N. 1975. Quantitative Untersuchungen zur Nahrungsausnutzung durch *Eudiaptomus gracilis*. Symp. Biol. Hung. 15: 225–231.
Poulet, S.A. & Marsot, P. 1978. Chemosensory grazing by marine calanoid copepods (Arthropoda,Crustacea). Science 200: 1403–1405.
Pourriot, R. 1965. Recherches sur l'écologie des rotifères. Vie et Milieu, Suppl. 21: 1–224.
Pourriot, R. 1977. Food and feeding habits of Rotifera. Arch. Hydrobiol. Beih. Ergebn. Limnol. 8: 243–260.
Proft, G. 1980. Der Kohlenstoffgehalt und die Kohlenstofffracht der mittleren Saale. Limnologica 12: 195–206.
Richman, S. 1958. The transformation of energy by *Daphnia pulex*. Ecol. Monogr. 28: 273–291.
Rigler, F.H. 1971. Methods for the measurement of assimilation of food by zooplankton. In: Secondary productivity in fresh waters. IBP Handbook No 17, W.T. Edmondson & G.G. Winberg (ed.). Oxford and London: Blackwell, pp. 264–269.
Ross, P.E. & Munawar, M. 1981. Preference for nanoplankton size fraction in Lake Ontario zooplankton grazing. J. Great Lakes Res. 7: 65–67.
Romigger, K.L. & Pollmann, W. 1972. Vergleichende Pharmakokinetik von Fenoterolhydrobromid bei Ratte, Hund und Mensch. Arzneimittelforschung 22: 1190–1196.
Ryther, J.H. 1954. Inhibitory effects of phytoplankton upon the feeding of *Daphnia magna* with reference to growth, reproduction and survival. Ecology 34: 522–533.
Saunders, G.W. 1969. Some aspects of feeding in zooplankton. In: Eutrophication – causes, consequences, corrections. Nat. Acad. Sci., Washington D.C., pp. 556–573.
Schiller, H., Spittler, P. & Heerkloss, R. 1977. Bestimmung von Assimilations-, Fress- und Abgaberaten planktischer Crustaceen mit Hilfe von Radiokohlenstoff. Wiss. Z. Univ. Rostock 26: 223–228.
Schindler, J.E. 1971. Food quality and zooplankton nutrition. J. Anim. Ecol. 40: 589–599.
Simon, H. & Rauschenbach, P. 1974. Messung von ratioaktiven und stabilen Isotopen. In: Anwendung von Isotopen in der organischen Chemie und Biochemie, Bd. 2, H. Simon (ed.). Berlin-Heidelberg-New York: Springer.
Starkweather, P.L. & Gilbert J.J. 1977. Feeding in the rotifer *Brachionus calyciflorus*. I. Regulatory mechanisms. Oecologia (Berlin) 28: 125–131.
Uhlmann, D. 1954. Zur Kenntnis der natürlichen Nahrung von *Daphnia magna* und *Daphnia pulex*. Z. Fisch. Hilfswiss. 3: 449–478.
Uhlmann, D. 1971. Influence of dilution, sinking and grazing rate on phytoplankton populations in hyperfertilized ponds and mikroecosystems. Mitt. Int. Verein. Limnol. 19: 100–124.
Unger, R. 1979. Ermittlung von Stoffwechselgrößen von Daphnia magna mit Hilfe kontinuierlicher Kulterverfahren und Untersuchungen zur Statistik, Struktur und Dynamik von Zooplanktonpopulationen in der Talsperre Bautzen. Dipl.-Arb., Techn. Univ. Dresden.
Ward, F.J., Wong, B. & Robinson, G.C. 1970. A liquid scintillation procedure for determining the effect of size on the self-absorption in *Daphnia pulex*. Limnol. Oceanogr. 15: 648–651.
Winberg, G.G., 1971. Symbols, units and conversion factors in studies of freshwater productivity. IBP Control office, London, pp. 1–23.
Wright, J.C. 1959. The Limnology of Canyon Ferry Reservoir. II. Phytoplankton standing crop and primary production. Limnol. Oceanogr. 4: 235–245.

12. Microbial activities at Lake Stechlin

H.-D. Babenzien & Christina Babenzien

The heterotrophic potential*

Introduction

According to Ohle (1962) up to 90% of the organic material produced is remineralized in the epilimnion. To understand the nutrient recycling in the lake it is very important to quantify these decomposition processes.

The preconditions for such in situ estimations include short exposure time and low substrate concentration. The heterotrophic potential technique (Parson & Strickland 1962; Wright & Hobbie 1966) is especially suitable for this purpose. In our investigations the parameters – maximum uptake velocity (V_{max}), turnover time (T_t), and the sum of transport constants and natural substrate concentration (K_{t+Sn}) – were determined by this method (Gocke 1974). Substrate concentrations between 0.5 and 25 $\mu g\ C \cdot l^{-1}$ (14C-glucose) were used. The samples were incubated at $\pm 1°$ C of the in situ temperature for three hours in the dark. Calculations were made using the Lineweaver-Burk transformation. If the correlation coefficient was below that required for the 95% probability level, the data were rejected. All the given V_{max} and T_t values are data for net assimilation.

In several cases the uptake rate was measured by the single concentration method (Griffith *et al.* 1977). For determination of mineralization rates we used the method described by Cole & Likens (1979).

The glucose uptake did not follow the Michaelis-Menten kinetics in every case. We interpret this finding as a result of heterogeneity of the bacterial population. About 10% of our experiments did not show Michaelis-Menten kinetics, especially in the metalimnion and near the sediment.

Orientating determinations of V_{max}, T_t, and K_{t+Sn} for the substrates acetate, glutamic acid, and glucose showed that acetate is incorporated with the highest

* We thank Dr. G. Haubold for the bacteriological data from 1976 to 1978.

Casper, S. Jost (ed.), Lake Stechlin. A temperate oligotrophic lake. ISBN 978-94-010-8930-2.

and glucose with the lowest speed. On average the turnover time for glucose is longer than for other substrates.

The most comparable results are known for glucose as a substrate. Therefore, we used this substrate for our investigations. According to Overbeck (1979) about 10% of the bacterial turnover of the organic production are determined by glucose.

The estimation of the CO_2–dark fixation is also widely employed for measuring the metabolic activity of bacteria. The biochemical background of this method is the anaplerotic reaction for completing the TCA-cycle by CO_2-fixation. Primarily pyruvate or phosphoenolpyruvate are carboxylated to oxaloacetate. This method is used with respect to two differing purposes: measurement of bacterial production (Kuznetzow & Romanenko 1966) and measurement of 'overall' heterotrophic activity (Overbeck 1981). We used the CO_2-dark assimilation described by Sorokin & Kadota (1972) mainly for determining the bacterial production.

Distribution pattern of heterotrophic activity

The intention of these studies was to obtain information about the seasonal and spatial distribution pattern of the kinetic data and the correlation with other environmental parameters.

Fig. 12.1 shows typical pictures of the vertical fluctuations of V_{max} for Lake Stechlin 1982. Pronounced variations are apparent in the epilimnion, while in the hypolimnion the pattern of activity is more similar. The oscillations ranges from 0.001 to 0.06 $\mu g\ C \cdot l^{-1} \cdot h^{-1}$ in the epilimnion and from 0.0014 to 0.017 $\mu g\ C \cdot l^{-1} \cdot h^{-1}$ in the hypolimnion. These values are extremely low in comparison to other waters (Table 12.1). Wide variations of activity occur in the warmed epilimnion between May and September. This indicates that the bacterial population, concerning their physiologic state, is strongly spatially structured. The bacterial activity runs parallel to the primary production during this time. The frequently observed increase of activity in the metalimnion can be explained by the nutrient enrichment in this barrier zone. The detectable dissolved organic carbon concentration in the range of 2–6 $mg \cdot l^{-1}$ does not reflect this dynamic. As reported (see p. 198), this fraction appears not to include the easily degradable compounds, which are rapidly mineralized by the bacterial population.

The main source for organic material is the exudation process by algae. Depending on the trophic level of a water (Larsson & Hagström 1979, Watanabe 1980, Wolter 1980, Bell 1980, Straškrabová & Desertová 1981, Chrost 1981, Itturiaga 1981), the concentration of the release varies between 0 and 54% of the primary production. For nutrient poor waters values for extracellular release of 50% are considered typical (Fogg 1983). Most of the released substances are compounds with molecular weights $\geq$1500 (Herbst 1976). About 10–20% are carbohydrates. The concentration of the well-investigated glycollate ranges from 15 to 76 $\mu g\ C \cdot l^{-1}$ (Billen *et al.* 1980 in Fogg 1983).

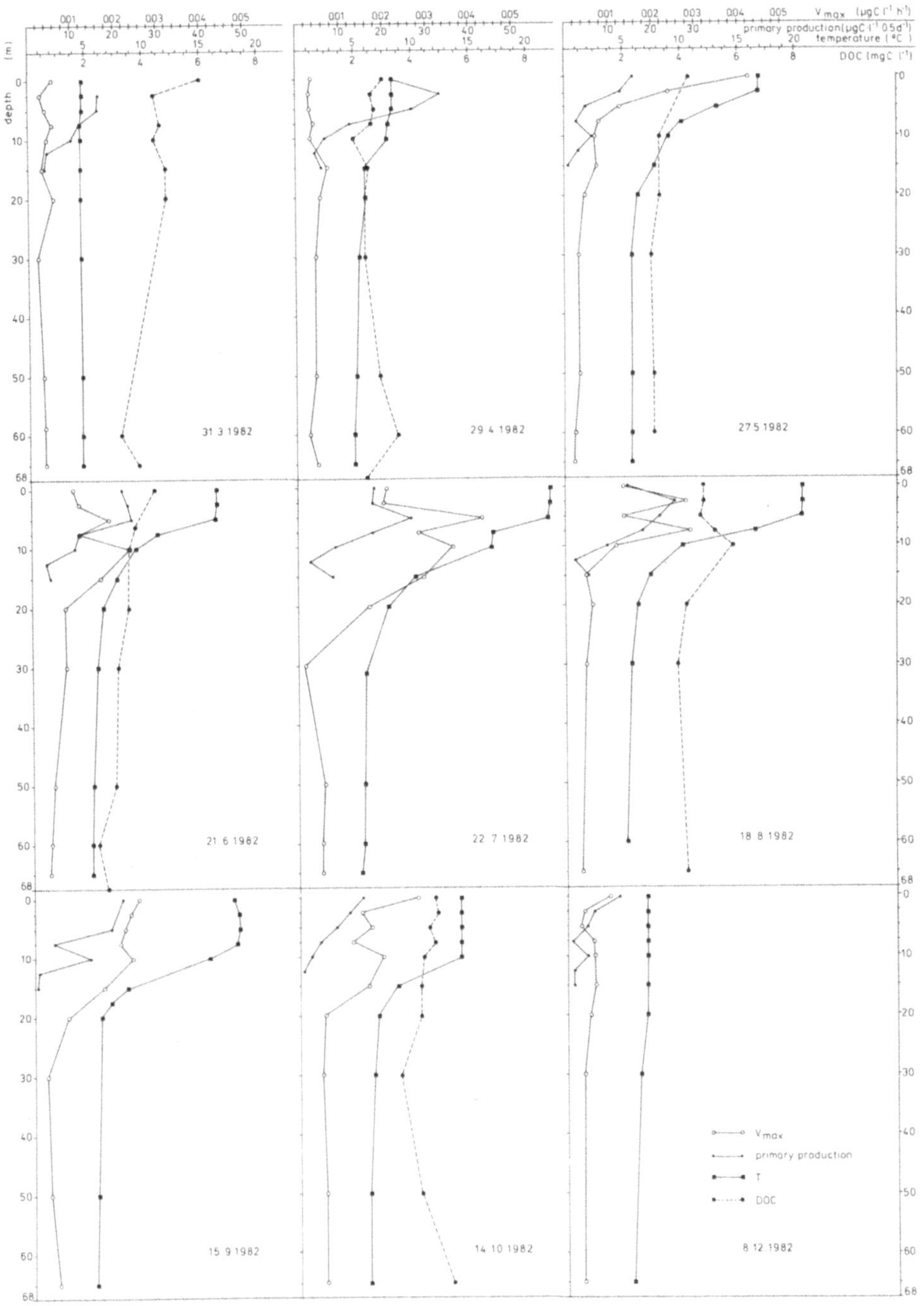

Figure 12.1 Vertical distribution of V_{max}, primary production, temperature, and DOC; Lake Stechlin 1982.

In our investigations with natural phytoplankton populations in Lake Stechlin, release rates of 20–80% of the primary production were obtained. The released substances are taken up by bacteria very rapidly. We measured incorporation rates up to 100%. Obviously, there is a highly effective mechanism, in which the consumption often is equal to the production rate. The close relation between phytoplankton DOC excretion and bacterial utilization is noticeable.

In Fig. 12.2 annual variations of the relationship between primary production and V_{max} are presented for various water depths. Pronounced fluctuations from year to year are visible. It is difficult to describe a 'typical rhythm'. Peaks of bacterial activity are closely related to high primary production rates. Observing the epilimnion as a whole, two phases of high activity are apparent, which in principle can be attributed to spring and autumn. But the results from 1981 indicate that this rhythm is not always present. In that year, in spite of normal primary production, a very low level of bacterial activity was found.

The low V_{max} at a depth of 30 m shows that only small amounts of easily assimilable substrates reach the deeper hypolimnion. From the calculation of K_{t+Sn} it can be concluded that concentration of glucose may be smaller than $5\,\mu g \cdot l^{-1}$.

A very instructive value for the description of nutrient turnover is the turnover time (Fig. 12.3). From June to October we noted the shortest T_t in the range of 10–600 h. In the winter months T_t varied from 200 to 5000 h. Comparing our

Table 12.1 Heterotrophic potential parameters V_{max} and T_t for different lakes.

Water	V_{max} (μg glucose $\cdot l^{-1} \cdot h^{-1}$)	T_t (h)	Source
Char Lake	0.001–0.008	43–1700	Morgan & Kalff 1972
Meretta Lake	0.01–0.075	5–175	Morgan & Kalff 1972
Lake Kinneret	0.095–1.94	1–87	Berman *et al.* 1979
Lake Biwa	0.46–0.49	–	Kadota & Tanaka 1975
Lake Erken	0.007–0.5	17–560	Hobbie & Wright 1968
Lapland Lakes	0.001–0.009	–	Morgan & Kalff 1972
Lake Lötsjön	1.0–35.0	0–300	Morgan & Kalff 1972
Slapy Reservoir	0.1–1.5	5–20	Fuksa 1977
Plußsee	0.2–1.2	6–201.8	Overbeck 1979
Kellersee	0.13–0.36	70–216	Overbeck 1979
Großer Plöner See	0.02–0.18	84–186	Overbeck 1979
Schöhsee	0.02–0.09	166–440	Overbeck 1979
Lake Stechlin	0.001–0.06*	10–5000	Babenzien & Babenzien, unpubl.
Lake Nehmitz-N	0.003–0.4*	7–2200	Babenzien & Babenzien, unpubl.
Lake Dagow	0.01–0.55*	4–292	Babenzien & Babenzien, unpubl.

* $\mu g\ C \cdot l^{-1} \cdot h^{-1}$

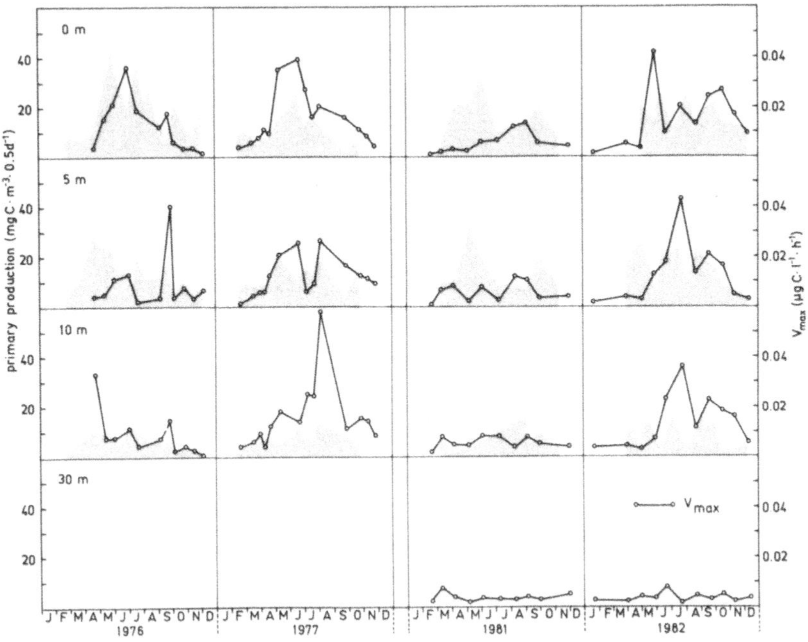

Figure 12.2 Seasonal fluctuations of V_{max} and primary production (background) in different depths; Lake Stechlin.

results with values of other waters, the low level of bacterial activity in Lake Stechlin again is obvious (Table 12.1).

In our investigations concerning the CO_2-dark assimilation values between 0.04 and 2.3 mg $C \cdot m^{-3} \cdot d^{-1}$ were obtained. In the hypolimnion the values were rather low (0.04–0.69 mg $C \cdot m^{-3} \cdot d^{-1}$) throughout the year (Fig. 12.9) while in the epilimnion the heterotrophic CO_2-assimilation increases between May and November. Seasonal variations show that the heterotrophic CO_2-assimilation coincides with metabolic activity described in terms of heterotrophic potential. But the V_{max} and dark assimilation are not parallel in every case, because the latter indicates higher metabolic activity due to the consumption of substrates other than glucose.

Our results are also in the range quoted for other oligotrophic waters (Gorlenko *et al.* 1983).

In the epilimnion of Lake Nehmitz-N, which is integrated in the cooling water cycle, the slightly higher V_{max} can be found in comparison to Lake Stechlin. Values range from 0.003 to 0.07 $\mu g\ C \cdot l^{-1} \cdot h^{-1}$. Seasonal variations are not apparent in the euphotic zone (Fig. 12.4). Vertical fluctuations show that surface, metalimnion, and the anaerobic hypolimnion are biotopes with increasing uptake rates (Fig. 12.5). A strong increase of V_{max} in connection with declining oxygen concentration is visible.

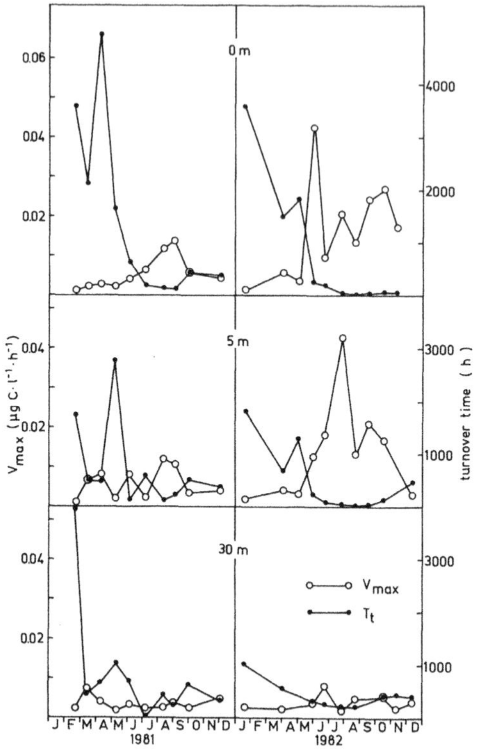

Figure 12.3 Relationship between T_t and V_{max} in different depths; Lake Stechlin.

The same effect was found in the anaerobic zone of Lake Dagow. Beginning with the decrease of oxygen concentration already in May, the heterotrophic carbon uptake strongly increases and reaches extremely high values (0.54 μg $C \cdot l^{-1} \cdot h^{-1}$ in 5 m depth) during the summer stagnation. It can be assumed that the lower efficiency of the anaerobic metabolism is compensated by greater substrate uptake.

As a result of the considerable phytoplankton production in Lake Dagow, the heterotrophic potential in the epilimnion was nearly one power higher than in Lake Stechlin and Nehmitz-N (0.01–0.27 $\mu g\ C \cdot l^{-1} \cdot h^{-1}$). Caused by the increased level of nutrient production, the organic substrate may not always have a limiting effect. Therefore, pronounced variations of the DOC concentration are possible (Figs. 12.6, 12.7).

For calculating the real carbon cycle it must be considered that a certain ratio of the uptake is respired immediately. The determination of this fraction is very important. Results of our respiration measurements are presented in Table 12.2. The percentage ratio of the respiration varies between 5 and 60%. Differences in connection with lake type, depth, and season are not remarkable. On average

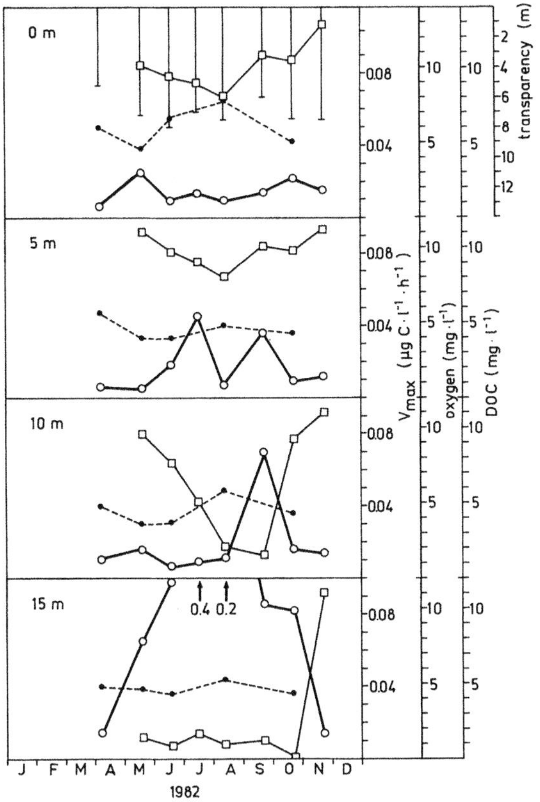

Figure 12.4 Seasonal and spatial distribution pattern of V_{max}, oxygen, DOC, and transparency; Lake Nehmitz-N. o—o V_{max}; □—□ oxygen; ●--● DOC; transparency.

Table 12.2 Respiration rates (%) of ^{14}C-glucose at ambient temperature.

Date	Lake Stechlin					Lake Nehmitz				Lake Dagow		
	0 m	5 m	10 m	30 m	60 m	0 m	5 m	10 m	15 m	0 m	2.5 m	7.5 m
1982												
Apr	38	43	31	6	21	15	16	8	13	–	–	–
May	51	–	–	–	–	27	35	29	11	27	–	–
Jun	49	36	21	19	26	24	56	21	17	30	31	14
Jul	36	37	32	32	9	32	8	16	6	27	12	14
Aug	28	55	24	17	6	49	47	39	14	32	29	20
Sep	22	13	22	32	16	22	21	23	10	37	43	15
Oct	23	22	24	55	47	21	20	21	9	16	23	24
Nov	5	17	9	12	36	22	35	22	41	34	28	26
Dec	6	20	20	13	20	10	12	17	21	17	27	12

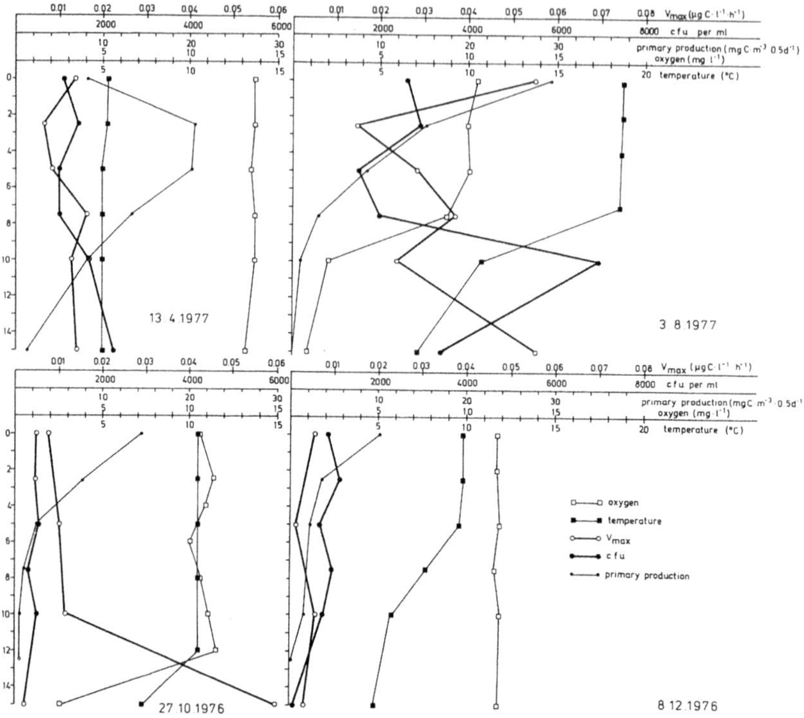

Figure 12.5 Vertical profiles of V_{max}, primary production, oxygen, temperature, and cfu (peptone-agar); Lake Nehmitz-N.

25% of the 14C-glucose uptake was released simultaneously as 14C-CO_2. Hence, for obtaining the gross uptake, V_{max} and T_t have to be corrected by this proportion.

Table 12.3 Heterotrophic potential parameters and trophic status.

Lake	V_{max} ($\mu g \cdot C \cdot l^{-1} \cdot h^{-1}$)	T_t (h)	Trophic status
Lake Stechlin	0.001–0.06	10–5000	oligotrophic
Lake Nehmitz			oligo- to mesotrophic
epilimnion	0.003–0.07	11–2200	
hypolimnion	0.002–0.4	7–500	–
Lake Dagow			eutrophic
epilimnion	0.01 –0.27	4–292	
hypolimnion	0.012–0.55	5– 68	

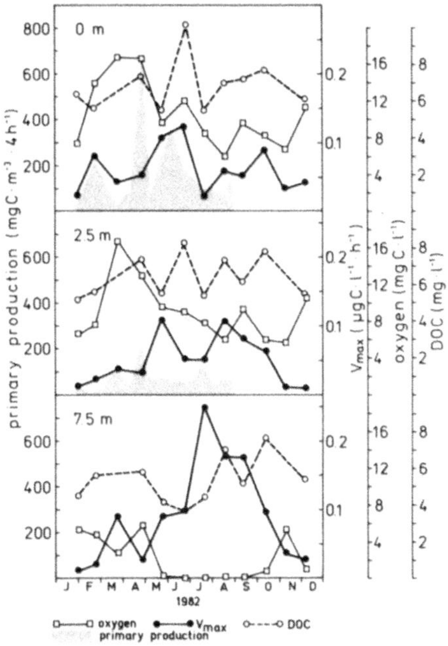

Figure 12.6 Seasonal changes of V_{max}, primary production, oxygen, and DOC; Lake Dagow.

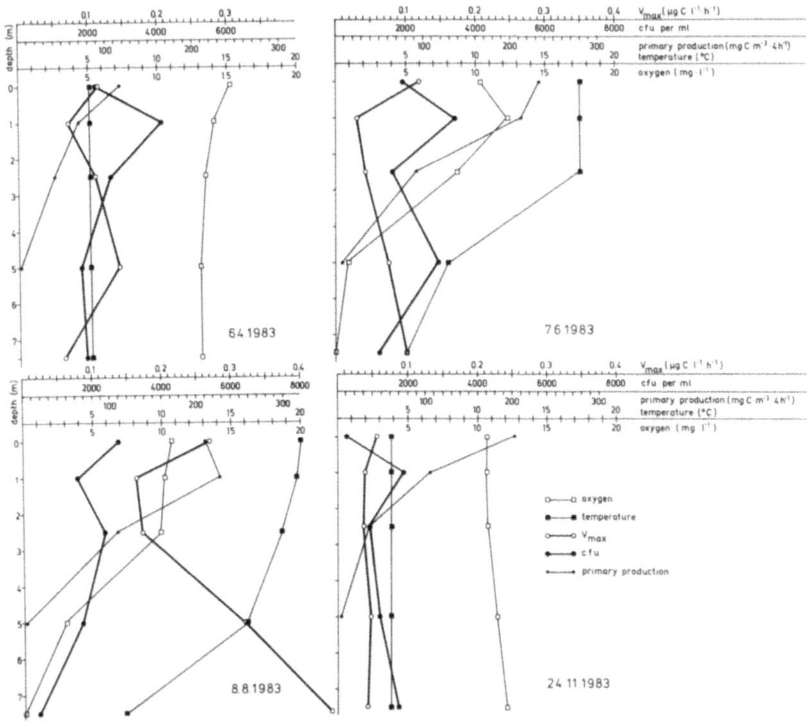

Figure 12.7 Vertical profiles of V_{max}, primary production, oxygen, temperature, and cfu (peptone-agar); Lake Dagow.

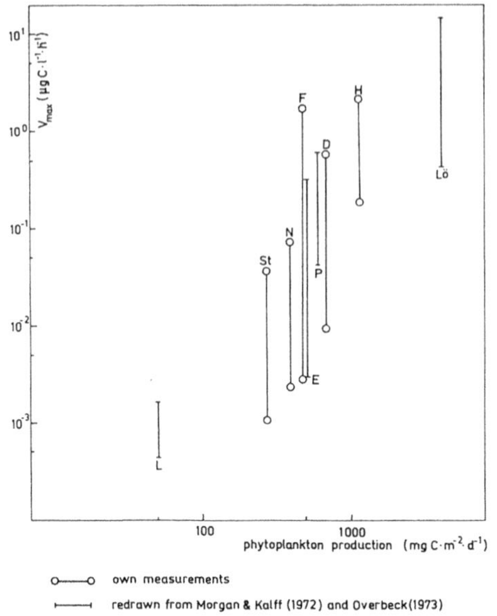

Figure 12.8 Seasonal range in maximum uptake velocity of glucose versus average daily primary production in several lakes. Graph redrawn from Morgan & Kalff (1972) and Overbeck (1973) with additions from own measurements (L – Lapland lakes, St – Lake Stechlin, N – Lake Nehmitz-N, F – Lake Große Fuchskuhle, E – Lake Erken, P – Lake Plußsee, D – Lake Dagow, H – Lake Haussee, Lö – Lake Lötsjön).

Heterotrophic activity and trophic status

The method of heterotrophic potential is recommended for the characterization of trophic level of lakes by many authors (Morgan & Kalff 1972; Overbeck 1973; Hobbie & Rublee 1977; Rheinheimer 1981; Gorlenko *et al.* 1983). According to Overbeck (1973), V_{max} can be understood as an expression of the 'heterotrophic capacity' of a lake.

Tables 12.1 and 12.3 express the distinct relationship between types of lakes and the heterotrophic potential parameters. It is evident that the oligotrophic status of Lake Stechlin is reflected in values of V_{max}.

Observing the V_{max} in relation to the primary production (Fig. 12.8) enables lakes to be classified according to their trophic level (Overbeck 1973).

Thermal pollution and microbial activity

Temperature is a fundamental environmental parameter controlling life processes at the stage of single enzymatic stages, the behaviour of individuals, as well as the whole population level (Krenkel & Parker 1969). Therefore, the phys-

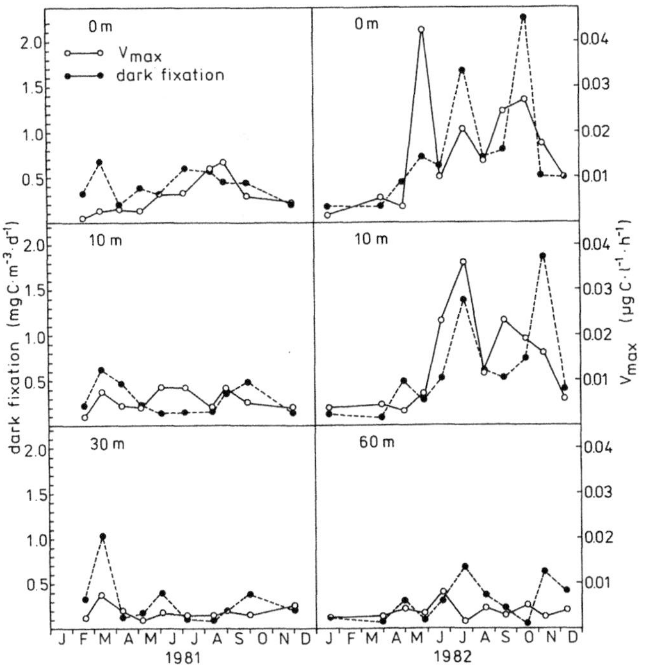

Figure 12.9 Seasonal and spatial fluctuations of V_{max} and dark fixation; Lake Stechlin.

iological activities of organisms, their reproduction rates, and the stability of the ecosystem are influenced by the ambient temperature.

Because of the potential hazard to life and to the balance of nature, the discharge of waste heat into natural waters is coming to be called thermal pollution (Clark 1969). For this resaon it is of great interest to understand the effects of discharges of the nuclear power plant (npp) at Lake Stechlin. Most biological investigations concern the impact on macroorganisms, such as fish and benthic invertebrates (Humpesch 1981), and likewise on phytoplankton, periphyton, and macrophytes; microorganisms, however, have been neglected (Buck 1973; Bott 1975; Koschel & Mothes 1976; see p. 421). Accordingly, we chose to examine the effects on selected bacterial parameters.

The starting point is the fact that an amount of 20 000 $m^3 \cdot h^{-1}$ water is taken from Lake Nehmitz and, having passed through the nuclear power plant, flows through a canal into Lake Stechlin. On average the output temperature is 10 degrees higher than the ambient temperature. The warming up process takes only a few minutes. As a maximum, the organisms are confronted in condensor tubes with temperatures up to 8° C higher than in the discharge canal. This short thermal stress is the only artifical change in the water.

In Fig. 12.10 the results of comparative measurements are shown concerning temperature and oxygen from the starting point of the nuclear power plant. They

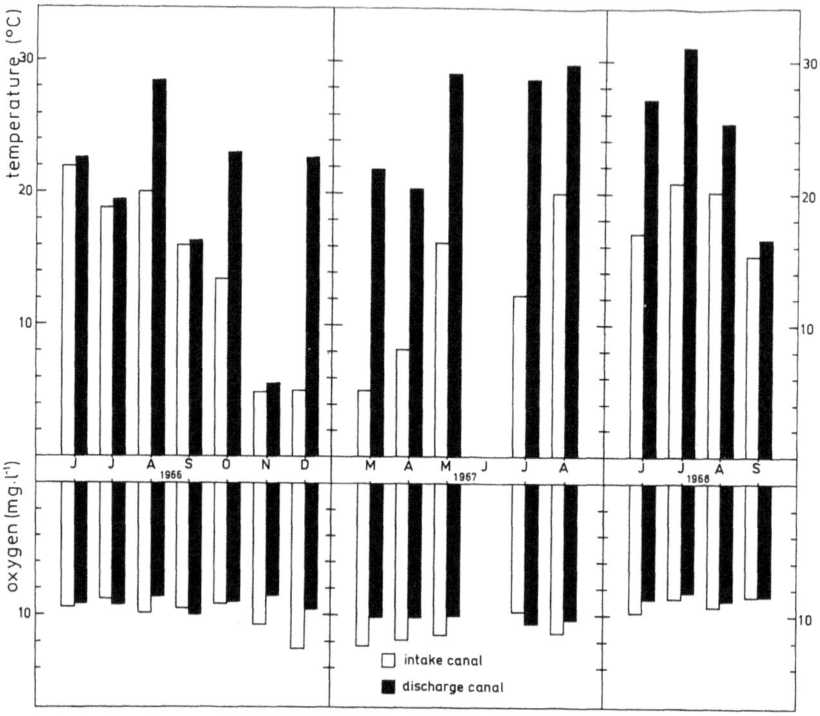

Figure 12.10 Changes of temperature and oxygen content due to nuclear power plant operation (nppo).

are representative for the working regime with maximum temperatures of 32° C in the discharge canal. But it must be remembered that a difference of 10 K in the water plume represents temperatures higher than the natural maximum for six months in a year. However, this gradient of temperature is completely dissipated in the west bay of Lake Stechlin (see p. 49; Fig. 3.3). Therefore, planktonic organisms are influenced for only a short time. Because of the lower saturation concentration at increasing temperatures, the oxygen content in the thermal effluent decreases rapidly. This effect is very critical in rivers, but in the well-oxygenated Lake Stechlin it is without any consequences.

In comparing the numbers of heterotrophic bacteria (colony forming units) of intake and discharge canal (Fig. 12.11), there were no pronounced variations. Commonly we can observe a tendency to higher counts in the effluent canal. This is probably due to a breakdown of microcolonies and detritus by mechanical damage from passing through the condensor tubes.

To understand whether or not a stimulation or depression of metabolic activity takes place, we determined the parameter of heterotrophic potential and the CO_2-dark fixation in the intake canal and immediately after the npp in the discharge canal. In addition we estimated the influence upon primary production.

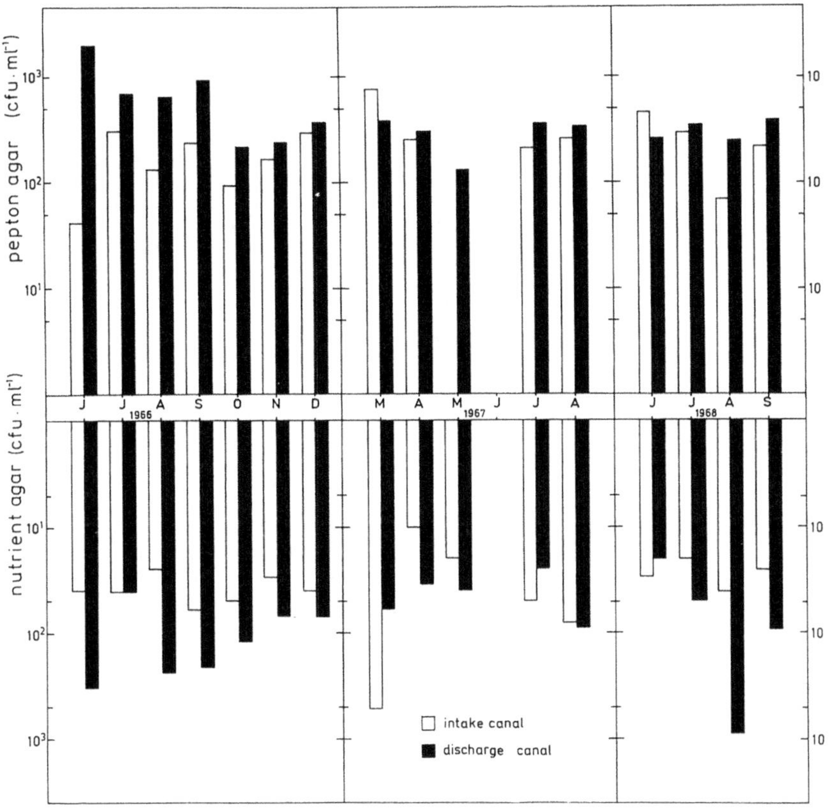

Figure 12.11 Comparison of bacterial colony forming units (cfu) before and after npp, incubation: 25° C, 10 d, 2 kinds of nutrient media.

The pronounced alterations of the parameters V_{max} and T_t are shown in Fig. 12.12. In these examples the short thermal stress causes a considerable shortening of turnover rate and higher uptake rate (Fig. 12.13).

This is the dominant behaviour throughout the year, although deviations – depressions – occur. Control experiments showed that only a part of the population could be stimulated while the other part must be inhibited. If water from the intake canal was incubated in the outflow without passing the condensors, we obtained information about the real temperature effect. In this case we noticed mostly higher values of activity (Fig. 12.13). Results of synoptic investigations are given in Fig. 12.14. In June an increase of temperature stimulated phytoplankton activity and the metabolic activity of bacteria. The higher ambient temperatures found in July were coupled with maximum values of 32° C in the discharge canal. During this time inhibitory effects in view of CO_2-light and dark fixation are apparent, while heterotrophic activity yielded still rising mineralization rates. For a better understanding of the plankton response to increasing temperatures,

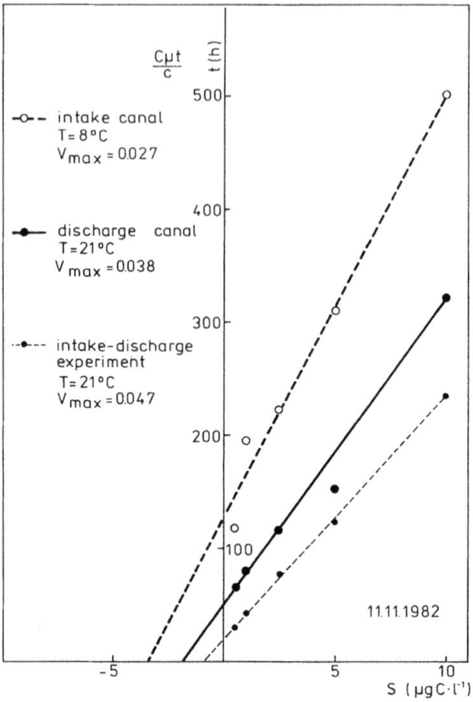

Figure 12.12 Lineweaver-Burk regression line of 14C-glucose uptake by natural microbial population as function of nppo.

some experiments have been carried out. The findings suggest a maximum permissible temperature for mostly mesophilic bacteria of 28–30° C. Temperatures above this critical threshold were followed by a rapid loss of activity (Fig. 12.15). At low temperatures the uptake velocity of 14C-glucose by the natural microbial communities decreases significantly. But a distinct correlation between temperature optima and different seasons was not observed.

Witzel (1980) also noted different substrate affinity in experimental studies in a river with a fluctuating thermal regime. He concluded: 'since the microorganisms in their natural environment are permanently competing for life – essential substances at suboptimum concentrations, an increase or decrease in substrate affinity may cause a distinct adventage of one over another organisms'.

Our results agree well with those of Miller *et al.* (1976), who used similar or identical methods. They found that, in general, any absolute temperature (ambient temperature + ΔT) less than 25° C caused stimulation of primary production. At ambient temperatures greater than 25° C, however, all temperature increases on the average caused inhibition of plankton production. The maximum uptake rate with 14C-glucose occurred at an absolute temperature between 20 and 25° C. Saltzmann (1980) reported data for a river power plant with marked

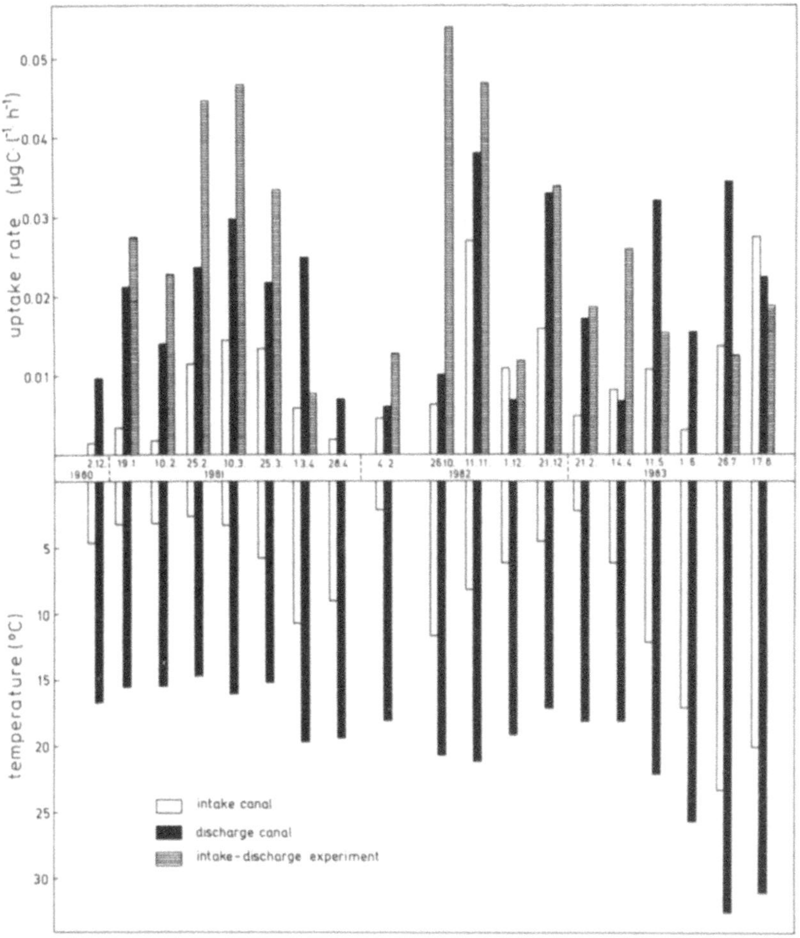

Figure 12.13 Effect of temperature elevation on uptake rate of 14C-glucose.

increase (22%) of the heterotrophic potential after the condensors.

It can be stated that community metabolism was increased substantially by passing the nuclear power plant. These effects have been restricted only to a small area. The low level of organic substances, balanced oxygen content, and the rapid exchange of energy with the atmosphere prevent any harmful consequences to the whole lake system (see p. 419).

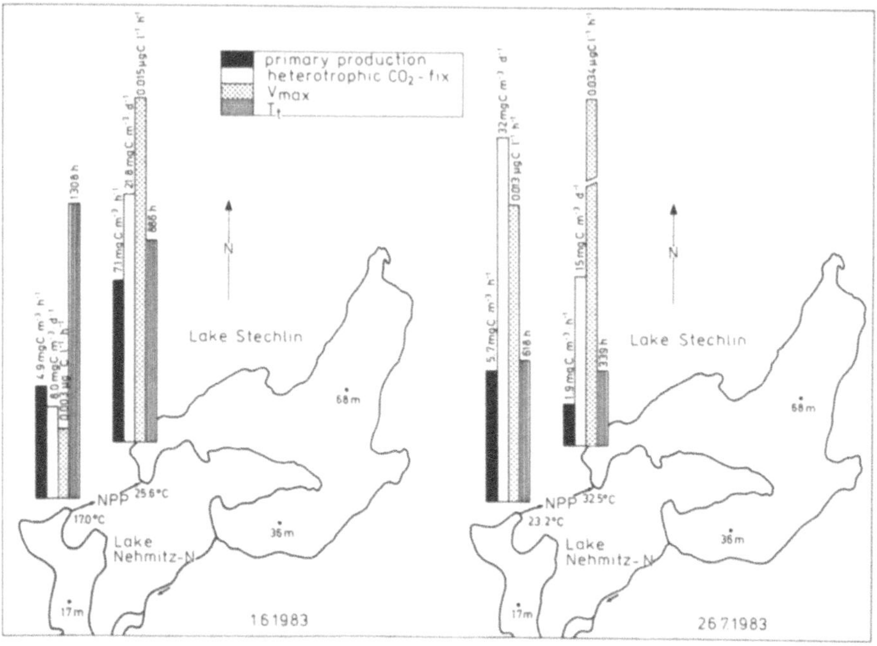

Figure 12.14 Synoptic analysis of some microbial activity parameters showing temperature depending stimulatory and inhibitory effects.

Microbial transformation processes in sediments

General characteristics

Mineralization processes in profundal sediments are primarily a domain of microorganisms. Furthermore, the physico-chemical environment is affected by their metabolism, by which the direction of abiotic transformations is determined. Finally, microbes play an important role in diagenetic conversion of settled substances. Since bacterial metabolism in profundal sediments is almost entirely dependent on inputs of organic matter from sedimentations, changes in the composition of particulate matter during settling can have significant impact on mineralization.

Composition and particle size of this 'detritus rain' depend on the autochthonous or allochthonous origin, e.g., algae, leaf litter, and macrophyte litter, because of their different degradation rates. On the other hand, morphometric peculiarities influence the extent of destruction by differing settling times and oxygen budgets. As has been shown by Olie & Cappenberg (1982), when seston is dying, roughly 20% POC is leached out of the cells. The leachate is almost completely assimilated by microbes in the same place. Organic matter entering

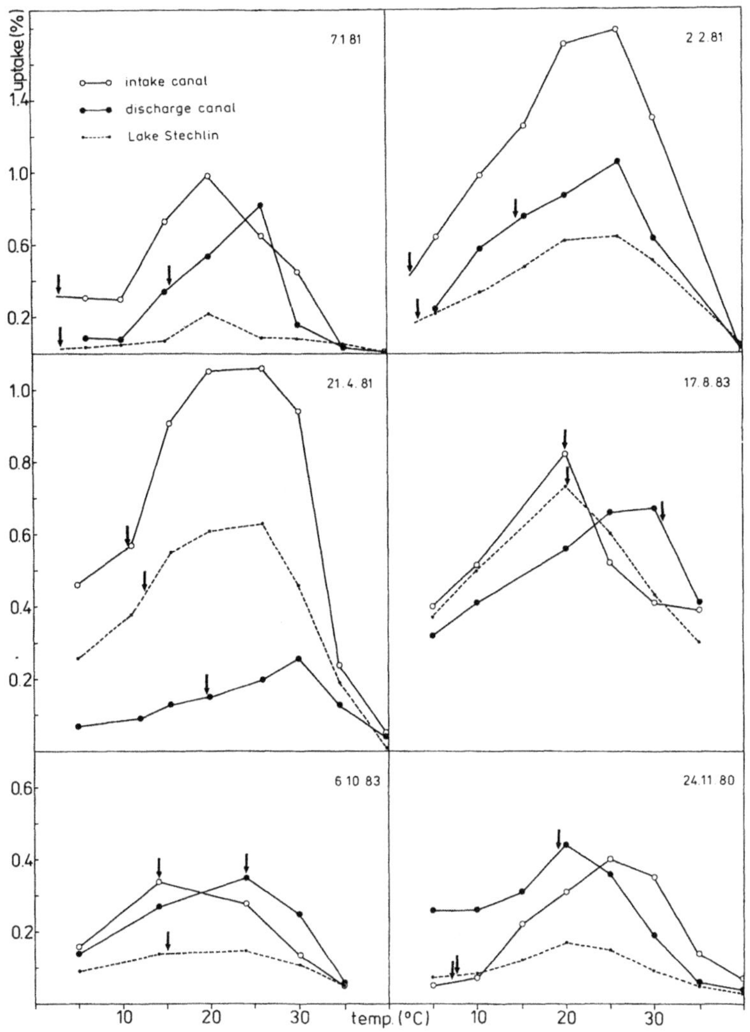

Figure 12.15 Uptake rates of 14C-glucose by 6 natural non-adapted microbial communities from temperature shift experiments. Arrows show activity at ambient temperature.

mud deposits, therefore, is enriched with cell debris, and thus more resistant to bacterial breakdown.

In Lake Stechlin the low values of V_{max} in the aerobic hypolimnion are explained by the loss of easily assimilable organic substances while sinking. Formation and nutritive characteristics of these mud sediments are affected by the different origins of residues. Assuming a 10% daily sedimentation rate, we can calculate an input of 30 mg $C \cdot m^{-2} \cdot d^{-1}$ from primary production. Macrophytes, which constitute a great part of the whole carbon flow (700 t DM = 107 t OC; see p. 146), are perhaps already mineralized in the littoral region. But they will contribute refractory plant residues, i.e., cellulose and lignin.

When studying macrozoobenthos in Lake Stechlin, Mothes (1965) came to the conclusion that sediments receive a remarkable supply of nitrogen from chitinic debris from zooplankters and insects.

Another source of carbon-input is airborne precipitation, but quality and composition are unknown. Considering the results of Ohle (1982), a yearly influx of 15–20 t organic carbon can be assumed (see p. 407).

Airborne litterfall from the catchment area of the lake plays an important role within the carbon budget too. In Lake Stechlin this source of allochthonous matter amounts on average to 23 t DM (= 11 t $C \cdot a^{-1}$; cf. p. 407). But is must be stressed that detritus derived from Beech leaves is very resistant to degradation (Kaushik & Hynes 1971). Studies of decay (Kunze & Bonnert 1982) show that after 84 d exposure the dry matter of Beech leaves consists of 50% lignin and cellulose, of which 68% was lignin. The percentage decomposition of these materials was only 0.2% per day.

Intensity of leaf litter decomposition is influenced by various conditions: mechanical breakdown in the littoral region, colonization by hyphomycetous fungi (Bengtsson 1982) and feeding by invertebrates (Iverson & Madsen 1977). Marinucci *et al.* (1983) emphasized the controlling (limiting) effect of the chemical composition of leaves and the availability of nitrogen compounds to the miccrobial colonization and decomposition rate. In small bays with leaf litter accumulation we could observe the heterotrophic activity to be strongly increasing, with up to 20–100 fold higher values than in the pelagic zone. Analysing the degradation process of beech leaves, it has been found an initial leaching phase with high concentration of DOC and high heterotrophic potential within 2–4 days (see p. 365). After that we observed only weak oscillations. Obviously, the depletion of available energy and resources of the substrate during decomposition did not allow any significant activity (Fig. 12.16).

All these findings indicate that the organic substances in the mud sediments of Lake Stechlin must be very immobilized. As a simple criterium of these circumstances the 'modified sediment activity' (Caspers 1962) was performed. The biological oxygen consumption of 10 ml mud (15° C, 24 h incubation, 1 l lake water) reached only 6.6%, whereas with Lake Nehmitz-mud 40.8% against the control (pure lake water) 1.9%.

The content of nitrogen (2.4%/dry matter, Kjeldahl-method) and phosphorus (0.14–0.32%/dry matter, Proft 1964) of sediments in Lake Stechlin is within the range of eutrophic lakes (Babenzien 1968). This sediment is characterized by high portions of organic substances (53%), water (95%), and a ratio C : N<10 (see p. 376). These are proporties of sediments, described by Ungemach (1966) as 'humic sediments'. However, the low level of bioactivity does not permit noticeable concentrations of nutrients in the interstitial fluids of sediments – in the example mentioned above 3.5 mg $NH_4^+ \cdot l^{-1}$ and 0.6 mg $P \cdot l^{-1}$, respectively. A rapid flux out of sediments would be typical in the case of eudynamic lakes (Ohle 1964), which is not the case in Lake Stechlin.

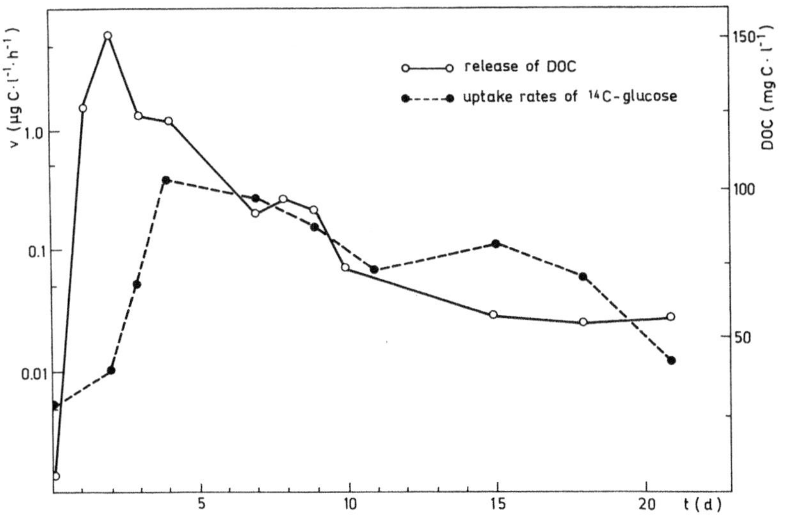

Figure 12.16 Leaching process of leaves of *Fagus silvatica* L. (50 g leaves, 2.5 l water, 15° C, aerobic conditions, data from Proft & Babenzien, unpublished).

Distribution of heterotrophic bacteria in sediment profiles

It could be expected that these characteristics are reflected also in the distribution of bacteria in sediment profiles. Therefore, we determined the number of aerobic heterotrophic bacteria in different sediment cores. The bacteria were enumerated by the standard spread plate method. Results for two samples are shown in Table 12.4. Generally, counts in the range of 10^3 cfu (per g dry matter) were observed. Only in the sediment-water-interface were higher numbers obtained. Although the bacterial number decreases with increasing depth, we did not find a sharp vertical stratification in the cores as it was described in other types of sediments (Collins 1977; Gorlenko *et al.* 1983). It can, then, be inferred that the high water content enables migration through recent sediments (Krumbein 1970). On the other hand, bacteria survive for a long time in a dormant stage. Many sediment particles failed to generate bacterial colonies; to some extent they are sterile.

In view of the general metabolic activity it must be considered that temperature in sediments reaches only a maximum of 5° C. Nevertheless, we were not able to isolate obligate psychrophilic strains. All of the organisms isolated and initially cultivated at 5° C grew better at 25–28° C, which is in accordance with the results of Boylen & Brock (1973). Temperature conditions, therefore, have a delaying effect on the turnover.

In addition we enumerated the number of putrefying bacteria producing hydrogen sulphide (MPN-method, nutrient solution modified after Lapage in

Table 12.4 Number of bacteria (per g of wet mud) in two sediment cores from Lake Stechlin (30 m water depth) as determined from plate counts (nutrient agar – NA – and peptone agar – PA –, 25° C, 10 d), content of dry matter (% DM, 105° C) and lost by ignition (in % of DM, 2 h, 550° C).

Core depth (cm)	Sample 29.6.1967				Sample 11.8.1983			
	bacteria/g · 10^3		% dry matter		bacteria/g · 10^3		% dry matter	
	NA	PA	105° C	550° C	NA	PA	105° C	550° C
0–10	13.0	7.6	2.9	62.7	138	180	4.0	41.7
20	4.4	6.0	3.3	60.9	3.9	9.0	4.7	50.5
35	2.4	7.5	3.1	60.4	10.1	4.4	5.2	38.7
50	2.4	1.6	3.3	67.7	59.7	7.2	9.0	52.3
65	3.5	6.0	3.7	64.5	4.9	5.1	4.2	23.6
80	1.6	3.1	5.3	47.8	5.9	11.0	4.0	15.5
100	4.2	7.3	5.5	52.5	1.6	5.4	3.8	15.4
130	–	–	–	–	1.0	0.3	4.0	13.5
150	0.1	0.2	6.3	62.7	1.6	2.1	4.3	14.2
190	0.4	0.6	7.1	68.3	0.2	0.4	4.5	15.2
220	–	–	–	–	0.3	0.05	5.0	19.3

Pfennig *et al.* 1981). Likewise, the number of sulphate-reducing bacteria was determined (Starkey-medium, in Babenzien 1964). The number of this important bacterial group was low. In detail, at 30 m-sample point the following results were found (per g dry mud): 0–10 cm: $3 \cdot 10^2$; 50 cm: $2.1 \cdot 10^2$; 65 cm: $0.2 \cdot 10^2$; 190 cm: $0.2 \cdot 10^2$. It is quite obvious that hydrogen sulphide producing processes do not possess any decisive significance.

Phosphate mobilization

Sediments play an important role in the overall phosphorus metabolism of lakes. In oligotrophic lakes, sediments possess the ability to retain a major proportion of phosphorus input. Phosphorus release from sediments is primarily affected by bacterial metabolism in different ways (Boström *et al.* 1982).

Firstly, phosphate is an end product of mineralization whereby organic phosphate esters are hydrolyzed by alkaline phosphatases (Reichardt 1978). Secondly, bacteria are able to solubilize phosphate precipitates. Solubilization is attributed to the formation of organic acids that function as chelating agents (Harrison *et al.* 1972). Lastly, bacterial activity changes the physicochemical conditions by oxygen consumption, sulphide and methane formation etc., thus affecting the redox potential that is responsible for exchange between water column and sediment (Ohle 1954).

The potential ability for solubilization of anorganic phosphates has been detected by the plate method (Rodina 1965) using tribasic calcium phosphate. Of 100 isolated strains 36% produced haloes around their colonies, i.e., showed dissolving ability. This process was thoroughly investigated using natural mixed populations. Intensity of phosphorus solubilization depended on available suitable carbon and energy sources, in these experiments glucose (Table 12.5). Our findings are in accordance with the results of Harrison *et al.* (1972) who analysed this phenomenon in detail.

Hence it can be assumed that phosphorus release from sediments in this way in aerobic Lake Stechlin sediments plays an important role. The classical model – describing phosphorus flux from sediments which is based upon the interaction between iron and phosphorus under aerobic and anaerobic conditions – probably does not fit Lake Stechlin. From the above it can be concluded that the low level of microbial activity prevents the nutrient outflow from sediments. As regards the whole lake, it functions as a protection of the oligotrophic status.

The methane cycle

Rudd & Hamilton (1978) established that 55% of the organic carbon input into the sediment of a eutrophic lake leaves it as methane. The steps of the biodegradation chain of organic molecules are performed by a metabolically diverse population. In this chain methane is produced only in anaerobic environments, requiring CO_2, H_2. and the main precursor acetate. Therefore, methane is a typical final product of these microbial interactions in anoxic sediments (Rudd & Taylor 1980). Further on it is suggested that methane oxidation in hypolimnion has a significant impact on the oxygen content of a lake.

Due to the low level of organic production and the general characteristics of their sediments, the methane cycle in oligotrophic lakes seems to be without significance. Observations about gas bubbles with a content of 50–60% methane

Table 12.5 Solubility of inorganic phosphates by natural bacterial communities and by 2 isolated strains in batch cultures.

	$Ca_3\ (PO_4)_2$	$Fe_3\ (PO_4)_2$		
Lake water + P_i	67	400	88	140
Lake water + P_i + C + N	1590	1120	1314	1050
Lake water + bacteria + C + N	83	150	129	250
Lake water + bacteria + C + N + P_i	1220	1235	645	1220
Lake water + bacteria + C + N	5	55	128	35
Lake water + bacteria + C + N + P_i	1270	1140	625	952

Phosphorus concentrations in filtrates after 3 d incubation (20° C) in μg P per 50 ml. P_i – $Ca_3\ (PO_4)_2$ resp. $Fe_3\ (PO_4)_2$, 0.1%; C – 0.5% glucose; N – 0.1% NH_4Cl. Values from 4 sets of experiments.

that escape from littoral sediments to the atmosphere prompted investigations of methane production and oxidation in Lake Stechlin (Heyer & Babenzien 1984).

Methods. Methane concentrations in water and interstitial fluids were estimated according to the modified gas chromatographic method of Naguib (1978). The number of methane-oxidizing bacteria was determined with the MPN-method in a gas atmosphere (20% CH_4, 75% air, 5% CO_2) after 6 weeks incubation at 30° C. Inorganic nutrient broth and details of quantification of methane oxidation or production are described by Heyer & Suckow (1984). Rates of release of methane from the sediment as bubbles (ebullition) were measured using enclosed funnels (0.1 m^2) placed on the sediment surface (Ohle 1958).

Results. Our results show the high intensity of the methane cycle and the importance of these processes at various locations in Lake Stechlin.

Generally, methane concentration in lake water (littoral and pelagic zone) is very low at all depths. Also during stagnation periods there was no hypolimnic accumulation as is typical in eutrophic lakes (Anagnostidis & Overbeck 1966). The lowest concentrations were found in water above the profundal sediments (0.1 μmol $CH_4 \cdot l^{-1}$), while above littoral sediments the values were higher (0.1–0.7 μmol $CH_4 \cdot l^{-1}$). Hence the number of methanotrophs in the water was negligible (1–95 cells per ml) and the rate of methane consumption was extremely low (0.01 μmol $CH_4 \cdot l^{-1}$).

On the other hand in littoral and profundal sediments surprisingly high concentrations were found (91–565 μmol $CH_4 \cdot l^{-1}$ interstitial fluid). Also the number of methylotrophic bacteria was high with differences between littoral and profundal sediments. In littoral sediments the values ranges from 1.6×10^4 to 6.2×10^5 per g dry matter, in profundal sediments (36 m and 68 m depth) from 8.1×10^3 to 7.1×10^4.

Differences between these two biotopes were also detectable in methane oxidation rates. The rates depend on CH_4-concentration according to the Michaelis-Menten-kinetics. Therefore, it was possible to calculate the natural consumption rate. Assuming an active uppermost layer of 1 cm in littoral sediments of 2 m depth, 475–961 μmol $CH_4 \cdot m^{-2} \cdot h^{-1}$ were calculated and in profundal regions of 36 m depth 169 μmol $CH_4 \cdot m^{-2} \cdot h^{-1}$.

Methane oxidation is probably the dominant oxygen consuming process at the sediment surface. Usually it takes place in the hypolimnion of lakes. Because of the well-oxygenated conditions in Lake Stechlin, it takes place in the immediate surface layer of mud. Intensities of methane oxidation obtained here are extremely high in comparison with eutrophic lakes. Note the values of Harrits & Hanson (1980; Lake Mendota: 1.4 μmol $CH_4 \cdot l^{-1} \cdot h^{-1}$), Howard *et al.* (1971; Lake Erie: 1.47 μmol $CH_4 \cdot l^{-1} \cdot h^{-1}$), Ivanov *et al.* (1976; Black Kichier Lake: 1.21 μmol $CH_4 \cdot l^{-1} \cdot h^{-1}$).

Methane oxidation in mud deposits has already been noted by Zaiss (1981) with

23 μmol CH_4 per kg sediment · h^{-1} (running water), respectively, 2.5 μmol CH_4 per kg sediment · h^{-1} (pond sediment). Additionally Bossard & Gächter (1981) carried out similar measurements (Lake Lungern, 4458 and 6375 μmol CH_4 · m^{-2} · h^{-1}, 0.7 cm layer, batch culture).

Enrichment and pure culture experiments have resulted in a taxonomical survey about different types of methanotrophs and their special properties that set them apart from the majority of other microbes. The most common strains of methane utilizers in Lake Stechlin belong to the *Methylocystis*-group with various morphological types. In addition the *Methylosinus*-group was observed (*M. trichosporium* and *M. sporium*). Among the other types *Methylomonas* was observed occasionally, while *Methylococcus* and *Methylobacter* seem to be absent.

High methane concentrations or oxidation rates are only understandable if high formation rates are also presumed. Measurements of methane evasions in littoral areas ranged 50–300 μmol CH_4 · m^{-2} · h^{-1}. Because of the heterogeneous environment and dependence upon meteorological factors it is impossible to generalize the results. But the extent of these processes is apparent. The loose consistency of sediments is probably due to the intensive evasion.

In littoral sediments (1.5 m) we obtained a methane production rate of 232–786 μmol CH_4 · m^{-2} · h^{-1}). In profundal sediments we were unable te demonstrate this process (18 d incubation, 6° C).

Apart from methodological difficulties, we assume the process to be limited to deeper layers (we used only the uppermost 20 cm). Moreover, the anaerobic breakdown takes place here very slowly. In comparison with littoral sediments in other lakes methane production rates are in the middle range (Table 12.6).

According to the relationship between input of organic carbon and methane release (Robertson 1979), the sedimentation ought to be in the range 150–500 mg C · m^{-2} · d^{-1}. In fact, this magnitude is not reached by sedimentation of planktonic organisms (30 mg C · m^{-2} · d^{-1}). Therefore, we have to consider allochthonous litter fall and macrophyte biomass. The important role of leaves in connection with methane production has already been noted by Tam *et al.* (1981). For understanding methane production in littoral areas we may assume that the

Table 12.6 Methane production rates (μmol CH_4 · m^{-2} · h^{-1}) in littoral sediments of different lakes.

Lake	Methane production rates	Source
Lake Lungern	50–154	Bossard & Gächter 1981
Lake 227	450	Rudd & Hamilton 1978
Third Sister Lake	230	Robertson 1979
Francis Lake	800	Robertson 1979
Lake Erie	4580	Howard *et al.* 1971
Lake Stechlin	232–786	Heyer & Babenzien 1984

intensive destruction of large amounts of organic matter (leaves, macrophytes) is responsible for local anaerobic circumstances. A two-stage process takes place consisting of organic acid-formation by heterogeneous microorganisms and gas-formation by the methanogens (Cappenberg 1982; Zeikus 1983). In profundal sediments these processes are limited by a depletion of suitable substrates. According to Cappenberg *et al.* (1982), the rate-limiting factor is the breakdown into acetate. Probably quantitative oxidation of methane in the uppermost layer of sediments causes the barrier against oxygen diffusion and thus the necessary conditions for anaerobic destruction. The channeling of breakdown processes into anaerobic paths seems to be unusual in an oligotrophic lake. However, the occurrence and restriction of the methane cycle to sediments represents a characteristic of the carbon flow of Lake Stechlin.

References

Anagnostidis, K. & Overbeck, J., 1966. Methanoxydierer und hypolimnische Schwefelbakterien. Studien zur ökologischen Biocönotik der Gewässermikroorganismen. Ber. Deutsch. Bot. Ges. 79: 163–174.

Babenzien, H.-D., 1964. Hydrobakteriologische Untersuchungen im Stechlinsee. Limnologica 2: 9–34.

Babenzien, H.-D., 1968. Studien zur Bakteriologie eines oligotrophen Gewässers (Stechlinsee). Limnologica 6: 207.

Bell, W.H., 1980. Bacterial utilization of algal extracellular products. I. The kinetic approach. Limnol. Oceanogr. 25: 1007–1020.

Bengtsson, G., 1982. Patterns of amino acid utilization by aquatic hyphomycetes. Oecologia 55: 355–363.

Berman, T., Hadas, O. & Marchaim, U., 1979. Heterotrophic glucose uptake and respiration in Lake Kinneret. Hydrobiologia 62: 275–282.

Billen, G., Joiris, C., Wijnant, J. & Gillain, G., 1980. Estuar. Coastl. Mar. Sci. 11: 279–294 (cited according to Fogg, G.E., 1983. The ecological significance of extracellular products of phytoplankton photosynthesis. Bot. Mar. 26: 3–14).

Bossard, P. & Gächter, R., 1981. Methan- und Sauerstoffhaushalt im mesotrophen Lungernsee. Schweiz. Z. Hydrol. 43: 219–252.

Boström, B., Jansson, M. & Forsberg, C., 1982. Phosphorus release from lake sediments. Arch. Hydrobiol. Beih. Ergebn. Limnol. 18: 5–59.

Bott, T.L., 1975. Bacterial growth rates and temperature optima in a stream with a fluctuating thermal regime. Limnol. Oceanogr. 20: 191–197.

Boylen, C.W. & Brock, T.D., 1973. Bacterial decomposition processes in Lake Wingra sediments during winter. Limnol. Oceanogr. 18: 628–634.

Buck, J.D., 1973. Effects of thermal effluent on the microbiology and chemistry of the Connecticut river. In: Stevenson, H.L. & Colwell, R.R. (eds.), 1973. Estuarine microbial ecology. Univ. S. Carolina Press, USA, 283–297.

Cappenberg, T.E., 1982. Carbon flow across the sediment-water interface in Lake Vechten, The Netherlands. Hydrobiologia 91: 161–168.

Caspers, H., 1962. Die Bestimmung der Sedimentaktivität. Int. Revue ges. Hydrobiol. 47: 581–586.

Chrost, R.J., 1981. The composition and bacterial utilization of DOC released by phytoplankton. Kieler Meeresforsch. Sonderh. 5: 325–332.

Clark, J.R., 1969. Thermal pollution and aquatic life. Sci. American 220: 19–27.

Cole, J.J. & Likens, C.E., 1979. Measurements of mineralization of phytoplankton detritus in an oligotrophic lake. Limnol. Oceanogr. 24: 541–547.

Collins, V.G., 1977. Methods in sediment microbiology. Adv. Aquatic Microbiol. 1: 219–272.

Fogg, G.E., 1983. The ecological significance of extracellular products of phytoplankton phytosynthesis. Bot. Mar. 26: 3–14.

Fuksa, F., 1977. Die Aktivitätsmessungen der heterotrophen Bakterien mittels markierter organischer Substrate. In: Daubner, J. (ed.). II. Internationales Hydromikrobiologisches Symposium, Smolenice. Bratislava: VEDA Verlag der Slowakischen Akademie der Wissenschaften, 301–313.

Gocke, K., 1974. Methodische Probleme bei Untersuchungen zur mikrobiellen Stoffaufnahme in Gewässern. Kieler Meeresforsch. 30: 12–23.

Gorlenko, V.M., Dubinina, G.H. & Kuznetsov, S.I., 1983. The ecology of aquatic micro-organisms. (Die Binnengewässer 28). Stuttgart: E. Schweizerbart'sche Verlagsbuchhandlung. 252.

Griffith, R.P., Hayasaka, St. S., McNamara, Th. M. & Morita, R.Y., 1977. Comparison between two methods of assaying relative microbial activity in marine environments. Appl. Environ. Microbiol. 34: 801–805.

Harrison, M.J., Pacha, R.E. & Morita, R.Y., 1972. Solubilization of inorganic phosphates by bacteria isolated from upper Klamath Lake sediments. Limnol. Oceanogr. 17: 50–57.

Harrits, S. & Hanson, R.S., 1980. Stratification of aerobic methane oxidizing organisms. Limnol. Oceanogr. 25: 412–421.

Haubold, G., 1982. Untersuchungen zur bakteriellen Aktivität in der euphotischen Zone des Stechlinseegebiets. Limnologica 14: 243–261.

Herbst, V., 1976. Physiologische Untersuchungen zur Kopplung des Stoffwechsels von *Oscillatoria Redekii* van Goor und Begleitbakterien. Kiel: Thesis, VIII + 133.

Heyer, J. & Babenzien, H.-D., 1985. Untersuchungen des Methankreislaufes in einem oligotrophen See (Stechlinsee). Limnologica 16: 267–276.

Heyer, J. & Suckow, R., 1985. Ökologische Untersuchungen der Methanoxydation in einem sauren Moorsee. Limnologica 16: 247–266.

Howard, D.L., Frea, J.I. & Pfister, R.M., 1971. The potential for methane carbon cycling in Lake Erie. Proc. 14 th Conf. Great Lakes Res. 236–240.

Hobbie, J.E. & Rublee, P., 1977. Radioisotope studies of heterotrophic bacteria in aquatic ecosystems. In: Cairns, J. Jr. (ed.). Aquatic microbial communities. New York – London: Garland Publ., Inc. 441–476.

Humpesch, U.U., Dokulil, M., Elliott, J.M. & Herzig, A., 1982. Ökologische Auswirkungen der thermischen Gewässerbeeinflussing. Oesterr. Wasserwirtschaft 34: 122–136.

Itturiaga, R., 1981. Phytoplankton photoassimilated extracellular products, heterotrophic utilization in marine environment. Kieler Meeresforsch. Sonderh. 5: 318–324.

Ivanov, M.V., Belyaev, S.S. & Laurinavichus, K.S., 1976. Methods of quantitative investigation of microbial production and utilization of methane. In: Schlegel, H.S., Gottschalk, G. & Pfennig, N. (eds.), 1976. Microbial production and utilization of gases. Göttingen: E. Goltze KG. 63–67.

Iverson, T.M. & Madson, B.L., 1977. Allochthonous organic matter in streams. Folia Limnol. Scandinavia 17: 17–20.

Kadota, H. & Tanaka, N., 1975. Bacterial production and decomposition of organic matter. In: Mori, S. & Yamamoto, G. (eds.), 1975. Productivity of communities in Japanese inland water. JIBP Synthesis 10: 13–15.

Kaushik, N.K. & Hynes, H.B.N., 1971. The fate of the dead leaves that fall into streams. Arch. Hydrobiol. 68: 465–515.

Koschel, R. & Mothes, G., 1976. Änderung der biologischen Produktivität in einem Seensystem durch den Kühlwasserzufluß eines Kernkraftwerkes. Materialien des Internationalen Symposiums über Eutrophierung und Sanierung von Oberflächengewässern. 20.–25.9.1976 Karl-Marx-Stadt, DDR, Bd. III: 219–233.

Krenkel, P.A. & Parker, F.I. (eds.), 1969. Biological aspects of thermal pollution. Proceed. Symp. Thermal Poll. Vanderbilt: Univ. Press.

Krumbein, W.E., 1970. On the behaviour of pure cultures of marine micro-organisms in sterilized and re-inoculated sediments. Helgol. wiss. Meeresunters. 20: 17–28.

Kunze, Chr. & Bonnert, R., 1982. Blattabbau in Wasserproben aus einem belasteten Fließgewässer. Angew. Bot. 56: 237–242.

Kusnetzow, S.I. & Romanenko, V.I., 1966. Produktion der Biomasse heterotropher Bakterien und die Geschwindigkeit ihrer Vermehrung im Rybinsker Stausee. Verh. Int. Verein. Limnol. 18: 176–187.

Larsson, U. & Hagström, A., 1979. Phytoplankton exudate release as an energy source for the growth of pelagic bacteria. Mar. Biol. 52: 199–206.

Marinucci, A.C., Hobbie, J.E. & Helfrich, J.U.K., 1983. Effect of litter nitrogen on decomposition and microbial biomass in *Spartina alterniflora*. Microb. Ecol. 9: 27–40.

Miller, M.C., Hater, G.R., Federle, Th.W. & Reed, J.P., 1976. Effects of power plant operation on the biota of a thermal discharge channel. In: Esch, G.W. & McFarlane, R.W. (eds.), 1976. Thermal Ecology II. ERDA-Symposium Series 40: 251–258. Springsfield, USA.

Morgan, J.J. & Kalff, J., 1972. Bacterial dynamics in high Arctic lakes. Freshwater Biol. 2: 217–228.

Mothes, G., 1965. Ein Beitrag zur Sediment-Charakterisierung des Stechlinsees und zur biologischen Typisierung von Seen. Limnologica 3: 381–388.

Naguib, M., 1978. A rapid method for the quantitative estimation of dissolved methane and its application in ecological research. Arch. Hydrobiol. 82: 66–73.

Ohle, W., 1954. Sulfat als 'Katalysator' des limnischen Stoffkreislaufs. Vom Wasser 21: 13–32.

Ohle, W., 1958. Die Stoffwechseldynamik der Seen in Abhängigkeit von der Gasauscheidung ihres Schlammes. Vom Wasser 25: 127–149.

Ohle, W., 1962. Der Stoffhaushalt der Seen als Grundlage einer allgemeinen Stoffwechseldynamik der Gewässer. Kieler Meeresforsch. 18: 107–120.

Ohle, W., 1964. Interstitiallösungen der Sedimente, Nährstoffgehalt des Wassers und Primärproduktion des Phytoplanktons in Seen. Helgol. Wiss. Meeresunters. 10: 411–429.

Ohle, W., 1982. Nährstoffzufuhren des Grebiner Sees durch atmosphärische Niederschläge und Oberflächenabschwemmung des Einzugsgebietes. Arch. Hydrobiol. 95: 331–363.

Olie, J.J. & Cappenberg, Th. E., 1982. Aspects of aerobic mineralization during spring in Lake Vechten with special reference to the 14 C-labelling technique. Hydrobiologia 95: 181–190.

Overbeck, J., 1973. Über die Kompartimentierung der stehenden Gewässer – ein Beitrag zur Struktur und Funktion des limnischen Ökosystems. Verh. Ges. Ökologie. Saarbrücken. 211–223.

Overbeck, J., 1979. Studies on heterotrophic functions and glucose metabolism of microplankton in Plußsee. Arch. Hydrobiol. Beih. 13: 56–76.

Overbeck, J., 1981. A new approach for estimating the overall heterotrophic activity in aquatic ecosystems. Verh. Int. Verein. Limnol. 21: 1355–1358.

Parson, T.R. & Strickland, J.D.H., 1962. On the production of particulate organic carbon by heterotrophic processes in sea water. Deep Sea Res. 8: 221–222.

Proft, G., 1967. Beiträge zur Sedimentchemie von Binnengewässern. I. Die Verteilung des Phosphors in Sedimenten von Seen und Talsperren. Limnologica 5: 397–404.

Reichardt, W., 1978. Responses of phosphorus remobilizing *Cytophaga* species to nutritional and thermal stress. Verh. Int. Verein. Limnol. 20: 2227–2232.

Rheinheimer, G., 1981. Mikrobiologie der Gewässer. Jena: VEB Gustav Fischer Verlag. 251.

Robertson, C.K., 1979. Quantitative comparison of the significance of methane in the carbon cycles of two small lakes. Arch. Hydrobiol. Beih. Ergebn. Limnol. 12: 123–135.

Rodina, A.G., 1965. Metody vodnoj mikrobiologii: praktičeskoe rukovodstvo. Moskow-Leningrad: Nauka. 362.

Rudd, J.W.M. & Hamilton, R.D., 1978. Methane cycling in a eutrophic shield lake and its effect on whole lake metabolism. Limnol. Oceanogr. 23: 337–348.

Rudd, J.W.M. & Taylor, C.D., 1980. Methane cycling in aquatic environments. Adv. Aquatic Microbiol. 2: 77–150.

Saltzmann, H.A., 1980. Untersuchungen über die Veränderungen der Mikroflora beim Durchgang von Brackwasser durch die Kühlanlagen von Kraftwerken. Kiel: Thesis, III + 124 (+36).

Sorokin, Y.I. & Kadota, H. (eds.), 1972. Techniques for the assessment of microbial production and decomposition in fresh waters. IBP Handbook Nr. 23. Oxford – London – Edinburgh – Melbourne: Blackwell Sci. Publ. 112.

Straškrabová, V. & Desertová, B., 1981. Bakterielle Ausnutzung der extrazellulären Algenprodukte. III. Internationales Hydromikrobiologisches Symposium (ed. I. Daubner). Bratislava. 423–434.

Tam, T.Y., Mayfield, C.I. & Inniss, W.E., 1981. Nitrogen fixation and methane metabolism in a stream sediment-water system amended with leaf material. Can. J. Microbiol. 27: 511–516.

Ungemach, H., 1960. Sedimentchemismus und seine Beziehungen zum Stoffhaushalt in 40 europäischen Seen. Kiel: Thesis, 1–420.

Watanabe, Y., 1980. A study of the excretion and extracellular products of natural phytoplankton in Lake Nakanuma, Japan. Int. Revue ges. Hydrobiol. 65: 809–834.

Witzel, K.-P., 1980. Temperature compensation of (U–14C) Glucose incorporation by microbial communities in a river with a fluctuating thermal regime. Appl. Environ. Microbiol. 39: 790–796.

Wolter, K., 1980. Untersuchungen zur Exsudation organischer Substanz und deren Aufnahme durch natürliche Bakterien-populationen. Report 45, Sfb 95, Universität Kiel, 1–124.

Wright, R.T. & Hobbie, J.E., 1966. Use of glucose and acetate by bacteria and algae in aquatic ecosystems. Ecology 47: 447–464.

Zaiss, U., 1981. Natural ebullition of mine gas and its microbial oxidation in the Netzbach brook, Saarland. Verh. Int. Verein. Limnol. 21: 1381–1385.

Zeikus, J.G., 1983. Metabolic communication between biodegradative populations in nature. In: Slater, J.H., Whittenburg, R. & Wimpenny, J.W.T. (eds.), 1983. Microbes in their natural environments. Symposia of the Society for General Microbiology 34: 423–462. Cambridge: University Press.

13. Sediments and sedimentation

G. Mothes & G. Proft

Introduction

The sediment of Lake Stechlin has been the subject of research many times, usually with the emphasis on biological aspects (Möller 1935; Schönborn et al. 1965; Vogt 1965).

Schönborn et al. (1965) attempted a characterisation of the sediment with the help of testaceen communities, and the results were based on the quantitative relation between carbon and nitrogen in the profundal sediment according to Hansen (1959, 1962). Uhlmann & Albrecht (1968) for their analysis relied on geochemical factors of the eutrophication process as well as on the C : N ratio (expressed as P : N : C ratio) and expressly referred to the conditions in Lake Stechlin.

Proft (1976, 1969) produced works on the spatial distribution of various chemical elements in the Lake's basins, and Mothes (1977; 1980; 1981a, b) studied the sedimentation process in detail and produced a balance of the phosphorus content.

Koschel et al. (1983) and Proft (1984) have reported on calcite precipitation and carbon content.

The following contribution summarizes the findings on sediments and the sediment process in the Lake Stechlin area, and compares them with other regions, discussing open questions.

Sediments (G. Proft)

The sediments of Lake Stechlin

Each deep zone in Lake Stechlin has its own typical sediment (Fig. 13.1).

In the eulittoral zone, the area of breaking waves and water movement, the bottom is sand, coarse sand or partially gravel. Autochthonous sediments are not

Casper, S. Jost (ed.), Lake Stechlin. A temperate oligotrophic lake. ISBN 978-94-010-8930-2.

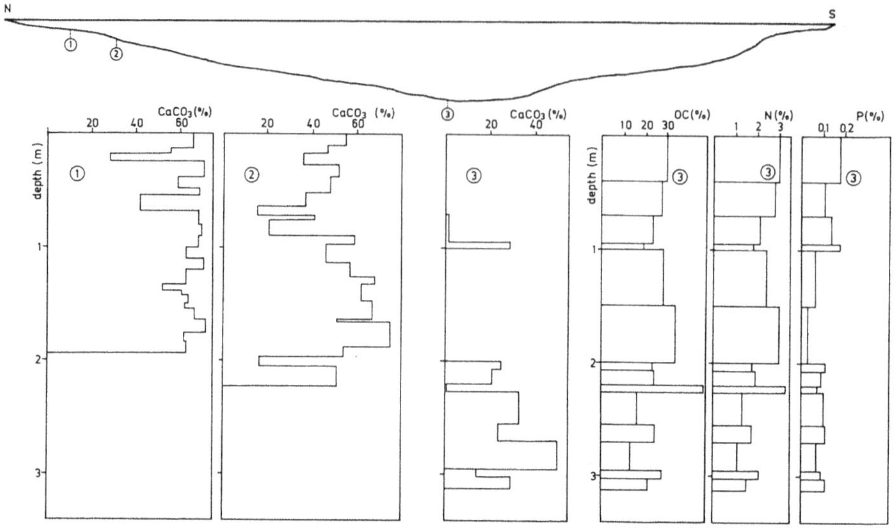

Figure 13.1 Lake Stechlin, S-basin, sediment profiles from different depths.

deposited in this relatively narrow zone.

Lake-wards, in the infralittoral zone, plant beds of increasing thickness and extensiveness are characteristic (p. 130). Here, from a depth of about 0.3 to 0.5 m, the first autochthonous lake sediments begin. The surface layer which can be obtained with an Ekman-Birge-grab, is black-grey in colour, rich in organic substances, but poor in calcite.

Beneath this layer, obtainable with a piston corer, is a lime rich gyttja of grey-white colour. We are here dealing with a lake-chalk-bank produced by biogenic decalcification of the littoral macrophytes and especially marked by a wealth of mollusc and ostracode shells. At a depth of 15–20 m it ends in an untypical formed shell deposit zone (littoriprofundal).

The sediments of the profundal are thin fluid in the uppermost layer and change gradually into a thick-fluid and finally a firm fine-particulate mass of dark olive to dark brown, partially brown-black colour. They are almost chalk-free (<0.1% $CaCO_3$) and rich in organic bound carbon (~30%). They consist mainly of organic detritus, coprogenic substances, chitin remains of plankton, shells from diatoms and chrysomonades, pollen grains as well as a small part of minerals such as quartz, feldspar and mica (<0.1 mm; cf. Vogt 1965). According to Mothes (1981b) the top-most sediment of the profundal 'bottom-grab' probes consists of 61.4% interstitial water, 38.6% soft sediment, 4.8% dry matter and 2.3% ash. In the interstitial water there is $0.4\,g \cdot m^{-3}$ orthophosphate and $2.6\,g \cdot m^{-3}$ of total phosphate.

Noteworthy is that the profundal sediment phenomenologically named gyttja, contains individual remains of leaves and brown humic-acid containing, muddy mass of dy-components (cf. p. 364, p. 401).

The thickness of the sediment is determined not only by the depth but by the morphometry of the lake and influenced by the internal waterstreams. Thus there are sediment-poor to sediment-free infralittoral regions, e.g., in the area of Fenchelberg and on the northern slopes of the north bay, with a slope of over five percent. Haregrave & Kamp-Nielsen (1976) refer to a similar difference in the rate of sedimentation being influenced by the slope and streams of water within the lake, with reference to Bedford Basin and Esrum Sø.

According to Håkanson (1982) the zones in Lake Stechlin which are subject to sediment movement by erosion and transport, can be calculated as follows:

$$a_{E+T} = 52.0 \times \left| \frac{\sqrt{a}}{D_{max}} - 0{,}2 \right| \times \frac{D_{max}}{3 \times \bar{D}} \times \log \left(\frac{60.6 \times \bar{D}}{\sqrt{a}} \right) + 19.3$$

a = lake surface area (Lake Stechlin = 4.3 km^2)
a_{E+T} = percentage area of lake bottom erosion and transportation
D_{max} = maximum depth (Lake Stechlin = 68 m)
$\bar{D}$ = mean depth (Lake Stechlin = 22.8 m)
According to this the a_{E+T}-area in Lake Stechlin covers 44 percent of the total lake bottom related to a depth of up to 16 m (cf. Håkanson 1982, p. 12, Fig. 3; see also Fig. 2.4 this volume).

Beds of macrophytes act as sediment 'catchers' (DeMarch 1978). In the vegetation period, in general, a fine detritus is deposited all over the infralittoral and additionally, in autumn, some leaves. Only in winter, when a large portion of the beds die off, can the sediment be carried and deposited in the direction of the deepest parts of the lake. The transport and enrichment of the sediments in the deepest parts of the lake-basin is called 'Trichtereffekt' (funneling-effect) by Ohle (1952) and 'focusing' by Likens & Davis (1975). Evans & Rigler (1983) established a linear relationship between depth and sedimentation accumulation for Lake Castello and Red Chalk Lake (Ontario).

Sediment chemistry and lake depth

Spatial distribution of phosphorus and uranium in lake sediments. Examinations of probes taken with an Ekman-Birge-grab at various depths allow the establishment of a relation between depths and concentration of the chemical elements in sediment.

In Lake Stechlin the profundal sediment is rich in organic substance. Organic-bound carbon, nitrogen and hydrogen are more abundant in the deepest parts of the basin (Fig. 13.2). To follow more exactly the relations between depth and concentration of chemical elements, we established 5 profiles through the lake (Figs. 13.3, 13.4) and examined a larger number of probes for their phosphorus and uranium content (Proft 1967, 1969).

The probes richest in phosphorus and uranium were found in the deepest parts

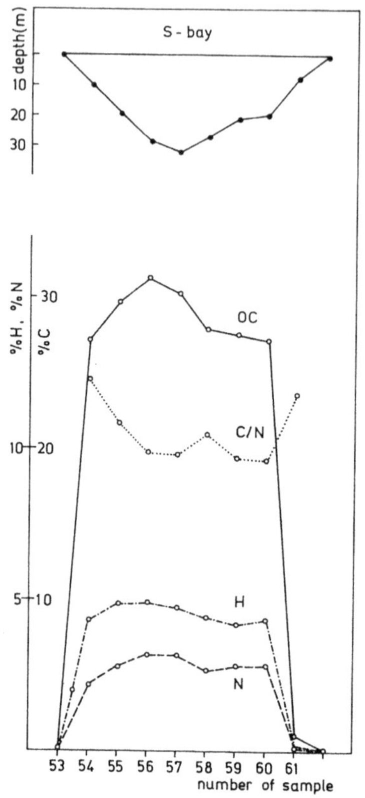

Figure 13.2 Lake Stechlin, S-basin, sediments from different depths (OC – organic carbon; H – hydrogen; N – nitrogen in per cent; C : N ratio: scale as N and H).

of the lake; with a reduction in water-depth there is a reduction in their concentration. A similar dependence of concentration on depth was found in Lake Nehmitz as well as other lakes and reservoirs. Each water-body examined seemed to have its own phosphorus: depth relationship (Fig. 13.5).

Evans & Rigler (1983) describe the phosphorus: depth relationship for Lake Castello with the equation $y = 0.1019 \times + 0.0558$, for the Red Chalk Lake with $y = 0.1464 \times - 0.1655$. According to this the relations in Lake Nehmitz coincide well with those of Lake Castello (Fig. 13.5).

The abundance of chemical elements in the profundal zone is not only the result of the accumulation of fine particles of slowly sedimenting sestons or sediment classification (Thomas et al. 1976), but also of chemical processes as is shown by examination of the carbonate content (Fig. 13.7) and the easily mobilized UO_2-ions in various waters. This phosphorus enriches itself independently of the trophic state of the water and the specific redox conditions in the hypolimnion in the deepest parts of the basin. In contrast the uranium distribution in sediment is dependent on the trophic state (Fig. 13.6). In the oligotrophic Lake Stechlin

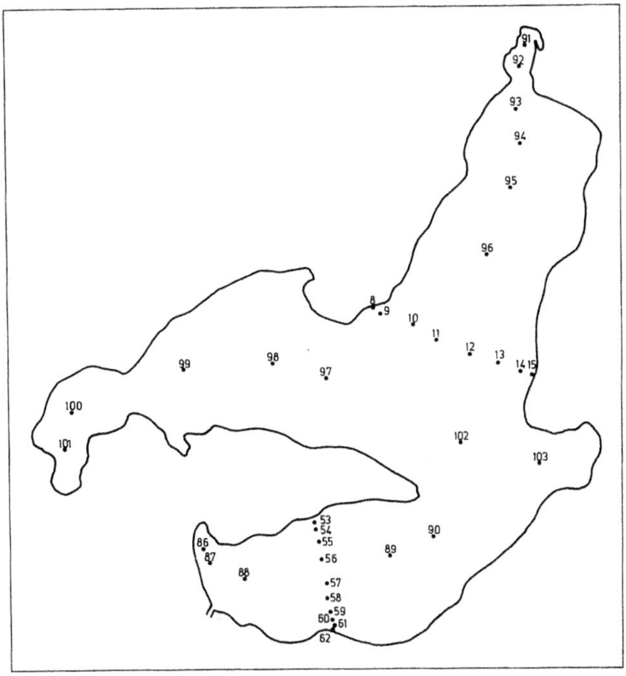

Figure 13.3 Sediment sample points at Lake Stechlin.

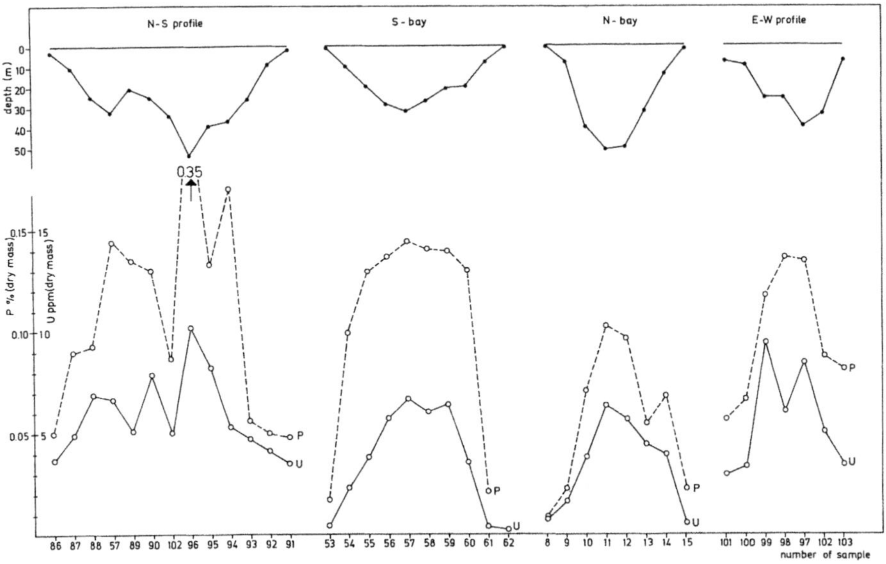

Figure 13.4 Lake Stechlin, uranium and phosphorus content in sediments from different depths.

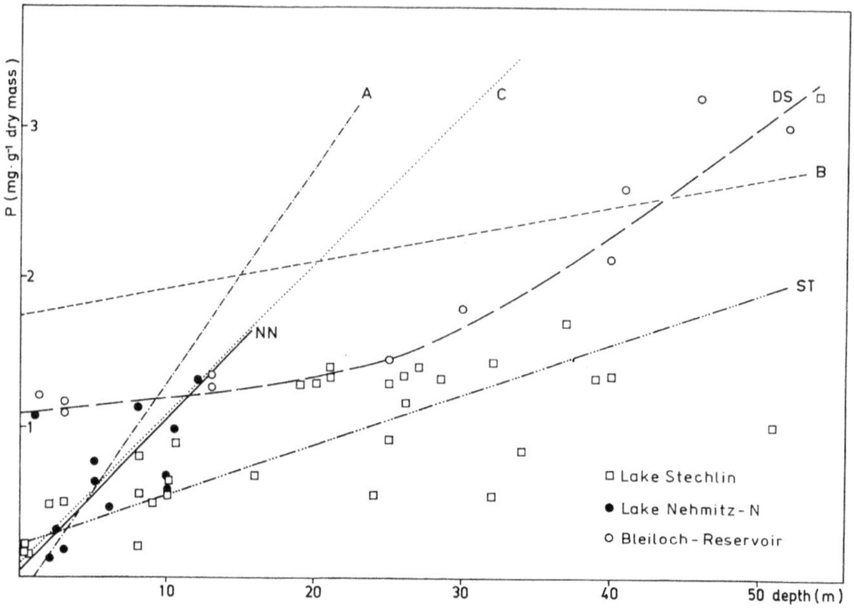

Figure 13.5 Correlation between phosphorus and depth in sediments (A – Red Chalk Lake, B – Bob Lake, C – Castello Lake, DS – Bleiloch Reservoir, NN – Lake Nehmitz-N, St – Lake Stechlin; A – C according to Evans & Rigler 1980, 1983).

uranium, like phosphorus, is more enriched in the deepest parts, whereas it is not so in the eutrophic Saale Valley Reservoir. In sediment from reservoirs it is even possible that inverse distribution patterns occur (Proft 1969).

On the basis of the redox conditions in the profundal zone sediment – interstitial water regime of the soft water lakes of the Schwarzwald Mountains, Tessenow (1975) developed a model for the enrichment of phosphorus, manganese and silicon, which should be applicable to Lake Stechlin to some degree. However, in addition to the redox conditions, a number of other factors have to be taken into account – such as retention of chemical elements, as well as biological-bacteriological degradation in the sediment itself.

Spatial distribution of carbonates in lake sediments. The lakes examined in our study have varying degrees of carbonate values in the profundal zone sediment. Thus the deep sediments in Lake Stechlin are calcium free (<0.1%; cf. p. 376). In the other hard water lakes the calcium-carbonate values fluctuate between 10% and 50% (Fig. 13.7). That the soft water acid bog-lakes have calcium-free sediments can only be mentioned in passing.

It is noticable that in all the lakes examined the sediments of the littoral-regions have been proven to have higher carbonate values than the profundal zone sediments (Fig. 13.7). Obviously, the distribution of the carbonate content is

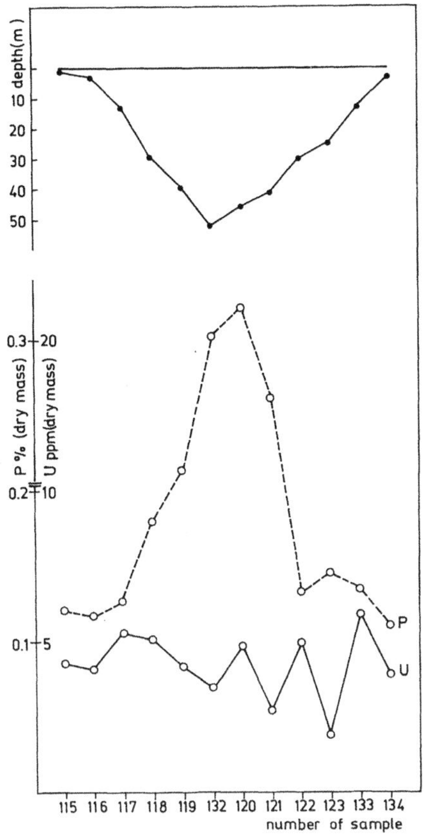

Figure 13.6 Bleiloch Reservoir, phosphorus and uranium in sediments of different depths.

dependent on the depth; the low calcium sediments are in the deepest parts.

An explanation for the decline in the carbonate content in the deepest parts of the lake basin can be found in the re-dissolving of the sedimenting and sedimented calcite (Lehmann 1983; Delorme 1983). With increasing depth the concentration of CO_2 grows (p. 104), and solution is mainly released by the biological decomposition of the organic detritus (p. 364). Thus the calcium content of the sediments is influenced by the size of the calcium sedimentation as well as the calcite solution potential of the profundal water and the interstitial water.

The pelagic calcite precipitation is bound with the primary production of the phytoplankton (pp. 412 ff.). Intensive calcite precipitation (whiting) has, so far, only been observed in mesotrophic lakes with a relatively high production (Minder 1923, 1926; Kelts & Hsü 1978; Strong & Eadie 1978; Koschel et al. 1983).

Examination of the sediment profiles from Lake Stechlin have shown that at certain deeper levels a noticeably higher carbonate content is found, as in upper levels, for instance, in 95–100 cm depth of sediment 29.8% $CaCO_3$ and in 270–

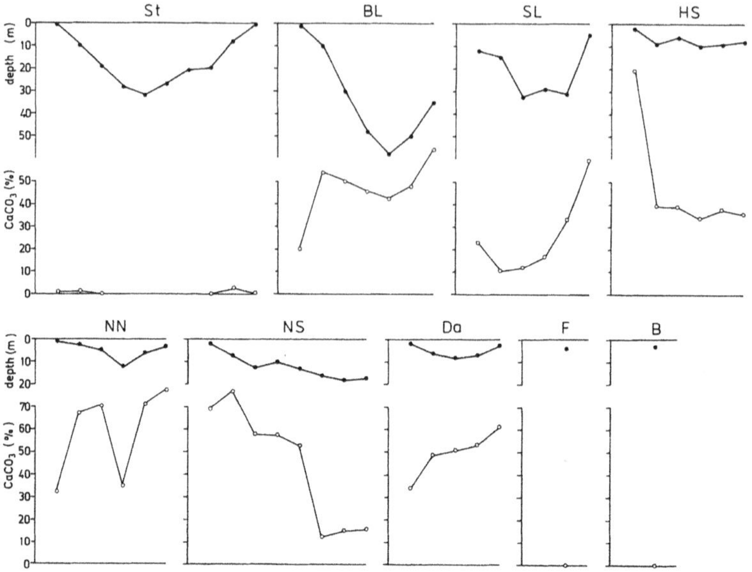

Figure 13.7 Calcite content of different depths (St – Lake Stechlin, BL – Lake Breiter Luzin, SL – Lake Schmaler Luzin, HS – Lake Haussee, NN – Lake Nehmitz-N, NS – Lake Nehmitz-S, Da – Lake Dagow, F – Lake Große Fuchskuhle, B – Lake Großer Barsch.

295 cm 50.3% $CaCO_3$ (Fig. 13.1). The conclusion might be made that in the development of Lake Stechlin there were phases with a higher productivity which led to a stronger calcite precipitation that was deposited as calcite in the profundal sediment. According to Müller (1979) $CaCO_3$ is an indicator of an anthropogenic determined eutrophic status. There are also signs that the change in carbonate content of sedimentary layers is connected with quaternary climatic changes, which could also be the cause of changing biotic interactions or higher production (Adelseck & Anderson 1978; Whiteside 1982).

The relations between pelagic calcite precipitation and the calcite content of sediments as researched here are not applicable to all standing waters. A particulate calcite introduced from the surrounding area can form a calcium sediment independently of the pelagic calcite precipitation.

C : N ratio in lake sediments

Introduction. In the classification of sediments from Baltic lakes Hansen (1959, 1962) used, among other factors, the relationship of carbon to nitrogen. In his research dy-sediments were shown to have a C : N ratio of >10, with gyttja-sediments having a ratio of <10. Dy-sediments are formed in dystrophic brown water lakes. Their relatively high carbon and humus content (v. Post 1862) arises from imports from its catchment area, and is, therefore, of allochthonous origin. The gyttja sediments arise in eutrophic lakes, mainly from the residue of the autochthonous production of the lakes.

In the profundal sediments of Stechlin C : N values of about and more than 10 were measured (Schönborn *et al.* 1965, 1966; Uhlmann & Albrecht 1968). These results caused discussion, since according to Hansen (1962) the sediment of an oligotrophic clear water lake such as Stechlin is not characterized as dy but as gyttja (Mothes 1965; Schräder 1965).

Lake Stechlin receives about 5.4 g C · m^{-2} as a result of the autumnal leaf-fall (see p. 407), according to Schönborn *et al.* (1965; see p. 217), which is the cause of the measured high rate of the C : N ratio. Beech leaves are seen as the source of the C-component which is so difficult to decompose in the profundal sediment. The lake is surrounded by a great many Beech (see p. 402). The neighbouring acid small bog-lakes (e.g. Kleiner and Großer Barsch, Große Fuchskuhle) have typical dy-sediments.

Latest research in other waters (Kemp & Mugrochova 1972;Premazzi & Ravera 1976; Flannery *et al.* 1982), analyses of the breakdown of nitrogen compounds in 'free' water and sediments (Sørensen 1978; Molongoski & Klug; Jones & Simon 1981; Jones *et al.* 1982; Blackburn & Hendriksen 1983), as well as a wealth of our own data from the lakes of the Mecklenburg Lake District led to a new understanding of the importance of the C : N ratio in lake sediments.

C : N ratio in sediments of lakes of the Mecklenburg Lake District. The lakes of this district, apart from the small bog-lakes, are hard-water lakes. Their sediment contains varying quantities of calcite (p. 382). In the analysis of the organic-bound carbon, therefore, the inorganic-bound carbon must be separately measured and deducted from the total carbon content. The results of the analysis show that the C : N ratio within a lake is not uniform, that the deepest profundal sediments show the lowest C : N values (Fig. 13.8).

The C : N ratio of the profundal sediments in hard-water lakes in the study area, on average, is between 8.5 and 12 and in bog-lakes, on average, between

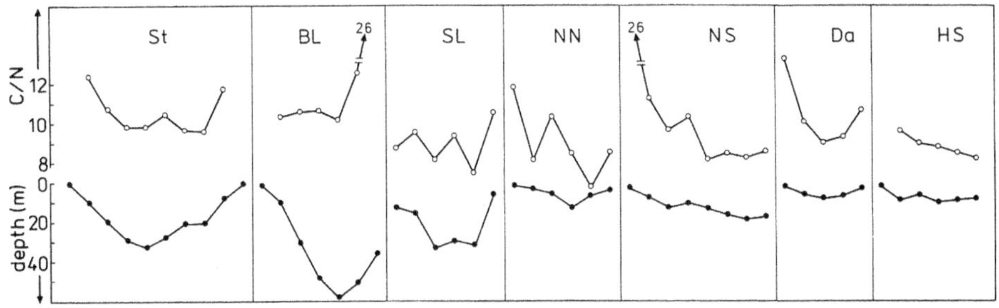

Figure 13.8 C : N ratio in sediments from different depths (symbols of lakes see Fig. 13.7).

13.5 and 15.3 (Tab. 13.1). The C : N ratio of the littoral sediments is higher than that of the profundal sediments and is mostly over 10 (10.6–26). The sediments with a higher C : N value were found in the oxygen-rich deep regions of the epilimnion.

Discussion. The lakes under examination in Mecklenburg can only be roughly categorized according to the C : N ratio of the profundal sediments. No linear correlation can be established with the trophic status of the water or the carbon content of the sediments as Flannery *et al.* (1982) found in the Florida lakes. In the eutrophic lakes we examined, the C : N ratio fluctuated between 8.5 and 9.5 independently of whether the water was strongly or weakly eutrophic ('eutrophic' according to Thienemann). A differentiation of the lakes according to the C : N ratio is not possible.

The deep oligotrophic lakes examined (Lakes Stechlin and Breiter Luzin; the Breiter Luzin has only become eutrophic in the last twenty years), have, with C : N ratios of 10 to 12, values above those of the eutrophic lakes, whereas the bog-lakes have even higher C : N values (about 15, see Table 13.1).

The C : N ratio of sediments is dependent on the combination of sedimenting materials and their decomposition. This follows directly with the sinking of the seston, and according to Jones *et al.* (1982) its C : N value increases with increasing depth from 8.1 to 10.7. In Lake Stechlin in the period of the summer stagnation in the sedimentation traps (Mothes 1981b) C : N ratios between 7.1 and 7.7 were found (Tab. 13.2). In winter and above all during the circulation, it is possible that through movement some of the disturbed sediment may have found its way into the sedimentation traps and led to false measurements of the C : N ratio.

The littoral and profundal sediments of oligotrophic lakes are often found to have high C : N values. According to Jones & Simon (1981) the nitrogen compound is reduced mainly to N_2 under aerobic condition. Gas bubbles rising from

Table 13.1 Average C : N ratios in lakes of the Mecklenburg Lake District.

Lake	C : N ratio
Haussee	8.8 ± 0.5
Dagow	9.5 ± 0.6
Schmaler Luzin	8.7 ± 0.9
Nehmitz-S	8.8 ± 0.9
Nehmitz-N	8.5
Breiter Luzin	10.9 ± 1.0
Stechlin	10.0 ± 0.5
Teufelssee	13.5
Fuchskuhle	15.0
Großer Barsch	15.3

Table 13.2 Comparison of C : N ratio-data according to Jones *et al.* (1982) and Mothes (1981).

Layer	C : N ratio Jones *et al.* (1982)	C : N ratio Mothes (1981)*	
Epilimnetic seston	8.1		
Sedimenting seston			
– below the thermocline	9.1	20 m	7.5 ± 1.4
		40 m	7.1 ± 1.4
– above the sediment	9.1	60 m	7.7 ± 2.55
Surface sediment	10.7		10.0 ± 0.5

* Lake Stechlin: carbon values corrected for carbonate-C.

the sediment contain 76% N and 15% CH_4. With a declining redox potential nitrogen compounds (e.g., in the hypolimnion of eutrophic lakes) are anaerobically reduced to ammoniac. In this case the gases released are 51% N and 46% CH_4. After that, under aerobic condition, a stronger degradation of the nitrogenous compounds follows, compared with the carbon compounds, which explains the higher C : N values in the littoral and profundal zones of oligotrophic lakes.

Züllig (1956) found even higher values in the Bodensee and alpine lakes. In these very deep lakes the sedimenting organic material is already heavily degraded due to oxidation, with C : N values of between 9.2 and 34 ($x = 20.4$). In the reservoirs of the Erzgebirge Mountains we also found extremely high C : N values in the sediments. These are soft-water lakes which received their water from a catchment area rich in acid rocks and covered with spruce forests (Tab. 13.3). Such high C : N values are within, or partially greater than those used by Hansen (1959, 1962), to characterize dy-sediments. The high C : N ratio of the dy-sediments is the result of the introduction of allochthonous carbon matter (humic acid, plant remnants such as fallen leaves) from the surrounding area. Sediments with a high C : N ratio also arise from the autochthonous production of a lake when the decomposition of the organic matter is caused by oxidation. Thus there are two causes of high C : N values in lake sediments which both operate in Lake Stechlin; by the leaf-fall which causes difficult degradable carbon material to be deposited in the sediment and the high oxygen-content of the total water-body creates the preconditions for an aerobic decomposition of the organic matter. It remains for future research to determine the importance of both causes quantitatively.

Table 13.3 C : N ratio of different temperate waters.

Lake*	Depth (m)	C : N ratio
Bodensee (Lake Constance)		
Konstanzer Trichter	55	19.5
near Utwil	150	17.7
Arbon	170	23.2
Mammern	29	15.1
Berlingen	40	25.1
Walensee	145	19.0
Pfäffikersee	34	34.2
Zellersee	14	13.7
Baldeggersee	66	26.9
Hallwilersee	47	28.3
Greifensee	32	23.5
Zugersee		
near Walchwil	197	9.2
near Oberwil	59	21.9
Gnadensee	17	14.3
Zürichsee		
near Meilen	120	18.1
near Oberrieden	140	15.1
Obere Neunzehnhainer Talsperre	27	139.0
	24	130.0
	8	247.0
Carlsfelder Talsperre	17	14.6
	10	17.4
	5	196.0
	3	219.0
Kranichsee Moor	0.5	24.9

* Lake Constance and other alpine lakes calculated according to Züllig (1956, p. 118, Table 1).

Sedimentation (G. Mothes)

Introduction

The present analyses are based on measurements of sedimentation in the period 1972–1975 (Mothes 1981a, b). In 1972 sedimenting masses and phosphorus were measured, and from 1974 analyses were extended to the elements C, N and H (the latter not dealt with here) and other measuring points added.

The measurements which extended over a whole year were carried out, as a rule, at intervals of 14 days with a total change of instruments used. Only at measuring point Swv (Lake Stechlin) were measurements carried out at intervals of 7 days. The sedimentation vessels (Fig. 13.9), which had been built on the basis of preparatory investigations lasting several years and following Ohle (1962),

Figure 13.9 Sedimentation vessels on the carrier. Photo W. Scheffler.

produced good results (Mothes 1983; cf. also Bloesch & Burns 1980; Blomquist & Håkanson 1981).

Sedimented masses during the summer stagnation period

The dry masses which sedimented from 1972–1975 each time during the vegetation period at the deepest point of Lake Stechlin (St) hardly differ at the 20-metres horizon (Table 13.4; see also the measured C values of 35 and 39 mg · $m^{-2} \cdot d^{-1}$). At the 40–60-metres horizon the sedimentation rate is about 1.7 times higher (with a maximum deviation of 20 per cent from the medium value at the 60-metres horizon in 1974). These results show that the supply of sedimenting material measurably increases in the depth region between 20 and 40 m, but does not continue to do with increasing depth. Sedimentation rates at depths of 40 and 60 m are largely identical.

The sedimentation of carbon and nitrogen corresponds to the one of the dry mass (at a C : N ratio – by atoms – of about 11).

In comparison, the measured values of sedimented phosphorus are different every year (Table 13.4). In 1974 a particularly large quantity of phosphorus sedimented at a depth of 40 and 20 m (at a lower C : P or N : P ratio), while at a depth of 60 m a relatively small amount of phosphorus was measured. The high medium P sedimentation rate at depths of 20 and 40 m results from 2 maxima with rates over 3 mg $P \cdot m^{-2} \cdot d^{-1}$, a 14-day maximum in April/May and a 14-(depth of 20 m) and 42-day (40 m) maximum from late July until early September (Fig. 13.10), the latter with a 28-day maximum value of 9 mg $P \cdot m^{-2} \cdot d^{-1}$. Except for another individual value in December 1972, sedimentation rates in Lake Stechlin lie below 2.5 mg $P \cdot m^{-2} \cdot d^{-1}$. Perhaps the high phosphorus content of the year 1974 could be explained by correspondingly high phosphorus inputs into the lake. They were in 1974 considerably higher than in 1975 (146 kg P from late June until mid-July from Lake Nehmitz, 46 kg P from Lake Dagow; cf. also Mothes 1981 b and Fig. 13.10).

A considerably larger amount of dry mass sediments in the west bay of Lake Stechlin (Swv; depth of 10 m) than at measuring point St. The material is poorer in N and P (with a higher C:N and C:P ratio, Table 13.4).

A larger amount of dry mass sediments in the north basin of Lake Nehmitz than in the south basin (Table 13.4). The sedimented mass in the south basin corresponds to the one in Lake Stechlin at measuring point St. In both lake basins

Table 13.4 Medium sedimentation rates* during the summer stagnation period in Lake Stechlin (measuring points St, Swv), Lake Nehmitz (Nn, Ns) and Lake Dagow (D) at different depths.

Period	No	Measuring point	Depth (m)	DM	C	N	P	C:N	C:P	N:P
19.4 –	14	St	20	181			0.40			
15.11.72	15	St	40	310			0.63			
	13	St	60	314			0.76			
4.4 –	14	St	20	166			0.44			
7.11.73	14	St	40	281			0.69			
	14	St	60	290			0.58			
10.4 –	15	St	20	167	35	3.85	0.81	11	112	11
6.11.74	15	St	40	295	58	6.21	2.08	11	72	7
	15	St	60	248	56	5.74	0.50	11	292	26
	30	Swv	10	548	111	10.20	0.92	13	310	24
9.4 –	15	St	20	186	39	4.01	0.43	11	235	21
5.11.75	15	St	40	324	61	5.77	0.60	12	263	21
	15	St	60	295	57	5.88	0.60	11	245	22
	30	Swv	10	662	128	10.87	0.93	14	354	26
3.4 –	16	Nn	10	379	84	9.03	0.84	11	257	24
13.11.74	16	Nn	15	643	143	17.15	1.69	10	219	22
	16	Ns	10	156	39	4.32	0.52	11	193	18
	16	Ns	15	228	54	6.23	0.96	10	145	14
24.4 –	14	D	5	490	146	19.34	3.64	9	104	12
6.11.74	14	D	7.5	474	145	21.59	5.94	8	63	8

No = number of exposures; DM = dry substance, C = carbon, N = nitrogen, P = phosphorus, all data in $mg \cdot m^{-2} \cdot d^{-1}$; C:N, C:P, N:P = atomic ratios in the sedimenting material.

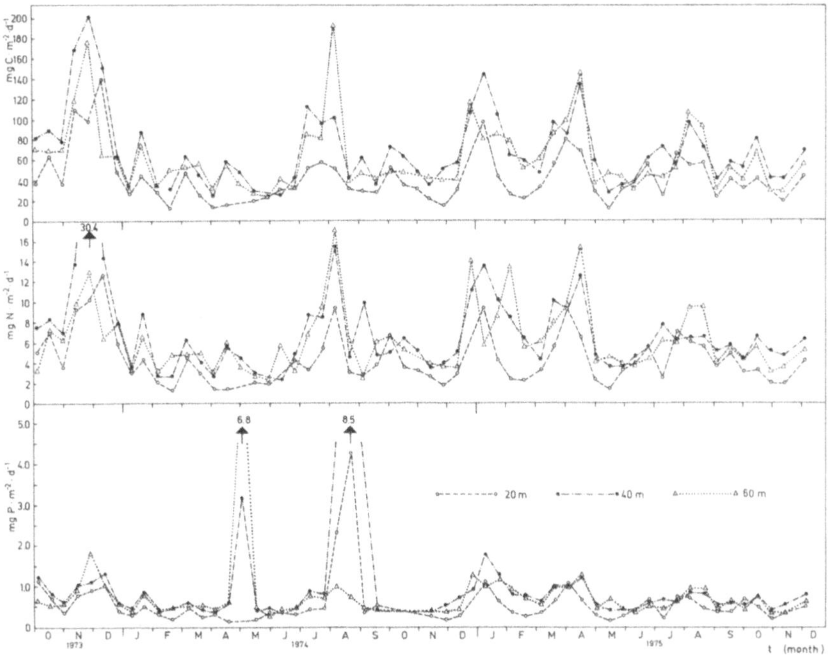

Figure 13.10 Variation of medium sedimentation rates of carbon, nitrogen, and phosphorus at Lake Stechlin, measuring point St (from Mothes 1981b).

material richer in phosphorus sediments at a depth of 15 m more than at 10 m. The material of the south basin is generally richer in phosphorus.

A relatively large amount of material, which is rich in nutrients, sediments in Lake Dagow (Table 13.4). The P content is clearly higher at a depth of 7.5 m over the 5 m horizon.

Generally we can say that in the eutrophic to mesotrophic lakes material richer in P sediments in deep regions more than in shallower horizons.

Sedimented masses in the winter half-year

In the winter half-years of 1972 and 1972–73 (Table 13.5) a larger amount of dry mass sedimented at measuring point St. In the corresponding periods of 1975, respectively 1974–75, however, about the same amount of dry mass sedimented as during the summer stagnation period. Sedimentation at the 20-metres horizon is again clearly lower than at 40 or 60 m. The composition of the material is relatively homogenous (C : N ratio about 12; P content of the dry mass about 0.2 per cent).

At measuring point Swv the sedimented dry mass corresponds to the one at the

Table 13.5 Medium sedimentation rates* from the beginning to the end of the spring full circulation in Lake Stechlin (measuring points St, Swv), Lake Nehmitz (Nn, Ns) and Lake Dagow (D) at different depths.

Period	No	Measuring point	Depth (m)	DM	C	N	P	C:N	C:P	N:P
22.3 –	1	St	20	396			0.77			
19.4.72	1	St	40	510			1.12			
	1	St	60	617			1.43			
15.11.72 –	8	St	20	436			1.02			
4.4.73	10	St	40	550			1.08			
	10	St	60	794			1.78			
7.11.73 –	11	St	20	279	55	5.33	0.51	12	279	23
10.4.74	11	St	40	426	84	8.92	0.69	11	313	29
	11	St	60	362	69	6.33	0.70	13	252	20
	14	Swv	10	382	86	7.02	0.55	14	400	28
6.11.74 –	10	St	20	193	42	4.32	0.51	11	214	19
9.4.75	11	St	40	365	78	7.82	0.86	12	235	20
	11	St	60	332	71	7.55	0.78	11	235	21
	11	Swv	10	307	60	5.45	0.50	13	314	24
5.11 –	2	St	20	155	34	3.37	0.40	12	218	18
17.12.75	2	St	40	267	59	5.63	0.66	12	232	19
	2	St	60	219	47	4.56	0.52	12	233	19
	2	Swv	10	241	49	3.91	0.49	15	259	18
31.10.73 –	6	Nn	10	460	113	10.98	0.94	12	313	26
3.4.74	4	Nn	15	611	165	14.95	1.01	13	422	33
	5	Ns	10	388	94	9.00	0.71	12	343	28
	3	Ns	15	474	119	10.84	0.92	13	334	26
13.11 –	3	Nn	10	159	37	3.82	0.30	11	315	28
30.12.74	3	Nn	15	238	50	5.87	0.55	10	238	24
	3	Ns	10	268	64	6.44	0.52	12	316	27
	3	Ns	15	281	70	7.28	0.57	11	319	28
6.11 –	3	D	5	36	7	0.98	0.13	8	141	17
8.12.74	3	D	7.5	36	7	1.33	0.13	6	136	22

* See Table 13.4.

40–60-metres horizon of measuring point St; however, the material is poorer in P and N (Table 13.5).

It is remarkable that in the winter half-year the sedimented dry masses in Lake Stechlin may vary from one year to another.

As for Lake Nehmitz, no significant differences in sedimentation between the two lake basins were found (Table 13.5). The sedimented materal is somewhat poorer in P than in Lake Stechlin (St).

In Lake Dagow a relatively small amount of material sediments in the winter half-year. It is poorer in nutrients than in the summer months (Table 13.5).

The oscillations of sedimentation

Given the chosen test conditions, the sedimentation process can be considered as a desegregation process of the seston which is randomly distributed in the water column over the sedimentation vessel. Obviously, there is no direct interaction between the measured seston concentration at the measuring horizon and the sedimentation result achieved in the funnel (Mothes 1977).

The elements C, N and P do not sediment to the same extent but rather differently (Fig. 13.10). Maxima of all the elements can be found simultaneously at all depths (St. Fig. 13.10; November–December 1973, January and April 1975). They were found exclusively in the winter half-year. On the other hand, there were high phosphorus values early in May 1974 at the 40– and 60-metres horizon and early September 1974 at a depth of 40 metres, i.e., at a time when N and C show only low sedimentation values.These facts (cf. Tables 13.4 and 13.5) can be explained as follows:

- The sedimentation maxima observed in the winter half-year are probably due to the wind-induced disturbance of sedimented detritus (De Groot 1981; Mothes 1982). In Lake Stechlin there was no winter stagnation period in the period under review (Richter *et al.* 1979; cf. p. 56).
- Moreover, in wind-exposed lakes such as Lake Stechlin, this disturbance of sedimented material is continuously taking place. The disturbed material, under the given morphometric conditions, is transported mainly to the depth horizon between 20 and 40 m. The increase in sedimentation rates at this horizon in Lake Stechlin as well as between 10 and 15 m in Lake Nehmitz may be explained in this way (funneling effect of organic sedimentation; Ohle 1962; cf. also p. 377).
- The seston of the epilimnion which is mostly exposed to high turbulence is mainly sedimented in terminal bays which are less turbulent: high sedimentation rate at measuring point Swv in contrast to St (20 m; Table 13.4). Transfers from the terminal bays with stormy weather seem to be possible (hardly any differences in sedimentation rates in Table 13.5). Any influence by the nuclear power plant on the high sedimentation rate at measuring point Swv must be excluded, since the values of 1974 (with the nuclear power plant operating) and 1975 (without operation) would then have to be very different (Fig. 13.11; Table 13.4).
- The phosphorus maxima may be so large that they will determine the medium sedimentation result during an entire period (Table 13.4, St 40 m 1974; Fig. 13.11). They are regularly found in mesotrophic and eutrophic lakes at greater depths when the phosphate dissolved in the anoxic hypolimnion is precipitated

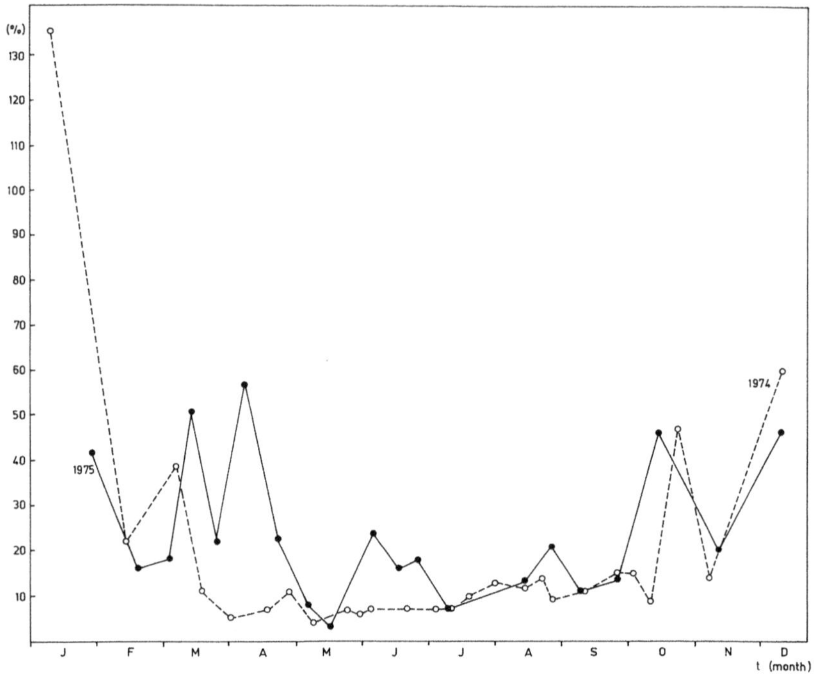

Figure 13.11 Percentage of medium carbon sedimentation rates (average of 14 days) at Lake Stechlin, measuring point St (from Mothes 1981b).

by oxidation (Mothes 1981b, Figs. 6 and 7). This is the reason for the relatively high P sedimentation at the deeper horizons of Lakes Nehmitz and Dagow (Table 13.4).

- Although the sedimentation rates measured in Lake Stechlin also include in their scope the maxima and minima of sedimentation in the other lakes concerned, it can generally be said that in an oligotrophic lake material sediments which, on average, is poorer in nutrients than in an eutrophic lake.

Phytoplankton and sedimentation

The phytoplankton is the main source of the seston in the pelagic zone. Thus it is not logical at first sight that there should be no significant connection between the appearance and disappearance of certain algae populations and the sedimentation rate concerned (Mothes 1981b). The decomposition rate of the seston is hardly algae-specific.

The following explanation could be given as regards the lack of correlation between algae development and sedimentation.

The organic substance produced by phytoplankton in the pelagic zone is partly used by the algae themselves, partly by other organisms (herbivorous or carnivorous zooplankters, bacteria) for producing their own substances. Part of the decomposition products becomes available again for the primary producers (short-circuited circulation). Only a small part of the primary product sediments (Burns & Rosa 1980; Reynolds *et al.* 1982).

The sedimentation rates determined in the measuring traps are composed of sedimenting particular remnants of primary and secondary productions and the disturbed, sedimented material. As a measuring horizon for determining the share of the short-circuited circulation, only the horizon immediately below the trophogenous zone and the summer stagnation period (minimization of the influence by disturbance) can thus be chosen. As a result, 80–90 per cent of the primary product remain in the short-circuited circulation (Table 13.6). During the summer stagnation period only some 10 per cent of the daily primary product sediments (Fig. 13.11).

Sediment balance

A sediment balance for Lake Stechlin as a whole cannot be exactly derived from the measuring values available. This is due to the following reasons:

- The sedimentation rates measured constitute the sum of the particular seston current of the body of water above the sedimentation vessel and of the disturbed mass stemming from other parts of the basin. Accordingly, the thickness of the sediment is less in the shallow regions than in the deeper ones.
- Sedimentation rates differ at various depth horizons and between the individual measuring points (Table 13.4; Fig. 13.10).
- The nutrient content of the sediment is dependent on depth (cf. p. 379). The P concentration is lower in shallow parts of the basin than in deeper ones.

A first balance of sedimentation, i.e., a rough calculation, on the basis of the data available, therefore, only is valid for the deepest point of Lake Stechlin. Since

Table 13.6 Comparison between annual primary production and sedimentation amount in the period under review and share of sedimentation ($g\ C \cdot m^{-2}$) in primary production ($g\ C \cdot m^{-2}$) in per cent. Locations as in Table 13.4.

Location	Depth (m)	Period of sedimentation	Production	Sedimentation	Share
St	20	7.3.73– 7.11.73	115	12	10
St	20	2.1.74– 2. 1.75	97	13	13
St	20	2.1.75–17.12.75	68	14	21
Nn	10	6.3.74–30.12.74	139	22	16
Ns	10	6.3.74–30.12.74	143	13	9
D	5	24.4.74–18.12.74	292	30	10

from among the nutrients C, N and P only phosphorus is conservative, i.e. is not in exchange with the atmosphere, this balance can only be established for phosphorus. The respective values for carbon and nitrogen result from the C : N : P ratios.

The sedimentation measurement is derived from the P input into the sediment. The P output necessary for the balance can only be calculated. The calculation is based on the P input into the summer hypolimnion and epilimnion and the P input into the whole pelagic zone during the circulation period, compensating for the P loss through sedimentation during the stagnation period.

In the hypolimnion of Lake Stechlin only an orthophosphate increase can be measured during the summer stagnation period (see p. 90). This increase, which also applies to the total phosphate, may be based both on a release from the sediment and on the decomposition of organic substance in the hypolimnion. The latter is calculated with the aid of the hypolimnetic orthophosphate balance. It is based on the following assumptions and measurement results (for details cf. Mothes 1981b).

Firstly it is assumed that the decomposition in the sedimentation vessel during exposition takes place to the same extent as it does in the hypolimnion (the release of P was measured in stagnant water situated above the sedimented material and calculated as mass $\cdot$ m^{-2}).

Secondly it is assumed that the C : P ratio of the sedimented material corresponds to the one decomposed in the hypolimnion.

Thirdly it is assumed that in the hypolimnion of Lake Stechlin which contains O_2, the O_2 deficit during the summer stagnation period corresponds, at least at the upper horizons, to the composition of organic substance (with a proportion of 2 C atoms for 3 O_2 molecules, valid for primary production).

From the second and third assumptions there results the phosphorus equivalent

$$\frac{1\,\mathrm{mg\ P}}{31} \mathrel{\hat{=}} \frac{1.5\,\mathrm{mg\ O_2}}{32} \cdot (\mathrm{C:P\ ratio})$$

The release of 1 mg P (with C : P = 200) is equivalent to a consumption of some 300 mg O_2 or, inversely, a deficit of 1 g O_2 to a release of 3 mg P.

To convert the area-related P release (first assumption) into a volume-related one, the measured summer O_2 deficit of the hypolimnion was divided, for all depth horizons of Lake Stechlin and 4 stagnation periods, by the O_2 consumption calculated from the third assumption and calculated for the same period by the measured P release. In all cases the quotient was about 1 rather than 0.5 or 2.

As a result, the mass calculated as mg P $\cdot$ m^{-2} according to the first assumption can be indicated in mg P $\cdot$ m^{-3} as an increase in concentration in terms of magnitude.

Figs. 13.12 and 13.13 compare the orthophosphate increase measured and calculated as above in the hypolimnion, taking Lakes Stechlin and Nehmitz as examples. Integrating all the years (including those which are not represented here) we get the following result.

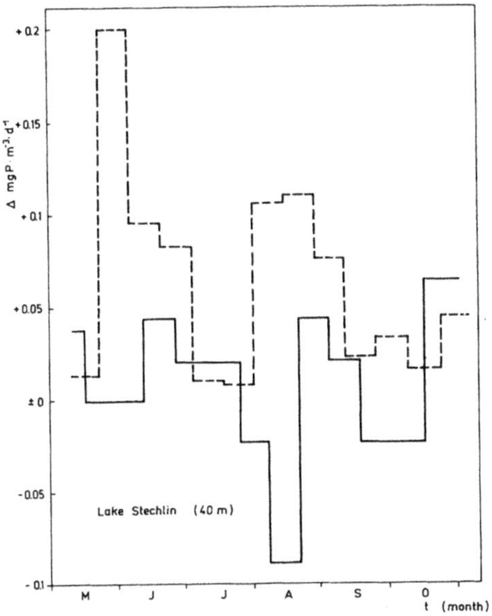

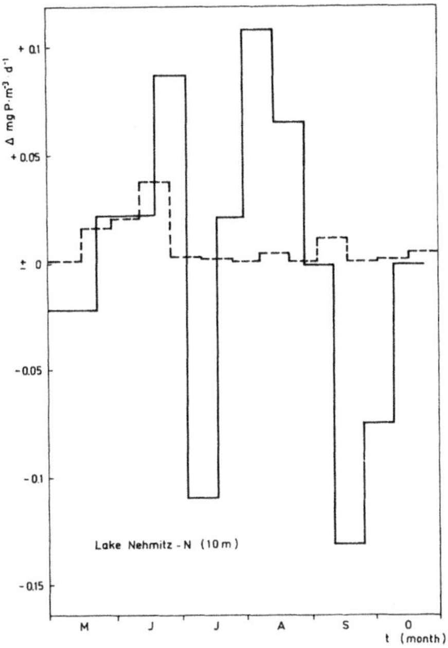

Figure 13.12 Comparison between the actual orthophosphate change with the one calculated from the orthophosphate release at selected measuring horizons (St 40 m; measuring points as in Table 13.4) of Lake Stechlin in 1974 (from Mothes 1981b).

Figure 13.13 Comparison between the actual orthophosphate change with the one calculated from the orthophosphate release at selected measuring horizons (Nn 10 m; measuring points as in Table 13.4) of Lake Nehmitz-N in 1974 (from Mothes 1981b).

In the hypolimnion containing O_2 (Lake Stechlin) the orthophosphate increase during the summer stagnation period can mostly be explained by the decomposition of sedimenting seston as a total supply of phosphorus from the euphotic zone. The increase calculated is temporarily higher than the measured orthophosphate consumption caused by living algae poor in P from the euphotic zone during the sedimentation process. A P outflow from the sediment was found in November and December 1974 at a depth of 60 m with values around 0.03 mg $P \cdot m^{-2} \cdot d^{-1}$. In the other years there was no indication of a phosphate release from the sediment during the summer stagnation period. A substantial P release from the profundal sediment of Lake Stechlin into the hypolimnion during the summer stagnation period is not likely to take place.

In the summer anoxic hypolimnion (Lakes Nehmitz and Dagow) the P release in the pelagic zone only makes a small contribution to the increased orthophosphate concentration. The larger part stems from the sediment.

In the mesotrophic and eutrophic lakes maximum release rates of 10 (Lake Nehmitz) to over 30 mg $P \cdot m^{-2} \cdot d^{-1}$ (Lake Dagow) were calculated for the deepest points. From this, about 1 (Lake Nehmitz) to 10 mg $P \cdot m^{-2} \cdot d^{-1}$ (Lake

Dagow) pass from the hypolimnion to the euphotic zone (Mothes 1981b; Osborne & Phillips 1978; Tanaka & Fukuhara 1981).

During the circulation period the P input from the sediment is calculated from the pelagic total phosphate increase plus the sedimentation and minus the import. With this a total phosphate balance can be established for the deepest points of the lakes.

Table 13.7 contains the calculation for Lake Stechlin in the years 1974 and 1975. Sedimentation at the 60-metres horizon (Tables 13.4, 13.5) is regarded as import, the calculated P output into the pelagic zone as export (see p. 100). Simplifying assumptions are disregard of seston currents near the sediment and consistence of the ratio C : N : P = 107 : 11 : 1 in the sediment.

In 1974 there was a positive balance of nutrients, in 1975 a negative one. In 1975, as in 1970, the pelagic total phosphate accumulation was particularly high until the beginning of the summer stagnation period. while from 1971–1974 the phosphorus output from the sediment was considerably lower.

If we assume that a year with a high total phosphate accumulation is followed by four years with a considerably lower one, the phosphorus retention of the sedimenting material in the sediment amounts to an average of over 30 per cent, the carbon retention, however, only to 12 per cent. The 30 per cent phosphorus retention equals about 100 per cent of the net phosphorus import into the lake.

Due to the composition of the raw sediment, an increase in the surface sediment, which still contains plenty of water, may be calculated at the deepest point of Lake Stechlin from the phosphorus remaining in the sediment. This leads to an amount of 0.2 mm $\cdot$ a^{-1}. At the transition of this surface sediment (average of the upper 160 mm) into the deeply situated permanent sediment a consolidation and reduction of the interstitial water share takes place. The real sediment increase was thus 0.1 mm for each year under review. This increase corresponds in its magnitude to the low thickness of the sediment in Lake Stechlin and the estimate by Vogt (1965) of less than 1 mm a year.

Table 13.7 Sediment balance for carbon (C), nitrogen (N) and phosphorus (P) and atomic relations of material currents in per cent (data of currents in mg $\cdot$ m^{-2} $\cdot$ d^{-1}) in Lake Stechlin.

	C	N	P	C:P	N:P	C:N
Sedimentation 1974	54.6	5.5	0.5	277	24	12
Release 1974	44.9	4.3	0.3	419	34	12
% release of sedimentation	82	79	54			
Remainder in the sediment 1974	9.7	1.2	0.2	107	11	10
Sedimentation 1975	61.3	6.4	0.7	237	21	11
Release 1975	69.6	7.4	0.9	207	19	11
% release of sedimentation	114	115	130			
Remainder in the sediment 1975	− 8.3	− 1.0	− 0.2	107	11	10
% release of sedimentation 1971–75	88	86	69			
Remainder in the sediment 1971–75	6.1	0.8	0.12			

From calculations of the year 1974 there result similar values for the sediment increase for Lake Nehmitz. In the north basin there are 0.5–0.3 mm · a^{-1}, in the south basin 0.3–0.2 mm · a^{-1}. Despite the short period of investigation and the mistakes possible due to this, the magnitude might well be correct.

The 1974 balance shows no sediment increase in Lake Dagow, but a sediment decrease of 20 mm · a^{-1}. Such a result which seems to be irrational may occur if a lake, after intense eutrophication, changes into an oligotrophizing phase, e.g., due to a redevelopment of the area. Then the decomposable carbon which did not oxidize in the anaerobic milieu will oxidize, and more phosphorus is absorbed in the sediment. There will be a decrease in the C : P ratio in the sediment, and thus a decrease of the sediment layer deposited. We consider this likely to happen in Lake Dagow (see p. 395). Whether the actual size of the sediment decrease corresponds to the value calculated is not yet known.

Calculations of the 'internal phosphorus balance' (Table 13.7) lead to the following conclusions.

In the oligotrophic Lake Stechlin the lake-internal P flows into the euphotic zone are twice the imports stemming from outside (Table 4.7). Nevertheless, these relatively small imports cause high primary production performances. In 1975, e.g., a total supply of some 150 mg P · m^{-2} · a^{-1} into the euphotic zone resulted in a primary production of 60 g C · m^{-2} · a^{-1}. Another import from the catchment area of 47 mg P · m^{-2} · a^{-1} in the summer 1974 led almost to a doubling of the primary production.

Eutrophic lakes such as Lake Dagow release plenty of phosphorus in the sediment and accumulate carbon. When the catchment area of a eutrophized body of water is being redeveloped, the expected oligotrophication will become visible only when the decomposable carbon has largely oxidized. A number of measures for sediment treatment are based on this fact (Klapper 1980).

References

Adelseck, Ch. G. Jr. & Anderson, Th. F., 1978. The late Pleistocene record of productivity fluctuations in the eastern equatorial Pacific Ocean. Geology 6: 388–391.

Blackburn, T.H. & Henriksen, K., 1983. Nitrogen cycling in different types of sediments from Danish waters. Limnol. Oceanogr. 28: 477–493.

Bloesch, J. & Burns, H.M., 1980. A critical review of sedimentation trap technique. Schweiz. Z. Hydrol. 42: 15–55.

Blomquist, S. & Håkanson, L., 1981. A review on sediment traps in aquatic environments. Arch. Hydrobiol. 91: 101–132.

Burns, N.M. & Rosa, F., 1980. In situ measurement of the settling velocity of organic carbon particles and 10 species of phytoplankton. Limnol. Oceanogr. 25: 855–864.

Davis, B.M.D. & Ford, M.S., 1982. Sediment focusing in Mirror Lake, New Hampshire. Limnol. Oceanogr. 27: 137–150.

De Groot, W.T., 1981. Phosphate and wind in a shallow lake. Arch. Hydrobiol. 91: 475–489.

De March, L., 1978. Permanent sedimentation of nitrogen, phosphorus and organic carbon in a high arctic lake. J. Fish. Res. Board Can. 35: 1089–1094.

Delorme, L.D., 1982. Lake Erie oxygen; the prehistoric record. Can. J. Fish. Aquat. Sci. 39(7): 1021–1029.

Evans, R.D. & Rigler, F.H., 1980. Measurement of whole lake sediment accumulation and phosphorus retention using lead-210 dating. Can. J. Fish. Aquat. Sci. 37: 817–822.

Evans, R.D. & Rigler, F.H., 1983. A test of lead-210 dating for the measurement of whole lake soft sediment accumulation. Can. J. Fish. Aquat. Sci. 40: 506–515.

Flannery, M.S., Snodgrass, R.D. & Whitmore, T.J., 1982. Deepwater sediments and trophic conditions in Florida lakes. Hydrobiologia 92: 597–602.

Håkanson, L., 1982. Bottom dynamics in lakes. Hydrobiologia 91: 9–22.

Hansen, K., 1959. The terms Gyttja and Dy. Hydrobiologia 13: 309–315.

Hansen, K., 1962. The dystrophic lake type. Hydrobiologia 19: 183–191.

Hargrave, B.T. & Kamp Nielsen, L., 1976. Accumulation of sedimentary organic matter at the base of steep bottom gradients. In: Interactions between sediments and fresh water. Proceedings of an international symposium held at Amsterdam, the Netherlands, September 6–10, 1976 (ed.: H.L. Golterman). The Hague: Dr. W. Junk Publishers. 168–173.

Jones, J.G. & Simon, B.M., 1981. Differences in microbial decomposition processes in profundal and littoral lake sediments. J. General Microbiol. 123: 297–312.

Jones, J.G., Simon, B.M. & Horsley, R.W., 1982. Microbiological sources of ammonia in freshwater lake sediments. J. General Microbiol. 128: 2823–2831.

Kamp Nielsen, L. & Hargrave, B.T., 1978. Influence of bathymetry on sediment focusing in Lake Esrom. Verh. Int. Verein. Limnol. 20: 714–719.

Kelts, K. & Hsü, K.J., 1978. Freshwater carbonate sedimentation. In: Lakes Chemistry, Geology, Physics (ed. A. Lerman). New York-Heidelberg-Berlin: Springer-Verlag. 295–323.

Kemp, A.L.W. & Mudrochowa, A., 1975. Nitrogen in sedimented organic matter from Lake Ontario. In: Humic substances. Their structure and function in the biosphere. Proceedings of an international meeting held at Nieuwersluis, the Netherlands, May 29–31, 1972 (ed.: D. Povoledo & H.L. Golterman). Wageningen: Centre for Agriculture Publishing Documentation. 137–157.

Kimmel, B.L., 1978. An evaluation of recent sediment focusing in Castle Lake (California) using a volcanic ash layer as a stratigraphic marker. Verh. Int. Verein. Limnol. 20: 393–400.

Klapper, H., 1980. Wassergütewirtschaft. 5. Bewirtschaftung von Standgewässer-Ökosystemen. Magdeburg: Ingenieurschule für Wasserwirtschaft.

Koschel, R., Benndorf, J., Proft, G. & Recknagel, F., 1983. Calcite precipitation as a natural control mechanism of eutrophication. Arch. Hydrobiol. 98: 380–408.

Lastein, E., 1976. Recent sedimentation and resuspension of organic matter in eutrophic Lake Esrom, Denmark. Oikos 27: 44–49.

Lehmann, R., 1983. Untersuchungen zur Sedimentation in einem oligotrophen Alpensee (Königsee) während der sommerlichen Schichtung. Arch. Hydrobiol. 96(4): 486–495.

Likens, G.E. & Davis, M.B., 1975. Post-glacial history of Mirror Lake and its watershed in New Hampshire, USA: an initial report. Verh. Int. Verein. Limnol. 19: 982–993.

Minder, L., 1923. Die biogene Entkalkung im Zürichsee. Verh. Int. Verein. Limnol. 1: 20–32.

Minder, L., 1926. Biologisch-chemische Untersuchungen im Zürichsee. Z. Hydrol. 3: 1–69.

Möller, L., 1935. Stechlinsee und Sakrower See. Ein Beitrag zur Charakteristik eutropher und oligotropher Seetypen. Arch. Hydrobiol. 29: 137–156.

Molongoski, J.J. & Klug, M.J., 1980. Quantification and characterization of sedimenting particulate organic matter in a shallow hypereutrophic lake. Freshwater Biol. 10: 497–506.

Mothes, G., 1965. Ein Beitrag zur Sediment-Charakterisierung des Stechlinsees und zur biologischen Typisierung von Seen. Limnologica 3: 381–388.

Mothes, G., 1977. Sedimentationsvorgänge im Stechlinsee-Gebiet. Acta hydrochim. hydrobiol. 5: 269–281.

Mothes, G., 1980. Die Nährstoffverteilung in Sedimenten norddeutscher Seen und Hinweise auf den Austausch Sediment – Pelagial. Acta hydrochim. hydrobiol. 8: 35–46.

Mothes, G., 1981a. Physikalische und chemische Parameter der Wasserbeschaffenheit in Seen des Stechlinseegebietes. Limnologica 13: 1–53.
Mothes, G., 1981b. Sedimentation und Stoffbilanzen in Seen des Stechlinseegebietes. Limnologica 13: 147–194.
Mothes, G., 1982. Sauerstoff- und Phosphorgehalte im Müggelsee als Indikatoren für Austauschprozesse. Acta Hydrophys. 27: 347–356.
Mothes, G., 1983. Sedimentation und Sedimentanalytik im Müggelsee. Acta hydrochim. hydrobiol. 11: 75–99.
Müller, G., 1979. Eutrophication changes sedimentation in part of Lake Constance. Naturwissenschaften 66: 261–262.
Ohle, W., 1962. Der Stoffhaushalt der Seen als Grundlage einer allgemeinen Stoffwechseldynamik der Gewässer. Kieler Meeresforschung 18: 107–120.
Osborne, P.L. & Phillips, G.L., 1978. Evidence for nutrient release of two shallow and productive lakes. Verh. Int. Verein. Limnol. 20: 654–658.
Post, H. v., 1862. Studier öfver nutidans koprogena jordbildningar, gyttja, torf, mylla. Kgl. SV. Vetensk. Akad. Handl. 4 (cit. according to Hansen 1959).
Premazzi, G. & Ravera, O., 1976. Chemical characteristics of Lake Lugano sediments. In: Interactions between sediments and fresh water. Amsterdam 1976 (ed.: H.L. Golterman). The Hague: Dr. W. Junk b. v. Publishers. 121–124.
Proft, G., 1967. Beiträge zur Sedimentchemie von Binnengewässern. I. Die Verteilung des Phosphors in Sedimenten von Seen und Talsperren. Limnologica 5: 397–404.
Proft, G., 1969. Chemische Schichtungen einiger Elemente im Sediment norddeutscher Seen. Mitt. Int. Verein. Limnol. 17: 252–255.
Proft, G., 1984. Die pelagische Calcitfällung und der Carbonatgehalt von Sedimenten pleistozäner Seen. Acta hydrochim. hydrobiol. 12: 319–324.
Reynolds, C.S., Morison, H.R. & Butterwick, C., 1982. The sedimentary flux of phytoplankton in the south basin of Windermere. Limnol. Oceanogr. 27: 1162–1175.
Schönborn, W., Flößner, D. & Proft, G., 1965. Die limnologische Charakterisierung des Profundals einiger norddeutscher Seen mit Hilfe von Testaceen-Gemeinschaften. Limnologica 3: 371–380.
Schönborn, W., Flößner, D. & Proft, G., 1966. Die limnologische Charakterisierung des Profundals einiger norddeutscher und masurischer Seen mit Hilfe von Testaceen-Gemeinschaften. Verh. Int. Verein. Limnol. 16: 251–257.
Schräder, Th., 1965. Das Sediment des Stechlinsees. Limnologica 3: 367–369.
Sørensen, J., 1978. Capacity for denitrification and reduction of nitrate to ammonia in a coastal marine sediment. Appl. Environm. Microbiol. 35: 301–305.
Strong, A.E. & Eadie, B.J., 1978. Satellite observations of calcium carbonate precipitation in the Great Lakes. Limnol. Oceanogr. 23: 877–887.
Tanaka, T. & Fukuhara, H., 1981. Behaviour of nutrients at the sediment-water interface in a eutrophic lake: laboratory and field experiments. Verh. Int. Verein. Limnol. 21: 609–610.
Tessenow, V., 1975. Akkumulationsprozesse in der Maximaltiefe von Seen durch postsedimentäre Konzentrationswanderung. Verh. Int. Verein. Limnol. 19: 1251–1262.
Thomas, R.L., Jaquet, J.M., Kemp, A.L.W. & Lewis, C.F.M., 1976. Surficial sediments of Lake Erie. J. Fish. Res. Board Canada 33: 385–403.
Uhlmann, D. & Albrecht, E., 1968. Biogeochemische Faktoren der Eutrophierung von Trinkwassertalsperren. Limnologica 6: 225–245.
Vogt, P., 1965. Aktuogeologische Untersuchungen am Großen Stechlinsee. Limnologica 3: 97–138.
White, W.S. & Wetzel, R.G., 1975. Nitrogen, phosphorus, particulate and colloidal carbon content of sedimenting seston of a hard-water lake. Verh. Int. Verein. Limnol. 19: 330–339.
Whiteside, M.C., 1983. The mythical concept of eutrophication. Hydrobiologia 103: 107–111.
Züllig, H., 1956. Sedimente als Ausdruck des Zustandes eines Gewässers. Schweiz. Z. Hydrol. 18: 5–143.

14. Fallen leaves in Lake Stechlin

S.J. Casper, L. Krey & G. Proft

Introduction

In larger eutrophic lakes autochthonous (i.e., dissolved and particulate organic substances produced within the water-body) are the main sources of energy. But in smaller eutrophic lakes, for instance in those of the 'Waldhumussee'-type (Utermöhl 1974), or in oligotrophic lakes, allochthonous organic matter can play an important ecological role difficult to estimate accurately (Ohle 1972; Casith & Hasler 1976; Overbeck 1982). Its main source is autumn-shed leaves (Kaushik & Hynes 1968; Petersen & Cummins 1974).

Within the extensive woodland of the Lake Stechlin area fine litter such as fallen leaves, twigs, barks, pollen, fruits, etc. and coarse wood litter from trees around the lakes strongly influence the organic nutrient cycle. Lake Stechlin does not belong to the eutrophic 'Waldhumussee'-type, although it looks like Lake Plußsee and although 96 per cent of its shore line is settled by trees. Its characteristic morphometry, oxygen budget and bioproduction, for instance, characterize Lake Stechlin as a water-body of the oligotrophic type (p. 463). But Schönborn *et al.* (1965), partly following former investigations of Möller (1935), showed that the profundal sediments are rich in humic substances, that the organic C/N-ratio is about 10, and that the micro- and meiozoobenthonic population density in the profundal zone is strikingly low. According to Hansen (1959, 1962) they postulated a dy-like profundal mud as a result of fallen leaves deposited on the bottom of the profundal region (p. 217). Uhlmann & Albrecht (1968) confirmed the high carbon content of the sediment and its interpretation according to the Hansen-thesis but Mothes (1965) rejected the conclusions.

In view of these contradictory view points it is regrettable that it has not been possible to carry out intensive research into leaf-fall, especially of the Beech-tree, into Lake Stechlin. The various experiments with catching cages and decomposition cages between 1962 and 1964 (Casper & Proft, unpublished) were the victim of deliberate destruction. Only figures on the occurrence and distribution of certain species of Hyphomycetes could be established (Casper 1965). Thus in this

Casper, S. Jost (ed.), Lake Stechlin. A temperate oligotrophic lake. ISBN 978-94-010-8930-2.

study only the production and decomposition of leaf litter are calculated and their effects on the lake ecosystem are cautiously discussed.

Estimations of production rates of litter in temperate lakes are given by Szczepanski (1965), Jordan & Likens (1975), Kubota & Shiraishi (1975) and Gasith & Hasler (1976). Data on the promotion of decay of autumn-shed leaves by associated microorganisms (bacteria, protozoa, fungi) are relatively numerous (Minshall 1967; Kaushik & Hynes 1971; Fisher & Likens 1972; Iversen 1973; Bärlocher & Kendrick 1974; Cummins 1974; Iversen & Madsen 1977). Information on the physical and chemical changes in litter quality during the process of decomposition are also available (Warnke 1968; Pieczynska 1972; Kaushik & Hynes 1971; Howarth & Fisher 1976; Rau 1976, 1978; Reed 1979; Dahm 1981; Hanlon 1981, 1982; Tam *et al.* 1981; Bengtsson 1982, 1983; Binder & Naiman 1983).

Vegetation surrounding Lake Stechlin

Almost all the shores of Lake Stechlin (~96 per cent) are occupied by trees. The total number of trees within a shore-hinterland of about 5 m is about 2900. 84 per cent are deciduous trees (*Fagus sylvatica* ~54 per cent, *Alnus glutinosa, Betula pendula, Sorbus aucuparia, Quercus petraea, Salix alba*) and 16 per cent are conifers, mainly *Pinus sylvestris*. Many of their crowns (~46 per cent) hang over the water (Fig. 14.1). The leaf-fall period begins at the end of September and ends in the middle of November although it differs slightly by species and individuals. *Quercus* often has its foliage still in February, but more than 90 per cent of the leaves fall during autumn. The peak of leaf-fall is in the period of the early frosts.

Methods

Cages of the size of 100 × 100 × 25 cm were set on the waterside by the east shore near the limnological laboratory in a continuous row. Usually the fallen leaves or needles in the cages were collected once a week during the whole leaf-fall periods from 1981–1983. The leaves were dried in the air, subsequently in an oven at 105° C, and weighed.

Some of the samples were pulverised and analysed elementarily into C, N, and H as well as being examined by the molybdat-vanadat-method for P.

To determine the soluble part of C and P a definite quantity of fallen leaves (10 g dry matter of Beech-leaves) were immersed in water from Lake Stechlin and the C and P content of the water were determined after ever longer standing periods (0–480 hours; see Table 14.4).

Figure 14.1 Lake Stechlin: crowns of trees hanging over the water. (Photo W. Scheffler)

Fig. 14.2 Lake Stechlin: tree fallen into the lake. (Photo W. Scheffler)

Results and discussion

The amount of fallen leaves

Table 14.1 shows the amount of fallen leaves gathered in the cages. The amount of the leaves gathered differs greatly each year (± 20–50%). Under natural conditions the rate of leaves falling into the lake decreases with increasing distance from the shore depending on the prevalent wind speed during the leaf-fall period. Assuming that fallen leaves are found up to 30 m away from the shore and that the density decreases according to an *e*-function (Szczepanski 1965; Kubota & Shiraishi 1975), we can calculate the total amount Q of fallen leaves by

$$Q = f\frac{a}{b}l \tag{1}$$

where *a* is the average total amount of the fallen leaves per one gathering cage through the leaf-fall period; *f*, the ratio of the amount of total fallen leaves in the lake to that within 1 m from the shore (taken from Kubota & Shiraishi 1975; f = 5.4); *b*, the area of the cage (m^2); *l*, the length of the wooded shore (m; Lake Stechlin 15, 450 m).

The calculated total amount of fallen leaves is shown in Table 14.2. The total amount of fallen leaves gathered per m shore line of Lake Stechlin with 330 g DM

Table 14.1 Amount of fallen leaves (in g dry weight) in cages 1981–1983, Lake Stechlin.

Year	1981	1982	1983
Total amount of dried leaves ($g \cdot 5\,m^{-2}$)	2180	1614	1121
Average weight in one cage ($g \cdot m^{-2}$)	436	323	224

Table 14.2 Calculated total amount Q (t) of fallen leaves; Lake Stechlin 1981–1983.

Year	1981	1982	1983
f	5.4	5.4	5.4
a_{dec} (g)[1]	431.64	319.11	221.6
a_{con} (g)[2]	4.36	3.23	2.24
b (m^2)	1	1	1
l_{dec} (m)[3]	12 978	12 978	12 978
l_{con} (m)[4]	2 472	2 472	2 472
Q (t)	30.3	22.4	15.6

[1] a_{dec} = proportion of deciduous leaves in the cage.
[2] a_{con} = proportion of coniferous needles in the cage.
[3] l_{dec} = shore line settled by deciduous trees.
[4] l_{con} = shore line settled by coniferous trees.

($\bar{x}$ 1981–1983) corresponds well with values of the estimation of the leaflitter of poor Beech woods of the Solling (Heller 1971; 300–400 g DM · m^{-2}).

Comparisons between the total amount of fallen leaves and airborne plant litterfall respectively per m shore line of Lake Stechlin and that of other lakes (Table 14.3) show that its values are the highest known. The high values are favoured by the trees growing immediately on the shore line, by the high proportion of deciduous trees, by the lack of dense belts of reed or other water-plants, and by the high wind speeds during the autumnal leaf-fall period.

Breakdown and decomposition of fallen leaves

The decay of the fallen leaves may be weight loss resulting from environmental factors such as wave action, from fragmentation by animal (for instance by *Gammarus*) feeding, and from microbiological activity (breakdown) or decay of the plant material into its constituent elements by microbiological activity and digestion by animals (decomposition). Only a few leaves are carried by the wind far away from the shore and sink undestroyed into the plant-covered littoral zone or the deep profundal zone. Some of them are carried out of the lake, passing with the cooling water along the canal between Lakes Stechlin and Nehmitz. Small quantities of leaves, difficult to estimate, are carried into little shallow bays (Feenbucht) protected from wind, where they collect in thick submerged layers.

Table 14.3 Comparison of airborne plant litterfall and phytoplankton productivity for lakes of different sizes (Lake Mikolajka: Szczepanski, 1965; Pieczynska, 1972b; Lake Wingra: Gasith & Hasler, 1976; Mirror Lake: Jordan & Likens, 1975; Lake Yunoko: Kubota & Shiraishi, 1975; Llyn Frongoch: Hanlon 1982).

Lake	Area (ha)	Shore-line length (km)	Afforestation (%)	Phyto-plankton productivity (gC · m^{-2} · yr^{-1})		Airborne litter (gC · m^{-2} · yr^{-1})		Airborne litter (gC · m^{-1} shore line)
Lake Mikolajka	459	14.4	52	330		0.36		225
Lake Wingra	131	5.0	75	400		0.916		320
Llyn Frongoch	7.2	1.45	23	15		0.625		135
Mirror Lake	15	2.0	90	78		4.3		354
Lake Yunoko	35.3	3.8	90.5	136		4.674*		433*
Lake Stechlin	425	16.1	96	95	1981:	3.4*		903*
					1982:	2.5*		668*
					1983:	1.8*		465*
					$\bar{x}$:	2.6*	$\bar{x}$:	680*

* Only fallen leaves.

Most of the fallen leaves are near the shore on the windward sides, driven by the wind-formed surface currents. Here in the zone of high wave action the leaves are reduced to small pieces and form layers of dark brown and black masses often covered by sand or are carried away into the littoral zone of submerged wintergreen plants or into the profundal zone. It is not possible to estimate accurately the proportion of the different ways in which the fallen leaves have been dispersed into the lake.

A rapid leaching of water soluble, low molecular weight substances (e. g. sugars, amino acids) has been observed from leaves (Nykvist 1963; Wetzel & Manny 1972).

The water-soluble parts of the fallen leaves are dissolved within a few hours of water contact. The easily soluble C-content is small and is quickly decomposed. The decomposition depends on many factors and can be biotic or abiotic (Dahm 1981). Freshly fallen Beech leaves from Lake Stechlin yielded 2–9% (p. 364).

P is also released from fallen leaves (Table 14.4). Within a few hours 20 μg P per g of Beech-tree leaves are dissolved; that is about 5 per cent of the total amount of P of the fallen leaves (fallen leaves 0.038 per cent).

The further decomposition of the fallen leaves is a very slow process. In the 'wind-free' bays leaf-mass remains for years before the leaves are biologically completely reduced. The early stages of leaf decay are dominated by Hyphomycetes producing extracellular enzymes for decomposition (Casper 1965; Kaushik & Hynes 1968, 1971; Suberkropp & Klug 1976; Bengtsson 1982, 1983). The mycelium on and in the leaves is grazed by animals, for instance by leaf-dy indicators such as *Arcella gibbosa, Nebela collaris,* etc. (see p. 217), which probably select mycelium – or bacterium – rich leaves due to their higher nutritional value.

For the decomposition due to bacteria see p. 364.

Own figures are not available. Experiments to determine leaf decomposition with the aid of cages made from net-cloth produced K-values of between $0.01 \cdot d^{-1}$

Table 14.4 P released from fallen leaves (Beech) submersed in water from Lake Stechlin.

Sample	Time of submersion (h)	P-content ($mg \cdot m^{-3}$)
Water *without* fallen leaves	–	6.1
Water *with* submerged fallen leaves	1	24.1
	2	26.1
	24	18.8
	48	9.1
	144	4.0
	480	4.6

and $0.002 \cdot d^{-1}$ (Petersen & Cummins 1974; Reed 1979; Meyer 1980; Hanlon 1982; Binder & Kaiman 1983). The K-values are dependent on the water currents, the mesh-size of the cages and the organism settlements which can block the cage-nets within a short time.

The amount of fallen leaves in the nutrient cycle of Lake Stechlin

Most of the fallen leaves are from Beech-trees (~54 per cent). Their breakdown and decomposition last relatively long, depending upon the biotic and abiotic environmental conditions. They are more persistent than Alder leaves and their fungal flora is poorer (Casper 1965; Bengtsson 1983).

The elementary analysis of individual mixed tests of fallen leaves as well as selective tests of individual leaftypes produced results showing 48% carbon, 1–2% nitrogen and 6% hydrogen, which gives a C : N : H ratio of 40 : 1 : 5.

A once-off determination of the P-content resulted in 0.038% from dry material.

Changes in the percentage content of chemical components arose when the leaves had been physically pulverised and sunk to the lake bottom. The C-content sank to about 35% of the previous values in such cases.

Assuming a mean annual import of fallen leaves of about 22.8 t (5.4 g $DM \cdot m^{-2}$) into Lake Stechlin (Table 14.2), the corresponding values for C and N are 10.9 t and 0.27 t, respectively. For the whole lake area the values are 2.6 g $C \cdot m^{-2} \cdot a^{-1}$ and 63.5 mg $N \cdot m^{-2} \cdot a^{-1}$.

Additional pollen, seeds, fruits (i.e. in some years greater amounts of Beech-nuts and fruit-pod), twigs, barks, and stems (Fig. 14.2) are carried into the lake, but their amount is negligible. Big Beech- or Pine-trees falling into the lake may bring this up to 1.5 t C, but as a rule they are removed by fishermen for use as firewood or timber. Thus we can estimate that fallen leaves are the most important allochthonous C-source of Lake Stechlin.

However, it should be taken into account that Ohle (1982) gives C_{org}-input values from Lake Grebin arising from rainfall (6 g $C_{org} \cdot m^{-3}$), which, assuming they can be transposed to our research area, lead us to expect considerable additional C-import into Lake Stechlin. Thus in 1980 (total precipitation 702 $l \cdot m^{-2}$), 20.3 t C_{org}; 1981 (total precipitation 690.5 $l \cdot m^{-2}$) 20.0 t C_{org} and 1982 (total precipitation 511.6 $l \cdot m^{-2}$) 14.8 t C_{org} was transported into the lake from the atmosphere (Koschel, personal communication.

References

Bärlocher, F. & Kendrick, B. 1973. Fungi and food preferences of *Gammarus pseudolimnaeus*, Arch. Hydrobiol. 72: 501–516.

Bengtsson, G. 1982. Patterns of amino acid utilization by aquatic hyphomycetes. Oecologia 55: 355–363.

Bengtsson, G. 1983. Habitat selection in two species of aquatic hyphomycetes. Microb. Ecol. 9: 15–26.

Binder, B.J. & Naiman, R.J. 1983. Decomposition of paper birch and speckled alder leaf litter in a sea water environment. Arch. Hydrobiol. 97: 163–179.

Casper, S.J. 1965. Hyphomyceten-Studien. I. Die Süßwasser-Hyphomyceten des Stechlinsee-Gebietes. Limnologica 3: 257–270.

Cummins, K.W. 1974. Structure and function of stream ecosystems. Bioscience 24: 631–641.

Dahm, C.N. 1981. Pathways and mechanisms for removal of dissolved organic carbon from leaf leachate in streams. Can. J. Fish. Aquat. Sci. 38: 68–76.

Daubenmire, R. & Prusso, Don C. 1963. Studies of the decomposition rates of tree litter. Ecology 44: 589–592.

Fisher, S.G. & Likens, G.E. 1972. Stream ecosystem: organic energy budget. Bioscience 22: 33–35.

Gasith, A. & Hasler, A.D. 1976. Airborne litterfall as a source of organic matter in lakes. Limnol. Oceanogr. 21: 253–258.

Hanlon, R.D.G. 1981. Allochthonous plant litter as a source of organic material in an oligotrophic lake (Llyn Frongoch). Hydrobiologia 80: 257–261.

Hanlon, R.D.G. 1982. The breakdown and decomposition of allochthonous and autochthonous plant litter in an oligotrophic lake (Llyn Frongoch). Hydrobiologia 88: 281–288.

Hansen, K. 1959. The terms gyttja and dy. Hydrobiologia 13: 309–315.

Hansen, K. 1962. The dystrophic lake type. Hydrobiologia 19: 183–191.

Heller, H. 1971. Estimation of biomass of forest. In: Integrated experimental ecology (Ecological studies 2. Ed. H. Ellenberg). Berlin-Heidelberg-New York: Springer Verlag, pp. 45–47.

Howarth, R.W. & Fisher, S.G. 1976. Carbon, nitrogen and phosphorous dynamics during leaf decay in nutrient enriched stream microecosystems. Freshwater Biol. 6: 221–228.

Iversen, T.M. 1973. Decomposition of autumn-shed beech leaves in a spring brook and its significance for the fauna. Arch. Hydrobiol. 72: 305–312.

Iversen, T.M. & Madsen, B.L. 1977. Allochthonous organic matter in streams. Folia Limnol. Scandinav. 17: 17–20.

Jordan, M. & Likens, G.E. 1975. An organic carbon budget for an oligotrophic lake in New Hampshire. Verh. Int. Verein. Limnol. 19: 994–1003.

Kaushik, N.K. & Hynes, H.B.N. 1968. Experimental study on the role of autumn-shed leaves in aquatic environments. J. Ecol. 56: 229–243.

Kaushik, N.K. & Hynes, H.B.N. 1971. The fate of the dead leaves that fall into streams. Arch. Hydrobiol. 68: 465–515.

Kubota, H. & Shiraishi, L.Y. 1975. Fallen leaves. In: Productivity of communities in Japanese inland waters. Eds. S. Mori & P.G. Yamamoto. Kyoto JIBP-synthesis 10: 69–72.

Mathews, C.P. & Kowalczewski, A. 1969. The disappearance of leaf litter and its contribution to production in the River Thames. J. Ecol. 57: 543–552.

Meyer, J.L. 1980. Dynamics of phosphorous and organic matter during leaf decomposition in a forest stream. Oikos 34: 44–53.

Minshall, G.W. 1967. Role of allochthonous detritus in the trophic structure of a woodland spring-brook community. Ecology 48: 139–149.

Möller, L. 1935. Stechlinsee und Sakrower See. Arch. Hydrobiol. 29: 137–156.

Mothes, G. 1965. Ein Beitrag zur Sedimentcharakterisierung des Stechlinsees und zur biologischen Typisierung von Seen. Limnologica 3: 381–388.

Naiman, R.J. & Sibert, J.R. 1979. Detritus and juvenile salmon production in the Nanaimo estuary. III. Importance of detrital carbon to the estuarine ecosystem. J. Fish. Res. Bd. Can. 36: 504–520.

Nykvist, N. 1963. Leaching and decomposition of watersoluble organic substances from different types of leaf and needle litter. Stud. For. Suec. 3: 3–29.

Ohle, W. 1972. Gelöste organische Stoffe. Aufnahme und Abgabe durch Planktonorganismen im See. Gewässerschutz-Wasser-Abwasser 8: 1–56.

Ohle, W. 1982. Nährstoffzufuhren des Grebiner Sees durch atmosphärische Niederschläge und Oberflächenabschwemmungen des Einzugsgebietes. Arch. Hydrobiol. 95: 331–363.

Olson, J.S. 1963. Energy storage and balance of producers and decomposers in ecological systems. Ecology 44: 322–330.

Overbeck, J. 1982. Bakterien und Kohlenstoffumsatz im Plußsee – Probleme und Möglichkeiten der Mikrobenökologie. Forum Mikrobiol. 5: 292–303.

Paul, R.W. Jr., Benfield, E.F. & Cairns, J. Jr. 1978. Effect of thermal discharge on leaf decomposition in a river ecosystem. Verh. Int. Verein. Limnol. 20: 1759–1766.

Petersen, R.C. & Cummins, K.W. 1974. Leaf litter processing in a woodland stream. Freshwater Biology 4: 343–368.

Pieczynska, E. 1972. Role of allochthonous matter in lakes. Wiad. Ekol. 18: 131–140.

Rau, G.H. 1976. Dispersal of terrestrial plant litter into a subalpine lake. Oikos 27: 153–160.

Rau, G.H. 1978. Conifer needle processing in a subalpine lake. Limnol. Oceanogr. 23: 356–358.

Reed, F.C. 1979. Decomposition of *Acer rubrum* leaves at three depths in a eutrophic Ohio lake. Hydrobiologia 64: 195–197.

Schönborn, W., Flößner, D. & Proft, G. 1965. Die limnologische Charakterisierung des Profundals einiger norddeutscher Seen mit Hilfe von Testaceen-Gemeinschaften. Limnologica 3: 371–380.

Short, J.R. & Ward, J.V. 1980. Leaf litter processing in a regulated Rocky Mountains stream. Can. J. Fish. Aquat. Sci. 37: 123–127.

Szczepanski, A. 1965. Deciduous leaves as a source of organic matter in lakes. Bull. Acad. Pol. Sci. Cl. II, 13: 215–217.

Suberkropp, K. & Klug, M.J. 1976. Fungi and bacteria associated with leaves during processing in a woodland stream. Ecology 57: 707–719.

Tam, T.-Y., Mayfield, C.I. & Inniss, W.E. 1981. Nitrogen fixation and methane metabolism in a stream sediment-water system amended with leaf material. Can. J. Microbiol. 27: 511–516.

Uhlmann, D. & Albrecht, E. 1968. Biogeochemische Faktoren der Eutrophierung von Trinkwasser-Talsperren. Limnologica 6: 225–245.

Warnke, P. 1968. Zur natürlichen Verunreinigung kleiner stehender Gewässer durch Fallaub. Fortschr. Wasserchem. 8: 143–152.

Welton, J.S. & Ladle, M. 1979. Two sediment trap designs in small rivers and streams. Limnol. Oceanogr 24: 588–592.

Wetzel, R.G. & Manny, B.A. 1972. Decomposition of dissolved carbon and nitrogen compounds from leaves in an experimental hard-water stream. Limnol. Oceanogr. 17: 927–931.

15. The ecosystem Lake Stechlin
The life of the lake

R. Koschel, G. Mothes & S.J. Casper

The ecological status of Lake Stechlin (S.J. Casper)

A typical lake of our north-temperate region is, physically spoken, a permanent standing water more or less sunk deeply into its environ without direct communication to the sea and filled with fresh water poor in electrolytes. But from the ecological point of view a lake is more than a simple water body; it is the biotope of many distinctive communities of plants and animals. The physical environment and the total ensemble of organisms together with their interactions comprise the ecosystem lake, the continuous dynamic state of which is maintained by the creation, transformation and decomposition of organic matter within different biogenic trophic levels.

The ability of phytoplankton and -benthos (see Chapter 5) to capture the energy of sunlight and utilize it to transform carbon dioxide and water into organic molecules (see Chapter 10) forms the basis of the lake metabolism as a whole and creates the observed orderliness of the ecosystem. The captured energy dissipates during predation by animals (see Chapter 11) or mineralization by bacteria (see Chapter 12) at rapid rates. Chemical elements cycle in regular and distinctive ways (see Chapter 4).

When the lake is studied as a whole and its organisms are counted out, rules emerge that could not have been predicted from a knowledge of the structure and function of single organisms, species or communities alone. The life of a lake is something else and more than the sum of all life in the lake (Thienemann 1955; cited according to Rodhe 1979). Ecology cannot fully explain the relation of organisms to their environment without substantial recourse to many distinctive scientific branches (for instance to as far distant branches as molecular biology, population dynamics, evolutionary history or social sciences). But even if we had all the knowledge necessary the result about the exact prediction of the development of water quality – one of the most important goals of all limnological research – might be dubious (see Chapter 16). Nevertheless, limnology as an inherent part of ecology of high complexity has to deal with the biogenic deter-

Casper, S. Jost (ed.), Lake Stechlin. A temperate oligotrophic lake. ISBN 978-94-010-8930-2.

mined structure and function of inland waters combining analytical and synthetical approaches.

By nature, Lake Stechlin is a temperate dimictic calcareous oligotrophic lake (see p. 55). Due to the dominating autotrophic organisms it may be designated a complete ecosystem. Its specific morphometry (see Chapter 2), its thermal regime, distinguished by well developed full-circulation periods in spring and autumn (see p. 55) together with a stable summer stagnation period (see p. 56) and an ice-dependent winter stagnation period (see p. 56), and its relatively isolated geographical situation (see Chapters 1 and 2) are the basis of an ecosystem, the structure of which is highly sensitive to external impacts but well-fitted to 'buffer' them.

The sediment does not significantly affect the hypo- and epilimnetic layers of the pelagic zone regarding nutrient release and oxygen consumption (see p. 393). The high hypolimnetic oxygen concentration at the end of the summer stagnation period guarantees the oligotrophic state of the lake (see p. 122). The yearly long-term development of distinctly separated vertical water layers prevents the unlimited exchange of the hypolimnetic and epilimnetic matter and promotes the inhomogeneity of the vertical distribution of the phytoplankton (see p. 189). Its primary production – but not its biomass – responds proportionally to the external phosphorus (see p. 101) load. The quantity of organic carbon in terms of DOC: POC is averaged at 16.8 (see p. 107), i.e., it is mainly of autochthonous origin and typical for unpolluted hard-water lakes (Wetzel 1975). The observed shifts in populations of communities (see pp. 140, 190, 213) may be the result of changes in the nutrient regime (p. 87), but they can also be linked to biotic interactions (p. 347), or as in the case of aquatic macrophytes to the changing incoming global radiation. In any case, the species diversity (see p. 277), the food-web complexity, and the production to biomass ratio (see p. 307) show that Lake Stechlin has the status of the trophic equilibrium.

Lake Stechlin is, historically spoken, a young lake, but no younger than any other north-temperate lake. As to its ontogeny, it is a mature lake ecosystem. As matters stand, the oligotrophy of Lake Stechlin represents the climax of its biocoenosis (Whiteside 1983).

To demonstrate the 'self-governing' powers of Lake Stechlin, we present firstly the calcite precipitation as a model of autochthonous natural control of lake metabolism by the ecosystem itself and secondly the reaction of Lake Stechlin to the impact by the nuclear power plant Rheinsberg I.

Calcite precipitation and its role in the life of the lakes (R. Koschel)

Calcite precipitation in Lake Stechlin and other lakes

The autochthonous calcite precipitation is a widespread phenomenon in hard-water lakes (Minder 1923; Ohle 1952; Kelts & Hsü 1978). We also observed calcite precipitations in the pelagic zone of most hard-water lakes of the Lakes Stechlin and Breiter Luzin areas (Koschel *et al.* 1983; cf. p. 380). For example, the concentration of calcite averaged 230 mg $CaCO_3 \cdot m^{-3}$ in Lake Stechlin, 290 mg $CaCO_3 \cdot m^{-3}$ in Lake Dagow or 470 mg $CaCO_3 \cdot m^{-3}$ in Lake Breiter Luzin during the summer stagnation period (May–September) in the epilimnion in 1982. Maximal calcite concentrations were 1 g $CaCO_3 \cdot m^{-3}$. Such high concentrations were measured in Lake Stechlin only in August and in Lake Dagow only in June, but in Lake Breiter Luzin more or less during the whole summer (Figs. 4.9–4.10; p. 103). In 1978 the calcite precipitation of Lake Breiter Luzin expanded, moreover, to phases which are known as 'seasonal cloudings' or 'whitings' (e.g. Brunskill 1969; Schäfer & Stapf 1972; Strong & Eadie 1978). The concentration of calcite increased near the surface to 15 g $CaCO_3 \cdot m^{-3}$ during the 'whiting' in June in Lake Breiter Luzin (Koschel *et al.* 1983).

Interrelation between calcite precipitation and ecological parameters

In recent years some investigations of calcite precipitation have shown that it is connected with coreactions which can play a fundamental role in self-purification (Otsuki & Wetzel 1972; Rossknecht 1977, 1980; Murphy *et al.* 1983; Koschel *et al.* 1983). But what efficiency these mechanisms have against eutrophication (or how calcite especially acts as a factor) is not clear. We have attempted to answer these questions through investigations of calcite precipitation and their interrelations with productivity, nutrient concentrations and the trophic state in different lakes of our study area. Important effects of calcite precipitation were analysed by scenario technique, using the dynamic ecological model SALMO (Koschel *et al.* 1983; p. 435).

Five coupled effects between calcite precipitation and lake metabolism can be derived from our and other observations on calcite precipitation in lakes (Table 15.1).

1. Calcite precipitation leads to a decrease of Secchi depth and a corresponding increase of the vertical extinction coefficient. During high intensities of calcite precipitation the Secchi depth decreases rapidly (Fig. 15.1). The period of limited Secchi depth through high concentration of calcite precipitates is short, as a rule only 1–3 weeks.
2. Most authors have assumed that co-precipitation of orthophosphate may be important in connection with calcite formation and sedimentation (Rossknecht 1980; Gloss *et al.* 1980; Otsuki & Wetzel 1972; Murphy *et al.* 1983). A marked decrease of orthophosphate during or after the maximum calcite precipitation in our lakes could not be significantly measured. The orthophosphate begins to decrease rapidly in late winter and early spring (see p. 87). During the autochthonous calcite precipitation the orthophosphat concentration is already less

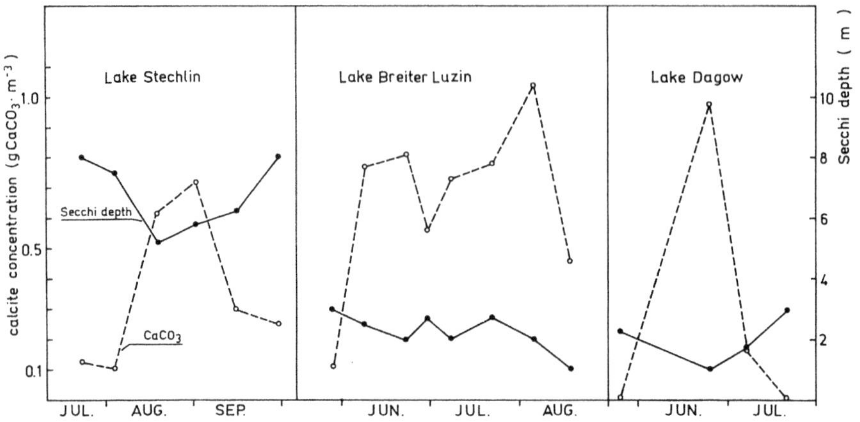

Figure 15.1 Epilimnetic calcite concentration and Secchi depth in Lakes Stechlin, Breiter Luzin and Dagow in 1982.

than 1 mg P · m^{-3} in Lake Stechlin and less than 5 mg P · m^{-3} in Lakes Breiter Luzin and Dagow. But a marked decrease, especially in Lake Breiter Luzin, of total phosphorus occurs when the calcite precipitation increases. The decrease of total phosphorus parallels an increase of sedimented material (Fig. 15.2).

3. Sedimentation rates always show a maximum after 2–4 weeks of maximum of calcite precipitation. After intensive calcite precipitation the primary production of phytoplankton increases (Fig. 15.3). A comparison of the seasonal dynamics of the primary production parameter and the calcite precipitation

Table 15.1 Ecological effects of calcite precipitation and calcite sedimentation on the metabolism of lakes.

Coupled reaction	Ecological effect	Source
Decrease of carbon dioxide	Limitation of primary production through CO_2	Ohle 1952
Co-precipitation of inorganic and organic compounds in the pelagic zone	Decrease of availability of dissolved nutrients (especially phosphorus)	Rossknecht 1980; Otsuki & Wetzel 1972; Murphy *et al.* 1983
Rise in turbidity	Light limitation of primary production	Minder 1923
Rise in sedimentation velocity (self-flocculation)	Elimination of particulate nutrients (Phosphorus, nitrogen, carbon etc.) and loss of biomass	Koschel *et al.* 1983
Co-precipitation of inorganic and organic compounds in sediments	Decrease of re-solution of nutrients	Jones & Bowser 1978

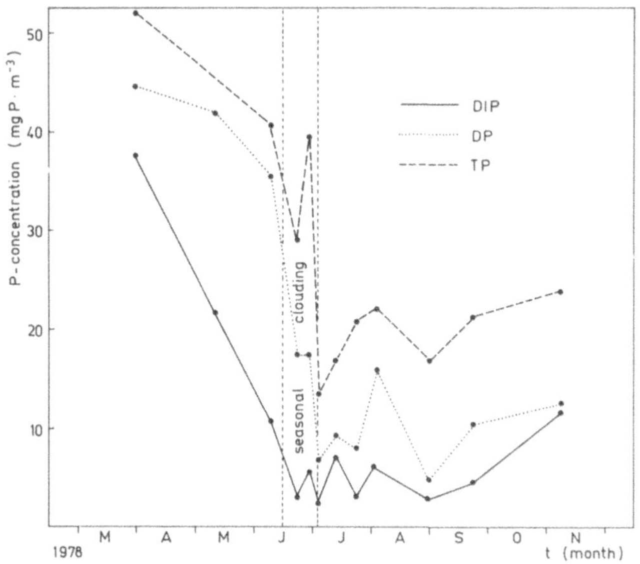

Figure 15.2 Orthophosphate (DIP), total soluble phosphate (DP), and total phosphate (TP) from March to December 1978 in Lake Breiter Luzin (mean values from 0 m, 2.5 m, 5 m, 7.5 m, and 10 m depth).

indicated that the maximum of specific carbon-incorporation of algae (P/B) is reached first, the maximum of primary production second and the maximum of biomass third. The maximum of calcite precipitation appears between the maximum of primary production and biomass. The calcite precipitation takes effect in a phase of reduced P/B and primary production values and during a relatively high phytoplankton biomass (Koschel *et al.* 1983). The increase of sedimentation rate through the calcite precipitation decreases the particular nutrients in epilimnion and thus limits the primary production. These relations are very significant for understanding ecological effects and the function of calcite precipitation for metabolism in stratified lakes.

4. We also observed that free CO_2-concentration decreases markedly in periods with high intensities of calcite precipitation (cf. Figs. 4.9–4.11; p. 103 ff.). Decrease of carbon dioxide can limit the productivity for short periods (Ohle 1952). But we think that these processes play a subordinate role.
5. On the other hand, the calcite deposit in sediment with the coupled reaction (formation of carbonate minerals, re-solution of nutrients etc.) can affect the metabolism for long periods. We know relatively little about these processes (see p. 380; Jones & Bowser 1978). Generally, the calcite deposit rate seems to follow the previously discussed principles of relation between calcite precipitation and trophic state of hard-water lakes (p. 102): high rates of deposit in mesotrophic and slightly eutrophic stratified lakes and low deposit rates in oligotrophic as well as in highly eutrophic and hypertrophic stratified lakes.

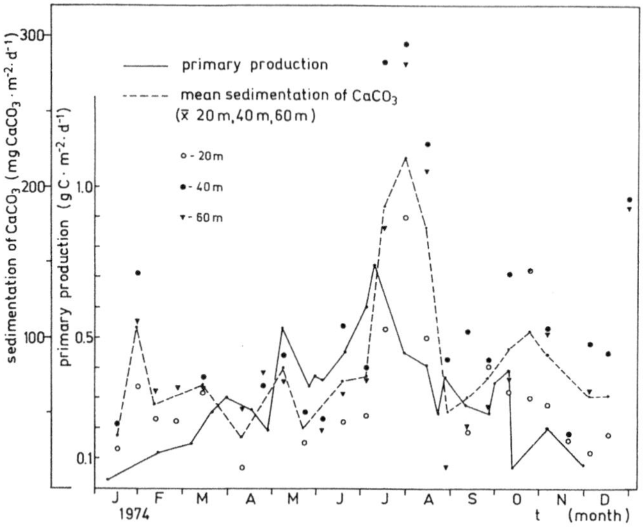

Figure 15.3 Primary production of phytoplankton and mean sedimentation rate of calcite (sedimentation rate in 20 m, 40 m, and 60 m depth of Lake Stechlin; values of sedimenting material according to Mothes 1981b).

Model simulation of ecological effects of calcite precipitation

We used the ecological lake model SALMO for scenario analyses to find out the dominant control mechanism of eutrophication linked with calcite precipitation in Lakes Breiter Luzin (Koschel *et al.* 1983) and Stechlin (see p. 435).

The three more important mechanisms – co-precipitation of dissolved orthophosphate, increasing of sediment velocity through self-flocculation processes, high turbidity under 'clouding' conditions – have been incorporated into five scenario analyses (Table 15.2).

Fig. 15.4 demonstrates the phosphorus and phytoplankton biomass concentrations of standard run, simulating what would happen if no calcite precipitation took place. The figure shows full accordance with the conclusion drawn by empirical models (Benndorf 1979) that the lake would produce higher phytoplankton biomass in the case of scenario 1 compared with its real state. Scenario 2 has been defined so that co-precipitation of orthophosphate may be important in connection with calcite formation and sedimentation. But the simulation results contradict this assumption. Although there is a high degree of conformity between observed and simulated phytoplankton biomass, the dissolved orthophosphate shows quite unrealistic behaviour, namely too fast a decrease in spring and a very low value at the end of the year. Consequently, scenario 2 cannot offer a satisfactory explanation of the problem.

Increased sedimentation velocity of phytoplankton algae due to flocculation by mineral particles is regarded as an important control mechanism in lake ecosystems. This assumption is confirmed by scenario 3 (Fig. 15.4). The simulated orthophosphate concentration as well as phytoplankton biomass coincide well with the observed values. Scenario 4, as a combination of scenarios 2 and 3, shows the worst agreement with the natural behaviour of the lake. All the stated variables are extremely low after April.

Scenario 5 is very similar to scenario 3 because the increased turbidity of the water due to 'clouding' at the maximum of the calcite formation is only of short duration. Fig. 15.4 demonstrates that the phytoplankton biomass is somewhat lower during the 'clouding' phase and higher in the following month when compared with scenario 3, but it can be concluded that increased turbidity exerts only a small influence on phytoplankton production.

The role of calcite precipitation in the metabolism of Lake Stechlin and other stratified hard-water lakes

From our observations in different lakes and from our scenario analyses, the overall conclusion must be drawn that self-flocculation of phytoplankton is the most important mechanism in connection with calcite formation, which has a

Table 15.2 Scenarios for evaluation of various mechanisms linked with calcite precipitation (according to Koschel *et al.* 1983).

Scenario	Mechanism involved	Simulated intensity of the mechanism	Duration of action of the mechanism
1	No calcite precipitation	KPSED = 0% · d^{-1} VS = 0.1 m · d^{-1} EPS_{min} = 0.4 m^{-1}	Whole year
2	Co-precipitation (KPSED) of dissolved orthophosphate	KPSED = 5% · d^{-1}	10.3–10.6
		KPSED = 15% · d^{-1}	10.6–10.7
		KPSED = 5% · d^{-1}	10.7–20.9
3	Increased sedimentation velocity (VS) of phytoplankton (self-flocculation)	VS = 1 m · d^{-1}	10.3–20.9
4	Co-precipitation of dissolved orthophosphate and increased sedimentation velocity of phytoplankton (self-flocculation)	as in Scenario 2 + 3	as in Scenario 2 + 3
5	Increased sedimentation velocity of phytoplankton (self-flocculation)	VS = 1 m · d^{-1}	10.3–20.9
	and increased light attenuation coefficient (EPS_{min})	EPS_{min} = 0.8 m^{-1}	10.6–10.7

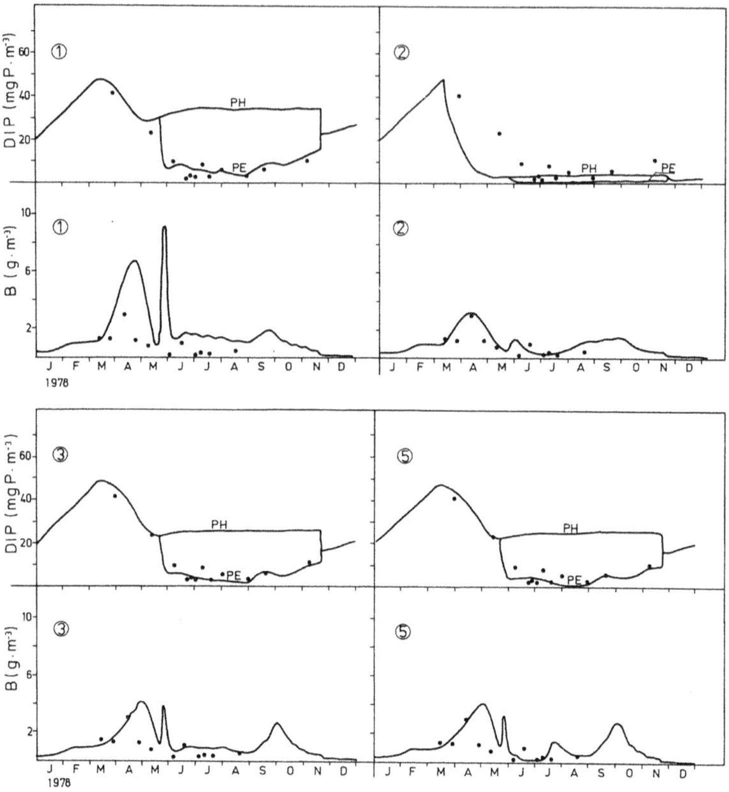

Figure 15.4 Simulation results of scenario 1 (no calcite precipitation); scenario 2 (co-precipitation of orthophosphate during spring and summer); scenario 3 (self-flocculation of phytoplankton with calcite during spring and summer), and scenario 5 (self-flocculation as well as increased attenuation coefficient during 'seasonal clouding'); dots = observed values in 1978; PE = DIP in the epilimnion; PH = DIP in the hypolimnion; B = phytoplankton biomass (g wet weight; details see Koschel *et al.* 1983).

considerable effect against the eutrophication of carbonate-rich lakes. This result is in full accordance with the assumption of some authors that increased sedimentation velocity of phytoplankton algae due to flocculation by mineral particles is regarded as being an important control mechanism in lake ecosystems (Uhlmann 1975; Ohle 1980). Sedimentation velocities of calcite crystals up to $3\,m \cdot d^{-1}$ have been found in Lake Constance (Rossknecht 1980). The investigations of calcite sedimentation in Lakes Stechlin, Dagow and Breiter Luzin have shown that the highest calcite sedimentation rates were coincident with the highest sedimentation rates of particulate organic carbon.

Obviously, the co-precipitation of dissolved orthophosphate is without any importance for the actual productivity of lakes. The co-precipitation rate of dissolved orthophosphate is slight, especially in phases of calcite precipitation, because the general content of dissolved orthophosphate is bound up as particulate organic phosphate. In addition to this, $CaCO_3$ has low sorption capacity for phosphate if pH is lower than about 9 (Sekoulov 1972; Griffin & Jurinek 1973; Fischer & Baumann 1978). It must be emphasized that adsorption of dissolved orthophosphate on clay minerals and the subsequent sedimentation does play a very important role (Goltermann 1973; Edzwald *et al.* 1978; Ohle 1980), but this mechanism is governed by allochthonous import of minerals or by sediment resuspension and is not a process like self-flocculation of algae by calcite, which provides an autochthonous self-protection of carbonate-rich lakes against eutrophication.

The calcite precipitation and its co-reactions play an important role in the lake metabolism of Lake Stechlin and could be a major mechanism of self-purification for the lake. A greater knowledge of the calcite precipitation and the utilization of this natural control mechanism of eutrophication is essential if the water management strategy of stratified hard-water lakes is ever to make progress.

The nuclear power plant and its role in the life of Lake Stechlin (R. Koschel, G. Mothes & S.J. Casper)

In 1966 the first nuclear power plant in the G.D.R. was put into operation with an external circulation of its cooling water which is taken from Lake Nehmitz. The heated water is then introduced into Lake Stechlin (see p. 50). Canal connections between both lakes close the circulation. The effects of the cooling water circulation on the physical, chemical, and biological conditions in a lake have been repeatedly investigated in the context of the Lake Stechlin system (e.g. Heitmann, Richter & Schumann 1969; Heitmann 1973, 1976; Richter, Neubert & Klämt 1979; Mothes *et al.* 1975; Koschel & Mothes 1976).

In this section we summarize the most important effects of the cooling water circulation and waste heat on the life of Lake Stechlin, partly discussed in various chapters of this volume.

Changes in water and heat balance

Significant changes were recorded from the water balance (Table 15.3). With a mean rate of flow of the cooling water (290 000 $m^3 \cdot d^{-1}$) its average retention time is 335 days for Lake Stechlin and 124 days for Lake Nehmitz-N for the total waterbody during the circulation periods. During the summer stagnation period the exchange of epilimnetic water is only 124 days for Lake Stechlin and 17 days for Lake Nehmitz-N with a mean rate of flow.

The retention times show that Lakes Stechlin and Nehmitz-N must be considered as one unit with respect to the cooling water circulation.

The difference between the temperature of the incoming and outgoing water is

Figure 15.5 In winter the heated water in the outflow canal of the nuclear power plant evaporates and forms a more or less dense mass of 'fog' on the canal. Photo: W. Scheffler.

by 10 K. The results are significant changes in the heat balance of Lake Stechlin (see p. 56; Heitmann 1976; Richter, Neubert & Klämt 1979).

The majority of the cooling water heat is transferred to the atmosphere (Fig. 15.5). The longwave reflexion of the lake and the evaporation under thermal load are considerably increased (see p. 65).

The residual amount of incoming heat leads to an increase in the temperature of the whole lake on an average of 1 K (see p. 51). From April to November the temperature increases from 1.4 K to 3.1 K (on average 1.9 K) in the surface layer and from 0.5 K to 1.3 K (on average 0.6 K) at a depth of 20 m (see Fig. 3.6; Heitmann 1976). The maximum increase of temperature occurs during complete circulation (November to March) in the deep water layers.

The development of the stability and position of the thermocline, influenced by the incoming cooling water, affects considerably the metabolism of Lake Stechlin (see p. 57; Fig. 3.8). Under thermal discharge, the thermocline decreases by 0.5 m in thickness. It becomes compressed (see p. 56), and its stability increases. Its position lies somewhat higher. The volume of the epilimnion decreases by 2–3 per cent.

During the winter months the cooling water circulation shortens the duration and reduces the thickness of the ice cover. The autumn full circulation period is lengthened and passes directly into the spring full circulation period. Only in very cold winters when, in spite of incoming cooling water, a complete ice cover is formed do stratification and destratification approach the natural conditions (see p. 55). By artificial heating Lake Stechlin shows a distinct trend to monomixis.

Changes in the matter balance

The operation of the cooling water circulation resulted in considerable changes in the load of matter in Lake Stechlin. We studied in detail the load and metabolism of phosphorus (Koschel 1976; Koschel & Mothes 1976; Mothes, Koschel & Scheffler 1976; Mothes 1981a; see p. 87).

The incoming cooling water led to an important increase of phosphorus inputs (see p. 98). As early as 1965 (during a trial run of the cooling water circulation without heat) the development of the macrozoobenthonic communities showed that there was an increase in matter (Mothes 1975).

Nearly 22 per cent of the input of total P into Lake Stechlin originates from the incoming cooling water of Lake Nehmitz-N (see Table 4.7). Orthophosphate, released from the plankton organisms damaged by the shock-like increase in water temperature up to ~28° C passing the nuclear power plant, is another source of phosphorus. With temperatures ~28° C a medium release of orthophosphate by 1.6 mg $P \cdot m^{-3}$ must be calculated (Mothes, Koschel & Scheffler 1976; see p. 424). Especially during the summer months additional amounts of orthophosphate become available for the planktonic community of Lake Stechlin. The input of orthophosphate by release due to the heating effect amounts to 22 per cent of the whole orthophosphate input into Lake Stechlin.

Under natural conditions, i.e., without incoming cooling water, the values of orthophosphate concentration in the upper epilimnion is 0. Under thermal load the value insignificantly lies above 0 (Mothes, Koschel & Scheffler 1976; Mothes 1981b).

From the phosphorus inputs due to the cooling water flow proceed one of the most important dangers for the oligotrophic state of Lake Stechlin.

The dimension of these imports is determined by the phosphorus concentration of and the phosphorus input into Lake Nehmitz-N. The phosphorus load of Lake Nehmitz-N is to a high degree the same as Lake Stechlin.

Since the middle of the seventies the phosphorus input into Lake Stechlin through the cooling water circulation shows a reduction due to various sanitation and environmental control measures in the catchment area of Lake Nehmitz-N (see p. 472).

Changes in the biocoenoses

The cooling water circulation of the nuclear power plant led to changes in the biotic interactions in Lake Stechlin (Koschel & Mothes 1976). The shifts observed in the phytoplanktonic, phytobenthonic, macrophytic, and macrozoobenthonic communities seem to indicate that Lake Stechlin has been 'eutrophized' at least for a short time and to a certain degree (cf. p. 156; p. 190; p. 213).

Macrophytes. Before the starting up of the nuclear power plant the favourable light conditions in Lake Stechlin led to the advance of the submerged macrophytic vegetation to depths of nearly 20 m. Nearly 40 per cent of the lake surface was covered with plant growth; the belts of Characeae-communities occupied 1.4 km^2, i.e., a third of the total lake area (see p. 141; cf. Fig. 5.6).

After the coming into operation of the nuclear power plant the translucence of the water was reduced by the increase in plankton and periphyton. Nearly all communities of Characeae receded, and in many places desolated areas without macrophytic vegetation arose (see p. 142; Fig. 5.7).

Chara filiformis, Ch. aspera, Stratiotas aloides and *Potamogeton rutilus,* adapted to conditions poor in nutrients, disappeared; *Ceratophyllum demersum, Potamogeton pectinatus, Ranunculus circinatus* and *Elodea canadensis,* adapted to conditions rich in nutrients, increased and formed dense stands, especially in the west bay. *Potamogeton crispus* began to expand. The growth and flowering of water- and reed-plants starts earlier; *Nymphaea alba* and *Potamogeton natans* flower at the end of April or the beginning of May, respectively (see p. 141).

In the course of time the macrophytic vegetation has adapted to the new conditions (see Fig. 5.7). The changes in the macrophytic vegetation of Lake Stechlin are shown in detail on Table 5.3.

Phytoplankton. The intensity and duration of the cooling water impact (see p. 192) led to spatial and temporal alterations in the structure and activity of the phytoplankton community in Lakes Stechlin and Nehmitz-N.

In 1966, when the nuclear power plant came into full operation, the cooling water disturbed the stability of the phytoplankton community of Lake Stechlin. It resulted in a distinct shifting of the previous structure and a distinct, higher level of biomass production.

Before 1966, Lake Stechlin was distinguished by a *Mallomonas-Dinobryon-Synedra-Asterionella-Oscillatoria rubescens* community (p. 164). After 1966, its phytoplankton was characterized by a *Fragilaria-Tabellaria-Mallomonas-Dinobryon* community (p. 177). In detail, formerly important members of the community disappeared totally (e.g. *Oscillatoria rubescens*; but see Küchler's 'Gattung *Oscillatoria*' in his paper of 1981, p. 96) or retreated temporarily (*Asterionella formosa, Synedra* spp.). Other taxa hitherto not observed amounted to large numbers (*Tabellaria fenestrata* var. *asterionelloides*) and dominated the community for a relatively long period. The development of microplanktonic diatoms (*Fragilaria crotonensis*) and nano- and ultraplanktonic green algae, especially phytoflagellates and coccoid forms, was favoured.

The biomass production (Table 5.12) amounted to an average of 200 to 400 per cent, with extraordinarily high peaks during fixed phases of the annual development (Figs. 5.18, 5.19). Under natural conditions the fluctuations in the composition of the phytoplankton community correspond to the seasonal alternation of meteorologic factors (see p. 190) and usually occur annually, but the period is also subject to the effect of random variables. During the period under consideration the cooling water circulation was the dominant environmental factor. Its 'input signal' (Uhlmann 1979) resembles a periodic oscillation (Figs. 5.19, 5.20). The observed alterations in the structure and activity of the phytoplankton community depend on the duration of the variations in the signal. Its pulse governs the periodicity of the community as well as its biomass level.

In 1966, the sudden increase of the inflow of external surface water into Lake Stechlin must have hit it like a hard blow. In the face of this shock the organismic society was thrown into disorder. Now, in view of the irregular alternation in intensity and time of the cooling water circulation, a new biological equilibrium has been established which, considering the composition of the phytoplankton community, partially approaches the previously existing equilibrium and preserves its diatom-chrysomonadean dominance.

Macrozoobenthos. The arrangement and make-up of the macrozoobenthonic communities changed considerably after the cooling water circulation began (see p. 241). In the lakes concerned similar but contrary changes took place.

In Lake Stechlin the medium abundance has become higher than during the initial conditions (see Table 7.5). The number of species increased from 406 to 468. The new immigrants include *Stictochironomus rosenschöldi* and *S. histrio* which indicate a higher standard of nutritious matter. *Pisidium conventus, Sergentia coracina,* and *Chaoborus,* typical for mesotrophic lakes, have spread perceptibly. *Micropsectra lindebergi* said to be an indicator of oligotrophic conditions decreased. In total the productivity of the macrozoobenthos in the depths increased by about 300% (Table 15.4, Fig. 15.6).

General reaction of biocoenoses and lake metabolism to the cooling water influence

Reaction of biocoenoses to heating. The cooling water from Lake Nehmitz, heated in the nuclear power plant by 6° C to 20° C (on average 10° C) is led into Lake Stechlin. In Lake Stechlin a small heated water region develops in the west bay near the nuclear power plant (Fig. 3.3). The rise in temperature is rapidly reduced toward the open lake (see p. 49; Fig. 3.3).

During the cold period of the year (<10° C) the rise in temperature in the heated water region has the following effects (Fig. 3.3):

- No damage to the organisms.
- Heightened productivity of periphyton and phytoplankton compared with the

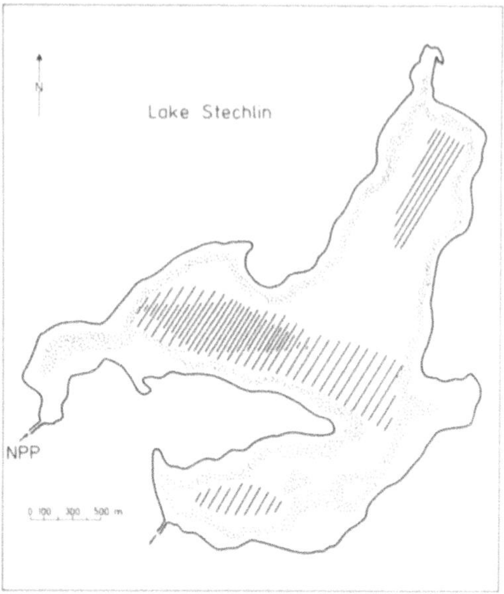

Figure 15.6 Changes in the distribution of macrozoobenthos in Lake Stechlin after starting cooling water circulation. – Wide slanting lines: zones with threefold increase in the abundance of macrozoobenthos; dotted parts: decline of macrophytic stands.

unheated lake region with roughly the same concentration of nutritious matter (see pp. 421, 317; Fig. 15.7).

- Increasing vitality of phytobenthos, e.g., premature development of macrophytes. In the heated water region *Nymphaea alba* already blossoms at the end of April, in the unheated part of the lake only at the beginning of June.
- Increasing vitality of macrozoobenthos, e.g., premature emergence of insects. In the heated water region many insects emerge regularly in winter, whereby two species of *Cricotopus* produce continuing sequences of summer and winter generations, whereas in the unheated lake region no emergence can be observed in winter (cf. p. 240).
- Heightened activity of bacteria compared with the unheated lake region (see p. 356).

During the warm months of the year (>15° C) with surface water temperatures up to 32° C in the heated water region it has the following effects (Fig. 3.3):

- Damage to phyto- and zooplankton.
- Greatly decreased productivity of periphyton and phytoplankton. With periphyton the decrease is lower because blue-green algaecoenoses adapted to warm water develop.
- Considerable increase of activity of heterotrophic bacteria coenoses (see p. 356; cf. Koschel & Mothes 1976).
- Release of nutritious matter following damage to organisms. The concentra-

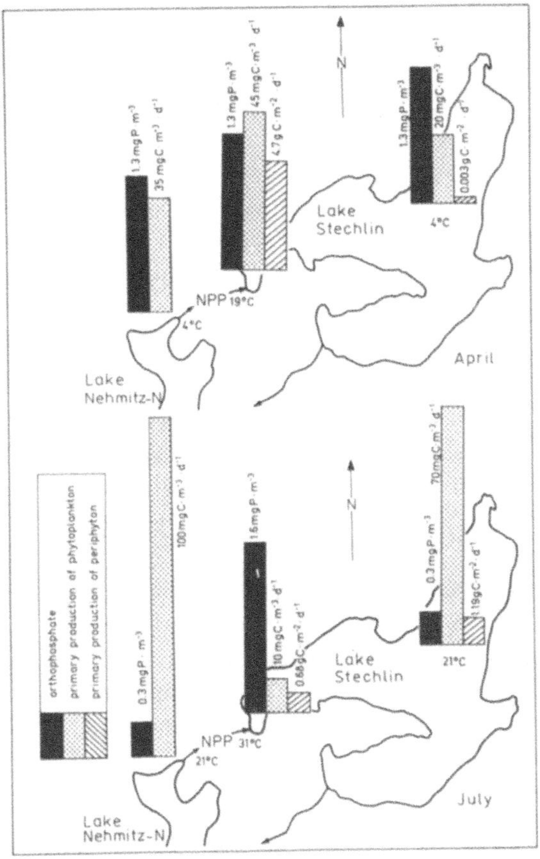

Figure 15.7 Comparison between primary production of phytoplankton and periphyton and orthophosphate concentration in close-to-surface layer of the Lake Stechlin-system cooling water circulation.

tion of orthophosphate can multiply considerably in the heated water region compared with other parts of the lake.

Tolerance of temperature of biocoenoses in Lake Stechlin. Outflow temperature of cooling water up to 25° C showed no considerable physiological damage; with temperature about 30° C phyto- and zooplankton were damaged irreversibly. It is clear from the carbon assimilation that damage at 30° C is only to its photoautotrophic portion (Koschel & Mothes 1976; see p. 424). Its heterotrophic portion (dark assimilation) is increased up to 35° C as a rule (see p. 424; Koschel & Mothes 1976).

Analyses of the release of orthophosphate and microscopic plankton observations show a damage to phyto- and zooplankton from 28° C onwards. Above this temperature, an above-average destruction and a decrease of primary produc-

Table 15.3 The cooling water circuit of the nuclear power plant.

	Lake Stechlin		Lake Nehmitz-N		Lake Gerlin	
Volume (Mio m^3) (total)	97		6.15		0.18	
Mean volume (Mio m^3) (epilimnion)	36		4.8		0.18	
Water exchange (days) (total)	202[a]	335[b]	13[a]	21[b]	0.4[a]	0.6[b]
Water exchange (days) (epilimnion)	75[a]	124[b]	10[a]	17[b]	0.4[a]	0.6[b]

[a] Water exchange for the maximum rate of flow of cooling water (480.000 $m^3 \cdot d^{-1}$).
[b] Water exchange for the mean rate of flow of cooling water (290.000 $m^3 \cdot d^{-1}$).

tion, a drastic deterioration in the oxygen regime and its negative consequences have to be noted. Heating above 27° C in comparable waters cannot be recommended for this reason. Nearly the same temperature limits have been discussed by other authors (e.g. Jaag 1972; Hornef 1972; Humpesch *et al.* 1981).

Reaction of lake metabolism to cooling water influence. Fig. 15.8 shows the effect of the different phases of cooling water circulation on the productivity of the whole lake. In the first phase the cooling water is taken from Lake Nehmitz. Then (phase 2) the cooling water is heated in passing through the nuclear power plant. The altered chemical, physical, and biological conditions have their first effect in the heated water region (phase 3; see p. 427). Their effects on the whole lake (phase 4) are as follows.

Heating in winter considerably reduces the ice cover of the lake and enables the vegetation to start earlier.

Outflow temperatures of 30° C mean a release of nutritious matter in the heated water region and thereby a continous supply of nutrients for the lake. This supply takes place at a time when the primary production normally is considerably limited by an impoverishment of nutritious matter, especially of phosphorus (see p. 89). It effects a considerable increase of carbon assimilation in horizons near the surface (see pp. 289 ff.). This rise in activity is accompanied by an increase in the efficiency of primary production in the 'kurzgeschlossenem Kreislauf'. Thus the effect of the supply of nutritious matter is multiplied.

The lengthening of the vegetation period and supply of nutritious matter in summer are responsible for the dangerously high rate of annual primary production. In addition, the dynamics of primary production and phytoplankton succession shift during the year (see pp. 172 ff.). In February and March there is already a first spring maximum of phytoplanktonic activity. The impoverishment of nutritious matter then leads to an early minimum (April). A second long-lasting maximum follows, normally beginning in autumn. The summer minimum (July, August), which is to be seen in comparable waters or in phases without cooling water influences, cannot be observed.

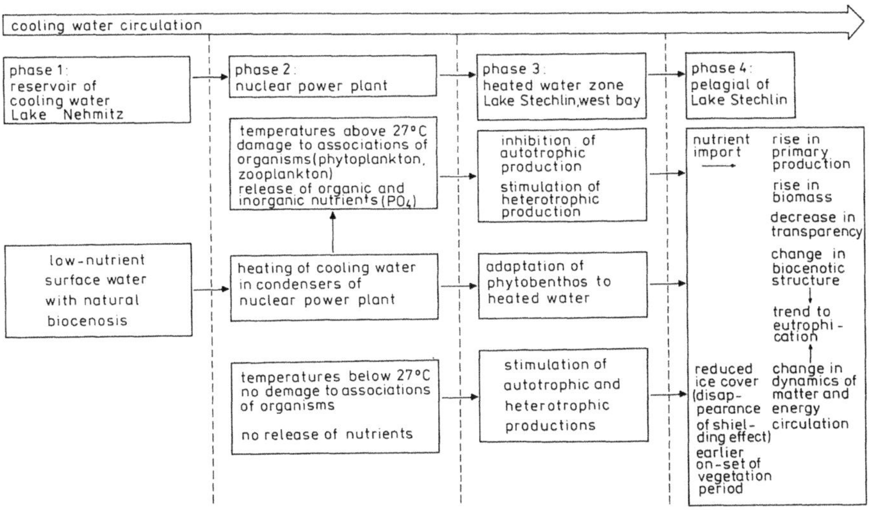

Figure 15.8 Nuclear power plant with external cooling water circulation, impact on surface waters.

Primary production of phytoplankton and phytoplankton density as a whole increased. The primary production of phytoplankton is at present about 100 g $C \cdot m^{-2} \cdot a^{-1}$ (see p. 290). This quantity is very high for on aligotrophic lake (see p. 312). Increased phytoplankton density is reflected in increased turbidity of water and causes a decrease in transparency (Table 15.5).

The effect of the nuclear power plant on the life of Lake Stechlin

Nuclear power plants which have an external circulation of cooling water significantly influence the waters by changing the water balance, the waste heat, and the

Table 15.4 Average production of macrozoobenthos per year before and after the nuclear power plant has become operational.

Macrozoobenthos	Before operation ($mg\ ww \cdot m^{-2} \cdot a^{-1}$)	After operation ($mg\ ww \cdot m^{-2} \cdot a^{-1}$)	Relative change (%)
Pisidium conventus	54	263	487
Oligochaeta	589	3106	527
Sergentia coracina	141	778	552
Tanytarsini	363	287	79
Total	1701	5280	310

* ww = wet weight

Table 15.5 Change of Secchi disc transparency in Lake Stechlin before and after the nuclear power plant has become operational.

	Secchi depth (m)	
	Before operation* 1958–1965	After operation 1970–1982
Spring		
average minimum	8.3	5.2
absolute minimum	5.5	4.0
Summer		
average maximum	10.0	6.0
absolute maximum	11.6	9.7
Winter		
absolute maximum	15.5	12.9

* according to Richter & Wiedrich 1967.

matter inputs (Fig. 15.9). In addition, the release of radionuclids is possible if defects occur in nuclear power plants. The last point plays in practice no role in Lake Stechlin. The radioactivity released by the nuclear power plant in our study area is very small (Ettenhuber *et al.* 1975; Winkler & Brune 1976). It doesn't affect the life of Lake Stechlin and of the surrounding area at all.

The effect of the cooling water circulation on the lake metabolism is ambivalent. The considerable increase in water temperature and, above all, nutritious matter are clear causes of tendencies towards eutrophication. However, it must be remembered that the various negative influences from Lakes Dagow and Wulwitz as well as the sewage purification plant of the nuclear power plant make an insight into the ecological balance more difficult. On the other hand, the change in the stability and position of the thermocline are directed against

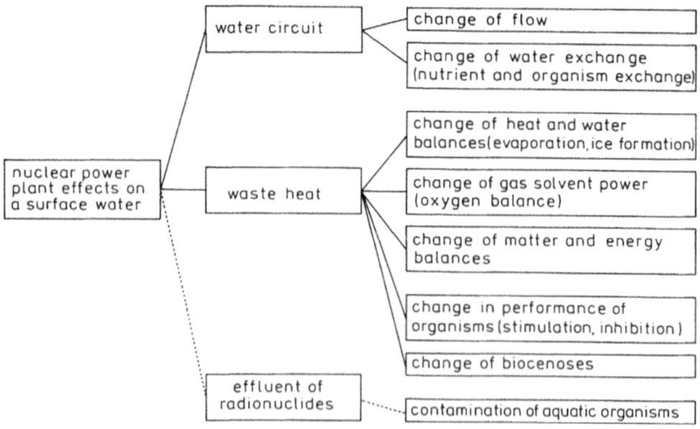

Figure 15.9 Effects of different phases of cooling water circulation on the productivity in the Lake Stechlin system.

eutrophication by reducing the load of the hypolimnion. There have been only small changes in the hypolimnetic oxygen capacity and in the hypolimnetic phosphate accumulation (see pp. 93, 122).

In spite of the observed tendencies towards eutrophication caused by the cooling water circulation now lasting for 18 years, Lake Stechlin has preserved its oligotrophic status.

References

Benndorf, J. 1979. A contribution to the phosphorus loading concept. Int. Revue ges. Hydrobiol. 64: 177–188.

Brunskill, G.J. 1969. Fayetteville Green Lake, New York. II. Precipitation and sedimentation of calcite in a meromictic lake with laminated sediments. Limnol. Oceanogr. 14: 830–847.

Busse, G. 1965. Biologische Untersuchungen im Gebiet des Stechlinsees. Teilprogramm 1: Floristisch-ökologische Bestandsaufnahme der Protophyten. In: Abschlußbericht . . . Jena: 1–36 (unpublished).

Busse, G. 1972. Phytoplankton. In: Grundlagen zur Modellierung des Stoffhaushaltes der Gewässer am Beispiel des Stechlin-Sees. Abschlußbericht . . . Jena: 1–44 (unpublished).

Edzwald, J.K., Toensing, D.C. & Chi-Yew Leung, M. 1976. Phosphate adsorption reactions with clay minerals. Environ. Sci. Technol. 10: 485–490.

Ettenhuber, E., Lönnig, M., Hoffmann, M.L., Röhnsch, W., Babenzien, Chr., Lehmann, R. & Schreiter, W.-D. 1975. Ergebnisse und Erfahrungen der Umgebungsüberwachung am KKW Rheinsberg. Staatliches Amt Atomsicherheit Strahlenschutz DDR. Report SAAS-190: 2–21.

Fischer, W.R. & Baumann, G., 1978. Die Sorption von Phosphat an carbonatreichen Unterwasserböden. Z. Pflanzenernähr. 141: 607–620.

Gessner, F. 1939. Die Phosphatarmut der Gewässer und ihre Beziehung zum Kalkgehalt. Int. Revue ges. Hydrobiol. 38: 203–211.

Gloss, S.P., Mayer, L.M. & Kidd, D.E. 1980. Advective control of nutrient dynamics in the epilimnion of a large reservoir. Limnol. Oceanogr. 25: 219–228.

Goltermann, H.L. 1973. Natural phosphate in relation to phosphate budgets: A contribution to the understanding of eutrophication. Wat. Res. 7: 3–17.

Griffin, R.A. & Jurinek, J.J. 1973. The interaction of phosphate with calcite. Soil. Sci. Proc. 37: 847–850.

Heitmann, M.-L. 1973. Stability of thermal stratification in a lake. Publ. Int. Assoc. Hydrol. Sci. 109: 321–326.

Heitmann, M.-L. 1976. Die thermische Belastung des Stechlinsees durch ein Kernkraftwerk. Materialien Int. Symp. Eutrophierung und Sanierung von Oberflächengewässern, 20.–25.9.1976. Karl-Marx-Stadt, DDR. Bd. III: 133–146.

Heitmann, M.-L., Richter, D. & Schumann, D. 1969. Der Wärme- und Wasserhaushalt des Stechlin- und Nehmitzsees. Abh. Meteor. Dienst DDR Nr. 96 (12).

Hornef. 1972. Thermische Belastung der Gewässer. Korrespondenz Abwasser 19: 271–272.

Humpesch, U.H., Dokulil, M., Elliot, J.M. & Herzig, A. 1981. Ökologische Auswirkungen der thermischen Gewässerbeeinflussung. Wien: Bundesministerium für Land- und Forstwirtschaft Österreich. 257.

Jaag, O. 1972. Zum Problem der Aufwärmung von Flußwasser durch Kühlwasser aus Kernkraftwerken. Verh. Int. Verein. Limnol. 18: 981–994.

Jones, B.F. & Bowser, C.J. 1978. The mineralogy and related chemistry of lake sediments. In: Lerman, A. (ed.): Lakes Chemistry, Geology, Physics. New York-Heidelberg-Berlin: Springer-Verlag. 179–235.

Kelts, K. & Hsü, K.J. 1978. Freshwater carbonate sedimentation. In: Lerman, A. (ed.): Lakes Chemistry, Geology, Physics. New York-Heidelberg-Berlin: Springer-Verlag. 295–323.

Koschel, R. 1974a. Einfluß der physikalischen und chemischen Umweltfaktoren auf die Primärproduktion des Phytoplanktons im Stechlinsee. Diss. Techn. Univ. Dresden. 1–72 (unpublished).

Koschel, R. 1974b. Primärproduktionsuntersuchungen im Stechlinsee. Limnologica 9: 143–156.

Koschel, R. 1976. Der Einfluß des Phosphorangebotes auf die Primärproduktion des Phytoplanktons in einem geschichteten Klarwassersee (Stechlinsee, DDR). Limnologica 10: 325–346.

Koschel, R. 1981. Bilanz der Primärproduktion des Phytoplanktons in den Gewässern des Stechlinseesystems. Limnologica 13: 55–81.

Koschel, R., Benndorf, J., Proft, G. & Recknagel, F. 1983. Calcite precipitation as a natural control mechanism of eutrophication. Arch. Hydrobiol. 98: 380–408.

Koschel, R. & Mothes, G. 1976. Änderung der biologischen Produktivität in einem Seensystem durch den Kühlwasserzufluß eines Kernkraftwerkes. Materialien Int. Sympos. Eutrophierung und Sanierung von Oberflächengewässern. 20.–25.9.1976, Karl-Marx-Stadt, DDR. Bd. III: 219–233.

Küchler, L. 1981. Phytoplanktonuntersuchungen im Stechlinseegebiet in den Jahren 1973–1975. Limnologica 13: 83–99.

Küchler, L. 1982. Phytoplanktonuntersuchungen im Stechlin und im Nordbecken des Nehmitzsees. Limnologica 14: 231–241.

Minder, L. 1923. Die biogene Entkalkung im Zürichsee. Verh. Int. Verein. Limnol. 1: 20–32.

Mothes, G. 1975. Der Einfluß eines Kraftwerkes auf das Makrozoobenthos des Stechlinseesystems. Acta hydrochim. hydrobiol. 3: 353–367.

Mothes, G. 1981a. Physikalische und chemische Parameter der Wasserbeschaffenheit in Seen des Stechlinseegebietes. Limnologica 13: 1–54.

Mothes, G. 1981b. Sedimentation und Stoffbilanzen in Seen des Stechlinseegebietes. Limnologica 13: 147–194.

Mothes, G., Koschel, R. & Scheffler, W. 1976. Der Einfluß der Aufwärmung eines Sees durch ein Kernkraftwerk auf Phosphorhaushalt und Stoffproduktion. Limnologica 10: 489–501.

Murphy, R.P., Hall, K.J. & Yesaki, I. 1983. Coprecipitation of phosphate with calcite in a naturally eutrophic lake. Limnol. Oceanogr. 28: 58–69.

Ohle, W. 1952. Die hypolimnische Kohlendioxyd-Akkumulation als produktionsbiologischer Indikator. Arch. Hydrobiol. 46: 153–285.

Ohle, W. 1980. Mineral impact of lakes as background for their biogenic dynamics. Hydrobiologia 72: 51–59.

Otsuki, A. & Wetzel, R.G. 1972. Co-precipitation of phosphate with carbonates in a marl lake. Limnol. Oceanogr. 17: 763–767.

Otsuki, A. & Wetzel, R.G. 1974. Calcium and total alkalinity budget and calcium carbonate precipitation of a small hard-water lake. Arch. Hydrobiol. 73: 14–30.

Richter, D. & Wiedrich, R. 1967. Die Sichttiefen im Stechlin- und Nehmitzsee. Limnologica 5: 299–305.

Richter, D., Neubert, W. & Klämt, A. 1979. Temperatur- und Wärmehaushalt des thermisch belasteten Stechlin- und Nehmitzsees. Abh. Meteorol. Dienst. DDR Nr. 123 (16): 1–40.

Rodhe, W. 1979. The life of lakes. Arch. Hydrobiol. Beih. Ergebn. Limnol. 13: 5–9.

Rossknecht, H. 1977. Zur autochthonen Calcitfällung im Bodensee-Obersee. Arch. Hydrobiol. 81: 35–64.

Rossknecht, H. 1980. Phosphatelimination durch autochthone Calcitfällung im Bodensee-Obersee. Arch. Hydrobiol. 88: 328–344.

Schäfer, A. & Stapf, K.R.G. 1972. Calcit-'whitings' im Bodensee-Untersee. Natur. Mus. 102: 307–311.

Sekoulov, I. 1972. Die Phosphatelimination mit Hilfe von kontinuierlich belichteten Blaualgen. Stuttgarter Berichte Siedlungs- Wasserwirtsch. 50: 1–186.

Strong, A. & Eadie, B.J. 1978. Satellite observation of calcium carbonate precipitation in the Great

Lakes. Limnol. Oceanogr. 23: 877–887.

Thienemann, A. 1955. Die Binnengewässer in Natur und Kultur. Eine Einführung in die theoretische und angewandte Limnologie (Verständliche Wissenschaft 55). Berlin-Göttingen-Heidelberg: Springer-Verlag. VIII + 156.

Uhlmann, D. 1975. Hydrobiologie. Ein Grundriß für Ingenieure und Naturwissenschaftler. Jena: VEB Gustav Fischer Verlag. 345.

Uhlmann, D. 1979. Hydrobiology. A text for engineers and scientists. Chichester. New York. Brisbane. Toronto: John Wiley & sons. 313.

Uhlmann, D. 1973. Störungen des biologischen Gleichgewichts in Gewässern. Abh. Sächs. Akad. Wissenschaften Leipzig, Math.-Naturwiss. Klasse *52* (2): 1–17.

Wetzel, R.G. 1975. Limnology. Philadelphia: Saunders, W.B. 743.

Whiteside, M.C. 1983. The mythical concept of eutrophication. Hydrobiologia 103: 107–111.

Winkler, R. & Brune, W. 1976. 10 Jahre Kernkraftwerk Rheinsberg. Kernenergie 19: 101–107.

16. The pelagic zone of Lake Stechlin
An approach to a theoretical model

J. Benndorf, R. Koschel & F. Recknagel

A strategy for development of dynamic ecological models as a method of limnological research

General principles

Dynamic ecological models are used frequently in theoretical and experimental ecosystem analysis (e.g. Patten *et al.* 1982; Straškraba 1982; Koschel *et al.* 1983) as well as in decision-making with respect to the optimum strategy of water quality management (e.g., Jørgensen 1980; Benndorf *et al.* 1981b; Ambühl 1982; Benndorf & Recknagel 1982). Although different specific requirements of the models must be met for both the different uses, there are some general principles which should be considered before a model can be properly used in either research or decision-making.

1. *The principle of the maximum possible simplicity.* The model should be as simple as possible with respect to the number of state variables. This requirement is important because of clarity for the user and because it has been found that models with a great number of state variables do not necessarily have a higher degree of reality than models having a relatively low number of state variables (Jørgensen *et al.* 1978; Benndorf & Recknagel 1982; Straškraba 1982).
2. *The principle of the maximum necessary complexity.* The model should exhibit the maximum necessary complexity with respect to the ecological control mechanisms. This requirement is inevitable to guarantee a degree of reality which allows the simulation of the effects of external and internal control factors. As indirect effects quite often predominate over direct effects, rather complex models are required (Patten *et al.* 1982), but it would be misleading to conclude that the increase of the number of state variables might be the only way to fulfil this requirement. There is also the more promising possibility to construct 'small' models (see above principle) which, nevertheless, can cover a great number of ecological control mechanisms. This can be achieved by improving the degree of reality of the description of the rate variables (e.g.,

Casper, S. Jost (ed.), Lake Stechlin. A temperate oligotrophic lake. ISBN 978-94-010-8930-2.

respiration rates, sedimentation rates, grazing rates) and by inserting appropriate auxiliary functions as 'replacements' for ecosystem components and processes not involved as state or rate variables.

3. *The principle of the maximum possible generalizability.* The model should contain 'generally valid' parameters, i.e., parameters that can be well explained in physical, biochemical or biological terms. The numeric values of these parameters should preferably be obtained through measuring the rates of the processes mainly in laboratory and field experiments. The decisive advantage of the determination of 'generally valid' parameters is that there is no need to reestimate the numeric values (with the exception of a few specific parameters) when the model is applied to different lakes or to different conditions in a particular lake. Considering the role of uncertainties (cf. Beck 1983; van Straten 1983 for details), the method of parameter determination recommended here might be an oversimplification, but it seems to be the only possibility to achieve predictive models, i.e., models which can be used for checking the validity of scientific hypothesis or for management decisions.
4. *The principle of the proper validation.* The validity of a model must be checked properly using water bodies having various trophic states and different hydrophysical and morphometric features. No change of the model structure or of the numerical values of the parameters should be made (see principle 3). Only such 'general' models, the validity of which has been checked in this way, can be applied with a sufficient degree of reliability to unknown situations.

There are, however, some doubts concerning the possibility of constructing 'general' models (e.g. van Straten *et al.* 1983); but, on the other hand, there have been some promising attempts (e.g. Walters *et al.* 1980; Benndorf & Recknagel 1982; Straškraba 1982). The accuracy of the existing 'general' models is, of necessity, still relatively low. This disadvantage can be compensated to a large extent by appropriate methods in the application of this model type, i.e., methods which minimize the effect of the simulation uncertainty on the evaluation of the simulation results (e.g., by the relative comparison of various scenarios with a standard scenario).

There are, of course, many dynamic ecological models the concepts of which do not follow the strategy outlined by principles 1 to 4 (e.g. Schellenberger *et al.* 1978; Jørgensen 1980; Somlyody 1983). The parameters of these 'lake-specific' models are estimated by calibration, using measured time series of the state variables and they are to be reestimated for each particular lake. This complicated procedure and the almost impossible validation for different lakes and conditions (see Patten *et al.* 1983) are great disadvantages for the application of this model type.

Since the aim of our model investigation consists in evaluating the effects of different control factors on the processes and components within the pelagic zone of Lake Stechlin, only such a model can be used, the applicability of which to unknown situations has been checked (principle 4) and the parameters of which are 'generally valid' in the sense of principle 3. The model SALMO meets these requirement and it corresponds also to principles 1 and 2, therefore, it has been chosen for the Lake Stechlin simulations.

The model SALMO

SALMO (Simulation by an Analytical Lake MOdel) was developed as a 'general' model at the Hydrobiological Laboratory of the Water Resources Department of the Dresden University of Technology and applied to some stratified lakes and reservoirs having quite different trophic states (see, e.g., Benndorf *et al.* 1981b; Benndorf & Recknagel 1982; Koschel *et al.* 1983).

In the present version of the model the water body is modelled as a two-layer system (Fig. 16.1). Both the epilimnion and the hypolimnion are assumed to be mixed homogeneously. The boundary between the two layers (the mixing depth) is variable in time. For both the epilimnion and the hypolimnion, the state variables – dissolved orthophosphate, inorganic nitrogen, phytoplankton (two functional groups), zooplankton, detritus and dissolved oxygen – are calculated. Fish and predatory zooplankton are only included in an indirect manner, by including their consumption behaviour in the zooplankton mortality rate. The interactions between sediment and water are considered by empirical relationships for nutrient release and oxygen consumption. An essential simplification consists in the fact that only two populations of phytoplankton are considered and that no separate calculations are made for the various functional groups of zooplankton, or for the various age classes within a certain zooplankton group. But time delay due to egg development is involved. The phytoplankton consist of (group 1) blue-green algae of the type *Oscillatoria redekei,* and (group 2) green algae and nanoplankton diatoms of the types *Scenedesmus quadricauda* and *Cyclotella meneghiniana.* The coexistence of the two phytoplankton groups, or dominance of one of the groups, is considered automatically as a consequence of the different parameter sets for the groups, and in addition as a result of the control of the food preference of the zooplankton by the phytoplankton composition. The zooplankton consist of herbivorous or omnivorous Cladocera and Copepoda of the types *Daphnia hyalina, Eudiaptomus gracilis* and *Cyclops vicinus.*

Despite the low number of state variables in SALMO, values for 117 parameters are required as a consequence of the inclusion of numerous control mechanisms. This is shown in Fig. 16.1 by the great number of arrows corresponding to the information flow which, in turn, controls the numerous rate variables. Because of the generalizability desired of the model, almost exclusively parameters were used that have been determined in accordance with principle 3.

The input variables are given as discrete values for a particular water body. In this context, for every ten days an average ten-day value is used to keep the

Table 16.1 List of symbols used in the diagrammatic representation of SALMO (Fig. 16.1) and in the scenario description (Table 16.2). H or (H) refers to the hypolimnion.

State variables (symbols in double squares) and auxiliary variables (symbols in single squares)	
D (H)	Detritus
F	Fish biomass
I, IHZ	Photosynthetic active radiation at a depth z_i of the mixed layer or zH_i of the hypolimnion
LOH	Load of the hypolimnion with respect to organic matter
N (H)	Inorganic nitrogen (sum of nitrate, nitrite and ammonia)
O (H)	Dissolved oxygen
P (H)	Dissolved orthophosphate
$X (H)_i$	Phytoplankton biomass of the functional group i
Z (H)	Biomass of the herbivorous (and omnivorous) zooplankton
Rate variables (symbols in circles)	
AE, AH	Rate of water exchange between epilimnion and hypolimnion
AZ (H)	Coefficient of the zooplankton assimilation (ratio of assimilation to ingestion)
BD (H)	Sedimentation rate of the detritus
BX (H)	Sedimentation rate of the phytoplankton (including the parameter VS = sinking velocity)
EGG	Factor of the development of zooplankton eggs
EPS (H)	Vertical extinction coefficient for photosynthetic active radiation
G, GHD	Grazing rate of the zooplankton
LSEZ (H)	Load of the mixed layer (LSEZ) or of the hypolimnion (LSEZH) due to oxygen consumption by the sediments
MIRAZ	Net rate of the vertical migration of the zooplankton
MORTZ (H)	Mortality rate of the zooplankton
NDENH	Rate of denitrification within the hypolimnion
NEXKR (H)	Rate of nitrogen remineralization due to the activity of the zooplankton
N (H) KONS	Rate of nitrogen assimilation by the phytoplankton
NMORT (H)	Rate of nitrogen remineralization by fish and from dead zooplankton
NSF (H)	Rate of nitrogen release from the sediments
PEXKR (H)	Rate of phosphate remineralization due to the activity of the zooplankton
P (H) KONS	Rate of phosphate assimilation by the phytoplankton
PHOTX (H)	Rate of the gross photosynthesis of the phytoplankton
PMORT (H)	Rate of phosphate remineralization by fish and from dead zooplankton
PSF (H)	Rate of phosphate release from sediments
RX (H)	Respiration rate of the phytoplankton
RZ (H)	Respiration rate of the zooplankton
WX (H)	Growth rate of the phytoplankton
WZ (H)	Growth rate of the zooplankton
Input variables (symbols without squares or circles)	
IIN	Incident photosynthetic active radiation at the surface
N (H) IN	Concentration of inorganic nitrogen (nitrate, nitrite and ammonia) in the water flowing into the mixed layer (NIN) or into the hypolimnion (NHIN)
P (H) IN	Concentration of dissolved orthophosphate in the water flowing into the mixed layer (PIN) or into the hypolimnion (PHIN)
POM (H) IN	Concentration of particulate organic matter in the water flowing into the mixed layer (POMIN) or into the hypolimnion (POMHIN)

Table 16.1 Continued.

Q (H) IN	Water flow to the mixed layer (QIN) or to the hypolimnion (QHIN)
Q (H) OUT	Water outflow from the mixed layer (QOUT) or from the hypolimnion (QHOUT)
T (H)	Water temperature
ZHM	Hypolimnion depth
ZMIX	Depth of the mixed layer

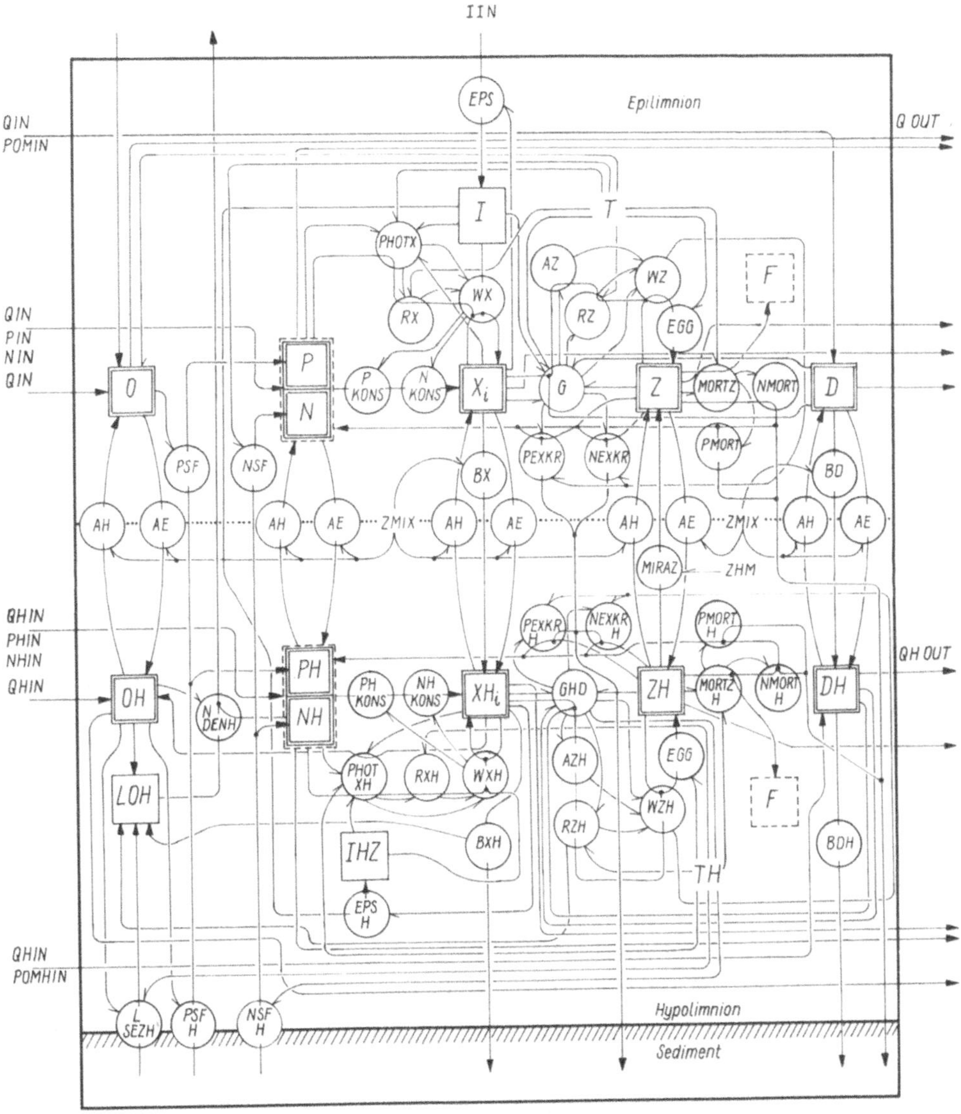

Figure 16.1 Diagrammatic representation of the structure of the model SALMO (filled arrows: flow of energy and matter; open half arrows: flow of information; see Table 16.1 for explanation of the symbols).

number of individual values within limits. The advantage of this type of input data compared with forcing functions consists above all in the possibility of modelling the effects of extraordinary meteorological and hydrological events or of control measures exerted by man. The following input variables are necessary: the depths and volumes of the two layers, the water temperature, the photosynthetic active radiation at the water surface, the water influx and outflow, and the concentrations of dissolved orthophosphate, inorganic nitrogen and particulate organic matter in the influx, as well as a dimensionless value for heavy-rain or melting snow in the drainage basin. All input data employed in the standard simulation of Lake Stechlin refer to the annual course of 1975 when the nuclear power plant was not in operation.

Hitherto the complete system of equations of the most recent version of SALMO shown in Fig. 16.1 has not been published. The development of the basic model (orthophosphate, phytoplankton and zooplankton state variables only) has been presented by Benndorf (1979) and Recknagel (1980). A brief description of this basic version was given by Benndorf & Recknagel (1982).

Checking the validity of SALMO

The most appropriate methods for checking the validity of a model seem to be: (1) retrospective comparisons of simulation results and observations, (2) sensitivity analysis, and (3) prospective scenario analysis. We have used all of these techniques in the validation of SALMO (Benndorf *et al.* 1981a; Benndorf & Recknagel 1982; Recknagel & Benndorf 1982; Koschel *et al.* 1983; Recknagel 1984). In order to achieve the main objective of this contribution – namely, to use a model as a reasonable method for the ecosystem's analysis – the method of retrospective comparisons is of primary interest. The first essential question in this context is whether the degree of conformity between simulated and observed data is within reasonable limits when the unmodified model (without calibration!) is applied to Lake Stechlin. A second problem consists of utilizing the disagreements between the simulation results and observations to find 'weak points' either in the theoretical assumptions involved in the model or in the observation methods. It should be emphasized that the second problem is at least as important as the first one, and, therefore, we will try to combine both the aspects in the following comparisons.

The phytoplankton biomass in the model grows significantly more slowly in spring than was actually observed (Fig. 16.2, top). Considering the low zooplankton biomass during late winter and early spring (Fig. 16.2, bottom) and the resulting low grazing pressure during this period, the reason for the disagreement must be sought mainly in the phytoplankton growth processes. This conclusion has also been confirmed by sensitivity analysis (Recknagel & Benndorf 1982). The most probable explanation seems to be that vertical inhomogeneity in the phytoplankton distribution (higher biomass in the upper water layers) occurs

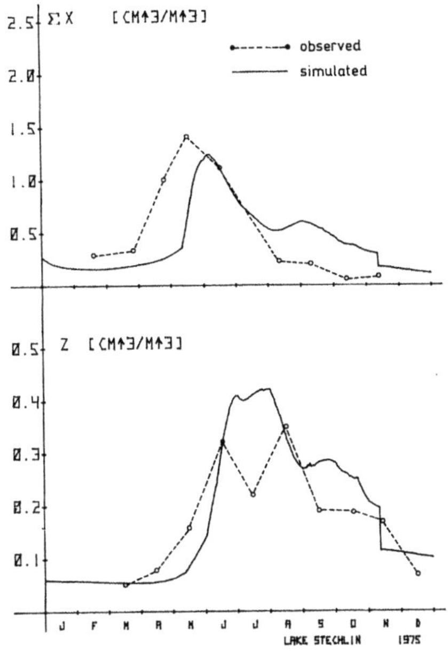

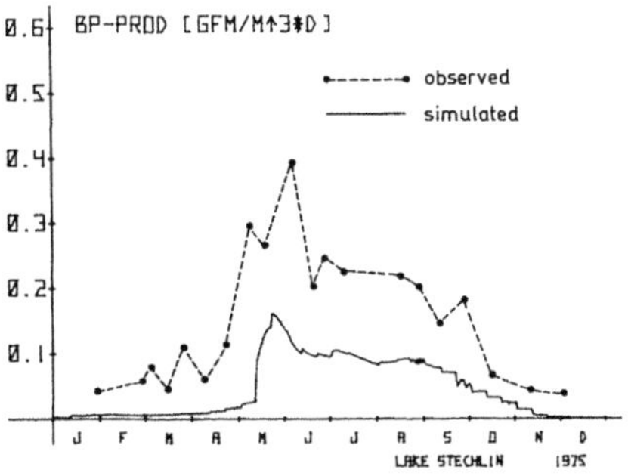

Figure 16.2 Comparison of simulated and observed annual courses of phytoplankton (top) and zooplankton (bottom) in Lake Stechlin. Both curves represent weighted averages for the mixed layer. Observations of phytoplankton taken from Küchler (1980, pers. comm.; see also p. 178 for details). Zooplankton observed values weighted averages from 1978 to 1980 (cf. p. 341).
Figure 16.3 Comparison of simulated and observed annual courses of the gross primary production of the phytoplankton in Lake Stechlin. Averages as Fig. 16.2 (cf. pp. 289 ff.).

during the spring circulation in Lake Stechlin. But in the model a homogeneous vertical distribution of phytoplankton during circulation is assumed. As Lake Stechlin has a mean depth of 23.4 m, such a homogeneous phytoplankton distribution leads, of necessity, to extremely low vertical averages of the simulated primary production (Fig. 16.3, January to early May). The extinction depth (i.e. the product: mixing depth × attenuation coefficient) during that time amounts to 9 . . . 10. In full agreement with findings by Oskam (1978) and Steel (1978), only extremely low phytoplankton production must be expected at this high extinction depth. If, on the other hand, the phytoplankton biomass is more abundant in the upper (well illuminated) water layers – as is actually the case in Lake Stechlin – the vertical average of primary production must be considerably higher. This is confirmed by the measured primary production (Fig. 16.3).

During summer and autumn the phytoplankton biomass in Lake Stechlin is found to be lower than the simulation results (Fig. 16.2, top). Since at the same time the observed primary production is higher than the simulated values (Fig. 16.3), the differences in the phytoplankton biomass during summer and autumn cannot be caused by the same reasons as discussed for the spring period (underestimation of phytoplankton growth by the model), but they must be initiated by an underestimation of the simulated loss processes of phytoplankton. The rather realistic simulation of zooplankton biomass (Fig. 16.2, bottom) indicates that this underestimation could hardly be explained by simulated grazing losses set too low (as has been assumed in an earlier investigation by Benndorf *et al.* 1981a). The most probable explanation seems to be that high sedimentation losses as a consequence of self-flocculation of the phytoplankton due to calcite precipitation play an important role in Lake Stechlin. Maximum rates of calcite sedimentation have actually been found only in July and August (see p. 413–417) when the observed phytoplankton curve decreases significantly more steeply than the simulated curve (Fig. 16.2, top). This explanation is also confirmed by simulation results which involve calcite precipitation (c.f. scenarios 5 and 6 in Figs. 16.9 and 16.10). It should be stressed that these simulations of the calcite scenarios show a decrease of the summer phytoplankton biomass as well as an increase of the photosynthesis rate in comparison with the standard simulation, which is in full accordance with the explanations derived from the comparisons between simulations and observations shown in Figs. 16.2 and 16.3.

In addition to the role of the vertical distribution of phytoplankton and of calcite precipitation, inaccuracies in the model parameters of the photosynthesis and respiration rates might also contribute to the differences between simulated and observed phytoplankton biomass as well as primary production. But we must also consider that the calculated values of phytoplankton biomass can be affected by various uncertainties (cf. p. 159).

The zooplankton state variable (Fig. 16.2, bottom) reveals a much more realistic behaviour then had been found in a first model investigation of Lake Stechlin (cf. Benndorf *et al.* 1981a; Koschel & Kasprzak 1984). The reason for this essential improvement must be sought in the following new constructs involved in the most recent model version:

- Temperature dependence of the zooplankton egg development is considered.
- In addition to phytoplankton, detritus is also used as a food component by zooplankton (cf. Fig. 16.5, bottom). Unfortunately, no observations are available with respect to detritus in 1975, so that the simulation results cannot be compared with measured data.
- As a result of more realistic experimental investigations, the numeric values of some zooplankton parameters have been re-calculated (e.g. parameters of the relationship between light and grazing rate, of the ratio of soluble and particulate phosphate and nitrogen in the zooplankton excrements, of the vertical migration).

The remaining differences between the two zooplankton curves in Fig. 16.2, bottom, refer most probably to indirect effects of calcite precipitation.

The spring decrease of the orthophosphate concentration occurs in Lake Stechlin earlier than calculated by the model (Fig. 16.4, top). This finding accords fully with the phytoplankton dynamics. If phytoplankton growth starts earlier in Lake Stechlin than in the simulation run, then phosphate consumption by the

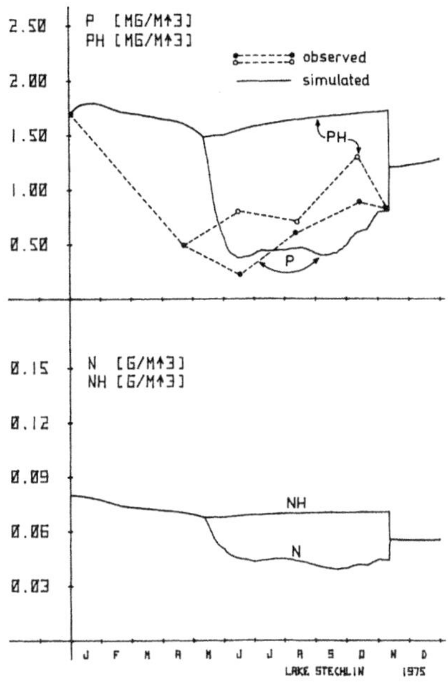

Figure 16.4 Comparison of simulated and observed dynamics of the dissolved orthophosphate (top) and demonstration of the simulated annual course of inorganic nitrogen (bottom) in Lake Stechlin. Averages as Fig. 16.2 (cf. p. 439).

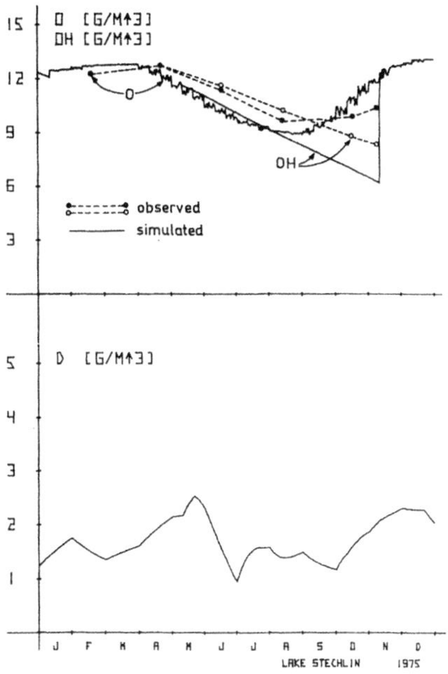

Figure 16.5 Comparison of simulated and observed dynamics of the dissolved oxygen (top) and demonstration of the simulated annual course of the detritus (bottom) in Lake Stechlin. Averages as Fig. 16.2 (cf p. 439).

algae must show the same behaviour, and so the reason for the difference in the spring decrease of phosphate must be identical with the reasons already discussed in connection with the phytoplankton dynamics (mainly vertical inhomogeneity in the phytoplankton distribution). As a direct consequence of the early spring decrease of orthophosphate in Lake Stechlin, a low concentration is already reached when thermal stratification is built up, so that the observed hypolimnetic concentration must be lower than the simulated values. The simulation results of the epilimnetic orthophosphate concentration accord well with the observations (Fig. 16.4, top). As no measured data of the inorganic nitrogen are available for the year 1975, only the simulation results of this state variable are demonstrated in Fig. 16.4, bottom.

The depletion of the hypolimnetic oxygen occurs more slowly in Lake Stechlin than calculated by the model (Fig. 16.5, top), but both of the curves show a high degree of conformity about the linear decrease from the beginning of the summer stagnation period in May to the start of autumn circulation in November. The most important criterion concerning the state variable oxygen – namely the hypolimnetic concentration at the end of the summer stagnation period – deviates to about 2 g $O_2 \cdot m^{-3}$ when the simulation results and the observations are compared. This deviation is not too serious if the long duration of the stratification period of six months is taken into account. By additional simulation runs, in which the oxygen consumption by the sediment has been excluded, the result was gained that the simulated oxygen concentrations in the hypolimnion are identical to the observed data. These additional simulations are not demonstrated in Fig. 16.5. This finding reveals the limits of the quite simple sediment submodel involved in SALMO, which overestimates the sediment impact in deep oligotrophic lakes. Obviously, the role of the sediment in this type of lake can be neglected as far as oxygen consumption (and nutrient release) is concerned.

Apart from the differences discussed above, a rather high degree of conformity is found between the simulation results and observations (Figs. 16.2–5). This optimistic evaluation of the model results is, of course, valid only because SALMO was applied without any adaptation to Lake Stechlin. Therefore, the conclusion is drawn that SALMO can be applied to unknown situations in this lake considering the limits of accuracy determined by the retrospective comparisons between the simulation results and observations.

Evaluation of ecological control mechanisms in the pelagic zone by SALMO

Scenario analysis as a method of research

The enormous complexity of ecosystems is one of the obstacles in the causal analysis of these systems. Neither an experienced limnologist – or even a group of them – nor simple empirical models (e.g. Vollenweider 1975) can consider quantitatively at the same time the numerous direct and indirect effects of the different control factors influencing a lake ecosystem. The aim of this section is to show that dynamic ecological models – assuming that their validity has been checked sufficiently – can provide an appropriate method to overcome this difficulty. As each model, even the most sophisticated one, represents only a simplification of the real world, a model cannot replace the judgement of the limnologist, but it can assist him in checking scientific hypotheses as well as in seeking optimum solutions for water quality management. The most appropriate method to achieve this goal is scenario analysis because the results from this technique are primarily founded upon a relative comparison of the various scenarios rather than using directly the absolute results of simulation runs. In this way, the uncertainties in the absolute simulation results are mitigated (Fedra 1980).

The flow chart of Fig. 16.6 demonstrates the general procedure of a scenario analysis. The procedure starts from the specification of some alternative scenarios. Each scenario represents a set of conditions, which reflects, in turn, a scientific hypothesis or a management strategy, respectively. The next step (scenario control) consists in assigning the different impacts, the effects of which

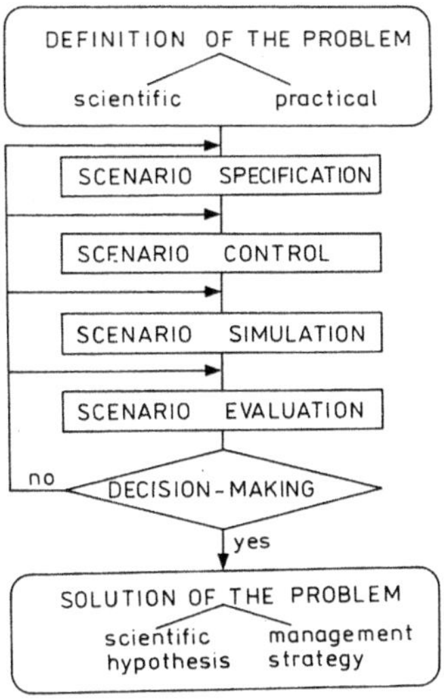

Figure 16.6 General procedure of a scenario analysis.

on the ecosystem are to be investigated, to the control variables which are the decisive links between the impacts and the effects on the lake ecosystem. After this, it can be clearly specified which term, parameter or input variable of the model is suitable to simulate the control variable. Table 16.2 shows an example how the specification and control of the scenarios can be conducted. After simulation of all of the specified scenarios the results are compared with the standard scenario, and the evaluation of these comparisons leads either to an acceptable solution of the problem or the procedure is repeated using modified assumptions.

In the following sections the direct role of the water temperature, the role of the external phosphorus loading, the importance of calcite precipitation and the impact of the fish stocks are subjected to scenario analysis where the standard run is represented by the simulation results shown in Figs. 16.2–5. Three state variables (phytoplankton [sum of the two groups], zooplankton and oxygen) and one rate variable (specific rate of photosynthesis [sum of the two groups]) have been selected for the evaluation of all of the scenarios (Figs. 16.7–11). In this way it is hoped to gain a deeper insight into some of the control mechanisms governing the Lake Stechlin ecosystem.

The role of temperature and thermal stratification

As Table 16.2 demonstrates, two scenarios have been specified with respect to the role of introducing warm cooling water from a nuclear power plant into Lake Stechlin. Scenario 1 refers to an indirect physical effect of the temperature

Table 16.2 Specification and control of the scenarios for the evaluation of the role of various control mechanisms in the metabolism of Lake Stechlin; (for explanation of the symbols see Table 16.1).

Scenario	Impacts and control mechanisms involved	Control variables	Variables or parameters of the model SALMO which are used for simulation of the scenario	Duration of action of the control mechanism or impact
1	Reducing the intensity of vertical mixing due to introducing warm water from a nuclear power plant	Mixing depth	Input variable ZMIX = values used in the standard run minus 1 m	20.5–10.9
2	Increasing the water temperature in the lake due to introducing warm water from a nuclear power plant	Water temperature	Input variables T = values used in the standard run plus 0.5 K	1.1–19.5, 11.9–31.12
			plus 1.7 K	20.5–10.11
			TH = values used in the standard run plus 0.5 K	20.5–10.11
3	Increasing the phosphorus import due to introducing cooling water from a nuclear power plant	External phosphorus load	Input variable PIN = values used in the standard run multiplicated by factor of 2	Whole year
4	Increasing the phosphorus import as in scenario 3 and additionally due to a higher load from the drainage area	External phosphorus load	Input variable PIN = values used in the standard run multiplicated by factor of 4	Whole year
5	Taking into consideration self-flocculation of the phytoplankton due to calcite precipitation	Phytoplankton sedimentation	Parameter VS = 0.5 m · d^{-1}	1.6–30.11
6	Increasing the phosphorus import as in scenario 4 and additionally taking into consideration calcite precipitation as in scenario 5	External phosphorus load; Phytoplankton sedimentation	Input variable PIN as in scenario 4;	Whole year
			Parameter VS = 1.0 m · d^{-1}	1.6–30.9

Table 16.2 Continued.

Scenario	Impacts and control mechanisms involved	Control variables	Variables or parameters of the model SALMO which are used for simulation of the scenario	Duration of action of the control mechanism or impact
7	Increasing the stocks of zooplankton-eating fishes	Zooplankton mortality	Parameters MOT = 0.018 MOMIN = 0.045	Whole year
8	Increasing the stocks of predatory fish	Zooplankton mortality	Parameters MOT = 0.002 MOMIN = 0.005	Whole year

increase, namely to the increasing vertical gradient of temperature which causes an epilimnion depth reduced by about one metre in comparison with the standard simulation run. Fig. 16.7 shows the simulation results. It can be seen that almost no differences occur between scenario 1 and reference senario. The decrease of the epilimnion thickness in the order of magnitude of one metre is obviously without any significant effect on the lake metabolism. (The epilimnion depth under standard conditions amounts to 5 metres in May and increases to 17 metres in November just before the beginning of the autumn circulation).

In contrast to these findings the direct physical effect of increasing the lake temperature by introducing cooling water (scenario 2) cannot be neglected with respect to the biological consequences (Fig. 16.7). From June to September lower zooplankton biomass is calculated in scenario 2 when compared with the standard run, and this fact leads to phytoplankton biomasses which are higher by about 20 per cent during summer in scenario 2. This result must be interpreted as a consequence of increased zooplankton mortality due to enhanced activity of the zooplankton-eating fish when water temperature is increased. The slightly reduced specific photosynthesis rate in scenario 2 precisely during the period of increased phytoplankton biomass indicates that no direct acceleration of the phytoplankton growth can be responsible for this increased phytoplankton biomass. This result might be a good example that quite often indirect ecological effects (in this case: temperature → fish → zooplankton → phytoplankton) predominate over direct effects (in this case: temperature → phytoplankton growth).

The rather small difference in the oxygen concentration in scenario 2 must be attributed to slightly higher respiration rates as well as a lower saturation value of oxygen at higher temperatures.

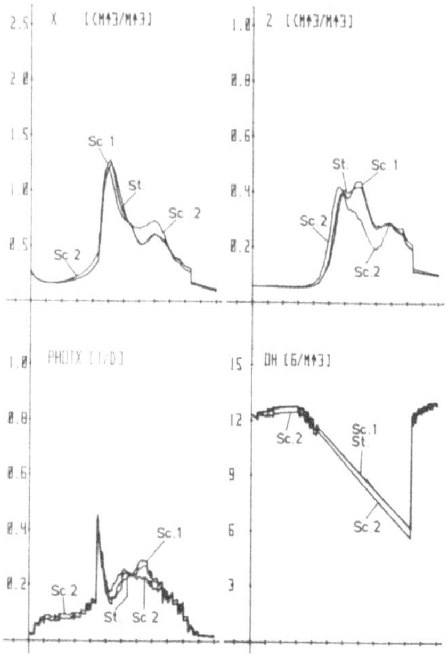

Figure 16.7 Simulation results of scenario 1 (reduced intensity of vertical mixing) and scenario 2 (increased water temperature) in comparison to the standard scenario. Sc. = scenario, St. = standard scenario, see Table 16.1 for explanation of symbols.

The role of the external nutrient load

The bioproduction of Lake Stechlin is primarily limited by phosphorus (see p. 114; p. 307). Nitrogen is available in excess throughout the year (see Fig. 16.4; cf. p. 112). There are four different sources of the external phosphorus loading, namely (1) the import from the drainage area, (2) the atmospheric fallout, (3) the direct input by recreation activities (swimming, boating, angling), and (4) the import through cooling water from a nuclear power plant (cf. p. 96). The causes of this latter loading component must be sought in the following phenomenon; the cooling water, which is taken from the oligotrophic to mesotrophic Lake Nehmitz, containe considerable amounts of plankton. When passing through the heat exchangers of the power plant, the plankton organisms are killed. These dead organisms release a great part of their phosphorus content into the cooling water which enters into Lake Stechlin (cf. p. 424). Consequently, the question arises: what effect does the power plant exert with respect to this increase of the phosphorus load of the lake? Scenario 3 was specified to give an answer to this question (Table 16.2). The results are shown in Fig. 16.8. A slight increase in the biomass both of the phyto- and zooplankton as well as in the photosynthesis rate is found. But it must be emphasized that the lake reveals the typical ability of oligotrophic lakes to compensate an impact as simulated in scenario 3. An increase in the P-load of 100% due to the introduction of cooling water leads only to an increase of 15 to 20% of phytoplankton biomass in spring and summer. The effects with respect to the other variables shown in Fig. 16.8 are even less.

As Table 16.2 indicates, an increase of the phosphorus load to 400% compared with the standard conditions is assumed in scenario 4 as a consequence of introducing cooling water as in scenario 3 and in addition to a threefold increase in other loading components (1) to (3). It could be expected from empirical models (cf. Straškraba 1978) that the 'buffering' capacity of oligotrophic lakes diminishes if the phosphorus loading increases. But this expectation is not confirmed by the simulation results of scenario 4 when compared with that of scenario 3 and of the standard scenario (Fig. 16.8). From Fig. 16.8 the conclusion can be drawn that all of the variables 'respond' more or less proportionally to the rise of the P-loading of oligotrophic Lake Stechlin. But it must be expected that this proportionality is lost if a lake becomes slightly eutrophic (hyper-proportional response) or hyper-eutrophic (hypo-proportional response). For practical eutrophication control, the findings from scenario 3 and 4 confirm the earlier conclusion (Benndorf *et al.* 1981b) that the 'classic' strategy of P-load reduction (Vollenweider 1968) represents the most effective control measure to prevent oligotrophic (and mesotrophic) lakes from eutrophication.

The role of calcite precipitation

In recent years some investigations of the widespread phenomenon of calcite precipitation have shown that it is connected with mechanisms which are able to play a fundamental role for the self-protection of hard-water lakes against eutrophication (e.g. Otsuki & Wetzel 1972; Rossknecht 1980; Koschel *et al.* 1983). The various hypotheses about the relative importance of the different mechanisms linked with calcite precipitation have been evaluated by Koschel *et al.* (1983), using scenario techniques and the model SALMO (cf. pp. 412 ff.). From this investigation the conclusion has been drawn that self-flocculation of the algae by forming aggregates, which consist of calcite crystals, algal cells, detritus and bacteria, predominates over the other mechanisms (e.g. co-precipitation of dissolved phosphates, light limitation due to rise in turbidity). Consequently, we must consider only the self-flocculation mechanism in the scenarios 5 and 6, which are intended for the investigation of the role of calcite precipitation in the pelagic zone of Lake Stechlin (Table 16.2).

Fig. 16.9 shows very clearly the diminution of the phyto- and zooplankton biomass during summer if calcite precipitation is involved in the simulation (scenario 5). The comparison of scenario 5 (Fig. 16.9) with the observed dynamics of phytoplankton and zooplankton (Fig. 16.2) proves that calcite precipitation does play an important role in Lake Stechlin, although the effects are even more considerable in mesotrophic and slightly eutrophic hard-water lakes (see Koschel

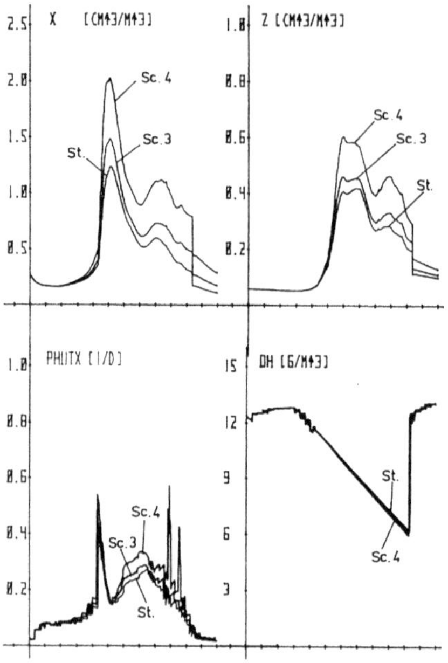

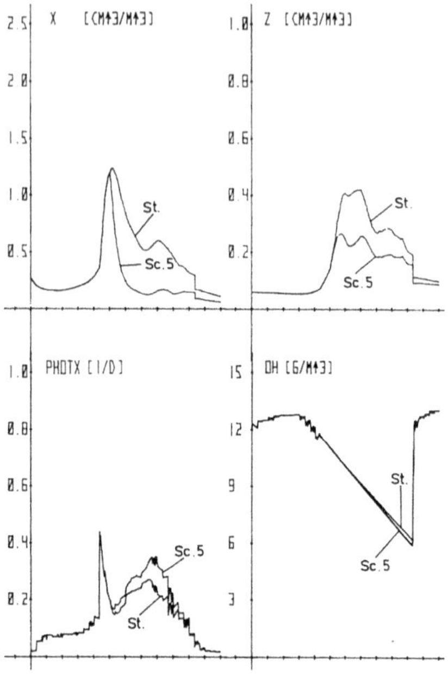

Figure 16.8 Simulation results of scenario 3 (rise in the external phosphorus load by a factor of 2) and scenario 4 (rise in the external phosphorus load by a factor of 4) in comparison with the standard scenario. Abbreviations and symbols as Fig. 16.7.

Figure 16.9 Simulation results of scenario 5 (self-flocculation of phytoplankton due to calcite precipitation) in comparison with the standard scenario. Abbreviations and symbols as Fig. 16.7.

et al. 1983; pp. 412 ff.). Note also that the specific photosynthesis rate is increased by the action of the calcite precipitation and that the hypolimnetic oxygen concentration is decreased. This decrease in oxygen in comparison with the standard run, which neglects calcite precipitation, is only very slight in the hypolimnion of oligotrophic Lake Stechlin, but it is considerably higher in mesotrophic and slightly eutrophic lakes. This latter finding is evident from unpublished simulation results of Lake Arend.

Scenario 6 analyses the effects of calcite precipitation under the conditions of a big rise in the external phophorus loading. The simulation results of scenario 6 are compared in Fig. 16.10 with that of scenario 4, in which a fourfold increase of the phosphorus load has been simulated (but no calcite precipitation was included). A higher intensity of calcite formation and sedimentation must be assumed in scenario 6 in comparison with scenario 5 (see Table 16.2) because of the higher intensity of primary production which is to be expected in scenario 6. It becomes evident from Fig. 16.10 that the calcite mechanism of the 'self-protection' of hard-water lakes acts very effectively after the spring maximum of phytoplankton, so that the summer average of phytoplankton biomass is found to be lower in scenario 6 than in the standard scenario, despite the fact that the phosphorus load is fourfold higher in scenario 6.

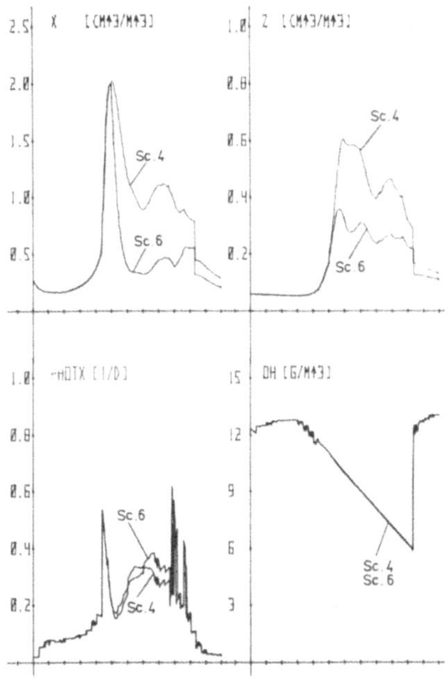

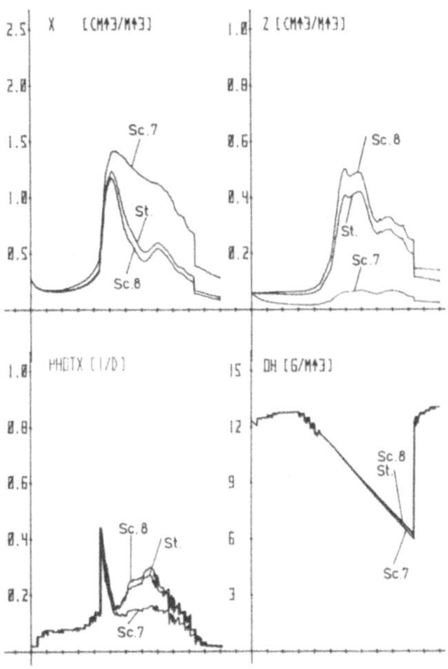

Figure 16.10 Simulation results of scenario 6 (fourfold increase of the external phosphorus load plus self-flocculation of phytoplankton as a consequence of calcite precipitation) in comparison with scenario 4. Abbreviations and symbols as Fig. 16.7.

Figure 16.11 Simulation results of scenario 7 (increased stocks of zooplankton-eating fish) and scenario 8 (increased stocks of predatory fish) in comparison with the standard scenario. Abbreviations and symbols as Fig. 16.7.

The role of the fish stocks

A lake-internal control variable of special limnological interest is the mortality of the herbivorous zooplankton. Its mortality depends mainly on the biomass of carnivorous zooplankton and of zooplankton-eating fish. As carnivorous zooplankton in turn is also eaten by fish, the control of the zooplankton mortality focuses on fish.

It has been known for at least twenty years that dense stocks of small fish drastically reduce zooplankton and may thus cause water quality degradation due to increasing biomass of phytoplankton (e.g. Hrbáček 1958; Merla 1967; Hurlbert *et al.* 1972). Later proposals were made to reverse this process for eutrophication control, namely to reduce plankton-eating fish by suitable management measures to a minimum stock. In experiments this can be done by poisoning all the fish or by maximizing the stock of predatory fish. Today, studies on this principle of 'biomanipulation' in large plastic containers (Shapiro *et al.* 1975; Andersson *et al.* 1978) or in water bodies (Björk 1978; Hrbáček *et al.* 1978; Stenson *et al.* 1978) are known. Also first results from long-term investigations have been gained (Benndorf *et al.* 1984). These studies show clearly an improved water quality as a consequence of biomanipulation (increased Secchi depth, reduced phytoplankton biomass, reduced pH, reduced total phosphate, dominance of larger herbivorous zooplankton organisms). But there are also some unsolved problems in connection with biomanipulation, namely (1) the role of the eutrophication degree and of the hydrophysical conditions (e.g. Benndorf *et al.* 1983), (2) the role of carnivorous zooplankton (e.g. Gliwicz *et al.* 1978; Benndorf *et al.* 1984), and (3) the possibility of changing the dominance structure of the phytoplankton composition (e.g. Stenson *et al.* 1978; McCauley & Briand 1979; Barthelmes 1981; Benndorf *et al.* 1983, 1984). Problem (1) only can be investigated at present by SALMO-simulations. The structure of the recent model version is still too simple for analysing problems (2) and (3) accurately.

Earlier model investigations of reservoirs having different trophic states have shown that biomanipulation is more effective in eutrophic than in mesotrophic waters (Benndorf & Recknagel 1983), but no simulations have been done for oligotrophic water bodies. Scenario 7 was specified to investigate the effects if the stocks of all zooplankton-eating fish were increased to 300 per cent of the standard run (cf. Table 16.2). In full agreement with the empirical findings cited above, a lower zooplankton biomass is calculated which causes in turn a remarkable increase in phytoplankton (Fig. 16.11; scenario 7). It seems to be interesting that the specific photosynthesis rate is reduced due to the lowered grazing pressure on phytoplankton in scenario 7.

A threefold rise of the stocks of predatory fish (Table 16.2; scenario 8) has only slight effects on the components of the pelagic zone of Lake Stechlin (Fig. 16.11). This fact indicates that the zooplankton in Lake Stechlin under the conditions of the standard run is more food-limited than influenced by predation. For eutrophication control, this state should be maintained by a sufficient quantity of predatory fish.

Conclusions

The following conclusions about 'weak points' in the model structure of SALMO can be drawn from the application of the model to Lake Stechlin:

- A more sophisticated consideration of the inhomogeneity of the vertical phytoplankton distribution is necessary.
- Activities to improve the accuracy of the determination of the model parameters by laboratory and field experiments should primarily be directed to the parameters of phytoplankton photosynthesis and respiration.
- A more sophisticated sediment submodel must be evolved, which covers the most essential sediment-water-interrelationships under conditions of all trophic states.
- The zooplankton submodel should be further developed with respect to the size-structure within the herbivorous zooplankton and to the explicit consideration of the carnivorous zooplankton.

Furthermore, the following conclusions about the importance of some control mechanisms for the functioning of the Lake Stechlin ecosystem can be derived from the comparisons between simulations and observations as well as from scenario analysis:

- The sediment of Lake Stechlin does not exert any significant effect on the pelagic zone regarding oxygen consumption and nutrient release.
- The limnological effects of a rise in water temperature due to introducing cooling water from a nuclear power plant (in the order of magnitude tested in the scenario analysis) are only slight. Indirect effects such as feedback control of phytoplankton by zooplankton (and zooplankton by fish) predominate over direct effects on phytoplankton growth.
- As long as the lake is characterized as oligotrophic by its hypolimnetic oxygen concentration at the end of the summer stagnation period, its bioproduction responds proportionally to an increase of the external phosphorus load.
- Calcite precipitation plays a role even in the oligotrophic Lake Stechlin (and most probably also in other oligotrophic hard-water lakes) in controlling phytoplankton biomass and acts, therefore, as an important factor stabilizing the oligotrophic state.
- Under the present conditions, which are typical for an oligotrophic lake, zooplankton is limited more by food than by predation. This state should be maintained by management measures which guarantee a sufficient stock of predatory fish.

References

Ambühl, H. 1982. Eutrophierungskontrollmaßnahmen an Schweizer Mittellandseen. Z. Wasser Abwasser Forsch. 15: 113–120.

Andersson, G., Berggren, H., Cronberg, G. & Gelin, C. 1978. Effects of planktivorous and benthivorous fish on organisms and water chemistry in eutrophic lakes. Hydrobiologia 59: 9–15.

Barthelmes, D. 1981. Hydrobiologische Grundlagen der Binnenfischerei. Jena: VEB Gustav Fischer Verlag. 1–252.

Beck, M.B. 1983. Uncertainty, system identification, and the prediction of water quality. In: Beck, M.B. & van Straten, G. (Eds.): Uncertainty and forecasting of water quality. Berlin: Springer-Verlag. 3–68.

Benndorf, J. 1979. Kausalanalyse, theoretische Synthese und Simulation des Eutrophierungsprozesses in stehenden und gestauten Gewässern. Diss. B. Technische Universität Dresden. 1–165.

Benndorf, J., Koschel, R. & Recknagel F. 1981a. Erste Ergebnisse der Ökosystemanalyse des Pelagials des Stechlinsees mit der Modellmethode. Limnologica 13: 195–202.

Benndorf, J., Uhlmann, D. & Pütz, K. 1981b. Strategies for water quality management in reservoirs in the German Democratic Republic. Water Quality Bulletin 6: 68–73 and 90.

Benndorf, J. & Recknagel, F. 1982. Problems of application of the ecological model SALMO to lakes and reservoirs having various trophic states. Ecol. Modelling 17: 129–145.

Benndorf, J. & Recknagel, F. 1983. Vorhersage der Beschaffenheit neu entstehender oder stark veränderter Ökosysteme. In: Busch, K.F., Uhlmann, D. & Weise, G.: Ingenieurökologie. Jena: VEB Gustav Fischer Verlag. 211–217.

Benndorf, J., Kneschke, H., Kossatz, K. & Penz, E. 1983. Manipulation der pelagischen Nahrungskette durch Raubfischbesatz in einem Experimentalgewässer. Veröff. Mus. Westlausitz Kamenz 7: 41–70.

Benndorf, J., Recknagel, F. & Schultz, H. 1984. Food-web manipulation for a man-made lake (Bautzen reservoir). Study material of the International post-graduate course in ecological approaches to resources development, land management and impact assessment in developing countries, Dresden. VI/3: 20–41.

Björk, S. 1978. Restoration of degraded lake ecosystems. Proceedings MAB Project 5, regional workshop 'Land use impacts on lake and reservoir ecosystems', Warsaw, May 26 to June 2: 196–219.

Fedra, K. 1980. Mathematical modelling – a management tool for aquatic ecosystems? Helgoländer Meeresunters. 34: 221–235.

Gliwicz, Z.M., Hillbricht-Ilkowska, A. & Weglеńska, T. 1978. Contribution of fish and invertebrate predation to the elimination of zooplankton biomass in two Polish lakes. Verh. Int. Verein. Limnol. 20: 1007–1011.

Hrbáček, J. 1958. Typologie und Produktivität der teichartigen Gewässer. Verh. Int. Verein. Limnol. 13: 394–399.

Hrbáček, J., Desertová, B. & Popovský, J. 1978. The influence of the fishstock on the phosphorus-chlorophyll ratio. Verh. Int. Verein. Limnol. 20: 1624–1628.

Hurlbert, S.H., Zedler, J. & Fairbanks, D. 1972. Ecosystem alteration by mosquito-fish (*Gambusia affinis*) predation. Science 175: 639–641.

Jørgenson, S.E., Mejer, M. & Friis, M. 1978. Examination of a lake model. Ecol. Modelling 4: 253–278.

Jørgensen, S.E. 1980. Lake management. Oxford: Pergamon Press. 1–167.

Koschel, R., Benndorf, J., Proft, G. & Recknagel, F. 1983. Calcite precipitation as a natural control mechanism of eutrophication. Arch. Hydrobiol. 98: 380–408.

Koschel, R. & Kasprzak, P. 1984. Model use and verification of ecological parameters for an oligotrophic lake (Lake Stechlin, G.D.R.). Ecol. Modelling 26: 97–101.

Küchler, L. 1980. Phytoplanktonuntersuchungen im Stechlinseegebiet in den Jahren 1973–1975. Limnologica 13: 83–99.

McCauley, E. & Briand, F. 1979. Zooplankton grazing and phytoplankton species richness: field tests of the predation hypothesis. Limnol. Oceanogr. 24: 243–252.

Merla, G. 1967. Wildfische im Karpfenteich – beachtliche Nahrungskonkurrenten für die Karpfen. Dt. Fischerei Ztg. 14: 377–382.

Oskam, G. 1978. Contribution to the discussion. DVGW-Schriften-Reihe Wasser Nr. 16 'Verminderung der Algenentwicklung in Speicherbecken und Talsperren': 112.

Otsuki, A. & Wetzel, R.G. 1972. Co-precipitation of phosphate with carbonates in a marl lake. Limnol. Oceanogr. 17: 763–767.

Patten, B.C., Richardson, Th.H. & Barber, M.C. 1982. Path analysis of a reservoir ecosystem model. Canadian Water Resources J. 7: 252–277.

Recknagel, F. 1980. Systemtechnische Prozedur zur Modellierung und Simulation von Eutrophierungsprozessen in stehenden und gestauten Gewässern. Diss. A. Technische Universität Dresden: 1–155.

Recknagel, F. & Benndorf, J. 1982. Validation of the ecological simulation model SALMO. Int. Revue ges. Hydrobiol. 67: 113–125.

Recknagel, F. 1984. A comprehensive sensitivity analysis for an ecological simulation model. Ecol. Modelling 26: 77–96.

Rossknecht, H. 1980. Phosphatelimination durch autochthone Calcitfällung im Bodensee-Obersee. Arch. Hydrobiol. 88:328–344.

Schellenberger, G., Kozerski, H.P., Behrendt, H. & Hoeg, S. 1978. A mathematical ecosystem model applicable to shallow water bodies. IAHS-Publ. 125: 128–136.

Shapiro, J., Lamarra, V. & Lynch, M. 1975. Biomanipulation. An ecosystem approach to lake restoration. Contribution 143 Limnol. Res. Centre Univ. Minnesota: 1–32.

Somlyody, L. 1983. Lake eutrophication management model. In: Somlyody, L., Herodek, S. & Fischer, J. (Eds.): Eutrophication of shallow lakes: Modelling and management. IIASA Collaborative Proceeding Series CP-83-S3: 207–250.

Steel, J.A. 1978. Contribution to the discussion. DVGW-Schriften-Reihe Wasser Nr. 16 'Verminderung der Algenentwicklung in Speicherbecken und Talsperren': 112.

Stenson, J.A.E., Bohlin, T., Henrikson, L., Nilsson, B.I., Nyman, H.G., Oscarson, H.G. & Larsson, P. 1978. Effects of fish removal from a small lake. Verh. Int. Verein. Limnol. 20: 794–801.

Straškraba, M. 1978. Theoretical considerations on eutrophication. Verh. Int. Verein. Limnol. 20: 2714–2720.

Straškraba, M. 1982. The application of predictive mathematical models of reservoir ecology and water quality. Canadian Water Resources J. 7: 283–318.

Straten, G. van, 1983. Maximum likelihood estimation of parameters and uncertainty in phytoplankton models. In: Beck, M.B. & van Straten, G. (Eds.): Uncertainty and forecasting of water quality. Berlin: Springer-Verlag. 157–171.

Straten, G. van, Shanahan, P. & Somlyody, L. 1983. State-of-the-art discussion on eutrophication modelling: summary and conclusions. In: Somlyody, L., Herodek, S. & Fischer, J. (Eds.): Eutrophication of shallow lakes: Modelling and management. IIASA Collaborative Proceedings Series CP-83-S3: 353–356.

Vollenweider, R.A. 1968. The scientific basis of lake and stream eutrophication, with particular reference to phosphorus and nitrogen as eutrophication factors. Tech. Rep. OECD Paris. DAS/CSI/68, 27: 1–182.

Vollenweider, R.A. 1975. Input-output models. With special reference to the phosphorus loading concept in limnology. Schweiz. Z. Hydrol. 37: 53–84.

Walters, C.J., Park, R.A. & Koonce, J.F. 1980. Dynamic models of lake ecosystems. In: Le Cren, E.D. & Lowe-McConnell, R.H. (Eds.): The functioning of freshwater ecosystems. Cambridge University Press. 455–479.

17. Lake Stechlin area and society

H. Klapper & R. Koschel

Introduction

Developing effective means of managing water quality is particularly necessary for the intensive utilization of lakes, including recreation.

Most lakes and reservoirs in the GDR are like 'islands' surrounded by intensively used agricultural, afforested and urbanized areas. High nutrient loads affect the water quality, and thus it is ever more difficult to maintain oligotrophic or mesotrophic lake ecosystems with a low level of bioproduction and high species diversity as required for recreational purposes or for drinking water supply (Uhlmann 1973).

It is important to find principles and general relationships applicable to water management (Forsberg 1982). Limnologists must meet the demands of our techniological century.

Management of water resources involving an ecological approach must not only produce descriptive documents but also create criteria that allow estimation of both prevailing and desired water quality (Lundquist 1982). In the first place the 'heavy-weight' limnologist himself must be able to solve the problem of lake management and must be able to bridge the gaps between controlling, using or restoring water resources and in some cases also to the economic rating of the ecosystem (Björk 1979). To demonstrate and achieve this point, we want to emphasize some guidelines for the lake quality management of Lake Stechlin and other lakes in its area. For that it is necessary not only to demonstrate limnological characteristics and man-made modifications of lakes and catchment areas but also to make some general remarks about the value of lakes for society.

Casper, S. Jost (ed.), Lake Stechlin. A temperate oligotrophic lake. ISBN 978-94-010-8930-2.

Lake classification and management

Parameters to be considered

Every lake is different and has its own distinct character. Even its detailed description will only portray the complex structure of the ecosystem which, compared with nature, is still simplified. Nevertheless, classification is the method suitable for arranging water data in simple categories.

Although the trophic state forms today the core of any classification of standing waters, it is insufficient to characterize the so classified waters as fitted for specific use. In countries where waters and their catchment areas are subject to intensive use, load and permissible-load have to be considered as essential parameters of their state. In addition, there is a series of other parameters which do not directly reflect the trophic state, but are very important for the various uses. These parameters include salinity and hardness, humic matter in brown water lakes as well as hygienically relevant criteria of toxicity, of bacterial load, of nitrate as water pollutants etc. We have summarized these parameters generally as pollutants.

Classification should satisfy the following elementary conditions:

- The lake should be seen as an inherent part of the catchment area.
- Relation between morphometry of the lake basin and water quality must be considered.
- Water quality must be envisaged as a unity of physical, chemical, and biological properties and components.
- The spatial and temporal dynamics of many of the criteria of the matter balance, annual vertical stratification and horizontal structure must be considered.
- The subdivision into classes should be made so that any changes in the quality due to load (eutrophication) and the efficiency of restorative measures (against eutrophication) are detectable.
- The classification should include information about the suitability of the water for various purposes and, if possible, it should also inform about the repercussions of these usages on the quality, i.e., about the load due to use.
- The manager responsible for water management must be supplied with understandable information.

A classification for standing inland waters meeting the above mentioned requirements has been declared a standard in the GDR since 1974 to classify lakes and since 1982 to classify lakes and reservoirs as well (TGL 27 885/01 1974, TGL 27 885/01 1982). Water quality is evaluated according to three criteria complexes: hydrographic and territorial criteria, trophic criteria, and pollutants. They are each further subdivided into three groups of features. Classification into a particular class requires several criteria from each group of features to be evaluated. Detailed evaluation of the three criteria complexes produces a survey of possible utilization.

We show how to make classification with the mentioned method, and – as far as data are available – Lakes Stechlin, Nehmitz, and Dagow are classified.

Hydrographic and territorial criteria. This complex of criteria is subdivided into following groups of features (Table 17.1): water body morphometry (I), hydrographic relations between catchment area and water body (II), and anthropogenic load (III). The class resulting from groups of features I and II describes the natural state; those from group III represent the anthropogenic load which changes the natural state. The class comprising hydrographic and territorial criteria represents an expectation value which, for the unloaded lake, corresponds with the trophic criteria complex. The comparison of the natural expectation value with the class according to trophic criteria allows conclusions to be drawn about loads and possible sanitation measures.

Among the morphometric criteria (I), the stratification behaviour is of prime importance. Water bodies in which trophogenic and tropholytic processes are separated from each other during the period of stagnation show lower circulation of matter and lower productivity (Patalas 1968; Richardson 1975). The quality of

Table 17.1 Classification of standing water bodies according to their hydrography (water body morphometry, hydrographic relations to the catchment area and load due to utilization).

Criteria	Quality class					
	1	2	3a	3b	4	5
I						
Morphometry						
Stratification	stratified holomictic or meromictic			no stable stratification polymictic		
Mean depth (m)	≥15	≥10	<10	2 to 10	1 to 2	<1
Maximum depth (m)	≥30	≥20	<20	non		
Volume-ratio[a]	≥1.5	≥1.0	<1.0	non		
Mean retention time (a)	≥10	≥1	≥0.2	≥0.1	<0.1	<0.1
II						
Drainage basin (DB)						
Volume quotient (V_q)[b]	≤3	≤5	≤10		≥10	non
Area quotient (F_q)	≤30	≤60	≤300		≥300	non
Share of forest (%)	≥80	≥50	≥20		≥10	<10
III						
Load[c]						
Load characteristic (L)[c]	≤50	≤500	≤2500		≤5000	>5000
PO_4-P-Import ($g \cdot m^{-2} \cdot a^{-1}$)			limits see Fig. 17.1			
N (inorg.)-Import[d] ($g \cdot m^{-2} \cdot a^{-1}$)	<5	<10	<15	non		

[a] Volume ratio = Volume hypolimnion: volume epilimnion, mean of stagnation period.

[b] For dams and reservoirs, the values refer to the normal damming aim.

[c] For load criteria L_1 from direct use see Table 17.2; for load criteria L_2 from uses in the catchment area see Table 17.3; for classification, multiply the population equivalents from direct use by 10 and add to the load from the catchment area.

[d] To be considered only if nitrogen acts as limiting nutrient.

V_q = Area of DB (km^2): water volume ($10^6 \cdot m^3$).

F_q = Area of DB: area of water body.

L = Population equivalent (PE): water volume ($10^6 \cdot m^{-3}$).

water improves with the increasing depth and stability of stratification. Polymictic shallow lakes tend, on the other hand, towards mass development of algae, vegetation colouring and water blooming.

Group II features describe the source of the load-causing substances, the drainage basin. It is related to lake volume (volume quotient) and lake area (area quotient). Among the vegetation factors which are significant to nutrient removal, only a percentage share of forest has been included in the evaluation because of its special protective effect. Disposition to erosion increases in the sequence forest, permanent grassland, forage crop cultivation, root and tuber crop cultivation.

In countries with intensive soil and water utilization, the point sources of phosphorus-load are of special significance (Müller 1978). They are summarized in the following group of features of anthropogenic load (III). Here two possible classifications are offered. The first possibility requires statistic ascertainment of the load-causing water uses (L_1) and uses in the catchment area (L_2). The population equivalents related to the water volume become the subject of classification (Tables 17.2, 17.3). Once sufficient data are available, it is possible to infer the trophic states from load per unit area of the water and the ratio of mean depth to mean retention time directly from the Vollenweider-scheme (Vollen-

Table 17.2 Load characteristics (L_1) for direct utilization of waters.

Type of load	Population equivalents[a] BOD[5]	total N	In terms of total P
Duck breeding			
with free access to the water 100 ducks per day	14	8 to 12	16
Duck farming			
on slatted floors with floating troughs[b] 100 ducks per day	11	7	10
Geese breeding			
with free access to the water 100 geese per day	42	24 to 36	48
Trout cultivation			
in channels, per 1000 kg trouts rate of stocking and day[b]	85	110	110
Fish intensive units			
average load per day at an annual production of 1000 kg (net cages, channels)	30	30	30
Recreation – use			
Loads caused by primary contact (100 bathing persons per day; seasonal average)	2	2	2

[a] 1 Population equivalent = 54 g BOD_5, 13 g N, 2 g P. For P-limited water bodies and N-limited water bodies use the values referring to P and those referring to N, respectively. For water bodies not nutrient-limited, use the population equivalents referring to the BOD_5.

[b] Approximate 50% of the nutrients are particle bound so that it is possible to reduce the load by almost 50% in settling basins.

weider 1976) which has been extended and modified with regard to orthophosphate (Benndorf 1979; Fig. 17.1).

Trophic criteria. This complex comprises those criteria which are essential for the level of primary production in standing waters and which are used for determination of the trophic classes subdivided into the following groups of features (Table 17.4): oxygen balance (I), nutrient balance (II), bioproduction (III).

The nutrient content is a prerequisite for possible production in the water body. Classification is impeded by seasonal variations. The conditions prevailing prior to phytoplankton production in early spring give relatively reliable information about the bioproduction in the pelagic zone to be expected. Since part of the phosphorus contained in lakes rich in inorganic carbon will be precipitated in the course of biogenic decalcification, the bioproduction remains under the value to be expected (Koschel *et al.* 1983; cf. p. 417, p. 447). Therefore, classification by the phosphate concentration should be performed separately for water with a high temporary hardness from that with a low temporary hardness.

The characteristics of primary production are suited best to provide a biological

Table 17.3 Load characteristics (L_2) for evaluating the impact of utilization on waters in the catchment area.

Type of load	Population equivalents[a] BOD$_5$ (1 = 54 g)	total N (1 = 13 g)	Referring to total P (1 = 2 g)
1 Inhabitant; no treatment	1.0	1.0	1.0
1 Inhabitant; septic tanks	0.7	0.8	0.8
1 Inhabitant; mechanical treatment	0.7	0.8	0.7
1 Inhabitant; biological treatment	0.2	0.4	0.4
1 Inhabitant; land treatment or advanced treatment with P-precipitation	0.1	0.3	0.1
1 Livestock unit; (LU)[b], diffuse after agricultural utilization (dry-farming)	1.0	5.0	1.0
1 Livestock unit; diffuse effluent after agricultural utilization (liquid manure)	2.0	12.0	5.0
1 Livestock unit; at pasture farming on permanent grassland[c]	0.5	3.0	0.6
Area effluent; from crop production (without the influence of organic manuring resulting from livestock production) per hectare of agricultural area[d]		9.0	0.5

[a] See Table 17.2.
[b] 1 LU = 1 cattle or horse, 6 pigs, 14 sheep or 150 head of poultry.
[c] 6 months at pasture, 6 months of dry feeding, no access to a water body, only slight slope.
[d] The annual N-removal rate depends on types of crops and respective amounts of fertilizers applied: for grassland approximate 10; clover 30; grain 40; root crops 70; vegetables 80 kg N · ha^{-1}.

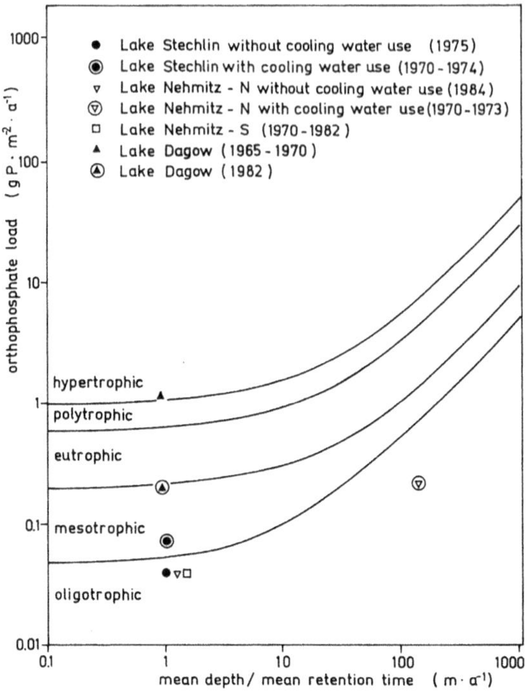

Figure 17.1 Graphic determination of the trophic degree of Lakes Stechlin, Nehmitz and Dagow to be expected on the basis of the orthophosphate-load per unit area.

basis for trophic classes (cf. p. 461). However, the user is more interested in the biogenic substances, i.e., in the plankton biomass of the water than in its photosynthetic capacity. Both the measurements of primary production and the counting of plankton are very time consuming. Seston determination by gravimetric methods is not reliable because of the allochthonous matter in shallow waters and in those with a strong current.

For practice, chlorophyll determination and the simple method of Secchi depth measurement are of prime importance. Frequent measurement of Secchi depth made for example by fishermen provides valuable information for classification.

But also zooplankton can undoubtedly play a major role in controlling phytoplankton biomass and, consequently, the trophic state of lakes. Summarizing zooplankton biomass observation, one can determine relations between the trophic levels. As a rule eutrophic or highly polluted lakes have higher average biomass values of crustacean plankton per cubic metre than meso- or oligotrophic lakes (cf. p. 461).

Pollutants. This complex comprises those groups of features which are also significant from the point of view of water utilization (Table 17.5): salinity (I), different universal pollutants (II), and hygienically relevant pollutants (III).

Salinization is important mainly with regard to industrial use (as boiler feeding water) and irrigation.

The group of different universal pollutants comprises mainly water conditions as existing in 'minerogenic', 'acidotrophic' or 'iron-containing' artificial lakes in mining areas as well as humic (formerly: 'dystrophic') water conditions. These are criteria which disturb the trophic relations between nutrient supply and bioproduction calibrated for neutral clear water lakes.

The group of hygienically relevant pollutants (e.g. nitrate, heavy metals,

Table 17.4 Classification by trophic criteria.

Criteria	Quality class 1	2	3a	3b	4	5
I						
Oxygen balance						
O_2-saturation (%)[a]	90 to 120	80 to 150	60 to 200		20 to 300	0 to 500
O_2-content ($g \cdot m^{-3}$)[b]	≥6	≥1	anaerobic	non		
II						
Nutrient balance[c]						
1. Spring circulation[d]						
Orthophosphate	≤0.005	≤0.01	≤0.03		≤0.05	>0.05
($g\ P \cdot m^{-3}$)[e]	≤0.005	≤0.015	≤0.2		≤1.2	>1.2
Total phosphorus ($g\ P \cdot m^{-3}$)[e]	≤0.015	≤0.025	≤0.04		≤0.06	>0.06
	≤0.015	≤0.045	≤0.3		≤1.5	>1.5
Dissolved inorganic nitrogen ($g\ N \cdot m^{-3}$)[f]	≤0.3	≤0.5	≤1.0		≤1.5	>1.5
2. Summer stagnation[g]						
Orthophosphate ($g\ P \cdot m^{-3}$)	0 to 0.002	0 to 0.005	0 to 0.1		>0.1	>0.5
Total phosphorus ($g\ P \cdot m^{-3}$)	≤0.015	≤0.04	0.04 to 0.3		>0.3	>0.5
Dissolved inorganic nitrogen ($g\ N \cdot m^{-3}$)	≤0.01	≤0.03	≤0.1		>0.1	>0.5
III						
Bioproduction						
1. Primary production phytoplankton						
Annual primary production ($g\ C \cdot m^{-2} \cdot a^{-1}$)	≤120.0	120 to 250	250 to 400		400 to 500	>500
Primary production-ratio[h] (%)	≤15.0	15 to 30	30 to 75		75 to 90	>90
2. Phytoplankton biomass[7]						
Phytoplankton volume ($cm^3 \cdot m^{-3}$)	≤1.5	≤5	5 to 10	10 to 20	20 to 30	>30
Chlorophyll a ($mg \cdot m^{-3}$)	≤3.0	≤10	10 to 20	20 to 40	40 to 60	>60
3. Transmissibility[g]						
Secchi depth (m)	≤6.0	≤4	≤1.0		≤0.5	<0.5
Extinct. coefficient [$E_{400-700}$] ($ln \cdot m^{-1}$)	≤0.5	≤0.6	≤1.5		>1.3	>2.5
4. Zooplankton biomass[g]						
Zooplankton ($g\ d.w. \cdot m^{-3}$)	<0.1	<0.3	<0.8		>0.8	0 to >0.8
Trophic state	oligotrophic	mesotrophic	eutrophic stratified	eutrophic unstratified	polytrophic	hyper-trophic

[a] At the surface (Summer, calm, day-night).

[b] O_2-measurements = 5 m above bottom at the end of summer stagnation.

[c] For lakes without a through current or retention times >1 year [for lakes and reservoirs with strong through current use the load per unit area ($P\ g \cdot m^{-2} \cdot a^{-1}$) only].

[d] Mean value of total water column involved in turnover at the beginning of the spring circulation.

[e] The lower values and higher values of the 'from-to-range' of the appropriate quality class are applicable to waters with a low alkalinity (many reservoirs in solid rock mountainous regions) and to those with a high temporary hardness (many lakes in lowland regions), respectively.

[f] To be considered only if nitrogen is clearly the limiting nutrient.

[g] Mean value for epilimnion (April–September).

[h] Percentage of a_{opt} in daily primary production.

Table 17.5 Pollutants.

Criteria[a]	Quality class					
	1	2	3a	3b	4	5
I						
Salinity						
Calcium Ca^{2+}	<60	<100	<150		<250	<250
Magnesium Mg^{2+}	<25	<50	<100		<150	>150
Sodium Na^{+}	<30	<70	<150		<300	>300
Chloride Cl^{-}	<50	<100	<250		<500	>500
Sulphate SO_4^{2-}	<100	<150	<350		<500	>500
Total hardness (as CaO)	≤100	≤150	≤300		≤500	>500
Temporary hardness (as CaO)	≤70	≤120	≤250			
Total salinity	350	750	1500		2500	2500
II						
Different universal pollutants						
Iron, total Fe	≤0.1	≤0.5	≤1.0		≤3.0	>3.0
Manganese Mn	≤0.02	≤0.1	≤0.2		≤0.5	>0.5
pH-value						
in the epilimnion of neutral waters	6.5 to 8	7 to 8.5	7 to 9	7 to 9.5	6.5 to 10	6 to 11
in minerogenic acidotrophic waters	>6	5 to 6	4 to 5		3 to 4	<3
Ammonium in the epilimnion NH_4	not detectable	not detectable	≤0.1		≤1.0	>1.0
Hydrogen sulphide in the epilimnion S^{2-}	not detectable	not detectable	not detectable		not detectable	>0.01
Hydrogen sulphide in the hypolimnion S^{2-}	not detectable	not detectable	>0.01			
Humic compounds as chemical oxygen demand-Mn (humic acid) O_2	<3	<5	<10		<20	>20
Humic standard	oligo-humous	oligo-humous	oligohumous		meso-humous	polyhumous
III						
Hygienically relevant pollutants						
Nitrate NO_3^-						
Maximum concentration	<15	<30	<40		>40	
Mean annual concentration	<10	<20	<30		>30	
Fluoride F^-	≤1.0	≤1.0	≤1.2		<5.0	>5.0
Phenols, water vapour volatile Phe	not detectable	not detectable	≤0.005		≤0.5	>0.5
Detergents Ten	not detectable	≤0.1	≤0.2		≤2.0	>2.0
Dissolved heavy metals						
Copper Cu	not detectable	not detectable	≤0.05		≤1.0	>1.0
Chromium (III) Cr^{3+}	not detectable	≤0.1	≤0.2		≤1.0	>1.0
Chromium (VI) Cr^{6+}	not detectable	≤0.01	≤0.02		≤0.1	>0.1
Lead Pb	not detectable	≤0.03	≤0.05		≤0.5	>0.5
Arsenic As	not detectable	≤0.01	≤0.05		≤0.2	>0.2
Zinc Zn	not detectable	not detectable	≤0.01		≤0.1	>0.1
Cadmium Cd	not detectable	not detectable	≤0.001		≤0.01	>0.01
Cobalt Co	not detectable	≤0.01	≤0.1		≤1.0	>1.0
Nickel Ni	not detectable	≤0.05	≤0.2		≤1.0	>1.0
Mercury Hg	not detectable	not detectable	≤0.001		≤0.01	>0.01
Polycyclic aromatic hydrocarbons PAK	not detectable	not detectable	≤0.0001		≤0.001	>0.001
Organophosphoric hydrocarbons POK	not detectable	not detectable	≤0.001		≤0.01	>0.01
Bacteriological criteria						
Coliforms	10 ml[b]	1 ml[b]	≤100[c]		≤1000[c]	>1000[c]
Enterococci	100 ml[b]	10 ml[b]	≤10[c]		≤100[c]	>100[c]

[a] Mean value of the total water column (for Mn in stratified waters mean value of hypolimnion); all values of ions $g \cdot m^{-3}$.

[b] Not detectable in the mentioned volumes.

[c] Colonies per ml.

detergents) has been formed mainly with a view to utilization as drinking water, for bathing and for conservation of natural biocoenoses.

Classification of possible uses

For the classification of standing water bodies for types of utilization see, the evaluating summary of three quality classes determined for the individual complexes of characteristic features (Table 17.6).

For reservoirs and lakes, management objectives should be stipulated. In the case of multiple use priorities must be taken into account. The respective binding regulations should be consulted when evaluating particular types of use. If the utilization requirements to be satisfied are more demanding than expressed in the given class of utilization, suitable measures must be taken to improve the quality. In special cases the class of utilization can be determined by a sole criterion, e.g., nitrate with regard to drinking water.

Classification of Lakes Stechlin, Nehmitz and Dagow. The classification of these three lakes shows how a relatively complete picture of water quality can be made with the help of this method. As for the water body morphometry and drainage basin hydrography, only Lake Stechlin is in good accordance with the norms of class '1' (Table 17.7; Fig. 17.2). About the load, the higher level for Lake Stechlin is, first of all, a result of the cooling water circulation. The nutrient import from Lake Nehmitz and the release of nutrient through heating of cooling water are significant eutrophication factors (cf. p. 424, p. 446; Fig. 17.2).

The north and south basins of Lake Nehmitz are by morphometric features class '3a'. The drainage basin hydrography corresponds well to class '1'. The effect of relatively high phosphorus-load (cf. p. 89) in Lake Nehmitz-N is diminished by the short retention time of water (Fig. 17.2). So really the current of Lake Stechlin water improves the hydrographic class in Lake Nehmitz-N.

Lake Dagow has, compared with the other two lakes, the worst natural prerequisites for water quality, with only 9 m maximum depth and a very small hypolimnion. Also the nutrient load was very high some years ago (Fig. 17.2). Then there was serious loading from Carp and duck farming as well as from sewage, but both have been removed by sanitation measures. Of course there are nutrients accumulated in the water body and in the sediments, which may cause an internal fertilization for many years. So as for hydrographic criteria, including the load, the quality class for Lake Dagow is '3'.

With respect to trophic criteria Lake Stechlin takes the first place of all lakes in the GDR. Oxygen in the epilimnion is almost permanently saturated, with 55% saturation near the bottom, and at the end of the summer stagnation period (cf p. 119). The nutrient content is extremely low (cf. p. 114). During the spring and the summer stagnation orthophosphate is often not detectable because of the very

Table 17.6 Classification of possible uses.

Type of use	Quality class					
	1	2	3a	3b	4	5
Drinking water*	simple treatment	normal treatment	comprehensive treatment; treatment partly complicated	complicated treatment; treatment partly impaired	unusable	unusable
Industrial water	simple treatment	normal treatment	comprehensive treatment; treatment complicated at times	complicated treatment; treatment partly impaired	unusable with reservations for particular purposes	unusable
Cooling water	perfect	perfect	perfect	can be used	usable with reservations	usable with reservations or unusable
Irrigation						
Organic load	perfect	perfect	perfect	perfect	usable	usable with reservation
Mineral load	perfect	perfect	usable	usable	usable with reservations	unusable
Hygienic load	perfect	perfect	usable	usable	usable with reservations	unusable
Recreation	perfect	perfect	perfect	usable	usable with reservations	unusable
Bathing	inadmissible when used as drinking water; otherwise perfect	inadmissible when used as drinking water; otherwise perfect	inadmissible when used as drinking water; otherwise usable	inadmissible when used as drinking water; otherwise usable	to be quered	unusable
Fishing	on a natural feeding basis	on a natural feeding basis	when used as drinking water or for bathing on a natural feeding basis	when used as drinking water or for bathing on a natural feeding basis	for intensive use with artifical feeding usable with reservations (NH_4^+, pH)	intensive use with artificial feeding at risk (O_2, NH_4^+, pH)
Water-fowl	inadmissible	inadmissible	inadmissible	inadmissible	admissible	admissible
Navigation and boating with internal combustion engines	inadmissible	inadmissible	inadmissible when used as drinking water; usable with reservations when used for bathing	inadmissible when used as drinking water; usable with reservations when used for bathing	admissible	admissible

* Examples of drinking water treatment from raw water from reservoirs and lakes with increasing trophic status:

Simple: Rapid filtration, chlorination;

Normal: Aluminium sulphate precipitation, floc filtration if necessary with pulverized active carbon, lime dosage, chlorination;

Comprehensive: Micro screening, aluminium sulphate precipitation, floc filtration, active carbon filtration (or slow sand filtration), lime dosage, chlorination;

Complicated: Denitrification, aluminium sulphate precipitation when adding flocculation aids, active carbon filtration, ozonization.

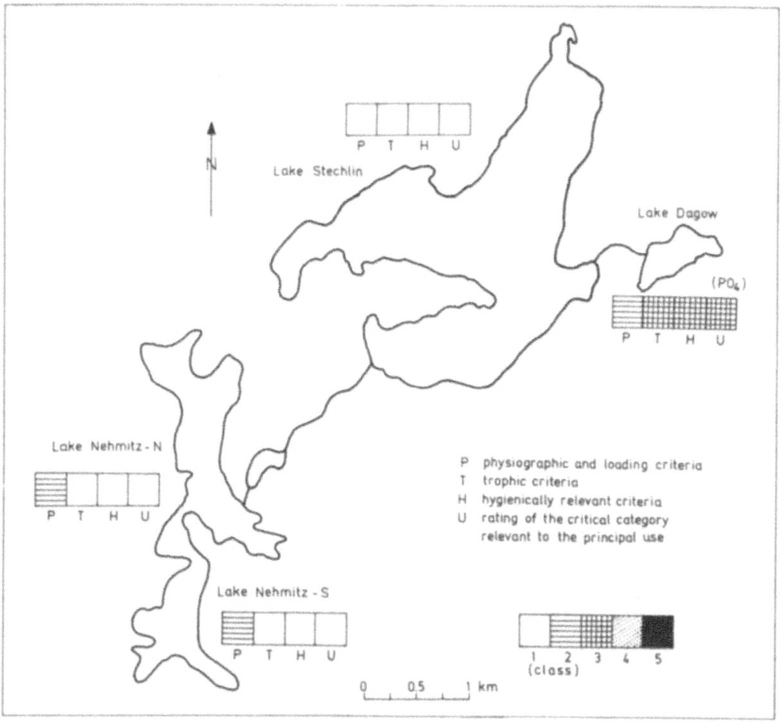

Figure 17.2 Mapping of water quality of Lakes Stechlin, Nehmitz and Dagow.

effective phosphorus incorporation by microplankton (cf. p. 311). With a low phytoplankton biomass a relatively high primary production is realized due to the high specific assimilation rate of the prevailing nanoplankton (cf. p. 304).

Lake Stechlin has to be classified as oligotrophic, i.e. class '1'.

The two basins of Lake Nehmitz are almost oxygen deficient in the deep water layer (cf. p. 120). Due to the cooling water circuit the nutrient content for stratified water column of Lake Nehmitz-N is nearly identical to that of Lake Stechlin. Nevertheless, the productivity in Lake Nehmitz-N is remarkably higher, for the morphometry of the lake basin is more conducive for bioproduction.

Lake Nehmitz has to be classified as oligotrophic to mesotrophic: class '1'–'2'.

Lake Dagow today belongs to trophic class '3' (eutrophic). But the drainage basin hydrography promises good chances to improve the water quality by protective measures (Fig. 17.2). Such improvement will be favourable also for Lake Stechlin, which receives some inflow from Lake Dagow (p. 31).

The general classification of water quality is completed through the complex of pollutants (Table 17.7). The water quality and the use of Lakes Stechlin, Nehmitz and Dagow are not affected by pollutants.

Table 17.7 Classification of Lakes Stechlin, Nehmitz and Dagow.

Criteria	Lake Stechlin		Lake Nehmitz-N		Lake Nehmitz-S		Lake Dagow	
	Value	Class	Value	Class	Value	Class	Value	Class
A. Hydrographic features								
Stratification	holomictic		holomictic		holomictic		holomictic	
Mean depth (m)	23.7	1	5.9	3a	7.1	3a	4.9	3a
Maximal depth (m)	68.0	1	18.6	3a	18.0	3a	9.0	3a
Volume-ratio	1.80	1	0.22	3a	0.25	3a	0.28	3a
Mean retention time (a)	>50 [a]		0.05[a]–>5		>5		5	
I								
Water body morphometry								
Volume quotient	0.1	1	0.6	1	0.6	1	4.2	2
Area quotient	2.9	1	3.6	1	3.6	1	21	1
Share of forest (%)	95	1	90	1	90	1	50	2
II								
Drainage basin hydrography		1		1		1		2
PO_4-P-Import ($gP \cdot m^{-2} \cdot a^{-1}$)	0.07[a] 0.04[b]	2–1	0.2	1[a] 2[b]	0.04	1	0.2[c]	2
III								
Load		2		2		1		2
Quality class as to hydrographic features		1		2		2		2
B. Trophic features								
O_2-saturation (%)	90–120	1	90–120	1	90–120	1	80–200	2
O_2-content (hypolimnion) ($g \cdot m^{-3}$)	7.0	1	<0.5	3a	<0.5	3a	0	3a
I								
Oxygen balance		1		2		2		3
1. *Spring circulation*								
Orthophosphate ($mg\ P \cdot m^{-3}$)	2.5	1	2.0	1	2.5	1	100	3
Total phosphorus ($mg\ P \cdot m^{-3}$)	10	1	10	1	12	1	140	3
Dissolved nitrogen ($g\ N \cdot m^{-3}$)	0.2	1	0.2	1	0.1	1	0.5	2
2. *Summer stagnation*								
Orthophosphate ($mg\ P \cdot m^{-3}$)	<1.0	1	<1.0	1	<1.0	1	5	3
Total phosphorus ($mg\ P \cdot m^{-3}$)	8	1	10	1	12	1	70	3
Dissolved nitrogen ($g\ N \cdot m^{-3}$)	<0.1	1	<0.1	1	<0.1	1	0.1	3
II								
Nutrient balance		1		1		1		3
Annual primary production ($g\ C \cdot m^{-2} \cdot a^{-1}$)	95	1	140	2	145	2	290	3
Primary production – ratio (%)	16	2	18	2	18	2	35	3
Phytoplankton volume ($cm^3 \cdot m^{-3}$)	0.5	1	0.8	1	0.8	1	10	3
Chlorophyll a ($mg \cdot m^{-3}$)	1	1	1.5	1	1.5	1	20	3
Secchi depth (m)	7.2	1	6.4	1	6.0	1	1.5	3
$E_{400-700}$ ($ln \cdot m^{-1}$)	0.4	1	0.5	1	0.5	1	1.2	3
Zooplankton biomass ($g\ d.w. \cdot m^{-3}$)	0.05	1	0.09	1	–	–	0.5	3
III								
Bioproduction		1		1		1		3
Trophic state	oligotrophic		oligo-mesotrophic		oligo-mesotrophic		eutrophic	
Quality class as to trophic features		1		1		1		3
C. Pollutants								
Magnesium ($g \cdot m^{-3}$)	6.0	1	6.0	1	6.0	1	6.0	1
Sodium ($g \cdot m^{-3}$)	10.0	1	10.0	1	10.0	1	14.0	1
Chloride ($g \cdot m^{-3}$)	16.0	1	15.0	1	15.0	1	20.0	1
Sulphate ($g \cdot m^{-3}$)	20.0	1	30.0	1	35.0	1	80.0	1
Temporary hardness ($g\ CaO \cdot m^{-3}$)	55.0	1	50.0	1	50.0	1	75.0	2
Total salinity value	2.6	1	2.8	1	2.8	1	4.0	1
I								
Salinity		1 (1.0)		1 (1.0)		1 (1.0)		1 (1.2)
Iron ($g \cdot m^{-3}$)	0.2	2	0.4	2	0.4	2	0.2	2

Table 17.7 Continued.

Criteria	Lake Stechlin		Lake Nehmitz-N		Lake Nehmitz-S		Lake Dagow	
	Value	Class	Value	Class	Value	Class	Value	Class
pH-value	8.1–8.3	2	8.0–8.2	2	8.0–8.2	2	8.0–9.0	3
Ammonium ($g \cdot m^{-3}$)	<0.05	1–2	<0.05	1–2	<0.05	1–2	0.1	3
Hydrogen sulphide (Epilimnion) ($g \cdot m^{-3}$)	n.n.	1	n.n.	1	n.n.	1	n.n.	1
Hydrogen sulphide (hypolimnion) ($g \cdot m^{-3}$)	n.n.	1	n.n.	1	n.n.	1	>0.01	3
II								
Different universal pollutants[d]		1 (1.5)		1 (1.5)		1 (1.5)		2 (2.8)
Nitrate								
Maxima ($g \cdot m^{-3}$)	0.1	1	0.1	1	0.1	1	0.3	1
Mean ($g \cdot m^{-3}$)	<0.05	1	<0.05	1	<0.05	1	<0.1	1
Chromium ($g \cdot m^{-3}$)	n.n.	1	n.n.	1	n.n.	1	n.n.	1
Lead ($g \cdot m^{-3}$)	n.n.	1	n.n.	1	n.n.	1	n.n.	1
Cadmium ($g \cdot m^{-3}$)	n.n.	1	n.n.	1	n.n.	1	n.n.	1
Nickel ($g \cdot m^{-3}$)	n.n.	1	n.n.	1	n.n.	1	n.n.	1
Coliforms colonies ($g \cdot m^{-3}$)	0	1	0	1	0	1	0	1
Enterococci colonies ($g \cdot m^{-3}$)	0	1	0	1	0	1	0	1
III								
Hygienically relevant pollutants[1]		1 (1.0)		1 (1.0)		1 (1.0)		1 (1.0)
Quality class as to salinity, different universal and hygienically relevant pollutants		1 (1.2)		1 (1.2)		1 (1.2)		2 (1.7)

[a] With cooling water influence.
[b] Without cooling water.
[c] At present.
[d] We thank Dr. Knaack for analyses of heavy metals and hygienic data and Dipl.-Biol. Rodhe for analyses of salinity datas.

General conclusions for lake management

Performance of classification leads to general conclusions about how to utilize lakes as natural resources to the best benefit for society.

Lakes Stechlin and Nehmitz are classified as utilization class '1', i.e., the water can be used, for instance, for drinking purposes with simple treatment and little expense. Lake Dagow with utilization class '3' cannot be subjected to multiple use as in Table 17.6.

Lakes such as Lakes Stechlin and Nehmitz, together with further clear-water lakes and reservoirs, are listed in the classification standard as 'waters especially worthy of being protected'. Protection concerns every kind of pollution, particularly nutrient load from diffuse and particular sources in the drainage basin and from water utilization.

In November 1978 the Regional Council of Potsdam made a decision on 'The effective utilization of standing waters in Potsdam Region'.

For Lakes Stechlin, Nehmitz and Dagow use as a nature reserve was given highest priority. Further permitted uses are the following (Rat des Bezirkes Potsdam 1978, 1982).

Lake Stechlin: (a) limited recreation with primary body contact (bathing, diving); (b) recreation without body contact (e.g. walking, hunting (Fig. 17.3); (c) traditional fishing (without feeding, or electrofishing, or pesticide application), with protection of reed belts and limitation of angleing.

Lake Nehmitz: (a) limited recreation with primary body contact, especially during the breeding period of the White-Tailed Eagle and Osprey; (b) limited recreation without body contact; (c) traditional fishing (as Lake Stechlin).

Lake Dagow: (a) recreation with primary body contact; (b) recreation without body contact; (c) traditional fishing (as Lake Stechlin).

The management directed towards a preservation of the oligotrophic state of Lake Stechlin has to consider that there are some established utilizations with fertilizing effects. The yearly use of 97 million cubic metres of cooling water from Lake Nehmitz by the nuclear power plant and its discharge into Lake Stechlin via an outlet canal has loading effects on the life of Lake Stechlin (p. 98). There have been many diffuse fertilizing sources in the drainage basin, the influence of which had to be curbed with help of a sanitation programme (cf. p. 472). It must be considered that body contact recreation is a loading factor too (p. 458; Table 17.2). Thus recreational use must be limited quantitatively, and loading has to be decreased by means of good sanitary equipment at the bathing sites (p. 472).

Lake Stechlin as a water resource

Strategy of water utilization and planned water quality

From an economic point of view water as a resource should be used with highest benefit to society. Water belongs to those resources which, to a certain degree, are able to regenerate their quality by natural powers after deterioration through use. This gratis effect of nature, called biological self-purification potential, has to be thoroughly planned, and the loading capacity should not be exceeded by overburdening. With classification and the loading concept developed by Vollenweider (1976; see p. 53–83) a reliable tool is available for estimating the permissible load.

The research on phosphorus budget (p. 89) has shown that at the end of the sixties the external load of Lake Stechlin already exceeded that permitted for oligotrophic lakes (Fig. 17.2). Planning water quality development, i.e., the regeneration to a former state and preservation of the oligotrophic conditions, it was differentiated between loads to be removed (e.g., from sewage inlet, garbage deposition, Carp-, Trout-, and duck breeding), permitted limited loads (e.g., cooling water circuit, bathing) and loads which cannot be controlled in practice (e.g., air pollution).

Lake Stechlin is a standard model of an oligotrophic temperate lowland clear-water lake. The natural prerequisites for its excellent water quality are small

Figure 17.3 Holiday-makers walking across ice-covered Lake Stechlin. Photo W. Scheffler.

drainage basin, lack of considerable surface inflows, favourable morphometry of lake basin, predominantly surrounding forests and nutrient-poor groundwater. These characteristics make Lake Stechlin unique in the GDR as well as in the whole lowland lake region south of the Baltic Sea. There is a broad interest, national and international, in preserving it for coming generations.

Suggestions for economic evaluation of the natural potential of the Lake Stechlin area

Paying no (or not enough) attention to the natural gratis effect, the economic value of lakes was many times underestimated in the past. In the last two centuries efforts have been made to reduce the lake water level and to expand usable agricultural land. Today the trend is directed to increasing the utilizable water supply by storage and raising the water level in order to satisfy the growing water demand, e.g., for irrigation. Here an effort is made to show the diversity of economic effects, including those which are not usually expressed in monetary figures. In these cases the natural potential is compared with the expenses necessary for substitutional activities – e.g., instead of bathing at a lake, the construction and operation of swimming pools, instead of biological self-purification, the construction and operation of sewage treatment plants. The aim is to show the order of magnitude and the importance of our study area within the political economy (Table 17.8).

The natural water utilization potential can be valued by economic aspects. The state utilization charges amount to compensation for average expenditure on water management to make the water available.

For power plants with a cooling water circuit drawn from surface waters, the water costs[1] $0.04\,M \cdot m^{-3}$, with water losses during the circulation $0.12\,M \cdot m^{-3}$. (To preserve ground-water as a drinking water resource, its employment for industrial purposes is at $0.24\,M \cdot m^{-3}$ made very expensive).

As to possible utilization for drinking water purposes, in the future it must be accounted that standing water mass with a volume of c. 100 million cubic metres is not identical with the possibilities for gaining raw water in such magnitudes (e.g., to satisfy the urgent demand of the capital, Berlin). Only the small runoff along the Polzow canal is usable (cf. p. 10). The declaration as a drinking water reserve is only of local interest in the cases of an emergency.

The economic result of fishing is influenced by the quality of fish species (Table 8.1). High quality fish are, e.g., Eel, Coregons, Pike and Pike-Perch. The result is further influenced by the expense of stock regulation with the use of fry and young fishes. Lakes Stechlin and Nehmitz are stocked with European Cisco, Eel

[1] Costs in Mark (M).

and Pike; Lake Dagow contains Pike, Eel, Pike-Perch and, since 1982 the Asian Silver Carp (cf. p. 261).

As to recreation potential, only the bathing places at Lake Stechlin were theoretically substituted by concrete basins of comparable capacity. In GDR conditions artificial bathing facilities specific costs run to 3.10 M per recreation day (Klapper 1972). Today the figure might be somewhat higher. This substitution expense includes depreciation, water treatment and operating costs, i.e., expenses which are not an equivalent to beauty of landscape and do not offer the possibilities for further activities in outdoor water recreation.

The self-purification potential may be employed to only a small degree compared with those of running waters. An oligotrophic lake changes to a higher trophic level already at very low nutrient loadings (Fig. 17.2). Overburdening has to be avoided because 'oligotrophication' of an eutrophized lake is most expensive and (if renewal is possible at all) most time consuming because of the very low

Table 17.8 Natural potential of Lakes Stechlin, Nehmitz and Dagow; utilization, usefulness and consequences of use (modified according to Graf & Klapper 1983; NPP = nuclear power plant).

Natural potential	Substitute	Consequence
1. Water utilization potential	Other water recources	Water management
Water utilization by NPP (100 million $m^3 \cdot a^{-1}$)	Wells, higher treatment expense	Saving of pumping, treatment costs, loading
Heat transport and cooling potential	Cooling towers	Increase of evaporation, changes in biological rhythms
Resource for drinking water demand	Drinking water reservoir	High costs per m^3 storage basin capacity
2. Potential of biologically productivity	Other kinds of bioproduction	Yield on biological products
Feed animals	Feed stuff	Saving food
Fishes ($6.5\,t \cdot a^{-1}$)	Fish breeding	Saving fish harvest, food and work
Reed	Other raw material	Relatively unimportant
3. Recreation potential	Building for recreation	Better life quality
Bathing possibility (500 000 recreation days $\cdot\, a^{-1}$)	Basins with water purification	Saving of costs for recreation
4. Self-purification potential	Sewage treatment	Water loading
Sedimentation	Mechanical treatment	Silting, oxygen depletion
Decomposition of organic matter	Biological sewage treatment	Oxygen depletion, nutrient accumulation
Nutrient elimination	Sewage treatment (3rd step)	Eutrophication, decreasing recreational value

renewal rate. The time factor must be involved in economic considerations because high social benefit should also be secured for coming generations and their claims to high quality usable water resources.

Man-made modifications in the Lake Stechlin area

In recent decades anthropogenic influences have altered external and internal factors in the lake ecosystems. The establishment of a cooling water circuit (Fig. 17.1) changed the basic figures for retention time as well as the temperature regime in Lakes Stechlin and Nehmitz (cf. pp. 59 ff.). The nuclear power plant discharged not only cooling water but also domestic sewage, cleaned by a biological treatment plant, and garbage, deposited at Lake Wulwitz (cf. p. 98). The villages Neuglobsow and Dagow have grown, and the 'recreation pressure' on the Lake Stechlin area increased (Figs. 17.4, 17.5). Fishing tried to increase productivity of Carp and Duck breeding in Lake Dagow, and Trout overwintering and feeding (in net containers placed in the heated west basin) in Lake Stechlin.

The increasing nutrient load (especially from 1965 to 1970) led to significant signs of eutrophication in Lake Stechlin – for example, decrease of light transmission and increase of bioproduction and biomass (cf. p. 423). Today the nutrient load shows signs of diminishing, with alterations from year to year due to different throughflow conditions. Phosphorus load amounts to 0.04–0.07 g $P \cdot m^{-2} \cdot a^{-1}$ (PO_4-P, cf. p. 101; Fig. 17.1). From 1965–1970 in cases of high run-off it exceeded 0.07 g $P \cdot m^{-2} \cdot a^{-1}$. That means that phosphorus input (not yet the actual concentration) was already typical for mesotrophic conditions. The intensive research work done by limnologists at Lake Stechlin (partly documented in this monography) produced a good insight and quantitative approach for the nutrient budget, nutrient sources and biological consequences of nutrient enrichment (cf p. 421). With this data as a basis, administration was ready to prepare and finance sanitation measures to protect the oligotrophic Lake Stechlin. In close cooperation various institutions (Water Management Board Oder-Havel, Potsdam Regional Council, Nuclear Power Plant Rheinsberg, the Academy of Sciences, the Nature Reserve Agency etc.) concluded a programme to combat the eutrophication of Lake Stechlin.

The main points in this programme are limiting the nutrient load from Lake Nehmitz (sewage treatment plant of nuclear power plant, rubbish dumping site at Lake Wulwitz) and from Lake Dagow with immediate catchment area, influenced by the village Neuglobsow.

These sanitation measures are now to a large extent in operation or are to be finished by 1985. Below are listed those measures which have been taken since 1965 to curb the nutrient load on Lake Stechlin.

1966: The camping site 'Sonnenbucht' was closed. The measure was necessary because of local loading in the north basin of Lake Stechlin due to inadequate sanitary equipment.

Figure 17.4 Restaurant 'Fontane-House' at Neuglobsow. Photo W. Scheffler.

Figure 17.5 Holiday-makers walking near Lake Stechlin. Photo W. Scheffler.

1970: Carp and duck breeding on Lake Dagow was ended to decrease the nutrient concentration and thus the nutrient import from Lake Dagow to Lake Stechlin. From 1970 to 1975 the mean phospate concentration decreased to about half compared those at the end of the sixties.

1970: Abolition of the temporary 1968/69 and 1969/70 permits given for Trout overwintering and feeding in the heated west basin of Lake Stechlin. In the immediate vicinity of the net containers the oxygen and nutrient conditions had deteriorated.

1972: The outfall of oxidation ponds (which served as a biological sewage treatment plant for the hotel complex 'Stechlin') to Lake Dagow was ended, resulting in no loadings on Lakes Dagow and Stechlin.

1973: Prohibition of garbage disposal by the nuclear power plant at Lake Wulwitz and of smaller sewage inlet from the nuclear power plant to Lake Nehmitz resulting in a decreased load on Lakes Wulwitz and Nehmitz.

1972: Building of sewage collection tanks without outfall for diverse restaurants, recreation facilities and for the camping site in Neuglobsow to decrease the diffuse nutrient input into the drainage basin.

1974: Solution of the phosphate elimination problem in the outfall of domestic sewage treatment plant of the nuclear power plant. The plant is situated near Lake Nehmitz. Originally there had been an inlet to Lake Nehmitz. 1st phase (1974): Phosphate elimination with help of iron shaving packed in a 100 m segment of the outflow ditch. 2nd phase (1977): Building of a discharge pipe for sewage diversion in a nearby marshy ground for natural phosphate elimination by micro- and macrophytes, and bank filtration. A direct outflow to Lake Nehmitz was stopped with a dam. The measures resulted in an about 50% elimination of nutrient input to Lake Nehmitz. 3rd phase (1976–1983): Building of a sewage discharge pipe followed by surface infiltration outside the drainage basin. This resulted in a complete removal of a sewage burdening of Lake Nehmitz.

1976: Outflow regulation and damming of Lake Dagow to minimize phosphorus load from Lake Dagow.

1978: Outflow regulation at Lake Wulwitz by constructing a dam with simultaneous diversion of storm and rinse water from the nuclear power plant to the River Havel. By this means the nutrient load from Lake Wulwitz was completely stopped.

1980: The construction of a central sewerage system with surface infiltration outside the catchment area and a drinking water supply for Neuglobsow was started. 1st phase (1981): Sewage diversion with provisional infiltration outside the drainage basin, thereby completely unburdening the oxidation ponds at Lake Dagow. 2nd phase (1982): Bringing into operation a screening vessel pumping station and the connection of further individual treatment plants to the community sewage system. 3rd phase (1983–1985): Full operation of the central sewerage system, the sewage diversion and the irrigation plant outside the drainage basin, completion of the central drinking water supply for Neuglobsow.

1980: Further decrease of phosphorus import from lake Dagow to Lake Stechlin by diversion of an outflow into swamp ground with about 30 days retention time.

1982: Start of deep water diversion at Lake Dagow. Nutrient-rich deep water is introduced in the sewage diversion pipe. The aim is restoration of Lake Dagow and minimization of outflow to Lake Stechlin.

Taken all together the measures should guarantee that in the future the orthophosphate load per area of Lake Stechlin will remain under $0.05\,g\,P \cdot m^{-2} \cdot a^{-1}$, i.e., in a range typical for oligotrophic conditions.

Lake Stechlin as a recreation area

The high quality of the Lake Stechlin area as a recreation area can be demonstrated by estimation of a recreation index (Klapper 1972; Klapper & Gringmuth 1976). The scale for judging the utilization effect of water for recreative purposes is subdivided into three main groups, in factors of environmental origin (A), in factors which characterize the degree of preparedness for the utilization of the waters for recreative purposes (B), and in factors reducing the use of waters for recreation (C; see Table 17.9).

The individual factors (seven in each case) of the main groups are evaluated with figures 0 to 2, in which case normal conditions are estimated at 1, conditions adverse to recreation at <1, and especially favourable conditions at >1. By means of the equation shown below (see Table 17.9) a complex numerical value (Rpr) is then obtained for all the individual values. This factor enables a comparison and judgement to be made of individual waters.

The restorative anti-eutrophication measures tending towards lowering the nutrient content of the waters to be evaluated, towards keeping out waste water, and the restriction of utilization adverse to recreation, find expression in the increased means of recreation and various restoration processes become comparable.

The Lake Stechlin area belongs to the best recreational landscapes of the GDR. The beauty of the landscape, the clear water with few aquatic weeds on the relatively steep littoral zone and the clean air in the large forests characterize excellent natural prerequisites for recreation (A-factors). The traffic connections are not suited for mass tourism. The nearest railway station in Fürstenberg is 10 km from Neuglobsow. But the growing use of cars makes the lakes a destination for weekend-visitors especially from the capital, Berlin. From there Lake Stechlin may be reached in about two hours by car.

Not only the fine bathing place should be mentioned, but also the possibilities for further water-based recreations (Fig. 17.6) canoeing (Fig. 17.7), surfing and sailing (Fig. 17.8). About 150 boats are available for hire. There are about 100 permanent permits for angling and 500 temporary ones issued annually. Of great importance in the clear water is sport-diving (Fig. 17.9). The training centre runs courses and competitions, with some international participation (Stechlin Diving Cup), for about a thousand divers each year.

The B-factor range between 1 and 2, i.e., between normal and excellent.

Adverse C-factors are almost unknown, except the limitation on recreational use in connection with the nature reserve (see p. 482).

The quantity of recreation at Lake Stechlin is shown in terms of recreation-days. One recreation-day is defined as 1 Rd = one overnight stay in private accommodation, hotels or at a camping site or one stay of 4 hrs or more in the lake region by a short-time visitor performing one or more water-based activities.

Table 17.9 Estimation of the recreation property of a lake landscape (according to Klapper 1972; S = Lake Stechlin, N = Lake Nehmitz, D = Lake Dagow).

	Range	Lake		
		S	N	D
A. Factors of environmental origin				
1. Largeness and multiformity	1.0 ... 2.0	2.0	2.0	1.2
2. Relief formation	1.0 ... 2.0	1.9	1.9	1.4
3. Recreational value of the surroundings	0.5 ... 2.0	2.0	2.0	1.5
4. Shore formation	0.5 ... 2.0	2.0	2.0	1.0
5. Trophic level (hypertrophic ... oligotrophic)	0.2 ... 2.0	2.0	1.9	1.2
6. Adverse aquatic weed	0.5 ... 2.0	2.0	1.8	1.5
7. Bioclimatic conditions	0.8 ... 1.5	1.5	1.5	1.5
Mean		*1.9*	*1.9*	*1.3*
B. Development of recreational conditions (man made)				
1. Recreational demand in vicinity	0.5 ... 2.0	1.0	1.0	1.0
2. Traffic	0.5 ... 1.5	1.0	0.8	1.0
3. Shore landscape	0.8 ... 2.0	2.0	2.0	1.2
4. Bathing possibilities	0.2 ... 2.0	2.0	2.0	0.4
5. Possibilities for further water-based recreation activities	0.2 ... 2.0	2.0	1.6	1.0
6. Camping sites, restaurants, hotels	0.8 ... 2.0	1.8	1.0	1.8
7. Complexity and equipment of water pathways for canoeing	0.8 ... 1.2	1.0	1.0	0.8
Mean		*1.5*	*1.1*	*1.0*
C. Factors adverse to recreational use				
1. Waterhygienic objections	0.2 ... 1.0	1.0	1.0	0.6
2. Water uses adverse to recreation	0 ... 1.0	1.0	1.0	0.5
3. Closing (full or in parts)	0 ... 1.0	0.8	0.5	0.5
4. Insects, noise, air pollution	0.2 ... 1.0	1.0	1.0	0.9
5. Overburdening by too many visitors	0.1 ... 1.0	1.0	1.0	0.9
6. Effects of falling water level	0.1 ... 1.0	1.0	1.0	1.0
7. Sewage irrigation or treatment in the immediate vicinity	0.1 ... 1.0	1.0	1.0	0.7
Mean		*1.0*	*0.9*	*0.7*
D. Judgement	Rpr*			
1. Not suitable	0 ... 0.1	–	–	–
2. Suitable with restrictions	0.1 ... 0.5	–	–	–
3. Suitable	0.5 ... 1.0	–	–	0.8
4. Favourable	1.0 ... 1.5	–	1.4	–
6. Excellent	1.5 ... 1.8	1.7	–	–

$$\text{Rpr} = \text{Recreational property} = \frac{\dfrac{\sum_{1}^{7} A}{7} + \dfrac{\sum_{1}^{7} B}{7}}{2} \cdot \frac{\sum_{1}^{7} C}{7}$$

Figure 17.6 Lake Stechlin: bathing in Neuglobsow bay. Photo W. Scheffler.

Figure 17.7 Canoeing on Lake Stechlin. Photo W. Scheffler.

Figure 17.8 Sailing on Lake Stechlin. Photo W. Scheffler.

Figure 17.9 Winter diving in Lake Stechlin. Photo W. Scheffler.

With regard to the development of the holiday accommodation overnight capacity, most striking is the extension of the season to the whole year (Fig. 17.10; Table 17.10). The healthy bioclimatic condition has led to an increasing use of recreation facilities for prophylactic physiotherapeutical cures in colder seasons too.

To sum up, the importance to the individual and society of active recreation in and on the waters of the lakes is constantly growing. Activities, with and without body water contact, are becoming increasingly indispensable for the physical and mental health of our citizens, for their recreation and recuperation. The lakes and their function as contributors to the well-being of individuals are of economic importance since they maintain or improve the ability of society to satisfy its needs through work. Thus any reduction in the ability of the lakes to satisfy these recreational needs (for instance as a result of eutrophication) is a loss to society which should be prevented as far as possible.

Lake Stechlin as nature reserve[2]

Lake Stechlin is said to be the most beautiful lake of Mark Brandenburg. Far from modern industry, agriculture and traffic, the landscape could preserve its

[2] Mainly by S.J. Casper and H.-D. Krausch.

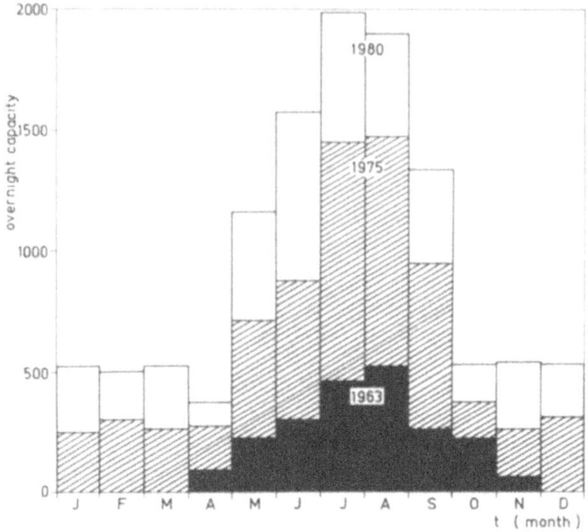

Figure 17.10 Lake Stechlin area: development of holiday accommodation 1963. to 1980.

undisturbed beauty. People interested in the preservation of places with beautiful scenery in their native country became anxious when, at the end of the twenties of our century, a railway was built connecting Berlin via Gransee with Neuglobsow. They were afraid building operations and holiday-makers would impair Lake Stechlin and its environs (Anonymous 1929).

In 1929 the 'Brandenburgische Provinzialkommission für Naturdenkmalpflege' applied for Lake Stechlin and its surrounding forests to be set up as nature reserve. But not until May 6th 1938 was the 'Naturschutzgebiet Großer Stechlin-, Nehmitz- und Großer Krukowsee' established. In addition to the lakes mentioned, the nature reserve included the surrounding forests and some little lakes and bogs (Lakes Glabatz, Breutzen, Kleiner Barsch, Teufel, Gerlin, Kleiner Stechlin). Its area was 1774 ha (cf. Verordnung über das 'Naturschutzgebiet . . .' 1938; Naturschutzgebiet Großer Stechlin 1938–1942).

In the beginning the only idea underlying the nature reserve Lake Stechlin was to preserve the given state of the protected area as a nature sanctuary. This

Table 17.10 Lake Stechlin area: trend in recreation-days.

Year	Holiday-makers overnight	Short-time visitors	Diving centre visitors
1963	60 000	10 000	–
1975	230 000	40 000	–
1980	350 000	60 000	5 000

Figure 17.11 Lake Stechlin, a water body like clear glass easy to see through. Photo W. Scheffler.

general notion changed fundamentally when the nuclear power plant Rheinsberg I was built, mainly on a tract of land not belonging to the nature reserve. Now the nature reserve 'Stechlin' was made larger to 2139.47 ha by including Lake Steutzen, and its character was transformed into an open-air laboratory.

Today the following tasks (cf. Handbuch der Naturschutzgebiete ... 1982) are set for the nature-reserve 'Stechlin':

1. Conservation of the natural potential of the Lake Stechlin area as a typical combination of wood-, bog- and lake-biogeocoenoses in their natural 'outfit' within the Lake District of the Baltic Land-Ridge.
2. Preservation of Lake Stechlin's unique character as a deep, oligotrophic clear-water lake rich in lime (Fig. 17.11).
3. Preservation of all protected species of native plants and animals.
4. Protection of the performance of ecological research on easy terms.

Human activities within the area have to preserve the undisturbed beauty of the landscape. Hunting has to provide protection for the nests of birds of prey. Fishing is allowed only by extensive technologies. The number of boats for anglers and holiday-makers is limited. Motor-driven boats are only allowed for professional fishermen, researchers and Red Cross life-savers. Forests are to be preserved as recreation areas, and the use of the signed walking-ways (Fig. 17.5) in private cars is strictly forbidden. To avoid overburdening of the landscape, bathing facilities and waters by too many visitors, no further increase of the hotel and camping capacity in the village Neuglobsow is planned (Fig. 17.4).

The nature reserve 'Stechlin' satisfies the statements of the Convention on Wetlands of International Importance from February 2nd, 1971 (see Übereinkommen ... 1971). It recognizes the fundamental ecological functions of wetlands as regulators of water regimes and as habitats supporting a characteristic flora and fauna and considers that wetlands constitute a resource of great economic, cultural, scientific and recreational values, the loss of which would be irreparable. Since 1974 it has been proclaimed as international biosphere reservation.

References

Anonymous, 1929. Vom Stechlinsee. Naturdenkmalpflege Naturschutz Berlin Brandenburg 1: 19–22, 41–43.

Benndorf, J. 1979. A contribution to the phosphorus loading concept. Int. Revue ges. Hydrobiol. 64: 177–188.

Björk, S. 1979. Lake management. Studies and results at the Institute of Limnology in Lund. Arch. Hydrobiol. Beih. Ergebn. Limnol. 13: 31–55.

Forsberg, C. 1982. Limnological research can improve and reduce the cost of monitoring and control of water quality. In: Ilmavitra, V., Jones, R.J. & Persson, P.-E. (Eds.): Lakes and Water Management. The Hague: Dr. W. Junk Publishers. 143–146.

Graf, D. & Klapper, H. 1983. Effektivitäts- und Entscheidungsprobleme bei der Nutzung von Naturelementen. In: Busch, K.-F., Uhlmann, D. & Weise, G. (Hrsg.): Ingenieurökologie. Jena: VEB Gustav Fischer Verlag. 426.

Handbuch der Naturschutzgebiete der Deutschen Demokratischen Republik. Bd. 2: Bezirke Potsdam, Berlin – Hauptstadt der DDR, Frankfurt/O und Cottbus. 1982. Leipzig, Jena, Berlin: Urania-Verlag (3. Aufl.) (D 5 Stechlin, 25–32).

Lundquist, I. 1982. The limnological basis for planning water quality management. In: Ilmavitra, V., Jones, R.J. & Persson, P.-E. (Eds.): Lakes and Water Management. The Hague: Dr. W. Junk Publishers. 147–152.

Klapper, H. 1972. Vorschlag zur Darstellung des aktuellen Erholungswertes einer Seenlandschaft. Wasserwirtschaft Wassertechnik 22: 141–144.

Klapper, H. & Gringmuth, W. 1976. Water-tied recreation and eutrophication. In: Institut für Wasserwirtschaft Berlin (Ed.): Proc. Eutrosym, Karl-Marx-Stadt, Vol. 2: 187–189.

Koschel, R., Benndorf, J., Proft, G. & Recknagel, F. 1983. Calcite precipitation as a natural control mechanism of eutrophication. Arch. Hydrobiol. 98: 380–408.

Müller, H.E. 1978. Belastung und Belastungsdynamik in See-Umland-Systemen. In: Tagungsber. wiss. Abh., 41. Deutsch. Geographentag Mainz. Wiesbaden: Franz Steiner Verlag. 499–516.

Naturschutzgebiet Großer Stechlin 1938–1942. Staatsarchiv Potsdam, Pr. Br. Rep. 6 C, Kreisausschuß Ruppin Nr. 414.

Rat des Bezirkes Potsdam (Hrsg.), 1978. Die effektive Nutzung der stehenden Gewässer des Bezirkes Potsdam. Beschluß vom 2.10.1978 (unpublished).

Rat des Bezirkes Potsdam (Hrsg.), 1982. Behandlungsrichtlinie für das Naturschutzgebiet Stechlin. Potsdam. 14 (unpublished).

Richardson, J.L. 1975. Morphometry and lacustrine productivity. Limnol. Oceanogr. 20: 661–667.

Patalas, K. 1968. Landschaft und Klima als Faktoren der Massenproduktion von Algen. Fortsch. Wasserchem. 8: 21–31.

TGL 27 885, 1974. Fachbereichstandard Wassergütebewirtschaftung Seen. Berlin: Ministerium für Umweltschutz und Wasserwirtschaft der DDR. 18.

TGL 27 885/01, 1982. Fachbereichstandard Nutzung und Schutz der Gewässer. Binnengewässer. Klassifizierung. Berlin: Ministerium für Umweltschutz und Wasserwirtschaft der DDR. 16.

TGL 27 885/03, 1983. Fachbereichstandard Nutzung und Schutz der Gewässer. Wassergütebewirtschaftung Seen. Berlin: Ministerium für Umweltschutz und Wasserwirtschaft der DDR. 25.

Übereinkommen vom 2. Februar 1971 über Feuchtgebiete, insbesondere als Lebensraum für Wasser- und Watvögel, von internationaler Bedeutung (Convention on Wetlands of International Importance especially as Waterfowl Habitat). Ges. Bl. 81. der DDR 1979, Teil II, Nr. 1, 10. Jan. 1979.

Uhlmann, D. 1973. Störungen des biologischen Gleichgewichtes in Gewässern. Abh. Sächs. Akad. Wiss. Leipzig Math.-Nat. Kl. 52: 1–17.

Verordnung über das 'Naturschutzgebiet Großer Stechlin-, Nehmitz- und Großer Krukowsee', Preußisches Forstamt Menz, Kreis Ruppin. 1938. Der märkische Naturschutz 4: 88–90.

Vollenweider, R.A. 1976. Advances in defining critical loading levels for phosphorus in lake eutrophication. Mem. Ist. Ital. Idrobiol. 33: 53–83.

Appendix

Taxa observed in the Lake Stechlin area

Taxa observed in the waters of the Lake Stechlin area made up by S.J. Casper, D. Flößner, H.-D. Krausch, G. Mothes, D. Ronneberger, W. Scheffler, and W. Schönborn (for Bacteria see Babenzien & Babenzien, p. 208).

Taxa[1]	S	N	D	W
PLANTS				
Cyanophyceae				
(S.J. Casper)				
Anabaena				
circinalis Rabenhorst	–	–	E	+
flos-aquae (Lyngbye) Brébisson in Brébisson et Goday	E	E	E	+
lemmermannii P. Richter in Lemmermann	E	E	–	–
Aphanizomenon flos-aquae (L.) Ralfs	–	–	E	+
Aphanocapsa delicatissima W. et G.S. West	–	–	–	+
Aphanothece				
castagnei (Kützing) Rabenhorst	–	–	–	+
microscopica (Meneghini) Naegeli	–	–	–	+
nidulans P. Richter	–	–	–	+
stagnina (Sprengel) Boye-Petersen	–	–	–	+
Calothrix				
ramenskii Elenkin	–	–	–	+
stellaris Bornet et Flahaut	–	–	–	+
Chamaesiphon cylindricus Boye-Petersen	–	–	–	+
Chroococcus				
limneticus Lemmermann	E	E	–	–
minimus (v. Keissler) Lemmermann	–	–	–	+
minutus (Kützing) Naegeli	T	–	–	–
turgidus (Kützing) Naegeli	T	T	–	+
Coelosphaerium				
kuetzingianum Naegeli	–	E	–	+

[1] The abbreviations stand for: S = Lake Stechlin, N = Lake Nehmitz (incl. Lakes Breutzen and Gerlin), D = Lake Dagow, W = Other lakes and waters; – = no observations, + = observations made, B = Benthonic, E = Euplanktonic, T = Tychoplanktonic.

Casper, S. Jost (ed.), Lake Stechlin. A temperate oligotrophic lake. ISBN 978-94-010-8930-2.

Taxa[1]	S	N	D	W
pallidum Lemmermann	–	E	–	–
Cylindrospermum stagnale (Kützing) Bornet et Flahaut	–	–	–	+
Dichothrix gypsophila (Kützing) Bornet et Flahaut	–	–	–	+
Diplonema rupicola Borge	–	–	–	+
Eucapsis alpina Clements et Shantz	–	–	–	+
Gloeotrichis pisum (C.A. Agardh) Thuret	–	–	–	+
Gomphosphaeria				
aponina Kützing	–	T	T	+
var. *limnetica* Virieux	+	–	–	–
compacta (Lemmermann) Ström	+	–	–	–
lacustris Chodat	E	E	E	+
naegeliana (Unger) Lemmermann	–	E	–	+
Homoeothrix juliana (Meneghini) Kirchner	–	–	–	+
Merismopedia				
glauca (Ehrenberg) Naegeli	–	+	–	+
punctata Meyen	–	–	–	+
Microcystis				
aeruginosa (Kützing) Kützing	E	E	–	+
elabens (Meneghini) Kützing	–	–	–	+
reinholdii (P. Richter in Reinhold) Forti	–	–	–	+
wesenbergii (Komárek) Starmach	–	E	–	–
Nostoc				
coeruleum Lyngbye	T	–	–	–
paludosum Kützing	–	–	–	+
sphaericum Vaucher	–	–	–	+
Oscillatoria				
acutissima Kufferath	–	–	–	+
angusta Koppe	–	–	–	+
chlorina (Kützing) Gomont var. *perchlorina* Lauterborn	T	T	–	–
geminata Meneghini	–	–	–	+
granulata Gardner	–	–	–	+
guttulata van Goor	–	–	–	+
irrigua Kützing	T	T	–	–
lauterbornii Schmidle	–	–	+	+
limosa C.A. Agardh ex Gomont	–	–	–	+
mougeotii Kützing	T	T	T	+
planctonica Wołoszyńska	–	–	–	+
pristleyi W. et G.S. West	–	–	–	+
pseudogeminata G. Schmid var. *unigranulata* Biswas	–	–	–	+
putrida Schmidle	–	+	–	+
rubescens DeCandolle ex Gomont	E	E	E	+
splendida Graville ex Gomont	T	–	–	–
tenuis C.A. Agardh ex Gomont f. *tergestina* (Kützing) Rabenhorst	T	T	T	+
Phormidium mucicola Huber-Pestalozzi et Naumann	–	–	–	+
Pseudanabaena catenata Lauterborn	T	T	T	+
Scytonema				
mirabile (Dillwyn) Bornet	–	–	–	+
myochrous (Dillwyn) Agardh	–	–	–	+

Taxa[1]	S	N	D	W
Synechococcus aeruginosus Naegeli	–	T	–	+
Tolypothrix				
distorte Kützing var. *penicillata* (Agardh) Lemmermann	–	–	–	+
helicophila Lemmermann	–	–	–	+
Chlorophyceae (incl. Chlamydophyceae) (S.J. Casper)				
Apiocystis				
brauniana Naegeli in Kützing	B	–	–	+
Ankistrodesmus				
bibraianus (Reinsch) Koršikov	–	T	–	–
falcatus (Corda) Ralfs	–	T	–	–
gracilis (Reinsch) Koršikov	–	E	–	–
spiralis (Turner) Lemmermann	–	–	–	+
Asterococcus				
superbus Scherffel	B	–	–	–
Botryococcus				
braunii Kützing	T	T	T	+
Chaetophora				
elegans (Roth) C.A. Agardh	+	–	–	+
incrassata (Hudson) Hazen	–	–	–	+
Chlamydocapsa planctonica (W. et G.S. West) Fott	E	E	E	–
Chlamydomonas dinobryonis G.M. Smith	E	–	–	–
Chlorella vulgaris Beijerinck	–	–	T	–
Coelastrum				
astroideum De-Notaris	E	–	–	–
microporum Naegeli in A. Braun	E	E	E	+
sphaericum Naegeli	–	–	–	+
reticulatum (Dangeard) Senn	E	E	–	+
Coleochaets				
irregularis Pringsheim	+	–	–	–
pulvinata A. Braun	+	–	–	–
Crucigenia				
quadrata Morren	–	E	E	–
tetrapedia (Kirchner) W. et G.S. West	E	E	–	+
Crucigeniella				
spiculata (Lemmermann) Komárek	–	E	–	–
rectangularis (Naegeli) Komárek	E	E	E	+
Dictyosphaerium				
ehrenbergianum Naegeli	T	T	T	+
pulchellum Wood	–	E	E	+
sphagnale Hindák	–	–	–	+
Dispore crucigenioides Printz	T	T	–	+
Elakatothrix				
acute Pascher	T	–	–	–
gelatinosa Wille	T	T	–	+
lacustris Koršikov	T	T	T	–
subacuta Koršikov	T	–	T	–
Eudorina elegans Ehrenberg	E	E	E	+

Taxa[1]	S	N	D	W
Kirchneriella				
lunaris (Kirchner) Moebius	T	T	T	+
obesa (W. West) Schmidle	T	T	T	+
Lagerheimia				
ciliata (Lagerheim) Chodat	–	E	–	–
longiseta (Lemmermann) Wille	E	–	–	–
Micractinium pusillum Fresenius	–	–	E	–
Monoraphidium				
griffithii (Berkeley) Komárová-Legnerová	–	E	E	–
minutum (Naegeli) Komárová-Legnerová	–	–	E	–
mirabile (W. et G.S. West) Pankow	E	E	E	–
contortum (Thuret) Komárová-Legnerová	–	–	E	–
convolutum (Corda) Komárová-Legnerová	–	–	E	–
Nautococcus mammilatus Koršikov	–	–	–	+
Nephrocytium				
agardhianum Naegeli	–	–	–	+
lunatum W. West	–	E	–	+
Oocystis				
borgei Snow	T	T	T	–
elliptica W. West	T	T	–	–
lacustris Chodat	–	T	–	–
marssonii Lemmermann	T	T	T	–
novae-semliae Wille	–	T	–	–
solitaria Wittrock in Wittrock et Nordstedt	–	–	–	+
Oonephris obesa (W. West) Fott	–	B	–	–
Pandorina morum (O.F. Müller) Bory	E	E	E	+
Paulschulzia tenera (Koršikov) Lund	E	–	–	–
Pediastrum				
biradiatum Meyen	–	E	–	–
boryanum (Turpin) Meneghini	E	E	E	+
var. *longicorne* Reinsch	–	E	–	–
var. *perforatum* (Raciborski) Nitardy	E	E	E	–
var. *brevicorne* A. Braun	–	E	–	–
duplex Meyen	–	E	E	+
simplex Meyen	–	–	E	–
tetras (Ehrenberg) Ralfs	–	E	E	+
var. *tetraodon* (Corda) Hansgirg	–	E	–	+
Phacotus lenticularis (Ehrenberg) Stein	E	E	E	+
Polyedriopsis spinulosa (Schmidle) Schmidle	E	–	–	–
Pseudosphaerocystis lacustris (Lemmermann) Nováková	E	E	–	+
Quadrigula pfitzeri (Schröder) G.M. Smith	T	T	–	+
Scenedesmus				
communis Hegewald	–	E	E	+
costato-granulatus Skuja	E	E	–	–
disciformis (Chodat) Fott et Komárek	E	E	–	+
hystrix Lagerheim	–	–	+	–
obliquus (Turpin) Kützing	–	T	–	+
obtusus Meyen	E	–	E	–

Taxa[1]	S	N	D	W
serratus (Corda) Bohlin	–	E	–	–
Sorastrum				
americanum (Bohlin) Schmidle	–	–	–	+
spinulosum Naegeli	T	T	–	+
Sphaerocystis schroeteri Chodat	E	E	E	–
Tetraedron				
caudatum (Corda) Hansgirg	–	E	–	+
minimum (A. Braun) Hansgirg	E	E	E	+
trigonum (Naegeli) Hansgirg	–	E	E	–
Tetrastrum triangulare (Chodat) Komárek	E	E	E	–
Treubaria triappendiculata	–	E	–	–
Volvox aureus Ehrenberg	E	E	–	E
Westella botryoides (W. West) De Wildeman	–	E	–	–
Codiolophyceae (S.J. Casper)				
Ulothrix zonata Kützing	B	–	–	+
Oedogoniophyceae (S.J. Casper)				
Bulbochaete varians Wittrock	+	–	–	–
Zygnemaphyceae (S.J. Casper)				
Closterium				
aciculare T. West	E	E	E	+
moniliferum (Bory) Ehrenberg ex Ralfs	T	T	–	+
pronum Brébisson	–	–	–	+
Cosmarium				
abbreviatum Raciborski[2]	T	T	T	+
depressum (Naegeli) Lundell var. *planctonicum* Reverdin	T	–	–	–
phaseolus Brébisson ex Ralfs[2]	T	–	–	–
punctulatum Brébisson[2]	T	T	T	–
reniforme (Ralfs) Archer	–	T	–	+
subtumidum Nordstedt in Wittrock et Nordstedt	T	–	–	–
synthlibomenum W. West var. *latum* (Nygaard) W. Krieger et Gerloff	–	–	–	+
Hyalotheca dissiliens (J.E. Smith) Brébisson	+	+	–	+
Mougeotia parvula Hassall	+	–	–	–
Spirogyra				
inflata (Vaucher) Kützing	–	–	–	+
longata (Vaucher) Kützing	–	–	–	+
majuscula Kützing em. Czurda	–	–	–	+
Spondylosium planum (Wolle) W. et G.S. West	–	–	–	+
Staurastrum				
anatinum Cook et Wills	+	–	–	–
cerastes Lundell	–	–	+	+
chaetoceras (Schroeder) G.M. Smith	–	+	+	+
cuspidatum Brébisson	+	+	–	+

[2] *C. abbreviatum* according to Busse (1965); Küchler (1982) only mentioned *C. punctulatum* (error?); Krieger determined *C. phaseolus* as dominant species of Lake Stechlin.

Taxa[1]	S	N	D	W
dejectum Brébisson	–	+	–	+
gracile Ralfs ex Ralfs	–	+	–	–
paradoxum Meyen	T	–	–	–
Chrysophyceae (incl. some colourless flagellates) **(S.J. Casper)**				
Bikosoeca multiannulata Skuja	–	+	–	–
Bitrichia chodatii (Reverdin) Chodat	E	E	–	–
Chrysamoeba radians Klebs	E	E	–	–
Chrysastrella				
aculeata (Dolgoff) Deflandre	E	E	–	–
furcata (Dolgoff) Deflandre	E	E	–	–
rasumowskoensis (Dolgoff) Deflandre	–	E	–	–
Chrysococcus minutus (Fritsch) Nygaard	–	E	–	–
Chrysopyxis inaequalis Fott	–	E	–	–
Codonomonas pascheri (Kisseliev) Van Goor	–	+	–	–
Dinobryon				
acuminatum Ruttner	E	E	–	–
bavaricum Imhof	E	E	–	+
crenulatum W. et G.S. West	E	E	–	+
cylindricum Imhof	E	E	–	+
var. *alpinum* (Imhof) Bachmann	E	–	–	+
divergens Imhof	E	E	E	+
var. *schauinslandii* (Lemmermann) Brunnthaler	–	E	–	–
pediforme (Lemmermann) Steinecke	E	–	–	+
sertularia Ehrenberg	E	E	E	+
sociale Ehrenberg	E	E	–	+
var. *americanum* (Brunnthaler) Bachmann	E	–	–	–
var. *stipitatum* (Stein) Lemmermann	E	E	–	+
suecicum Lemmermann var. *longispinum* Lemmermann	–	E	–	–
Erkenia subaequiciliata Skuja	E	E	–	–
Hyalobryon				
lauterbornii Lemmermann	E	B	–	–
ramosum Lauterborn	–	–	–	+
Kephyrion asper (Lackey) Bourrelly	–	E	–	–
Mallomonas				
acaroides Perty var. *striatula* Asmund	E	E	–	–
akrokomos Ruttner in Pascher	E	E	–	–
crassisquama (Asmund) Fott	E	E	–	+
fastigata Zacharias	E	E	E	+
heterospina Lund	–	–	–	+
tonsurata Teiling em. Krieger	E	E	–	+
Pachysoeca ruttneri (Bourrelly) Fott	B	B	B	+
Phaeothamnion confervicolum Lagerheim	–	–	–	+
Pseudokephyrion				
circumvallatum Bourrelly	–	B	–	–
entzii Conrad	–	E	–	–
Rhizochrysis gracillima Naumann	–	B	–	–

Taxa[1]	S	N	D	W
Salpingoeca frequentissima (Zacharias) Lemmermann	B	B	–	–
Synura				
sphagnicola Koršikov	–	–	–	+
uvella Ehrenberg	E	E	E	+
Uroglena				
americana Calkins	E	E	E	–
botrys (Pascher) Conrad	E	–	–	–
eustylis Skuja	E	–	–	–
volvox Ehrenberg	–	E	–	–
Xanthophyceae				
(S.J. Casper)				
Goniochloris				
fallax Fott	–	E	–	+
sculpta Geitler	–	T	–	–
Ophiocytium				
capitatum Wolle	–	T	–	+
cochleare (Eichwald) A. Braun	–	T	–	+
Vaucheria dichotoma (L.) Martius	B	B	–	–
Bacillariophyceae				
(W. Scheffler & S.J. Casper)				
Acanthoceras zachariasii (J. Brun) Simonsen	E	E	E	+
Achnanthes				
clevei Grunow	T	T	–	–
var. *rostrata* Hustedt	T	–	–	–
conspicua A. Mayer	T	T	–	–
delicatula (Kützing) Grunow	B	–	–	–
dispar Cleve	T	T	–	–
var. *angulata* Hustedt	T	–	–	–
exigua Grunow	T	–	–	–
var. *heterovalvata* Krasske	T	T	–	–
flexella Kützing	T	–	–	–
kolbei Hustedt	T	T	–	–
lanceolata (Brébisson in Kützing) Grunow	T	T	–	–
var. *elliptica* Cleve	–	B	–	–
var. *rostrata* Hustedt	–	B	–	–
minutissima Kützing (incl. *microcephala* Kützing)	T	T	–	–
Amphipleura pellucida Kützing	T	T	–	–
Amphora				
libyca Ehrenberg	T	T	–	–
ovalis Kützing var. *pediculus* Kützing	T	T	–	–
Anomoeoneis vitrea (Grunow) Ross	B	–	–	–
Asterionella formosa Hassall	E	E	E	+
Caloneis				
bacillum (Grunow) Cleve	T	–	–	–
schumanniana (Grunow) Gleve var. *biconstricta* Grunow	–	B	–	–
silicula (Ehrenberg) Cleve	–	T	–	–

Taxa[1]	S	N	D	W
zachariasii Reichelt	T	T	–	–
Campylodiscus hibernicus Ehrenberg	T	T	–	+
Cocconeis				
diminuta Pantocsek	T	–	–	–
disculus Schumann	T	T	–	–
placentula Ehrenberg	T	T	–	–
var. *euglypta* (Ehrenberg) Cleve	T	T	–	–
var. *lineata* (Ehrenberg) Cleve	B	–	–	–
Cyclotella				
comta (Ehrenberg) Kützing	E	E	–	–
var. *oligactis* (Ehrenberg) Grunow	E	E	–	–
glomerata Bachmann	T	–	–	–
kützingiana Thwaites	T	T	–	–
var. *planetophora* Fricke	T	T	–	–
ocellata Pantocsek	T	T	–	–
operculata (Agardh) Kützing	T	T	–	–
quadrijuncta (Schröter) em. Hustedt	E	–	–	–
Cymatopleura				
elliptica (Brébisson) W. Smith	E	E	–	–
solea (Brébisson) W. Smith	E	E	–	+
var. *apiculata* (W. Smith) Ralfs	E	E	–	+
Cymbella				
affinis Kützing	T	–	–	–
amphicephala Naegeli	T	–	–	–
aspera (Ehrenberg) Cleve	B	–	–	–
cistula (Hemprich) Grunow	T	–	–	–
cuspidata Kützing	T	–	–	–
cymbiformis (Agardh? Kützing) Van Heurck	–	T	–	–
ehrenbergii Kützing	–	T	–	–
helvetica Kützing	T	–	–	–
hustedtii Krasske	T	–	–	–
hybrida Grunow	B	–	–	–
lanceolata (Ehrenberg) Van Heurck	T	–	–	+
leptoceros (Ehrenberg) Grunow	T	–	–	+
microcephala Grunow	T	–	–	–
obtusiuscula (Kützing) Grunow	T	–	–	–
parva (W. Smith) Cleve	T	–	–	–
prostrate (Berkeley) Cleve	T	T	–	–
thumensis (A. Mayer) Hustedt	T	–	–	–
tumida (Brébisson) Van Heurck	T	–	–	–
turgida (Berkeley) Cleve	T	–	–	–
ventricosa Kützing	T	–	–	–
Denticula tenuis Kützing var. *crassula* (Naegeli) Hustedt	T	–	–	–
Diatoma				
elongatum (Lyngbye) C.A. Agardh	E	E	–	+
hiemale (Lyngbye) Heiberg	B	–	–	–
var. *mesodon* (Ehrenberg) Grunow	B	–	–	–
vulgare Bory	T	–	–	–

Taxa[1]	S	N	D	W
Diploneis				
domblittensis (Grunow) Cleve	T	–	–	–
elliptica (Kützing) Cleve	T	T	–	–
mauleri (J. Brun) Cleve	T	T	–	–
Epithemia				
argus (Ehrenberg) Kützing	T	–	–	–
hyndmannii W. Smith	T	–	–	–
intermedia Fricke in A. Schmidt	T	–	–	–
muelleri Fricke	T	–	–	–
sorex Kützing	T	–	–	–
turgida (Ehrenberg) Kützing	T	T	–	–
var. granulata (Ehrenberg) Grunow in Van Heurck	B	–	–	–
zebra (Ehrenberg) Kützing	T	–	–	–
var. *porcellus* (Kützing) Grunow	B	–	–	–
var. *saxonia* (Kützing) Grunow	T	T	–	–
Eucocconeis flexella (Kützing) Hustedt	T	–	–	–
Eunotia monodon (Ehrenberg) var. *major* (W. Smith) Hustedt	T	–	–	–
Fragilaria				
brevistriata Grunow in Van Heurck	T	T	–	–
capucina Desmaziéres	T	T	–	+
var. *mesolepta* (Rabenhorst) Grunow	T	–	–	–
construens (Ehrenberg) Grunow	T	T	–	+
var. *binodis* (Ehrenberg) Grunow	T	T	–	–
var. *subsalina* Hustedt	B	–	–	–
var. *triundulata* Reichelt	–	E	–	–
var. *venter* (Ehrenberg) Grunow	T	T	–	–
crotonensis (Edwards) Kitton	E	E	E	+
intermedia Grunow in Van Heurck	T	–	–	–
pinnata Ehrenberg	T	T	–	+
var. *lancettula* (Schumann) Hustedt	T	–	–	–
Gomphonema				
acuminatum Ehrenberg	T	–	–	–
var. *brebissonii* (Kützing) Cleve	T	–	–	–
var. *coronatum* (Ehrenberg) W. Smith	B	–	–	–
var. *elongatum* (A. Braun) Cleve	B	–	–	–
var. *turris* (Ehrenberg) Cleve	T	–	–	–
constrictum Ehrenberg	B	–	–	–
var. *capitatum* (Ehrenberg) Cleve	–	–	–	+
gracile Ehrenberg	T	T	–	–
intricatum Kützing	T	–	–	–
var. *pumilum* Grunow	T	–	–	–
olivaceum (Lyngbye) Kützing	T	–	–	–
var. *calcareum* Cleve	T	–	–	–
parvulum (Kützing) Grunow	T	–	–	–
Gyrosigma				
acuminatum (Kützing) Rabenhorst	T	T	–	–
attenuatum (Kützing) Rabenhorst	T	T	–	–
Mastogloia				

Taxa[1]	S	N	D	W
grevillei W. Smith	T	–	–	–
smithii Thwaites in W. Smith	T	–	–	–
var. *amphicephala* Grunow in Cleve et Möller	T	–	–	–
Melosira				
arenaria Moore	T	T	–	–
granulata (Ehrenberg) Ralfs in Pritchard	–	T	–	–
var. *angustissima* O. Müller	–	T	–	+
italica (Ehrenberg) Kützing var. *tenuissima* (Grunow) O. Müller	B	–	–	–
varians C.A. Agardh	T	T	–	+
Meridion circulare C.A. Agardh	E	–	–	–
Navicula				
bacillum Ehrenberg	T	–	–	–
cari Ehrenberg	T	–	–	–
var. *cincta* (Ehrenberg) Lange-Bertalot	T	–	–	–
costulata Grunow	T	T	–	–
cryptocephala Kützing	T	–	–	–
var. *exilis* (Kützing) Grunow	B	–	–	–
cuspidata Kützing	T	–	–	–
hungarica Grunow var. *capitata* (Ehrenberg) Cleve	T	–	–	–
jentzschii Grunow	T	–	–	–
oblonga Kützing	T	T	–	–
pupula Kützing	T	–	–	–
radiosa Kützing	T	T	–	+
var. *tenella* (Brébisson) Grunow	T	–	–	–
reinhardtii Grunow	T	E	–	–
scutelloides W. Smith ex Gregory	T	–	–	–
tuscula (Ehrenberg) Van Heurck	T	T	–	–
viridula (Kützing) Ehrenberg	B	–	–	–
var. *vulpina* (Kützing) Lange-Bertalot	T	–	–	–
Neidium				
dubium f. *constrictum* Hustedt	T	–	–	–
iridis (Ehrenberg) Cleve	–	T	–	–
var. *amphigomphus* (Ehrenberg) Van Heurck	T	T	–	–
Nitzschia				
acuta Hantzsch	T	–	–	–
angustata var. *acuta* Grunow	T	–	–	–
amphibia Grunow	T	–	–	–
angustata var. *acuta* Grunow	T	–	–	–
denticula Grunow	T	–	–	–
dissipata (Kützing) Grunow	T	–	–	–
microcephala Grunow	T	–	–	–
romana Grunow	B	–	–	–
sigmoidea (Nitzsch) W. Smith	T	T	–	–
vernicularis (Kützing) Grunow	T	–	–	–
Opephora martyi Héribaud	E	E	–	–
Pinnularia major (Kützing) Cleve	E	E	–	–
Rhizosolenia longiseta Zacharias	E	E	E	–
Rhoicosphenia curvata (Kützing) Grunow ex Rabenhorst	T	–	–	–

Taxa[1]	S	N	D	W
Rhopalodia gibba (Ehrenberg) O. Müller	T	–	–	–
Stauroneis				
acuta W. Smith	–	E	–	–
phoenicenteron Ehrenberg	T	–	–	+
Stephanodiscus				
rotula (Kützing) Hendey	E	E	E	–
var. *minutulus* (Kützing) Grunow in Van Heurck	T	–	–	+
hantzschii Grunow in Cleve et Grunow	E	–	–	+
spec. (*hantzschii*-group)	E	–	–	–
Surirella				
biseriata Brébisson var. *bifrons* (Ehrenberg) Hustedt	T	T	–	–
f. *punctata* Meister	T	T	–	–
caproni Brébisson in A. Schmidt	–	T	–	–
elegans Ehrenberg	T	T	–	–
Synedra				
acus Kützing	E	E	–	–
var. *angustissima* Grunow	E	E	–	–
var. *radians* (Kützing) Grunow	E	–	–	–
amphicephala Kützing	E	–	–	–
capitata Ehrenberg	T	T	–	+
parasitica (W. Smith) Hustedt	T	T	–	–
rumpens Kützing	T	–	–	–
var. *meneghiniana* Grunow	B	–	–	+
ulna (Nitzsch) Ehrenberg	T	T	–	+
var. *biceps* (Kützing) v. Schönfeld	B	–	–	–
vaucheriae Kützing	T	–	–	–
var. capitellata Grunow	T	–	–	–
Tabellaria				
fenestrata (Lyngbye) Kützing	E	E	–	+
var. *asterionelloides* Grunow in Van Heurck	E	E	–	+
flocculose (Roth) Kützing	T	T	–	+
Dinophyceae				
(S.J. Casper)				
Ceratium				
cornutum (E.) Claparede et Lachmann	E	E	–	+
hirundinella (O.F. Müller) Schrank	E	E	E	+
Gymnodinium lacustre Schiller	E	E	E	–
Peridinium				
cinctum (O.F. Müller) Ehrenberg	E	E	E	–
umbonatum Stein	E	–	–	–
Cryptophyceae				
(S.J. Casper)				
Chroomonas caerulea (Geitler) Skuja	–	–	E	–
Cryptomonas				
erosa Ehrenberg	E	–	–	–
obovata Skuja	E	–	–	–

Taxa[1]	S	N	D	W
ovata Ehrenberg	E	–	–	–
Rhodomonas minuta Skuja var. *nannoplanctica* Skuja	E	E	–	–
Euglenophyceae				
(S.J. Casper)				
Colacium				
simplex Huber-Pestalozzi	–	E	–	–
vesiculosum Ehrenberg	–	E	–	–
Euglena acus Ehrenberg	–	E	–	+
Rhodophyceae				
(S.J. Casper)				
Batrachospermum vagum (Roth) C.A. Agardh	–	–	–	+
Fungi				
(S.J. Casper)				
Alatospora acuminata Ingold	+	+	–	+
Anguillospora longissima (Sacardo et Sydow) Ingold	+	–	+	+
Articulospora tetracladia Ingold	–	–	–	+
Clavariopsis aquatica De Wildeman	+	–	–	+
Clavatospora				
longibrachiata (Ingold) S. Nilsson	+	–	–	–
stellata (Ingold et Cox) S. Nilsson	+	–	–	–
Dactyella				
aquatica (Ingold) Ranzoni	–	+	–	–
submersa (Ingold) S. Nilsson	+	+	–	+
Dimorphospora foliicola Tubaki	+	+	–	–
Flagellospora curvula Ingold	+	+	–	+
Lemonniera aquatica De Wildeman	+	–	–	+
Lunulaspora curvula Ingold	+	+	–	–
Margaritispora aquatica Ingold	+	+	–	–
Monotosporella microaquatica (Tubaki) S. Nilsson	–	+	–	–
Tetrachaetum elegans Ingold	–	–	–	+
Tetracladium				
marchalianum De Wildeman	+	–	–	+
setigerum (Grove) Ingold	+	–	–	–
Tricladium				
anomalum Ingold	+	–	–	–
splendens Ingold	+	–	–	+

Taxa[3]	S	N	D	W	M
Charophyceae					
(H.-D. Krausch)					
Chara aspera Detharding ex Willdenow	**c**rrC	rcC		rcC	
Chara aculeata Kützing in Reichenbach	cCP	cCP		cCP	
Chara contraria A. Braun ex Kützing	cC	cC		cC	
Chara filiformis Hertsch	**c**rCP	rcCP		rcCPF	
Chara deliculata C.A. Agardh	rcRFP	rcR		rcR	
Chara fragilis Desvaux	ccCPF	ccCPF		ccCPF	
Chara hispida L.	rFPC	cFPC		cFPC	
Chara tomentosa L.	**rr**rcCPF	rcCPF		rcCPF	
Nitella flexilis (L.) C.A. Agardh	rcC	rcC		rcC	
Nitella opaca C.A. Agardh	rrP	rrP			
Nitellopsis obtusa (Desvaux) J. Groves	ccCPF	ccCPF	**r**F	ccCPF	
Marchantiatae					
(H.-D. Krausch)					
Riccia fluitans L.				rP	
Bryatae					
(H.-D. Krausch)					
Drepanocladus aduncus (Hedwig) Moenkemeyer	rC			rR	
Drepanocladus fluitans (L. ap. Hedwig) Warnstorf					rcPF
Fontinalis antipyretica L. ap. Hedwig	**rc**rrCP	rcCSF		rR	
Platyhypnidium riparioides (Hedwig) Podpera	rcC				
Equisetatae					
(H.-D. Krausch)					
Equisetum fluviatile L. em. Erhart	rcR	rcR	rR	rcR	
Filicatae					
(H.-D. Krausch)					
Thelypteris palustris Schott	rcR	rcR	rR	rcR	
Monocotyledoneae					
(H.-D. Krausch)					
Sparganium erectum L. em. Reichenbach	rR	rR	rR	rR	

[3] In the following lists abbreviations stand for: S = Lake Stechlin, N = Lake Nehmitz (incl. Lakes Gerlin and Breutzen), D = Lake Dagow, M = bog-lakes, W = other lakes and waters of the region; R = Reeds, F-zone of nymphoids with floating leaves, P = zone of submerged pondweeds, C = zone of Characeae; rr = very rare, r = rare, rc = scaterred, rather common, c = common, cc = very common, frequent, + = frequent, – = no observations; **bold** printed letters: formerly frequent.

Taxa[3]	S	N	D	W	M
Sparganium minimum Wallroth		**rF**		rF	
Typha angustifolia L.	cR	cR	ccR	cR	
Typha latifolia L.	rrR	rrR		rR	
Potamogeton natans L.	rcF	cF	**rcF**	ccF	rF
Potamogeton lucens L.	cP	cP		cP	
Potamogeton x *nitens* Weber	rrP	rrP		rP	
Potamogeton perfoliatus L.	rcP	rcP		rcP	
Potamogeton crispus L.	**rrP**		rP		
Potamogeton praelongus Wulfen				crrP	
Potamogeton obtusifolius Mertens et Koch	**rP**	rP			
Potamogeton mucronatus Schrader ex Sonder	**rP**	rcP		rP	
Potamogeton rutilus Wolfgang	**rrP**	rrP		rrP	
Potamogeton panormitanus Bivona-Bernardi		**rrP**			
Potamogeton acutifolius Link		**rrP**			
Potamogeton gramineus L.	rrR			rR	
Potamogeton x *zizii* Koch ex Roth	**rP**	**rP**		**rP**	
Potamogeton pectinatus L.	cP	cP		cP	
Potamogeton filiformis Persoon	rP	**rP**		rP	
Najas marina L.		**rP**		**rP**	
Najas marina subsp. *intermedia* (Wolfgang) Casper		**rP**		**rP**	
Najas minor Allioni		**rrP**		**rrP**	
Zannichellia palustris L.				rrP	
Scheuchzeria palustris L.					rcR
Sagittaria sagittifolia L.	rrR				
Alisma plantago-aquatica L.	rrR			rR	
Butomus umbellatus L.			rrR		
Hydrocharis morsus-ranae L.		rF	rR	rcF	
Stratiotes aloides L.				cPF	
Stratiotes aloides 'submersa'	**rcF**	rcF		rcF	
Elodea canadensis Richard in Michaux	cF	cF	cF	cF	
Glyceria maxima (Hartman) Holmberg		rR			rcR
Glyceria fluitans (L.) R. Brown	rrR			rR	
Phragmites australis (Cavanilles) Trinius ex Steudel	cR	ccR	cR	ccR	rR
Schoenoplectus lacustris (L.) Palla	cR	cR	cR	cR	
Schoenoplectus lacustris subsp. *glaucus* (Smith) Becherer				rR	
Eleocharis palustris (L.) Roemer ex Schulte	rcR	rcR		rcR	
Cladium mariscus (L.) Pohl	rcR	rcR	rrR	rcR	
Rhynchospora alba (L.) Vahl					cR
Carex appropinquata Schumacher		rR		rcR	
Carex paniculata Juslenius		rR		rR	
Carex elata Allioni	rR	rR		rcR	
Carex limosa L.					cR
Carex pseudocyperus L.				rR	rR
Carex rostrata Stokes in Withering	rcR	rcR		rcR	cR
Carex vesicaria L.				rrR	
Carex acutiformis Ehrhart	cR	cR	cR	cR	
Carex riparia Curtis	rR	rR	rR	rR	

Taxa[3]	S	N	D	W	M
Carex lasiocarpa Ehrhart	rR	rR		rR	ccR
Acorus calamus L.				rR	
Calla palustris L.					rcR
Lemna trisulca L.	rrR	rP	rcP	cP	
Lemna minor L.	rrF	rrF	rR	cF	
Spirodela polyrrhiza (L.) Schleiden	rrF			rF	
Juncus subnodulosus Schrank	rrR	rR		rrR	
Iris pseudacorus L.	rR	rR	rR	rcR	
Hammarbya paludosa (L.) O. Kuntze					rrR
Dicotyledoneae (H.-D. Krausch)					
Rumex hydrolapathum Hudson	rR	rR	rR	rR	
Polygonum amphibium L.	rF	rF		rF	
Nymphaea alba L.	rcF	cF	ccF	cF	rcF
Nuphar lutea (L.) J.E. Smith	rcF	cF	rcF	cF	rcF
Ceratophyllum demersum L.	rcP	cP	ccPF	cP	
Caltha palustris L.	rrR	rR	rR	rR	
Ranunculus sceleratus L.	rrR			rrR	
Ranunculus flammula L.	rR	rR	rR	rcR	
Ranunculus lingua L.				rcR	
Ranunculus circinatus Sibthorp	cP	cP	rcP	cP	
Cardamine palustris (Wimmer et Grabowski) Petermann	rR	rR		rR	
Aldrovanda vesiculosa L.				rrP	
Drosera rotundifolia L.					cR
Drosera anglica Hudson					rcR
Potentilla palustris (L.) Scopoli				rcR	rcR
Hypericum tetrapterum Fries	rrR			rR	
Lythrum salicaria L.	rR	rR	rR	rcR	
Epilobium hirsutum L.	rrR		rcR		
Epilobium roseum Schreber			rR	rrR	
Epilobium palustre L.	rrR	rR	rR	rR	
Hippuris vulgaris L.	rR			rR	
Myriophyllum alterniflorum DC.	cP	rcP		rcP	
Myriophyllum spicatum L.	ccFP	ccFP	cP	ccFP	
Myriophyllum verticillatum L.	rrF	rs	rF	rcFP	
Hydrocotyle vulgaris L.					cR
Peucedanum palustre (L.) Moench	rR	rR		rcR	rcR
Cicuta virosa L.		rR	rR	rcR	rR
Berula erecta (Hudson) Coville	rrR			rR	
Lysimachia thyrsiflora (L.) Reichenbach	rR	rcR	rcR	rcR	rR
Lysimachia vulgaris L.	rR	rR	rR	rcR	
Lysimachia nummularia L.	rR	rR	rR	rcR	
Menyanthes trifoliata L.	rR	rR			rcR
Scutellaria galericulata L.	rR	rR	rR	rcR	
Stachys palustris L.				rcR	
Lycopus europaeus L.	rcR	rcR	rcR	rcR	
Mentha aquatica L.	rrR		rR	rR	

Taxa[3]	S	N	D	W	M
Scrophularia umbrosa Dumortier	rrR				
Veronica scutellata L.	rR	rR	rR	rcR	
Veronica beccabunga L.	rrR			rR	
Veronica anagallis-aquatica L.				rRP	
Utricularia vulgaris L.	rcP	cP	rcP	cP	
Ultricularia intermedia Hayne				rR	rP
Utricularia minor L.		rP		rP	rP
Galium palustre L.	rR	rR	rcR	rcR	
Senecio congestus (R. Brown) DC.	**rrR**			**rrR**	

Taxa[4]	S	N	W	M	D
ANIMALS					
Protozoa					
(W. Schönborn)					
Rhizopoda					
Amoebina (incl. Aconchulina and Athalamia)					
Amoeba beryllifera Penard	rL				
Amoeba proteus Leidy	cL				
Astramoeba radiosa (Dujardin)	cL				
Astramoeba radiosa var. *granulifera* (Penard)	cL				
Biomyxa cometa (Cienkowsky)	cL				
Dactylosphaerium vitraeum Hertwig et Lesser	cL				
Elaeorhanis cincta Greeff	cL				
Hartmanella hyalina (Dangeard)	cL				
Penardia cometa (Penard)	cL				
Penardia granulosa De Saedeleer	cL				
Penardia mutabilis Cash	cL				
Thecamoeba striata (Penard)	cL				
Vahlkampfia limax-group	cL				
Heliozoa					
Acanthocystis turfacea Carter	ccP				
Raphidocystis tubifera Penard	rP				
Testacea					
Arcella arenaria Greeff	rL				
Arcella artocrea (Leidy)	rL				
Arcella catinus Penard	cL	+			
Arcella catinus var. *sphaerocysta* Deflandre	rL				

[4] In the following lists abbreviations stand for: S = Lake Stechlin, N = Lake Nehmitz, D = Lake Dagow, M = bog-lakes, W = other lakes; P = pelagial, L = littoral, L(F) = litter in the littoral, B = profundal, I = interstitial space; r = rare, c = common, cc = very common.

Taxa[4]	S	N	W	M	D
Arcella catinus f. *undulata* Schönborn	rL				
Arcella dentata Ehrenberg	cL (F)				
Arcella discoides Ehrenberg	ccL	+	+		+
Arcella discoides var. *foveosa* Playfair	cL	+			+
Arcella discoides var. *pseudovulgaris* Deflandre	cL	+			+
Arcella gibbosa Penard	ccB	+		+	
Arcella gibbosa var. *laevis* Deflandre	rB				
Arcella hemisphaerica Perty	ccB	+			
Arcella hemisphaerica var. *angulata* Schönborn	rB				
Arcella hemisphaerica var. *intermedia* Deflandre	cB				
Arcella hemisphaerica f. *undulata* Deflandre	cB				
Arcella megastoma Penard	cL	+	+		+
Arcella polypora Penard	cL	+	+		+
Arcella rotundata Playfair	cL	+			
Arcella rotundata var. *aplanata* Deflandre	cL				
Arcella rotundata var. *stenostoma* Deflandre	rL				
Arcella rotundata f. *undulata* Stepánek	rL				
Arcella vulgaris Ehrenberg	ccL	+	+		+
Assulina muscorum Greeff	rL(F)				
Campascus minutus Penard	rB	+			
Campascus triqueter Penard	cB	+			
Centropyxis aculeata (Ehrenberg)	ccL,cB	+	+		+
Centropyxis aculeata var. *grandis* Deflandre	rB				
Centropyxis aculeata var. *oblonga* Deflandre	cL,rB				+
Centropyxis aerophila Deflandre	cL,cB				
Centropyxis aerophila var. *sphagnicola* Deflandre	cL(F)				
Centropyxis aerophila var. *sylvatica* Deflandre	cL(F)				
Centropyxis cassis (Wallich)	cL				
Centropyxis constricta (Ehrenberg)	cL				
Centropyxis delicatula Penard	cL,cB	+	+		+
Centropyxis discoides (Penard)	cL	+	+		+
Centropyxis ecornis (Ehrenberg)	rL				
Centropyxis gibba Deflandre	cL	+			
Centropyxis gibbosa Rampi	rL,B				
Centropyxis hirsuta Deflandre	ccL	+	+		+
Centropyxis minuta Deflandre	rL(F)				
Centropyxis orbicularis Deflandre	cL(F)				
Centropyxis platystoma (Penard)	cL(F)			+	
Chlamydophrys stercorea Cienkowsky	rL				
Cochliopodium bilimbosum Auerbach	cL	+	+		+
Cochliopodium echinatum Korotneef	rL				
Corythion dubium Taranek	cL(F)				
Corythion pulchellum Penard	rL(F)				
Cyclopyxis arcelloides (Penard)	rL				
Cyclopyxis eurystoma Deflandre	cL(F)				
Cyclopyxis grospietschi Schönborn					+
Cyclopyxis kahli Deflandre	rL,B				
Cyphoderia ampulla Ehrenberg	cB	+	+		+

Taxa[4]	S	N	W	M	D
Cyphoderia laevis Penard	rB				
Cyphoderia trochus Penard	cB	+	+		+
Difflugia acuminata Ehrenberg	cB	+	+	+	+
Difflugia bacillariarium Perty	rL				
Difflugia bacillariarium var. *teres* Wailes	rL				
Difflugia amphora Leidy	rB	+	+		+
Difflugia avellana Penard	rB				+
Difflugia avellana var. *inflata* Penard	rB				
Difflugia corona Wallich	rB	+	+		+
Difflugia elegans Penard	rL,cB	+	+	+	+
Difflugia elegans var. *teres* Penard	cB	+	+		+
Difflugia glans Penard	rB				
Difflugia globulosa Dujardin	cL,rB	+			+
Difflugia gramen Penard	rB	+			
Difflugia lanceolata Penard	cB				
Difflugia lebes Penard	cB	+			
Difflugia lemani Blanc	cB				
Difflugia limnetica (Levander)	ccP	+	+		+
Difflugia lithophila (Penard)	rB	+			
Difflugia lobostoma Leidy	cL	+			
Difflugia lucida Penard	rB				
Difflugia manicata Penard	cL	+	+		+
Difflugia mica (Frenzel)	cL	+	+		+
Difflugia mica var. *anulata* Schönborn	rL				
Difflugia oblonga Ehrenberg	cB	+	+	+	+
Difflugia oblonga var. *nodosa* Leidy	rB				
Difflugia oblonga var. *claviformis* Penard	rB				
Difflugia oblonga var. *lacustris* Penard	cB	+	+		+
Difflugia pulex Penard	cL	+			
Difflugia rubescens Penard	cL	+	+		+
Difflugia rubescens var. *brevicollis* Schönborn	rL				
Difflugia stechlinensis Schönborn	rB				
Difflugia tuberculata Wallich	rB	+	+		+
Difflugia urceolata Carter	rB	+	+		+
Difflugia urceolata var. *olla* Leidy	rB	+	+		+
Difflugia varians Penard	rB	+	+		+
Difflugia viscidula Penard	cB	+	+		+
Difflugiella angusta Schönborn	rL				
Difflugiella crenulata (Playfair)	cL	+	+		+
Difflugiella crenulata var. *glabra* Schönborn	rL				
Difflugiella crenulata var. *globosa* (Playfair)	rL				
Difflugiella compressa (Penard)	rL	+	+		+
Difflugiella horrida Schönborn	cL				
Difflugiella oviformis (Penard)	rL	+	+		+
Difflugiella splendida Schönborn	cL				
Difflugiella voigti (Schmidt)	rL				
Difflugiella sacculus (Penard)	cL	+	+		
Euglypha acanthophora (Ehrenberg)	cL	+	+		+

Taxa[4]	S	N	W	M	D
Euglypha compressa Carter	cL	+			
Euglypha laevis Perty	rL	+			
Euglypha rotunda Wailes & Penard	rL				
Euglypha strigosa (Ehrenberg)	rL				
Frenzelina reniformis Penard	cL				
Gocevia pontica Valkanov	rL				
Heleopera petricola Leidy	rB	+		+	
Heleopera petricola var. *amethystea* Penard	rB				
Heleopera rosea Penard	rL(F)				
Heleopera sylvatica Penard	rL				
Hyalosphenia cuneata Stein	cB				
Hyalosphenia elegans Leidy	rL				
Lecythium hyalinum (Ehrenberg)	cL	+	+		+
Lesquereusia modesta Rhumbler	cB	+			
Lesquereusia spiralis (Ehrenberg)	cB	+		+	
Microchlamys patella Cockerell	cL	+	+		+
Microcometes paludosa Cienkowsky	cL				
Microcorycia flava Greeff	rL				
Nebela collaris (Ehrenberg)	cL			+	
Nebela dentistoma Penard	rB			+	
Nebela militaris Penard	cL(F)				
Nebela minor Penard	rL(F)				
Nebela tincta Leidy	rL(F)				
Paraquadrula discoides (Penard)	ccL	+	+		+
Paraquadrula globulosa (Penard)	cL	+	+		
Phryganella acropodia (Hertwig & Lesser)	cL	+			
Phryganella microps Valkanov	rL				
Phryganella nidulus Penard	rB				
Phryganella paradoxa Penard	cL	+	+		+
Pontigulasia bigibbosa Penard	cB	+		+	+
Pontigulasia spectabilis Penard	cB	+		+	+
Pseudodifflugia fascicularis Penard	cL	+	+		
Pseudodifflugia virescens Penard	cL	+			
Pyxidicula cymbalum Penard	cL	+			
Quadrulella symmetrica Schulze	cL(F), B				
Tracheleuglypha dentata (Vejdowsky)	cL	+			+
Trigonopyxis arcula (Leidy)	cL(F), B				
Trinema complanatum Penard	rL(F)				
Trinema enchelys (Ehrenberg)	cL	+	+		+
Trinema lineare Penard	cL	+	+		+
Ciliata					
Aspidisca costata (Dujardin)	cL				
Aspidisca lynceus Ehrenberg	rL				
Codonella cratera var. *reticulata* Entz	cP				
Coleps hirtus Nitzsch					ccL
Dileptus anser (O.F. Müller)	cL				
Dileptus monilatus (Stokes)	rL				

Taxa[4]	S	N	W	M	D
Epistylis rotans Sveç	cP				cp
Lacrymaria olor (O.F. Müller)	cL				
Litonotus fasciola (Ehrenberg)	rL				
Litonotus lamella (Ehrenberg)	cL				
Ophrydium versatile (O.F. Müller)	cL				
Spirostomum ambiguum (O.F. Müller)	cL				
Staurophrya elegans Zacharias	cP				
Stentor amethystinus Leidy	ccP				
Stentor coeruleus Ehrenberg	cL				
Stentor polymorphus (O.F. Müller)	cL				
Tintinnidium fluviatile Stein	cP				
Trachelius ovum Ehrenberg	cL				
Vorticella campanula (Ehrenberg)	cL				
Vorticella nebulifera var. *similis* (Stokes)	cP,L				
Vorticella monilata Tatem	cL				
Vorticella picta (Ehrenberg)	rL				
Zoothamnium spp.	cL				
Parazoa					
(G. Mothes)					
Spongillidae					
Eunapius fragilis (Leidy)		rL			
Spongilla lacustris (L.)	cL	cL	rL		cL
Coelenterata					
(G. Mothes)					
Hydrozoa					
Hydra oligactis (Pallas)	cL	cL	cL		cL
Hydra vulgaris (Pallas)	cL				
Plathelminthes					
(G. Mothes)					
Tricladida					
Bdellocephala punctata (Pallas)	rL				
Dendrocoelum lacteum (Müller)	rL	rL			
Dugesia lugubris (Schmidt)	cL	cL			
Dugesia polychroa (Schmidt)	rL	rL			
Dugesia tigrina (Girard)	rL		rL		
Phagocata albissima (Vejdovsky)	rL				
Planaria torva (Müller)	cL				
Polycelis tenuis Ijima	cL				
Nemertini					
(G. Mothes)					
Prostoma sp.	rL				

Taxa[4]	S	N	W	M	D
Nemathelminthes					
Rotatoria					
(D. Ronneberger & W. Schönborn)					
Anuraeopsis fissa (Gosse)	rP	cP			
Argonotholoa foliacea (Ehrenberg)	rP, cL	cP			
Ascomorpha ecaudis Perty	cP				
Ascomorpha saltans Bartsch	ccP				
Aspelta aper (Harring)	rL				
Aspelta circinator (Gosse)	rL				
Asplanchna priodonta Gosse	cP	cP			ccP
Brachionus angularis Gosse	r	c			cc
Brachionus calyciflorus Pallas		r			c
Brachionus calyciflorus anuraeopsis (Brehm)		r			
Brachionus calyciflorus Pallas		r			
Brachionus dimidiatus (Bryce)		r			
Cephalodella apocolea Harring & Myers	rL				
Cephalodella auriculata (O.F. Müller)	rL				
Cephalodella catellina (O.F. Müller)	rL				
Cephalodella eva (Gosse)	rL				
Cephalodella forficula (Ehrenberg)	rL				
Cephalodella gibba (Ehrenberg)	ccL				
Cephalodella gibba subsp. *microdactyla* Koch-Althaus	rL				
Cephalodella gigantea Remane	rL				
Cephalodella glandulosa Koch-Althaus	rL				
Cephalodella gracilis (Ehrenberg)	rL				
Cephalodella hoodi (Gosse)	rL				
Cephalodella incila Wulfert	rL				
Cephalodella limosa Wulfert	rL				
Cephalodella megalocephala var. *rotunda* Donner	rL				
Cephalodella nana Myers (= *C. cuneata* Harring & Myers)	rL				
Cephalodella pachydactyla Wulfert	rL				
Cephalodella plicata Myers	rL				
Cephalodella psammophile (?) Koch-Althaus	L				
Cephalodella sterea f. *minor* Donner	rL				
Cephalodella tachyphora Myers	rL				
Cephalodella tenuis Koch-Althaus	rL				
Cephalodella theodora Koch-Althaus	rL				
Cephalodella ventripes (Dixon-Nuttall)	cL				
Chromogaster ovalis (Bergendal)	cP	cP			
Collotheca ambigua (Hudson)	rL				
Collotheca libera (Zacharias)	cP				
Collotheca mutabilis (Hudson)	rP				
Collotheca pelagica (Rousselet)	rP				
Colurella adriatica Ehrba	rL				
Colurella colurus (Ehrenberg)	rL				
Colurella denticauda Carlin	rL				
Colurella obtusa (Gosse)	ccL				
Colurella sulcata (Sternroos)	rL				

Taxa[4]	S	N	W	M	D
Colurella uncinata f. *bicuspidata* (Ehrenberg)	ccL				
Colurella uncinata f. *deflexa* (Ehrenberg)	rL				
Conochilus natans (Seligo)		rP	rP		cP
Conochilus unicornis Rousselet	cP	cP			cP
Dicranophorus cernuus Harring & Myers	rL				
Dicranophorus epicharis Harring & Myers	rL				
Dicranophorus hercules var. *capucinoides* Wiszniewski	rL				
Dicranophorus robustus Harring & Myers	cL				
Dicranophorus robustus var. *europaeus* Wulfert	cL				
Dissotrocha aculeata (Ehrenberg)	rL				
Dorystoma caudata (Bilfinger)	rL				
Encentrum mustela (Milne)	rL				
Encentrum spinosum Koch-Althaus	L				
Encentrum stechlinensis Koch-Althaus	L				
Euchlanis alata Voronkov	rL				
Euchlanis dilatata Ehrenberg	rL				
Euchlanis dilatata var. *lucksiana* Hauer	rP				
Euchlanis micropous (?) Koch-Althaus	rL				
Filinia longiseta (Ehrenberg)	cP	cP			cP
Filinia major (Colditz)	cP				
Filinia terminalis (Plate)	cP	cP			
Gastropus hyptopus (Ehrenberg)	rP	ccP			
Gastropus stylifer Imhof	cP	ccP			
Itura pseudoaurita (?) Koch-Althaus	rL				
Kellicottia longispina (Kellicott)	ccP	ccP			cP
Keratella cochlearis (Gosse)	ccP	ccP	cP	cP	ccP
Keratella quadrata (O.F. Müller)	cP	cP			ccP
Lecane bulla (Gosse)	rL				
Lecane closterocerca (Schmarda)	rL				
Lecane flexilis (Gosse)	rL				
Lecane furcata (Murray)	rL				
Lecane hornemanni (Ehrenberg)	rL				
Lecane inermis (Bryce)	rL				
Lecane intrasinuata (Olofsson)	rL				
Lecane levistyla (Olofsson)	rL				
Lecane ludwigi Eckstein	rL				
Lecane luna (O.F. Müller)	ccL				
Lecane lunaris (Ehrenberg)	rL				
Lecane nana (Murray)	rL				
Lecane ploenensis (Voigt)	rL				
Lecane quadridentata (Ehrenberg)	rL				
Lecane sympoda Hauer	rL				
Lecane tenuiseta Harring	rL				
Lecane thetis (Harring & Myers)	rL				
Lecane ungulata (Gosse)	rL				
Lepadella acuminata (Ehrenberg)	cL				
Lepadella ehrenbergi (Perty)	rL				
Lepadella heterostyla (Murray)	rL				

Taxa[4]	S	N	W	M	D
Lepadella ovalis (O.F. Müller)	cL				
Lepadella patella (O.F. Müller)	cL				
Lepadella patella subsp. *minor* Koch-Althaus	rL				
Lepadella pseudosimilis Koch-Althaus	rL				
Lepadella triptera Ehrenberg	cL				
Lindia euchromatica subsp. *europaea* Koch-Althaus	rL				
Lindia torulosa Dujardin	rL				
Lophocharis oxysternon (Gosse)	rP				
Macrochaetus subquadratus Perty	rL				
Monommata longiseta (O.F. Müller)	rL				
Mytilina ventralis (Ehrenberg)	cL				
Mytilina ventralis var. *brevispina* (Ehrenberg)	cL				
Notholca acuminata (Ehrenberg)	rP,rL			cP	
Notholca labis Gosse	rP,rL				
Notholca squamula (O.F. Müller)	rL				
Notommata contorta (Stokes)	rL				
Notommata copeus Ehrenberg	rL				
Notommata cyrtopus Gosse	rL				
Notommata glyphura Wulfert	rL				
Notommata omentata Wulfert	cL				
Notommata pseudocerberus De Beauchamp	rL				
Notommata tripus Ehrenberg	rL				
Notommata voigti Donner	rL				
Philodina megalotrocha Ehrenberg	rL				
Philodina roseola Ehrenberg	rL				
Platyias quadricornis (Ehrenberg)	rL				
Pleurotrocha petromyzon Ehrenberg	rL				
Ploesoma hudsoni (Imhof)	cP	rP			
Polyarthra dolichoptera Idelson	ccP	ccP	cP		ccP
Polyarthra minor Voigt	rP	rP			
Polyarthra vulgaris Carlin	cP	cP			
Pompholyx complanata Gosse	rP	rP			rP
Pompholyx sulcata Hudson	rP	cP			cP
Proales decipiens (Ehrenberg)	rL				
Proales fallaciosa Wulfert	rL				
Proales micropus Gosse	rL				
Proales sordida Gosse	rL				
Proalinopsis caudatus (Collins)	rL				
Resticula melandocus (Gosse)	rL				
Resticula nyssa Harring & Myers	rL				
Rotaria rotatoria (Pallas)	rL				
Scardium longicaudum (O.F. Müller)	rL				
Sinantherina socialis (Linné)	rP				
Squantinella rostrum (Schmarda)	rL				
Squantinella tridentata var. *mutica* (Ehrenberg)	rL				
Synchaeta asymmetrica Koch-Althaus	rP				
Synchaeta grandis Zacharias	cP				
Synchaeta oblonga Ehrenberg	cP	cP			cP

Taxa[4]	S	N	W	M	D
Synchaeta pectinata Ehrenberg	ccP	cP			cP
Synchaeta tavina Hood	cP				
Synchaeta tremula (O.F. Müller)	rP	rP			
Taphrocampa annulosa (Gosse)	rL				
Testudinella parva (Ternetz)	rL				
Testudinella parva var. *bidentata* (Ternetz)	rL				
Testudinella patina (Hermann)	cL				
Testudinella truncata (Gosse)	rL				
Trichocerca bicristata (Gosse)	rL				
Trichocerca brachyura (Gosse)	cL				
Trichocerca cavia (Gosse)	rL				
Trichocerca iernis (Gosse)	rL				
Trichocerca intermedia (Sternroos)	rL				
Trichocerca longiseta (Schrank)	rL				
Trichocerca minuta (Olofsson)	cL				
Trichocerca myersi (Hauer)	rL				
Trichocerca procellus (Gosse)	cL				
Trichocerca pusilla (Jennings)	rP,rL				
Trichocerca rattus (O.F. Müller)	rL				
Trichocerca rousseleti (Voigt)	rP				
Trichocerca sulcata (Jennings)	rL				
Trichocerca taurocephala (Hauer)	rL				
Trichocerca tenuior (Gosse)	rL				
Trichocerca tigris (O.F. Müller)	rL				
Trichocerca uncinata (Voigt)	cL				
Trichotria pocillum (O.F. Müller)	ccL				
Trichotria similis Sternroos	rL				
Trichotria tetractis (Ehrenberg)	rL				
Trochoceroa capucina Wierzejski & Zacharias	rP	rP			cP
Trochoceroa rousseleti (Voigt)	rP				
Nematomorpha					
(G. Mothes)					
Gordius albopunctatus Müller	rL				
Nematodes					
(D. Ronneberger)					
Achromadora terricola (De Man)	rI				
Dorylaimus stagnalis (Dujardin)	rI				
Dorylaimoides microdentatus Altherr	rI				
Eudorylaimus holsaticus (W. Schneider)	rI				
Ironus ignavus Bastian	rI				
Ironus ignavus var. *tenuicaudatus* (De Man)	rI				
Labronema stechlinensis Altherr	rI				
Monohystera agilis De Man	rI				
Paractinolaimus intermedius Altherr	cI				
Plectus granulosus Bastian	rI				
Plectus rhizophilus De Man	rI				

Taxa[4]	S	N	W	M	D
Prionchulus muscorum (Dujardin)	rI				
Prismatolaimus dolichurus De Man	rI				
Prodorylaimus longicaudatoides Altherr	rI				
Tobrilus aberrans (W. Schneider)	rI				
Tobrilus allophysis (Steiner)	cI				
Tobrilus gracilis (Bastian)	rI				
Tobrilus grandipapillatus (Brackenhaus)	rI				
Tobrilus husmanni Altherr	rI				
Tobrilus medius (G. Schneider)	rI				
Tobrilus steineri (Micoletzki)	cI				
Tripyla glomerans Bastian	cI				
Tripyla setifera Bütschli	rI				
Tentaculata					
(G. Mothes)					
Bryozoa					
Cristatella mucedo Cuvier	cL	cL	cL		cL
Fredericella sultana (Blumenbach)	cLB	cLB			
Paludicella articulata (Ehrenberg)	rL				
Plumatella fungosa (Pallas)			cL		
Plumatella repens (L.)	rL		rL		
Mollusca					
(G. Mothes)					
Gastropoda					
Acroloxus lacustris (L.)	cL	cL	cL		cL
Anisus vortex (L.)	cL	cL			
Bithynia leachi Sheppard		rL			
Bithynia tentaculata (L.)	ccL	ccL	cL		cL
Bulinus contortus Michaud	rL	rL			
Gyraulus acronicus Férussac	cL	cL	cL		
Gyraulus albus (Müller)	cL	cL			
Gyraulus crista (L.)	cL	cL	cL		
Gyraulus riparius Westerlund	cL	cL			
Lymnaea auricularia (L.)	rL				
Lymnaea palustris (Müller)	cL	cL			
Lymnaea peregra Müller	cL	cL	cL		cL
Lymnaea stagnalis (L.)	cL	cL	cL		cL
Physa fontinalis (L.)	ccL	cL	cL		cL
Planorbarius corneus (L.)	rL	rL			
Planorbis planorbis (L.)	cL	cL	cL		cL
Segmentina complanatus (L.)	cL	cL			cL
Segmentina nitida (Müller)	rL	rL			
Theodoxus fluviatilis (L.)	cL	cL			
Valvata cristata Müller	cL	cL	cL		cL
Valvata piscinalis (Müller)	cL	cL	cL		cL
Viviparus contectus (Millet)	cL	cL	rL		cL

Taxa[4]	S	N	W	M	D
Lamellibranchiata					
Anodonta cygnea (L.)	cL	cL			cL
Dreissena polymorpha (Pallas)	ccLB	ccLB	cL		ccL
Pisidium amnicum (Müller)	rL	rL			
Pisidium casertanum (Poli)	cLB	cL			cL
Pisidium conventus Clessin	cB				
Pisidium henslowanum (Sheppard)	cL	cL			
Pisidium hibernicum Westerlund	rL	rL			
Pisidium liljeborgi Clessin	rL	rL	rL		
Pisidium milium Held	rL				
Pisidium moitesserianum Paladilhe		rL			
Pisidium nitidum Jenyns	cL	cL	cL		cL
Pisidium obtusale (Lamarck)	rL				
Pisidium pseudosphaerium Bentheim-Jutting & Kuiper	cL				
Pisidium subtruncatum Malm	cL	cL	cL		
Sphaerium corneum (L.)		rL			
Sphaerium lacustre Müller	rL				
Unio pictorum (L.)		cL	cL		
Annelida					
(G. Mothes)					
Oligochaeta					
Chaetogaster diaphanus (Gruithuisen)	cL	cL	cL		
Dero digitata (Müller)	rL				
Dero obtusa d'Udekem	rL				
Nais pseudobtusa Piguet	cL	cL		cL	
Ophidonais serpentina (Müller)		rL			
Paranais litoralis (Müller)	cL	cL			
Pristina bilobata (Bretscher)	rL			rL	
Pristina longiseta Ehrenberg	rL				
Pristina rosea (Piguet)	rL			rL	
Slavina appendiculata (d'Udekem)	rL				
Specaria josinae (Vejdovsky)	rL				
Stylaria lacustris (L.)	ccL	ccL	cL	cL	ccL
Uncinais unicinata (Ørstedt)		rL			
Vejdovskyella comata (Vejdovsky)	rL				
Hirudinea					
Batrachobdella verrucata (Fr. Müller)	rL				
Dina leneata (Müller)	rL				
Erpobdella nigricollis (Brandes)	rL	rL			
Erpobdella octoculata (L.)	rL	rL			
Glossiphonia complanata (L.)		rL			
Glossiphonia heteroclita (L.)	cL	cL	cL		
Haemopis sanguisuga (L.)		rL			
Helobdella stagnalis (L.)	cL	cL	cL		cL
Hemiclepsis marginata (Müller)	rL				
Hirudo medicinalis L.	rL	rL			

Taxa[4]	S	N	W	M	D
Piscicola geometra (L.)	cL	cL	cL		cL
Theromyzon maculosum (Rathke)		rL			
Arthropoda					
Arachnida					
(G. Mothes)					
Aranea					
Argyroneta aquatica (Clerk)	rL	rL	rL	rL	rL
Acari					
Hydracarina					
Arrenurus adnatus Koenike	cL				
Arrenurus affinis Koenike	rL	rL			
Arrenurus albator (Müller)	cL	cL	cL		
Arrenurus batillifer Koenike		rL			
Arrenurus bicuspidator Berlese	cL	cL			
Arrenurus bifidicodulus Piersig		cL			
Arrenurus biscissus Lebert		cL			
Arrenurus brunsvivensis Viets		rL			
Arrenurus bruzelii Koenike		rL			
Arrenurus buccinator (Müller)	cL	cL			
Arrenurus claviger Koenike	cL	cL			
Arrenurus conicus Piersig		rL			
Arrenurus crassicaudatus Kramer		rL			
Arrenurus crenatus Koenike	rL	cL			
Arrenurus curvisetus Viets		rL			
Arrenurus cuspidator (Müller)	cL	cL			
Arrenurus cuspidifer Piersig	rL	rL			
Arrenurus cylindratus Piersig	rL	rL			
Arrenurus fimbriatus Koenike	rL	rL	rL		
Arrenurus forpicatus Neuman	cL	cL			
Arrenurus globator (Müller)	cL	cL	cL	cL	cL
Arrenurus inexploratus Viets	rL	rL			
Arrenurus integrator (Müller)		rL			
Arrenurus knauthei Koenike	rL	rL			
Arrenurus latus Barrois & Moniez	rL	rL			
Arrenurus maculator (Müller)	rL	rL			
Arrenurus mediorotundatus Thor		rL			
Arrenurus neumani Piersig		rL		cL	
Arrenurus nobilis Neuman		rL			
Arrenurus nodosus Koenike	cL	cL	cL	cL	
Arrenurus perforatus George	cL	cL			
Arrenurus pustulator (Müller)	rL				
Arrenurus radiatus Piersig		rL			
Arrenurus robustus Koenike	rL				
Arrenurus securiformis Piersig	cL	cL			
Arrenurus sinuator (Müller)	cL	cL			
Arrenurus stecki Koenike		rL			

Taxa[4]	S	N	W	M	D
Arrenurus tetracyphus Piersig	rL	rL			
Arrenurus tricuspidator (Müller)	cL	cL			
Arrenurus truncatellus (Müller)		rL			
Arrenurus virens Neuman	rL				
Arrenurus zachariae Koenike		rL			
Atractides ovalis Koenike	cL	cL			
Atractides pavesii Maglio	cL	cL			
Brachypoda versicolor (Müller)	cL	cL	cL		
Euthyas truncata (Neuman)	rL				
Eylais extendens (Müller)	cL	cL			
Eylais infundibulifera Koenike	cL				
Forelia brevipes (Neuman)		rLB			
Forelia curvipalpis Viets	cL	cL			
Forelia liliacea (Müller)	cL	cL			
Frontipoda musculus (Müller)	cL	cL			
Hydrachna globosa (Geer)				rL	
Huitfeldia rectipes Thor	cLB				
Hydrochoreutes krameri Piersig	cL	cL	cL		cL
Hydrochoreutes ungulatus (Koch)	rL	rL			
Hydrodroma despiciens (Müller)	cL	cL	cL		cL
Hydryphantes bayeri Pisarov		cL			
Hydryphantes crassipalpis Koenike		rL			
Hydryphantes dispar (Schaub)		rL			
Hydryphantes flexuosus (Koenike)		cL			
Hydryphantes hellichi Thon		rL			
Hydryphantes octoporus Koenike		cL			
Hydryphantes parmulatus Koenike		rL			
Hydryphantes placationis Thon	rL				
Hydryphantes ruber (Geer)		cL			
Hygrobates longipalpis (Herman)	ccL	cL	cL		cL
Hygrobates nigromaculatus Lebert	cL	cL			
Lebertia densa Koenike	cLB	cLB			
Lebertia dubia Thor	cLB	cLB			
Lebertia inaequalis (Koch)	cL	cL			
Lebertia porosa Thor	cL	cL			
Limnesia connata Koenike		cL			
Limnesia fulgida Koch		cL			
Limnesia maculata (Müller)	cL	cL	cL		cL
Limnesia polonica Schechtel	cL	cL			
Limnesia undulata (Müller)	cLB	cL			
Limnochares aquatica (L.)	cL	cL	cL		cL
Limnohalacarus wackeri (Walter)	rL				
Midea orbiculata (Müller)	cL	cL	cL		cL
Mideopsis crassipes Soar	cL	cL	cL		cL
Mideopsis orbicularis (Müller)	ccL	cL	cL		cL
Neumania callosa (Koenike)	cL				
Neumania deltoides (Piersig)	rL	cL			
Neumania limosa (Koch)	cL	cL			

Taxa[4]	S	N	W	M	D
Neumania spinipes (Müller)	rL	rL			
Neumania vernalis (Müller)	cL	cL			
Oxus longisetus (Berlese)	rL	rL			
Oxus ovalis (Müller)	rL	rL	rL		
Oxus strigatus (Müller)	cL	cL			
Piona alpicola (Neuman)	cL	cL			
Piona carnea (Koch)				cL	
Piona clavicornis (Müller)		cL			
Piona coccinea (Koch)	cL	cL	cL	cL	cL
Piona conglobata (Koch)	cL	cL	cL	cL	cL
Piona discrepans (Koenike)		rL			
Piona longipalpis (Krendowski)	cL	cL	cL		cL
Piona neumani (Koenike)	rL	rL			
Piona nodata (Müller)	rLB	rLB			
Piona obturbans (Piersig)	rL	rL			
Piona paucipora (Thor)		cL			
Piona pusilla (Neuman)	cL	cL	cL		cL
Piona variabilis (Koch)	cL	cL			
Pionacercus norvegicus Thor		rL			
Pionacercus uncinatus (Koenike)		cL			
Pionacercus vatrax (Koch)		rL			
Pionipsis lutescens (Herman)	cL	cL			
Porohalacarus alpinus (Thor)	rL				
Teutonia cometes (Koch)		rL			
Thyas pachystoma Koenike		rL			
Tiphys ensifer (Koenike)		rL			
Tiphys latipes (Müller)		cL			
Tiphys ornatus Koch		cL			
Tiphys scaurus (Koenike)		cL			
Tiphys torris (Müller)	rL				
Unionicola crassipes (Müller)	ccL	ccL	cL		cL
Unionicola minor (Soar)	cL	cL			
Oribatei					
Hydrozetes parisiensis Grandjean	cL	cL	cL		cL
Crustacea					
(D. Flößner)					
Branchiopoda					
Acantholeberis curvirostris (O.F. Müller)				+	
Acroperus harpae (Baird)	ccL	+	+		+
Acroperus angustatus Sars	cL	+	+		
Alona affinis (Leydig)	ccL	+	+	+	+
Alona costata Sars	cL	+	+		+
Alona guttata Sars	cL	+	+	+	+
Alona karelica Sternroos	rL				
Alona intermedia Sars				+	
Alona protzi Hartwig	rL				

Taxa[4]	S	N	W	M	D
Alona rectangula Sars	cL	+	+		
Alona rustica Scott				+	
Alona quadrangularis (O.F. Müller)	cL	+	+		
Alona weltneri Keilhack	rL				
Alonella excisa (Fischer)	cL	+	+	+	
Alonella exigua (Lilljeborg)	cL	+	+		+
Alonella nana (Baird)	cL	+	+	+	+
Alonopsis elongata Sars	ccL	+	+		
Anchistropus emarginatus Sars	cL	+	+		
Bosmina coregoni Baird	ccP, rL	+	+		+
Bosmina longirostris (O.F. Müller)	cP, rL	+	+	+	+
Bunops serricaudata (Daday)			+		
Bythotrephes longimanus Leydig	cP				
Camptocercus rectirostris Schoedler	cL	+	+	+	
Camptocercus lilljeborgi Schoedler	cL	+	+		+
Ceriodaphnia laticaudata P.E. Müller	rL	+	+		+
Ceriodaphnia megops Sars		+	+		
Ceriodaphnia pulchella Sars	rP, rL	+	+		+
Ceriodaphnia quadrangula (O.F. Müller)	cP,rL	+	+	+	
Ceriodaphnia reticulata (Jurine)	rL	+	+		
Ceriodaphnia rotunda Sars	rL	+	+		
Chydorus gibbus Sars	cL	+			
Chydorus latus Sars		+			
Chydorus ovalis Kurz				+	
Chydorus piger Sars	cL	+	+		
Chydorus sphaericus (O.F. Müller)	ccL	+	+	+	+
Daphnia cucullata Sars	ccP, rL	+	+		+
Daphnia curvirostris Eylmann	rL	+			
Daphnia galeata Sars		+			+
Daphnia hyalina Leydig	ccP	+	+		+
Daphnia longispina O.F. Müller			+		+
Diaphanosoma brachyurum (Liévin)	cP,rL	+	+	+	+
Disparalona rostrata (Koch)	cL	+			
Drepanothrix dentata (Eurén)	rL	+			
Eurycercus lamellatus (O.F. Müller)	ccL	+	+	+	+
Graptoleberis testudinaria (Fischer)	cL	+	+	+	+
Ilyocryptus acutifrons Sars			+		
Ilyocryptus agilis Kurz		+		+	
Ilyocryptus sordidus (Liévin)	rL	+		+	
Kurzia lattisima (Kurz)	rL			+	
Lathonura rectirostris (O.F. Müller)	cL	+	+		
Latona setifera (O.F. Müller)	rL	+	+		
Leptodora kindti (Focke)	cP	+	+		
Leydigia leydigi (Schoedler)	rL	+			
Macrothrix rosea (Jurine)	rL	+			
Monospilus dispar Sars	rL	+			

Taxa[4]	S	N	W	M	D
Oxyurella tenuicaudis (Sars)	rL		+		
Pleuroxus aduncus (Jurine)	rL	+	+		+
Pleuroxus laevis Sars	rL	+	+		
Pleuroxus trigonellus (O.F. Müller)	cL	+	+		+
Pleuroxus truncatus (O.F. Müller)	ccL	+	+	+	+
Pleuroxus uncinatus Baird	rL	+	+		
Polyphemus pediculus (L.)	cL,rP	+	+		+
Pseudochydorus globosus (Baird)	cL	+	+		+
Rhynchotalona falcata (Sars)	cL	+	+		
Scapholeberis microcephala Sars				+	
Scapholeberis mucronata (O.F. Müller)	cL	+	+	+	+
Sida crystallina (O.F. Müller)	ccL	+	+		+
Simocephalus exspinosus (Koch)			+		
Simocephalus serrulatus (Koch)			+	+	
Simocephalus vetulus (Schoedler)	cL	+	+	+	+
Streblocerus serricaudatus (Fischer)	rL			+	
Copepoda					
Calanoida					
Eudiaptomus gracilis (Sars)	ccP,rL	+	+	+	
Eudiaptomus graciloides (Lilljeborg)		+	+	+	+
Eurytemora lacustris (Poppe)	cP,rL	+			
Heterocope appendiculata Sars	rP,rL	+			
Cyclopoida					
Acanthocyclops robustus (Sars)				+	
Cyclops bohater Kozminski	rP		+		
Cyclops furcifer Claus	rL				
Cyclops insignis Claus	rL	+	+		
Cyclops kolensis Lilljeborg	rP,rL	+	+		
Cyclops strenuus Fischer	rL	+	+	+	+
Cyclops vicinus Uljanin					+
Cryptocyclops bicolor (Sars)	cL		+		
Diacyclops bicuspidatus (Claus)	rP	+	+		+
Diacyclops bisetosus (Rehberg)	rI		+		
Diacyclops languidus (Sars)	rL			+	
Diacyclops languidoides (Lilljeborg)	rL	+		+	
Diacyclops nanus (Sars)	rL			+	
Ectocyclops phaleratus (Koch)	rL		+	+	
Eucyclops denticulatus (Graeter)	rL		+	+	
Eucyclops macrurus (Sars)	ccL	+	+		
Eucyclops macruroides (Lilljeborg)	cL		+		
Eucyclops serrulatus (Fischer)	cL	+	+	+	+
Eucyclops speratus (Lilljeborg)	rL	+	+		
Macrocyclops albidus (Jurine)	ccL	+	+	+	
Macrocyclops distinctus (Richard)	rL		+		
Macrocyclops fuscus (Jurine)	cL	+	+	+	
Megacyclops gigas (Claus)	cL,cB				

Taxa[4]	S	N	W	M	D
Megacyclops viridis (Jurine)	ccL	+	+	+	+
Mesocyclops leuckarti (Claus)	ccP,ccL	+	+	+	+
Microcyclops rubellus (Lilljeborg)	cL	+			
Microcyclops varicans (Sars)	rL	+			
Paracyclops affinis (Sars)	rL		+	+	+
Paracyclops fimbriatus (Fischer)	rL		+		
Paracyclops poppei (Rehberg)	rL		+		
Thermocyclops dybowskii (Lande)		+	+		
Thermocyclops crassus (Fischer)			+		
Thermocyclops oithonoides (Sars)	ccP,rL	+	+		+
Poecilostomatoida					
Ergasilus spec.		+			
Harpacticoida					
Attheyella crassa (Sars)	cL	+	+	+	
Attheyella dentata (Poggenpol)	cL		+	+	
Attheyella trispinosa (Brady)	ccL		+		
Bryocamptus minutus (Claus)	cL				
Bryocamptus pygmaeus (Sars)	cL		+	+	
Bryocamptus vejdovskyi (Mrazek)	cL				
Canthocamptus staphylinus (Jurine)	ccL	+	+	+	
Elaphoidella gracilis (Sars)	rI			+	
Moraria brevipes (Sars)				+	
Moraria mrazeki Scott	rL				
Nitocra hibernica (Brady)	ccL		+		
Parastenocaris brevipes Kessler	cI	+		+	
Parastenocaris glacialis Noodt	rI				
Parastenocaris phyllura Kiefer	cI				
Ostracoda					
Candona acuminata Fischer	rL				
Candona albicans Brady	rL				
Candona candida (O.F. Müller)	cL		+		
Candona hyalina Brady et Robertson	rL		+		
Candona neglecta Sars	cB				
Candona protzi Hartwig	rL,cB		+		
Candona rostrata Brady et Norman		+			
Candona stagnalis Sars				+	
Cyclocypris laevis (O.F. Müller)	cL		+		
Cyclocypris ovum (Jurine)	ccL	+	+		
Cyclocypris serena (Koch)	cL				
Cypria exsculpta (Fischer)	rL		+		
Cypria lacustris Sars	cB				
Cypria ophthalmica (Jurine)	cL,rB	+	+		
Cypridopsis vidua (O.F. Müller)	ccL		+		
Cytherissa lacustris Sars	cB				
Darwinula stevensoni (Brady et Robertson)	rB				

Taxa[4]	S	N	W	M	D
Herpetocypris reptans (Baird)	rL				
Limnocythere inopinata (Baird)	rL				
Limnocythere psammophila Flößner	rL		+		
Notodromas monacha (O.F. Müller)	ccL				
Paracandona euplectella Brady et Norman				+	
Potamocypris villosa (O.F. Müller)	rL				
Malacostraca					
Gammarus lacustris Sars	cL				
Gammarus pulex L.	cL	cL	cL		
Gammarus roeseli Gervais	cL	cL	cL		
Asellus aquaticus (L.)	ccL	cL	cL	cL	cL
Orconectes limosus (Rafinesque)	cL	cL	cL		cL
Insecta					
(G. Mothes)					
Ephemeroptera					
Caenis horaria L.	ccL	ccL	cL		cL
Caenis lactea Pictet	rL				
Caenis luctuosa Burmeister	cL	cL			
Centroptilum luteolum Müller	cL	cL	cL		cL
Cloeon dipterum L.	cL	cL	cL		cL
Cloeon simile Eaton	cL	cL			
Ephemera vulgata L.	cL	cL			
Heptageniidae spec.				rL	
Leptophlebia marginata L.	cL	cL	cL		cL
Leptophlebia vespertina L.				cL	
Procloeon bifidum Bengtsson	cL	cL	cL		
Plecoptera					
Nemoura cinerea Retzius	cL	cL	cL		
Odonata					
Aeshna cyanea (Müller)	rL		rL	rL	
Aeshna grandis (L.)	rL		rL	rL	
Aeshna isosceles (Müller)	rL				
Aeshna juncea (L.)				rL	
Aeshna mixta Latreille	rL	rL	rL	rL	
Aeshna subarctica Walker				rL	
Brachytron pratense (Müller)	rL	rL			
Calopteryx splendens (Harris)	rL				
Calopteryx virgo (L.)	rL				
Coenagrion hastulatum (Charpentier)			rL	rL	
Coenagrion lunulatum (Charpentier)			rL	rL	
Coenagrion puella (L.)	cL	cL	cL	cL	cL
Coenagrion pulchellum (Linden)	cL	cL	cL	cL	cL
Cordulea aenea (L.)	rL			cL	
Enallagma cyathigerum (Charpentier)	ccL	cL	cL	cL	cL

Taxa[4]	S	N	W	M	D
Epitheca bimaculata (Charpentier)	rL				
Erythromma najas (Hansemann)	cL	cL	cL	cL	cL
Gomphus vulgatissimus (L.)	rL				
Ischnura elegans (Linden)	cL	cL	cL	rL	cL
Lestes dryas Kirby				rL	
Lestes sponsa (Hansemann)	cL	cL	cL	rL	
Lestes virens (Charpentier)		rL		cL	
Lestes viridis (Linden)	cL		cL	cL	
Leucorrhinia albifrons (Burmeister)				cL	
Leucorrhinia caudalis (Charpentier)				cL	
Leucorrhinia dubia (Linden)				cL	
Leucorrhinia pectoralis (Charpentier)				cL	
Leucorrhinia rubicunda (L.)				cL	
Libellula depressa L.	cL		cL		
Libellula fulva Müller	cL	cL	cL		
Onychogomphus forcipatus (L.)	rL				
Ophiogomphus serpentinus (Charpentier)	rL				
Orthetrum cancellatum (L.)	cL	cL	cL	cL	
Platycnemis pennipes (Pallas)	cL	cL	cL		
Pyrrhosoma nymphula (Sulzer)	cL	cL	cL	cL	cL
Somatochlora flavomaculata (Linden)	rL	rL			
Somatochlora metallica (Linden)		rL			
Sympecma fusca (Linden)	rL	rL		rL	
Sympetrum danae (Sulzer)	rL			rL	
Sympetrum flaveolum (L.)	cL	cL			
Sympetrum sanguineum (Müller)	cL	cL	cL	cL	cL
Sympetrum striolatum (Charpentier)	cL	cL			
Sympetrum vulgatum (L.)	rL	rL		rL	
Heteroptera					
Corixa dentipes (Thomson)				rL	
Cymatia coleoptrata (Fabricius)	rL		rL		
Gerris argentatus Schummel		cL			
Gerris lacustris (L.)	cL				
Gerris najas (De Geer)	cL	cL			
Gerris paludum (Fabre)	cL	cL			
Glaenocorisa propinqua (Fieber)				cL	
Hesperocorixa linnei (Fieber)	cL				
Hesperocorixa sahlbergi (Fieber)	cL	cL			
Hydrometra gracilenta Horion	rL				
Hydrometra stagnorum (L.)			rL		
Ilyocoris cimicoides (L.)	cL	cL			
Mesovelia furcata Mulsant et Reyche				cL	
Micronecta minutissima (L.)	cLB	cLB	cL		
Nepa cinerea L.	cL	cL	cL		
Notonecta glauca L.	cL	cL	cL	cL	
Notonecta maculata Fabricius				cL	
Ranatra linearis (L.)		rL			

Taxa[4]	S	N	W	M	D
Coleoptera					
Acilius canaliculatus Nicolaus	rL				
Acilius sulcatus L.	rL				
Anacaena limbata Fabricius		cL			
Berosus luridus L.		cL			
Berosus signaticollis Charpentier		cL			
Coelambus marklini (Gyllenthal)		rL			
Colymbetes (Larvae)	rL				
Cybister lateralimarginalis Degert				rL	
Dytiscus circumcinctus Ahrens		rL	rL		
Dytiscus dimidiatus Bergström			rL		
Dytiscus marginalis L.			rL		
Galerucella nymphaeae L.	rL	cL	cL		cL
Gaurodytes paludosus (Fabricius)		rL			
Gaurodytes undulatus (Schrank)	rL				
Gyrinus caspius Ménétriés	rL	cL	cL		
Gyrinus colymbus Erichson		rL			
Gyrinus natator L.	cL				
Gyrinus paykulli Ochs	cL				
Haliplus apicalis Thomson	rL				
Haliplus confinis Stephens	cL	cL			
Haliplus fulvicollis Erichson	rL				
Haliplus immaculatus Gerhard	rL				
Haliplus laminatus Schaller		rL			
Haliplus obliquus Fabricius	cL	cL			
Haliplus ruficollis Degert	rL				
Haliplus variegatus Sturm	rL	rL			
Hydaticus (Larvae)	cL	cL			
Hydrobius fuscipes L.	cL			cL	
Hydroporus elongatulus Sturm	cL				
Hydroporus erythrocephalus (L.)	cL	cL			
Hydroporus melanocephalus (Marshal)	rL	cL			
Hydroporus palustris (L.)	cL	cL			
Hygrotus decoratus Gyllenhal		cL			
Hygrotus versicolor Schaller	cL				
Hyphydrus spec.	rL				
Hyphydrus ovatus (L.)	cL				
Ilybius obscurus Marshal		rL			
Laccobius bipunctatus Fabricius		rL			
Laccornis oblongus (Stephens)	rL				
Noterus clavicornis Degert	cL	cL			
Oulimnius troglodytes Gyllenhal	cL	cL			
Peltodytes caesus Duftschmid	rL				
Porhydrus lineatus (Fabricius)		cL			
Potamonectes depressus (Fabricius)	cL				
Rhantus exsoletus (Förster)	rL				
Scirtes haemisphaericus L.	cL	cL			

Taxa[4]	S	N	W	M	D
Megaloptera					
Sialis lutaria L.	cL	cL	cL	cL	cL
Trichoptera					
Agrypina obsoleta Hagen			rL	cL	
Agrypina varia Fabricius	rL	cL			
Anabolia laevis Zetterstedt	cL	cL			
Apatania auricula Forsslund	cL				
Athripsodes aterrimus Stephens	cL	cL			
Athripsodes cinereus Curtis	cL				
Ceraclea fulvus Rambur		rL			
Ceraclea senilis Burgmeister			rL		
Colpotaulius incisus Curtis	cL				
Cyrnus flavidus McLachlan	cL	cL	cL		cL
Cyrnus trimaculatus Curtis	ccL	ccL	cL	cL	cL
Ecnomus tenellus Rambur	rL	rL			
Erotesis baltica McLachlan	cL	cl	cL	cL	
Glyphotaelius pellucidus Retzius	cL	cL			
Goera pilosa Fabricius	rL	cL			
Grammotaulius spec.			rL		
Halesus digitatus Schrank	rL	rL	rL		
Holocentropus dubius Rambur	cL		cL		
Holocentropus picicornis Stephens		cL		cL	
Hydroptila forcipata Eaton			cL		
Hydroptila pulchricornis Pictet	cL	rL			
Hydroptila tineoides Dalman	cL		cL		
Leptocerus tineiformis Curtis	cL	cL			
Limnophilus auricula Curtis	rL				
Limnophilus binotatus Curtis	rL	cL			
Limnophilus centralis Curtis			rL		
Limnophilus decipiens Kolenati	cL	cL			
Limnophilus elegans Curtis			rL		
Limnophilus flavicornis Fabricius		cL	cL		
Limnophilus germanus McLachlan				rL	
Limnophilus lunatus Curtis	rL				
Limnophilus marmoratus Curtis	cL	cL	cL	cL	cL
Limnophilus nigriceps Zetterstedt	cL		cL	cL	
Limnophilus picturatus McLachlan		rL			
Limnophilus politus McLachlan	cL	cl	cL		
Limnophilus rhombicus L.	cL	cL			
Limnophilus sparsus Curtis		rL			
Limnophilus stigma Curtis	rL		rL	cL	
Lypa phaeopa Stephens	cL	cL	cL		
Micropterna spec. (? *sequax* McLachlan)					rL
Molanna angustata Curtis	cL	cL			
Mystacides azurea L.	cL	ccL	cL		cL
Mystacides longicornis L.	cL	cL			
Mystacides nigra L.	cL	cL	cL		

Taxa[4]	S	N	W	M	D
Neureclipsis bimaculata L.			rL		
Oecetis furva Rambur	rL	rL			
Oecetis lacustris Pictet	cL	cL	cL		
Oecetis ochracea Curtis	cL	cL	cL		
Oecetis testacea Curtis	cL	cL			
Oligotricha striata l.	cL	cL	cL	ccL	
Orthotrichia angustella McLachlan		rL	rL		
Orthotrichia costalis Curtis	cL	cL	cL	cL	cL
Oxyethira flavicornis Pictet	cl	ccL	cL	cL	cL
Phacopteryx brevipennis Curtis	rL				
Phryganea bipunctata Retzius	cL		cL	cL	
Phryganea grandis L.	cL	cL		cL	
Plectrocnemia spec. (? *conspersa* Curtis)	rL				
Polycentropus flavomaculatus Pictet	cL	cL	cL		
Tinodes spec. (? *pallidula* McLachlan) (? *unicolor* Pictet)		rL			
Tinodes waeneri L.	cL	cL	cL		
Triaenodes bicolor Curtis	cL	cL	cL	cL	
Triaenodes conspersus Rambur	rL				
Tricholeiochiton fagesii Guinard	rL				
Trichostegia minor Curtis		rL			
Lepidoptera					
Acentropus niveus Oliver	rL				
Paraponyx stagnata Donovan	rL	rL			
Paraponyx stratiotata L.	rL	rL			
Diptera					
Chaoboridae					
Chaoborus flavicans (Meigen)	rPB	cPB	cPBL	cPL	cPB
Chaoborus obscuripes (v.d. Wulp)				rL	
Chaoborus pallidus (Fabricius)			rL	rL	
Mochlonyx martinii Edwards				cL	
Chironomidae – Orthocladiinae					
Ablabesmyia longistyla Fittkau	cL	cL			
Ablabesmyia monilis (L.)	cL	cL	cL		cL
Ablabesmyia phatta (Eggert)	cL	cL		cL	cL
Anatopynia plumipes (Fries)		rL			
Arctopelopia barbitarsis (Zetterstedt)	ccL	ccL			
Arctopelopia griseipennis (v.d. Wulp)	cL	cL			
Clinotanypus nervosus (Meigen)	rL	rL			
Conchapelopia melanops (Wiedemann)	rL	rL			
Guttipelopia guttipennis (v.d. Wulp)		rL			
Krenopelopia binotata (Wiedemann)	cL	cL			
Labrundinia longipalpis (Goetghebuer)	cL	cL			
Larsia spec.	rL	rL			
Monopelopia tenuicalcar (Kieffer)	cL	cL			

Taxa[4]	S	N	W	M	D
Natarsia punctata (Fabricius)		cL		cL	
Paramerina cingulata (Walker)	cL	cL	cL		cL
Procladius choreus (Meigen)	cLB	cL	cL		cL
Procladius crassinervis (Zetterstedt)	cL	cL			
Procladius fulvus (Kieffer)	rL				
Procladius fuscus Brundin	cLB			rL	
Procladius islandicus Goetghebuer	rL	rL			
Procladius ? nigriventris (Kieffer)					
(sensu Brundin)	cL	cL			
(sensu Goetghebuer)	rL	cL			
Procladius nudipennis Brundin		rL			
Procladius pectinatus Kieffer	cB				
Procladius sagittalis Kieffer	cL	cL			
Procladius scapularis Kieffer	rL	rL	rL		
Procladius suecicus Brundin	rL	rL			
Procladius spec.	cL			rL	
Psilotanypus flavifrons Edwards		rL			
Tanypus punctipennis (Meigen)	cL	cL			
Tanypus vilipennis (Kieffer)	rL	rL			
Telmatopelopia nemorum (Goetghebuer)				rL	
Thienemannimyia vitellina (Kieffer)	cL				
Trissopelopia spec.				rL	
Xenopelopia falcigera (Kieffer)		rL			
Xenopelopia nigricans Fittkau		rL			
Chironomidae – Orthocladiinae					
Acricotopus lucens (Zetterstedt)	cL				
Brillia longifurca Kieffer	cL		cL		
Bryophaenocladius spec.	r				
Bryophaenocladius aestivus Brundin	r				
Bryophaenocladius nidorum (Edwards)		r			
Chaetocladius spec.	r				
Chaetocladius spec.	r				
Corynoneura spec.	rL	rL			
Corynoneura celeripes Winnertz	cL	cL			
Corynoneura lacustris Edwards	cL	cL			
Corynoneura minuscula Brundin	cL	cL			
Corynoneura scutellata Winnertz	cL	cL	cL		cL
Cricotopus spec.	cL	cL			
Cricotopus bicinctus Meigen	rL	rL			
Cricotopus festivellus Kieffer	cL	cL	cL		cL
Cricotopus obnixus (Walker)	rL	rL	rL		rL
Cricotopus oscillator (Meigen)	rL	rL	rL		
Cricotopus pilitarsis (Zetterstedt)	cL	cL			
Cricotopus pilosellus Brundin	cL	cL			
Cricotopus relucens Hirvenoja	cL	cL			
Cricotopus reversus Hirvenoja	cL	cL			
Cricotopus sylvestris (Fabricius)	cL	cL	cL		cL

Taxa[4]	S	N	W	M	D
Cricotopus tricinctus (Meigen)	rL				
Epoicocladius flavens (Malloch)		cL			
Eukiefferiella hospita Edwards		rL			
Gymnometriocnemus brumalis Edwards		r			
Heterotanytarsus picalis Kieffer	rL				
Limnophyes spec.	rL	rL			
Limnophyes eltoni (Edwards)	cL	cL			
Limnophyes globifer Lundström	rL				
Limnophyes interruptus Goetghebuer	cL	cL	cL	cL	cL
Limnophyes laccobius (Kieffer)	rL				
Limnophyes smolandicus Brundin	rL			cL	
Metriocnemus atratulus (Zetterstedt)	cL	cL			
Metriocnemus atratus Kieffer	rL				
Metriocnemus fuscipes (Meigen)	rL				
Metriocnemus lacteolus Goetghebuer	rL			cL	
Metriocnemus longicornis Kieffer	rL				
Microcricotopus bicolor (Zetterstedt)	rL				
Orthocladius oblidens (Walker)		rL			
Orthocladius obtexens Brundin	rL				
Rheocricotopus spec.	rL				
Parakiefferiella coronata (Edwards)	rL				
Parakieferiella nigra Brundin	cL	cL	cL		
Paralimnophyes spec. [? *hydrophilus* (Goetghebuer)]		rL			
Parametriocnemus stylatus (Kieffer)	rL	rL			
Paraphaenocladius impensus (Walker)	cL	cL	cL	cL	
Paraphaenocladius ocularis (Kieffer)	cL	cL	cL		
Paraphaenocladius penerasus (Edwards)	cl	cL	rL		
Parorthocladius spec.			rL		
Potthastia longimana Kieffer		rL			
Pogonocladius consobrinus (Holmgren)	cL	rL			
Prodiamesa olivacea (Meigen)	rL				
Propsilocerus lacustris Kieffer			cL		
Psectrocladius barbatipes (Kieffer)	rL	rL			
Psectrocladius barbimanus Edwards	cL	cL			
Psectrocladius bisetus Goetghebuer				cL	
Psectrocladius edwardsi Brundin	cL	cL	cL		
Psectrocladius fennicus Stora	cL		rL		
Psectrocladius limbatellus (Holmgren)	rL				
Psectrocladius octomaculatus Wülker				rL	
Psectrocladius psilopterus Kieffer	rL	cL			
Psectrocladius sordidellus (Zetterstedt)	cL	ccL			
Psectrocladius zetterstedti Brundin	cL				
Pseudorthocladius spec.	cL	cL	cL		
Pseudorthocladius spec.				cL	
Pseudorthocladius spec.		rL			
Pseudorthocladius spec.	cL	cL	cL		
Pseudosmittia trilobata (Edwards)	rL	rL			

Taxa[4]	S	N	W	M	D
Smittia spec.	r				
Smittia spec.	r	r	r		
Smittia spec.	r				
Smittia aterrima (Meigen)	r	r			
Smittia edwardsi Goetghebuer	r	r	r		
Syncricotopus rufiventris (Meigen)	cL				
Chironomidae – Chironomini					
Camptochironomus pallidivittatus Malloch	rL				
Chironomus spec.				rL	
Chironomus spec.	rL	rL			
Chironomus annularius (Meigen)	rL		rL	rL	
Chironomus anthracinus Zetterstedt	cL	clB	cLB	ccL	cLB
Chironomus cingulatus (Meigen)	cL	cL	cL		
Chironomus macani Freeman	rL	rL	rL	rL	
Chironomus melanotus Keyl				cL	
Chironomus nuditarsis Strenzke	cL	cLB	cL		cLB
Chironomus plumosus (L.)		(ccL)?			
Chironomus pseudothummi Strenzke		cL			
Chironomus thummi Kieffer	rL				
Cryptochironomus albofasciatus (Staeger)			rL		
Cryptochironomus redekei Kruseman		rL			
Cryptochironomus supplicans (Meigen)	cL	cL			
Cryptocladopelma bicarinata Brundin	cL	rL	rL		
Cryptocladopelma edwardsi Kruseman	rL	rL			
Cryptocladopelma virescens (Meigen)	cL	cL	cL		
Cryptocladopelma viridula (L.)	cL	cL	cL		
Cryptotendipes usmaensis Pagast	cL	cL			
Demeijerea rufipes L.		rL			
Demicryptochironomus vulneratus (Zetterstedt)	cL	cL	cL		
Dicrotendipes lobiger Kieffer	cL	ccL	cL		cL
Dicrotendipes nervosus (Staeger)	cL	cL	cL		
Dicrotendipes notatus (Meigen)	rL	rL			
Dicrotendipes pulsus (Walker)	rL	rL		rL	
Dicrotendipes tritomus Kieffer	cL	cL	cL		
Einfeldia carbonaria (Meigen)		rL	rL		
Einfeldia dilatatus Goetghebuer	cL				
Einfeldia dissidens (Walker)			rL		
Einfeldia longipes (Staeger)	cL	cL	cL		
Einfeldia pagana (Meigen)	cL	cL	rL		
Einfeldia pectoralis Kieffer	rL		rL		
Endochironomus spec.	rL				
Endochironomus dispar (Meigen)		rL			
Endochironomus impar (Walker)	cL	cL	cL		
Endochironomus intextus (Walker)	cL	cL	cL		
Endochironomus tendens Fabricius	ccL	ccL	cL		cL
Glyptotendipes spec.		rL			
Glyptotendipes anomalus Kieffer		rL			

Taxa[4]	S	N	W	M	D
Glyptotendipes gripekoveni Kieffer	ccL	ccL	ccL		cL
Glyptotendipes pallens (Meigen)		cL	cL	cL	
Harnischia spec.	cL	cL			
Kiefferulus tendipediformis Goetghebuer	rL		rL		
Lauterborniella agrayloides (Kieffer)	cL		cL		
Leptochironomus tener Kieffer		rL			
Microtendipes caledonicus Edwards			rL		
Microtendipes chloris (Meigen)	ccL	cL	cL		cL
Microtendipes pedellus (De Geer)	ccL	ccL	ccL		ccL
Pagastiella orophila (Edwards)	cL	cL		cL	
Parachironomus arcuatus Goetghebuer	cL	cL	cL		cL
Parachironomus biannulatus (Staeger)	rL	rL			
Parachironomus danicus Lehman		cL			
Parachironomus monochromus (v.d. Wulp)	rL	rL			
Parachironomus parilis (Walker)	cL	cL	cL		cL
Parachironomus subalpinus Goetghebuer	rL				
Parachironomus vitiosus Goetghebuer	cL	rL			
Paracladopelma camptolabis Kieffer	rL				
Paratendipes albimanus (Meigen)	cL	cL			
Paratendipes hexatomus Kieffer	cL	cL			
Paratendipes nudisquama (Edwards)				rL	
Pentapedilum sordens (v.d. Wulp)	cL	cL	cL		cL
Pentapedilum stratiotis Goetghebuer		rL			
Pentapedilum tritum (Walker)	cL	cL			
Phaenopsectra albiventris Kieffer	cL	cL	cL		cL
Phaenopsectra flavipes (Meigen)	cL	cL	cL		cL
Phaenopsectra punctipes (Wiedemann)	rL	rL	rL		rL
Polypedilum spec.	cL	cL			
Polypedilum bicrenatum Kieffer	cL	cL			
Polypedilum cultellatum Goetghebuer	cL		cL		
Polypedilum nubeculosum (Meigen)	ccL	ccL	cL	cL	cL
Polypedilum pullum (Zetterstedt)	cL	cL			
Polypedilum scalaenum (Schrank)	cL	cL	cL		
Pseudochironomus prasinatus (Staeger)	cL	cL			
Sergentia coracina (Zetterstedt)	ccL	cL			
Stenochironomus fascipennis (Zetterstedt)		rL	rL		
Stenochironomus gibbus Fabricius	rL	cL			
Stenochironomus hibernicus Edwards		rL			
Stictochironomus histrio (Fabricius)	cL				
Stictochironomus pictulus (Meigen)	ccL	ccL			
Stictochironomus rosenschöldi (Zetterstedt)	cL	cL			
Xenochironomus xenolabis Kieffer		rL			
Chironomidae – Tanytarsini					
Cladotanytarsus spec.	rL		rL		
Cladotanytarsus atridorsum (Kieffer)	cL	cL	cL		
Cladotanytarsus difficilis Brundin	rL				
Cladotanytarsus dispersopilosus Goetghebuer	ccL	ccL	cL	cL	cL

Taxa[4]	S	N	W	M	D
Cladotanytarsus iucundus Hirvenoja	rL		rL		
Cladotanytarsus mancus (Walker)	rL	rL	rL		
Cladotanytarsus nigrovittatus (Goetghebuer)	rL	cL	cL		cL
Corynocera ambigua Zetterstedt		cL	cL		
Lauterbornia coracina Kieffer	cB				
Micropsectra lindebergi Säwedahl	ccB				
Micropsectra styriaca Reiss	cL	cL			
Micropsectra tenellula Goetghebuer			rL		
Parastempellina subglabripennis Brundin		cL			
Paratanytarsus handlirschi Goetghebuer	cL	cL	cL		cL
Paratanytarsus confusus Palmen	cL	cL	cL	cL	cL
Paratanytarsus intricatus Goetghebuer	cL	cL			
Paratanytarsus laetipes Zetterstedt		cL	rL	rL	rL
Paratanytarsus penicillatus Goetghebuer	cL				
Paratanytarsus tenuis (Meigen)	cL	cL	cL		cL
Stempellina spec.			rL		
Stempellina bausei (Kieffer)	rL	rL			
Stempellinella flavidula Edwards		rL			
Stempellinella minor Edwards		cL			
Tanytarsus chinyensis Goetghebuer	cL	cL	cL		rL
Tanytarsus curticornis Kieffer	cL	cL	cL	cL	cL
Tanytarsus debilis (Meigen)	cL	cL	cL		cL
Tanytarsus excavatus Edwards	cL	cL	cL		cL
Tanytarsus holochlorus Edwards	rL	cL	cL	cL	cL
Tanytarsus insequalis Goetghebuer	cLB	cLB	rL		
Tanytarsus lactescens Edwards	cL	rL	rL		
Tanytarsus lestagei-group	cL	cL	cL		cL
Tanytarsus lugens Kieffer	cL				
Tanytarsus medius Reiss et Fittkau	rL	rL			
Tanytarsus mendax Kieffer	rL	cL	cL	cL	
Tanytarsus miriforceps Kieffer	rL	cL	cL		
Tanytarsus multipunctatus Brundin		rL			
Tanytarsus niger Anderson	cB				
Tanytarsus nigricollis Goetghebuer		cL	cL		
Tanytarsus norvegicus Kieffer	ccB	rL			
Tanytarsus occultus Brundin	cL	cL	cL		cL
Tanytarsus pallidicornis (Walker)			rL		
Tanytarsus quadridentatus Brundin	rL				
Tanytarsus recurvatus Brundin	cL	cL			
Tanytarsus signatus v.d. Wulp	cL	cL	cL		cL
Tanytarsus sylvaticus v.d. Wulp	rL	cL	cL		cL
Tanytarsus usmaensis Pagast	cL	cL			
Tanytarsus veralli Goetghebuer	cL	cL	cL		
Thienemanniola ploenensis Kieffer		cL	cL		
Zavrelia pentatoma Kieffer			rL		
Ceratopogonidae					
Atrichopogon stribarba Kieffer	rL	rL			

Taxa[4]	S	N	W	M	D
Atrichopogon minutus Meigen	cL	rL			
Atrichopogon haesitans Kieffer	rL				
Bezzia bicolor Meigen				rL	
Bezzia bidentata Kieffer		rL			
Bezzia gracilis Winnertz	cL	cL			
Bezzia ploenensis Kieffer			rL		
Ceratopogon sociabilis Goetghebuer				rL	
Ceratopogon vitiosus Winnertz	rL				
Clinohelea unimaculata Macquart		rL			
Culicoides spec.		rL			
Culicoides chiopterus Meigen	cL	rL			
Culicoides cilipes Kieffer	rL	rL			
Culicoides obsoletus Meigen		rL			
Culicoides salinarius Kieffer		rL	rL		
Culicoides setosinervis Kieffer				rL	
Dasyhelea spec.	rL				
Dasyhelea longipalpis Kieffer	cL				
Dasyhelea modesta Winnertz	rL				
Dasyhelea semistriata Goetghebuer	rL	rL			
Dicrobezzia venusta Meigen	cL			cL	
Forcipomyia spec.	cL	cL	cL	cL	cL
Forcipomyia brevicubitus Goetghebuer	cL	cL	cL	cL	
Forcipomyia ciliata Winnertz				rL	
Forcipomyia fuliginosa Meigen	rL				
Forcipomyia lepida Winnertz	rL				
Forcipomyia pallida Winnertz	cL	cL			
Macropeza spec. (? *albitarsis* Meigen)		rL			
Mallochhelea inermis Kieffer		rL			
Palpomyia flavipes Meigen	rL	rL			
Palpomyia fulva Macquart	cL	cL	cL	cL	
Palpomyia tibialis Meigen				rL	
Serromyia atra Megerle	rL				
Vertebrata					
Pisces					
(D. Ronneberger)					
Abramis brama (L.)	c	cc	cc		
Acerina cernua (L.)	c	r	r		
Alburnus alburnus (L.)	c	c	r		
Anguilla anguilla (L.)	c	c	c		
Blicca bjoerkna (L.)	r	r	r		
Carassius auratus gibelio (Bloch)	r	r	r		
Carassius carassius (L.)	r	r	r		
Cobitis taenia L.	r	r			
Coregonus albula L.	c	c			
Cyprinus carpio L.	r	r			
Esox lucius L.	c	c	c		
Gasterosteus aculeatus L.	r	r	r		

Taxa[4]	S	N	W	M	D
Hypophthalmichthys molitrix (Valenciennes)			cc		
Leucaspius delineatus (Heckel)	r	r	r		
Rutilus rutilus (L.)	cc	cc	cc		
Lota lota (L.)	r	r	r		
Lucioperca lucioperca (L.)			r		
Misgurnus fossilis (L.)			r		
Perca fluviatilis L.	cc	cc	cc		
Rhodeus sericeus amarus Bloch			r		
Salmo gairdneri Richardson	r				
Scardinius erythrophthalmus (L.)	c	c	c		
Silurus glanis L.	r	r			
Tinca tinca (L.)	r	c	c		
Amphibia					
(D. Ronneberger)					
Bufo bufo L.	r	c	c		
Rana arvalis L.		c	c	c	
Rana 'esculenta' L.			c		c
Rana temporaria L.		c	c		c
Triturus vulgaris L.			r		
Reptilia					
(D. Ronneberger)					
Emys orbicularis L.	r				
Natrix natrix L.	c	c	c	c	c

Taxa[5]	Lake Stechlin area
Aves	
(D. Flößner)	
Gavia arctica (L.)	im; r
Podiceps cristatus (L.)	rb; cc
Podiceps griseigena (Boddaert)	og
Podiceps auritus (L.)	iw; r
Podiceps ruficollis (Pallas)	ib
Phalacrocorax carbo (L.)	og
Ardea cinerea L.	fb; r
Botauris stellaris (L.)	? ib
Cygnus olor (Gmelin)	rb; c
Cygnus cygnus (L.)	iw; r
Anser fabalis (Latham)	im; c

[5] The following symbols are used: rb = regular breeding bird, ib = irregular breeding bird, fb = formerly breeding bird, rm = regular migrant, im = irregular migrant, rw = regular wintering guest, iw = irregular wintering guest, og = occassional guest.

Taxa[5]	Lake Stechlin area
Anser albifrons (Scopoli)	og
Anas penelope L.	og
Anas crecca L.	ib
Anas platyrhynchos L.	rb; cc
Anas querquedula L.	og
Aythya ferina (L.)	ib
Aythya nyroca (Guldenst.)	ib
Aythya fuligula (L.)	rm; rw; c
Aythya marila (L.)	iw; r
Melanitta nigra (L.)	og
Melanitta fusca (L.)	og
Bucephala clangula (L.)	rb; c
Mergus albellus L.	og
Mergus merganser L.	fb; r
Milvus migrans (Boddaert)	rb; c
Haliaeetus albicilla (L.)	rb; r
Circus aeruginosus (L.)	ib
Pandion haliaetus (L.)	rb; c
Grus grus (L.)	rb; r
Rallus aquaticus L.	ib
Porzana porzana (L.)	ib
Gallinula chloropus (L.)	rb; r
Fulica atra L.	rb; cc
Charadrius dubius Scopoli	og
Gallinago gallinago (L.)	rb; c
Numenius arquata (L.)	og
Limosa limosa (L.)	og
Tringa ochropus L.	ib
Tringa glareola L.	og
Tringa hypoleucos L.	rm; r
Larus fuscus L.	og
Larus canus L.	og
Larus ridibundus L.	rm; rw; c
Chlidonias niger (L.)	og
Alcedo atthis L.	? ib
Motacilla cinerea (Tunstall)	rb; r
Cinclus cinclus (L.)	iw
Locustella luscinioides (Savi)	rb; r
Acrocephalus schoenobaenus (L.)	rb; c
Acrocephalus scirpaceus (Hermann)	rb; r
Acrocephalus arundinaceus (L.)	ib
Emberiza schoeniclus L.	rb; c

Taxonomic index*

Ablabesmyia 233
monilis 237
phatta 238
Abramis brama 263
Acanthoceras zachariasii 177
Acari 280
Achromobacter 208
Acrocephalus
arundinaceus 270
schoenobaenus 269
scirpaceus 269
Aconchulina 213
Acroladium cuspidatum 139
Acroloxus lacustris 236
Acroperus 227, 228, 229
harpae 219, 220, 222
Aegagrophiletum holsaticae 133
Aerobacter 208, 341
Aeshna 232
Aira
caryophyllaceae 13
praecox 13
Alcaligenes 208
Aldrovanda vesiculosa 13, 129
Alnetum 140, 219
Alnus 14
glutinosa 402
Alona
affinis 219, 220, 222, 225, 227, 229
costata 227, 229
guttata 221, 224
karelica 219, 226
protzi 219
quadrangularis 221, 222, 228, 229
rectangula 223, 224, 225, 228
weltneri 219
Alonella
excisa 222, 228, 229
exigua 223, 225, 227
nana 220, 221, 222
Alonopsis 223, 225, 226, 229
elongata 219, 221, 222, 225, 227, 229
Amoebina 213, 500
Amphipoda 234, 235, 239
Anabaena
hassallii 165
lemmermannii 162, 163, 165, 166, 169, 170, 171, 172, 176, 179, 182, 185, 187
Anas platyrhynchos 269
Anchistropus emarginatus 223
Anguilla anguilla 263
Ankistrodesmus 169, 184
pfitzeri 177
Anodonta 238
cygnea 232
Anuraeopsis fises 255
Apatania fimbriatum 236
Arcella
arenaria 217
artocrea 217
discoides 214, 216
gibbosa 217, 406
hemisphaerica 216, 217
rotunda 216
vulgaris 216
Arctopelopia
barbitarsis 233, 235
griseipennis 232
Argyroneta aquatica 232
Arrenurus perforatus 237
Arnoseris minima 13

* Taxa of appendix pp. 485–529 excluded; phytosociological categories included.

Ascomorpha
ecaudis 245
saltans 245
Asellus 234, 236
aquaticus 238
Aspidisca costata 217
Astramoeba radiosa var. *granulifera* 213
Asterionella 422
formosa 159, 161, 162, 163, 164, 165, 166, 169, 170, 172, 173, 176, 177, 178, 179, 180, 184, 185, 188, 189, 190, 283
Athalamia 213
Atractides 235
pavesii 236
Attheyella
crassa 222, 224, 229
dentata 220, 222, 224, 228
trispinosa 220, 222, 224
Aythya
fuligula 272
marila 272

Bacillariophyceae 280
Bacillus 208
Bacteria 197, 198, 199, 200, 201, 207
Beneckea 208
Betula 14, 16
pendula 402
pubescens 11
Biomyxa 214
Bithynia 236
tentaculata 236
Bitrichia chodatii 177
Bosmina 221, 266
coregoni 249
longirostris 221, 225, 227
obtusirostris 252
Botryococcus 180
braunii 159, 166, 170, 171, 173
Brachionus
angularis 255, 335, 336
calyciflorus 248, 336
Brevibacterium 208
Bryocamptus vejdovskyi 219, 221, 222, 225, 226, 229
Bryozoa 236, 238, 241, 280
Bucephala clangula 269
Bunops serricaudata 219
Bythotrephes longimanus 225

Caenis 236, 240
horaria 236, 237
Calluna vulgaris 13
Campascus triqueter 216
Camptocercus 224
lilljeborgi 220, 222, 223, 224, 225, 227
rectirostris 220, 221, 224, 228
Camptochironomus pallidivittatus 232
Candona 222
acuminata 228
albicans 222, 229
candida 221, 222
hyalina 228
neglecta 229
Canthocamptus 227
staphylinus 220, 221, 222
Carex
acutiformis 139, 144, 147
elata 139
lasiocarpa 11, 139, 140
limosa 11
rostrata 139
Caricetum ripario-acutiformis 13, 135
Carici-Agrostietum caninae 13
Carici elongatae-Alnetum glutinosae 12, 135
Carnivora 232, 233
Carpinus 14
Caulobacter 208
Cellvibrio 208
Centronella reicheltii 164
Centroptilum 236
Centropyxis 216
aculeata 214, 216
aerophila 217
var. *sphagnicola* 217
discoides 214, 216
gibba 216
orbicularis 217
Cephalodella
apocolea 218
gibba 218
glandulosa 215
tenuis 215
theodora 215
Ceratium 171
cornutum 177
hirundinella 162, 165, 166, 170, 171, 172, 173, 178, 179, 180, 182, 187, 188
Ceratophyllum 223, 227, 235
demersum 13, 132, 133, 135, 136, 137, 141, 143, 144, 148, 422
Ceriodaphnia 221

laticaudata 220
megops 225
pulchella 225, 227
quadrangula 225, 227
reticulatae 225, 227
Chaetogaster diaphanus 238
Chaoborus 242, 423
flavicans 240, 242
Chara 135, 136, 147, 148, 149, 214, 216, 218, 219, 220, 223, 224, 227, 228, 265, 275
aculeolata 132
aspera 130, 132, 141, 142
delicatula 138
filiformis 130, 132, 141, 143, 274, 422
fragilis 132, 133, 274
f. *hedwigii* 132
hispida 130, 143
subsp. *rudis* 133
rudis 132
tomentosa 130, 132, 133, 143
vulgaris 138
Characeae 130, 132, 133, 136, 140, 141, 143, 144, 145, 146, 219, 227, 228, 235, 237, 274, 422
Charales 18, 130, 132, 141–144
Charetum asperae 131, 140, 143, 144, 145, 233
Charetum filiformis 131, 132, 140, 143, 144, 145, 146, 147, 235, 237
Charetum tomentosae 132, 143, 144, 145
Charophyceae 277, 280
Chlamydomonas 337
Chironomidae 231, 232, 233, 236, 238, 240, 241
Chironomini 234, 235, 236, 238, 239
Chironomus 236, 238, 242
anthracinus 242
macani 232
nuditarsis 236
plumosus 236, 242
tenuistylus 243
Chlorella 324, 336, 342
Chlorophyceae 164
Chromobacterium 208
violaceum 207
Chroococcus 182, 185
limneticus 189
turgidus 171, 177
Chrysamoeba radians 161, 162, 163, 164, 170, 171, 177
Chrysomonadales 158
Chrysophyceae 164, 170
Chrysococcus 173, 176
Chydorus 224
gibbus 221, 222, 223, 225, 226, 228, 229
piger 220, 222, 223, 224, 225, 226, 228
sphaericus 219, 220, 221, 222, 224, 227
Ciliata 213, 217–218, 502
Cladietum 219
Cladietum marisci 138, 139
Cladium 139
mariscus 13, 139, 143, 144, 147
Cladocera(ns) 219, 220, 223, 224, 225, 243, 244, 266, 435
Cladonio-Pinetum 11
Cladophora-Fontinalis (*antipyretica-*) community 131, 132, 140, 143, 144
Cladotanytarsus 233, 234, 235, 241
dispersopilosus 234
Cochliopodium bilimbosum 214
Cocconeis
diminuta 150
disculus 150
Coelambus marklini 281
Coelastrum
microporum 173
reticulatum 177
Coenagrion
puella 236, 241
pulchellum 236, 241
Coleoptera 231, 234, 235, 238, 239
Collotheca libera 255
Colurella adriatica 218
Conochilus
natans 255
unicornis 245, 340
Copepode 219–223, 225–229, 243, 435
Coregonus 263, 266, 268
albula 262, 263, 265, 266
Corynephorus canescens 13
Corynocera 236
ambigua 242
Corylus 14
Corythion dubium 217
Cosmarium 173, 177
phaseolus 165
Cricotopus 236, 240, 242, 424
sylvestris 237
Cristatella mucedo 238
Crustacea 213, 233, 280, 323
Cryptochironomus 233
Cryptocyclops bicolor 221, 222
Cyanophyceae 164, 169, 170, 171
Cyclocypris 227
laevis 220, 223, 225

ovum 220, 222, 224
serena 220, 224, 228
Cyclops
furcifer 227
insignis 221, 227
kolensis 221, 246
vicinus 435
Cyclopyxis
arcelloides 216
eurystoma 217
kahli 217
Cyclotella 159, 161, 162, 163, 164, 166, 172, 173, 177, 179, 182, 186, 187, 188, 189, 192, 229, 299, 301
comta 173, 279
var. oligactis 173, 279, 281
var. *quadrijuncta* 164
glomerata 173, 279
kützingiana 162, 165, 169, 170, 176, 177, 190, 279
var. *planetophora* 279
meneghiniana 435
'nano' 162, 165, 169, 176, 184
ocellata 279
operculata 173, 279
Cygnus olor 269
Cymatopleura
elliptica 150
solea 173, 179, 283
Cymbella
hustedtii 150
hybrida 150
microcephala 150
obtusiuscula 150
thumensis 150
Cyphoderia trochus 216
Cyperia lacustris 222, 229
Cypridopsis vidua 220, 222, 228
Cyprinus carpio 263
Cyrnus
flavidus 236, 237
trimaculatus 236, 237
Cytherissa lacustris 222, 229

Dactylosphaerium vitraeum 213
Daphnia 221, 340
cucullata 244, 246, 247, 248, 249, 254, 324, 325, 326, 327, 328, 330, 331, 332, 333, 334, 335, 338, 339
curvirostris 227
galeata mendotae 331
hyalina 249, 331, 341, 435
longispina 331
magna 331
pulex 251, 325, 326, 327, 329, 331, 341
pulicaria 252
Darwinula 229
stevensoni 222
Dasyhelea 234
Demicryptochironomus 233
Desulfovibrio 208
Diacyclops nanus 226
Diaphanosoma 227
Diatoma elongatum 161, 164, 173, 283
Diatomeae 283
Dicranophorus
epicharis 218
robustus 218
Dicrotendipes 236
Dictyosphaerium sphagnale 280
Diaptomus gracilis 331
Difflugia 216
amphora 217
bacillaria 217
corona 217
elegans 214, 216
globulosa 216, 281
limnetica 243, 244
mica 214, 216
oblonga 216
pulex 214
stechlinensis 215, 217
urceolata 217
Difflugiella 214
horrida 215
sacculus 216
Dileptus anser 217, 218
Dinobryon 24, 160, 164, 166, 169, 170, 171, 173, 178, 179, 182, 184, 185, 186, 187, 188, 422
bavaricum 161, 162, 170, 176
crenulatum 176
cylindricum 161, 162, 169, 177
divergens 165, 169, 171
pediforme 324, 330, 332
sertularia 176, 180
sociale 165, 166, 169, 171, 172, 177
suecicum 177
Dinobryonaceae 280
Diptera 234, 235, 239
Disparalona 223, 228
rostrata 222
Dorylaimoides microdentatus 215

Dreissena polymorpha 231, 233, 236, 237, 238, 239, 240, 274
Drepanocladus
aduncus 132
fluitans 136
Drepanothrix 223, 226, 228
dentata 222, 225
Drosera rotundifolia 11
Dugesia polychroa 238
Dytiscidae 232
Dytiscus 232

Ectocyclops 223, 227
phaleratus 221, 222, 225
Einfeldia 236, 238
Elaeorhanis cincta 214
Elakatothrix gelatinosa 177
Eleoharis palustris 139, 144
Elodea canadensis 15, 133, 135, 136, 141, 143, 145, 148, 422
Emberiza schoeniclus 269
Enallagma cyathigerum 236, 237, 241
Endochironomus 240
Ephemera vulgata 234
Ephemeroptera 234, 235, 239
Epithemia
hyndmannii 150
mülleri 150
turgida 150
zebra var. *porcellus* 150
Equisetum fluviatile 133, 138
Erotesis baltica 237
Eriophorum angustifolium 11
Erkenia subaequiciliata 173
Erythromma naias 237
Esox lucius 263
Encentrum
spinosum 215
stechlinensis 215
Eucyclops 222, 228
macruroides 227
macrurus 221, 227
serrulatus 228
Eudiaptomus gracilis 244, 246, 248, 249, 254, 324, 325, 326, 327, 328, 329, 330, 331, 332, 333, 334, 338, 339, 341, 342, 435
Eudorina 172, 180
elegans 159, 165, 173, 176, 178, 179
Euglypha
acanthophora 216
compressa 214, 216
Eurycercus 223, 227
lamellatus 219, 220, 222
Eurytemora 221, 227
affinis 246
lacustris 220, 221, 225, 244, 246, 248, 249, 254, 324, 341

Fagus 14, 16
sylvatica 402
Festuca psammophila 13
Filinia
longiseta 245, 335, 340
terminalis 245, 249, 251, 335
Flavobacterium 208
Fontinalis antipyretica 132, 133, 136, 145, 148
Fragilaria 172, 422
brevistriata 283
capucina var. mesolepta 283
construens 283
var. *binodis* 283
var. *subsalina* 283
var. *triundulata* 283
var. *venter* 283
crotonensis 165, 166, 171–173, 176–180, 185, 188–191, 299, 303, 422
pinnata var. *lancettula* 283
Fraxinus 14
Fulica atra 269

Galium palustre 139
Gammaridae 233
Gammarus 234, 236
Gasterosteus aculeatus VII
Gastropoda 232
Gastropus 248
stylifer 335
Gastrotricha 213
Genista pilosa 13
Geranium pusillum 13
Gloeococcus schroeteri 164
Glyceria maxima 139
Glycerietum maximum 140
Glyptotendipes 236
Goera pilosa 233, 234
Gomphonema constrictum 150
Graptoleberis 223, 227, 228
testudinaria 222
Gymnodinium 159, 165, 173
Gyraulus
albus 236
riparius 236

Gyrinidae 233
Gyrinus natator 280

Haliaeetus albicilla 274
Haliplus apicalis 280
Harpacticoida 219
Heleidae 234, 235, 239
Heleopera sylvatica 217
Helichrysum arenarium 13
Helobdella stagnalis 233, 234
Hemiptera 232, 234, 235, 238, 239
Herpetocypris reptans 222, 228
Heterocope 221, 227
  appendiculata 225
Heteroptera 231
Hieracium 13
Hirudinea 234, 235, 238, 239, 240
Hyalosphenia
  cuneata 217
  elegans 217
Hydracarina 231, 234, 235, 236, 238, 239, 240
Hydrilla verticillata-Potamogeton-community 146
Hydrocharis morsus-ranae 137
Hydrocharitetum morsus-ranae 137
Hydrozoa 234, 235, 239, 280
Hydrozetes parisiensis 232
Hydrobates longipalpis 236
Hyphomicrobium 208
Hyphomycetes 401, 406
Hypnum 11
Hypochoeris glabra 13

Ilyocoris cimicoides 232
Ilyocryptus 223, 228
  sordidus 222
Insecta 280
Isopoda 234, 235, 239

Junco-Molinietum 13

Kellicottia longispina 245, 251, 327, 335, 336, 341
Kephyrion 173
Kephyriopsis 166
Keratella
  cochlearis 245, 251, 327, 335, 336, 337, 338, 340, 341, 342
  quadrata 335
Kirchneriella lunaris 324, 329, 330, 332, 333, 334
Koeleria glauca 13
Kurzia latissima 220, 227

Labronema stechlinensis 215
Lacornis oblongus 280
Lacrymaria olor 217
Larix 11, 16
Lathonura 227
  rectirostris 221
Latona 223, 228, 229
  setifera 222, 225, 229
Lauterbornia coracina 240, 242, 281
Lecane
  flexilis 218
  luna 218
  lunaris 218
  ovalis 218
Lecythium hyalinum 214
Ledo-Pinetum 11, 12, 14
Ledum palustre 11, 13
Lemna
  minor 137, 138
  trisulca 137, 138
Lemnetum trisulcae 137
Lepadella
  patella 218
  pseudosimilis 215
  triptera 218
Leptophlebia 236
Lesquereusia
  modesta 216
  spiralis 217
Leucorrhinia dubia 281
Leydigia 223, 229
  leydigi 229
Limnephilus 233, 237
Limnocythere psammophila 215, 219, 222, 223, 225, 229
Litonotus lamella 217
Locustella luscinioides 269
Lonicera periclymenum 13
Lucioperca lucioperca 263
Lutra lutra 280
Lycopodiella inundata 13
Lymnaea
  peregra 238
  stagnalis 233
Lysimachia vulgaris 139
Lythrum salicaria 139

Macrocyclops 222, 228

albidus 224
fuscus 223, 224, 225, 227
Macrothrix 223
rosea 222, 228
Mallomonas 160, 164, 165, 166, 171, 182, 185, 278, 422
acaroides var. *striatula* 278
crassisquama 177, 278
fastigata 159, 162, 163, 164, 165, 166, 169, 170, 171, 172, 177, 186, 189, 278
tonsurata 164, 177, 278
Mammalia 280
Mastogloia smithii var. *amphicephala* 150
Megalocyclops 228
gigas 220, 221, 229
viridis 220, 222, 225, 227, 228
Melosira 177
Mergus 273
merganser 269
Mesocyclops 227
leuckarti 220, 221
Methylobacter 369
Methylococcus 369
Methylocystis 208, 369
Methylomonas 208, 369
Methylosinus 208, 369
sporium 369
trichosporium 369
Micrococcus 208
Microcyclops
rubellus 222, 223, 225
varicans 221
Microcystis 165, 166, 171, 180
Micropsectra lindebergi 240, 242, 423
Microtendipes 236
chloris 238, 239, 240, 242
Midea orbiculata 236
Mideopsis orbicularis 236
Molanna angustata 233, 234
Mollusca 231, 233, 234, 235, 237, 239, 280
Monopelopia tenuicalcar 236
Monospilus 223, 228, 229
dispar 222
Moraria mrazeki 219, 221, 225, 226, 229
Myriophyllo-Nupharetum 133, 135, 136, 143, 144, 219, 224, 227
Myriophyllum 136, 149, 214, 218, 223, 225, 227, 235
alterniflorum 13, 133, 143, 144, 148, 280
spicatum 13, 132, 133, 136, 148
verticillatum 13, 133, 136, 144
Myrtillo-Pinetum 11
Mystacides 237
Mytilina ventralis 218

Naidida 238
Najas
marina 13, 129
subsp. *intermedia* 129, 280
minor 13
Navicula jentzschii 150
Nebela
collaris 217, 406
dentistoma 217
militaris 217
Nematoda 213, 215, 218, 234, 235, 239, 240, 241, 507
Nemoura cinerea 234, 236
Nepa cinerea 232
Nevskia 208
ramosa 208
Nitella 131, 238, 239
flexilis 132, 143, 145, 148, 228, 238, 274
Nitellopsidetum obtusae 131, 132, 135, 140, 143, 144, 145, 146, 147, 235, 237
Nitellopsis obtusa 130, 132, 133, 146, 148, 274
Nitello-Vaucherietum 131, 132, 140, 144, 147, 238
Nitocra 223, 228, 229
hibernica 220, 222, 228
Nitzschia 176, 178
sigmoidea 177
Notholca
acuminata 255
foliacea 255
Notodromas 223, 225, 227
monada 221, 222
Nuphar 136, 237
lutea 11, 13, 133, 136, 138
Nupharetum 227
Nymphaea 133, 134, 218, 222, 223, 237, 422, 424
alba 11, 133, 136, 138
'minor' 136

Odonata 231, 234, 235, 237, 238, 239
Oecetis ochracea 238
Oligochaeta 213, 233, 234, 235, 236, 239, 427
Oocystis 173, 189
Ophridium versatile 218
Orthocladiinae 231, 232, 234, 235, 236, 237, 238, 239

Oscillatoria 182, 185, 187, 190, 329, 422
redekei 435
rubescens 159, 161, 162, 163, 164, 165, 166, 169, 170, 171, 177, 180, 189, 190, 191, 422
utermöhliana 324, 328, 329, 330, 332
Osmunda regalis 13
Ostracoda 215, 219–223, 225, 227–229
Oulimnius 233
Oxalis 11
Oxyethira flavicornis 236, 237

Pagastiella orophila 236
Pandion haliaetus 270
Pandorina 160, 172, 180
morum 165, 177
Paractinolaimus intermedius 215, 218
Paracyclops fimbriatus 222, 228, 229
Parakiefferiella 236
Paranais litoralis 233, 234
Paraquadrula 217
discoides 214, 216, 217
globulosa 214, 216, 217
Parastenocaris 222, 225, 229
brevipes 229
glacialis 219, 225, 226, 229
phyllura 229
Paratanytarsus 236
Paratendipes albimanus 234, 235
Pediastrum
boryanum 177
duplex 177
tetras 173
Penardia 214
Pennatae 283
Perca fluviatilis 263
Peridinium 159, 165, 171, 177, 188
Peucedano-Pinetum 11
Phacotus lenticularis 161, 169, 176, 177
Phragmites 134, 139, 144, 146, 147, 214, 218, 227, 233, 234
australis 11, 13, 133, 136, 138, 139, 143, 145, 146
Phragmitetum 219, 227, 231, 234
Phragmitetum australis 138
Phryganea 233
Phryganella
acropodia 214, 216
nidulus 216
paradoxa 214
Physa fontinalis 236, 237, 238
Picea 11, 14, 16
Pinus 14, 16
sylvestris 11, 13, 402
Piscicola geometra 238
Pisidium 239
amnicum 233, 234
casertanum 234, 239
conventus 238, 240, 241, 242, 423, 427
hibernicum 281
lilljeborgi 281
Planaria torva 238
Planorbis
corneus 233
planorbis 233
Platyhypnidium rusciforme 132, 148
Plecoptera 234, 235, 239
Pleuroxus
aduncus 222, 225, 227
trigonellus 222, 224, 228, 229
truncatus 219, 222, 224, 225, 227, 228
uncinatus 222, 223, 228
Podiceps cristatus 269
Pogonocladius consobrinus 237
Polyarthra 248, 252
dolichoptera 245, 251, 252, 335
vulgaris 245, 251, 252, 327
Polygonum amphibium ('*natans*') 137, 139, 148
Polypedilum 236, 238
bicrenatum 234
pullum 234, 238, 239
scalaenum 236
subgenus *Tripodura* 233
Polyphemus 223, 225, 227
pediculus 221, 223
Polypodium vulgare 13
Pompholyx
complanata 255
sulcata 245, 255
Pontigulasia bigibbosa 216
Populus 14
Potameto-Nupharetum 136
Potametum filiformis 135
Potametum lucentis 133, 134, 135, 143, 144, 146, 147, 219, 224, 227
Potamogeton 136, 149, 214, 216, 218, 220, 222, 223, 224, 227
alpinus 13
crispus 141, 143, 422
filiformis 13, 135, 138, 280
gramineus 13, 132, 280
lucens 13, 133, 135, 148
mucronatus 133, 135, 144, 145

natans 13, 133, 136, 137, 138, 141, 148, 223, 422
nitens 13, 280
obtusifolius 133
pectinatus 13, 131, 132, 141, 147, 422
perfoliatus 133, 135, 144, 148
praelongus 13, 280
rutilus 13, 141, 144, 280, 422
Potamocypris villosa 222, 229
Procladius 235, 236, 238, 240
choreus 242
fuscus 242
nigriventris 238
Prodorylaimus longicaudatoides 215
Propsilocerus lacustris 242
Protaminobacter 208
Proteus 208
Protozoa 213–218, 500
Psectrocladius 236, 237, 238, 240, 242
Pseudanabaena catenata 180
Pseudochironomus 236
Pseudochydorus globosus 220, 223
Pseudodifflugia 216
fascicularis 216
Pseudokephyrion entzii 165, 166, 169, 173, 176, 178
Pseudomonas 208
Pseudosphaerocystis schroeteri 161
Pseudotsuga 11, 16
Pulsatilla pratensis 13
Pyrrhosoma nymphula 237

Quadrulella symmetrica 217
Quercus 14, 16, 402
petraea 402

Rallidae 272
Ramischia secunda 13
Ranunculus 11
circinatus 133, 135, 136, 141, 143,422
repens 11, 139
Rhizosolenia longiseta 177
Rhodomonas minuta var. *nanoplanktonica* 173
Rhodotorula 340
Rhopalodia gibba 150
Rhynchospora alba 11
Rhynchotalona 223, 228, 229
falcata 222, 225
Rotatoria 213, 215, 218, 243, 280, 343, 504
Rutilus rutilus 263

Sagittaria sagittifolia 138
Salix 14
alba 402
Salvia pratensis 13
Sarothamnus scoparius 13
Scapholeberis 223, 225, 227
mucronata 222
Scenedesmus 189, 329, 340
costato-granulatus 161–163, 173, 177, 189, 280
quadricauda 324, 328, 330, 332, 435
Scheuchzeria palustris 11
Schoenoplectus lacustris 135, 136, 138, 143
Scirpo-Phragmitetum 135, 138, 139
Scorpidio-Utricularietum 138
Scorpidium scorpidioides 138
Sergentia coracina 240, 241, 242, 243, 423, 427
Sida 223, 225, 227
crystallina 219, 222, 223, 225
Sieglingia decumbens 13
Simocephalus 227
vetulus 220, 221
Sinantherina socialis 255
Solanum dulcamara 139
Sorbus aucuparia 402
Sparganietum erecti 138
Sparganietum minimi 137
Sparganium
erectum 136, 138
minimum 13, 137, 280
Sphaerocystis schroeteri 164, 169, 172
Sphagnum 11, 37, 39, 139, 217
recurvum 11
Spirodela 137
polyrrhiza 137
Spirodeletum polyrrhizae 137
Spirostomum ambiguum 217
Spongillidae 280
Staurastrum paradoxum 177
Stauroneis phoenicenteron 150
Stentor
coeruleus 217
polymorphus 217
Stephanodiscus 166, 172, 185, 192
astraea 164
hantzschii 279, 303
minutus 159, 161, 163, 173
rotula var. *minutulus* 279, 281
Stictochironomus 233
histrio 241, 423
pictulus 234, 236, 238, 240, 241, 242
rosenschoeldi 241, 242, 423

Stratiotes 137, 223, 227, 235, 422
aloides 133, 136, 137, 141, 144, 148
f. *submersa* 133
Stylaria lacustris 234, 236, 238
Synchaeta 244, 248
assymmetrica 215
grandis 255
oblonga 245
pectinata 245, 249
tremula 245
Synechococcus elongatus 329
Synedra 161, 164, 169, 172, 186, 422
acus 159, 161, 162, 163, 164, 177, 283
capitata 283
'nano' 162, 163, 166
parasitica 150
var. *subconstricta* 150
Synura 24
sphagnicola 280
Synuraceae 280

Tabellaria 422
fenestrata 177, 299, 301, 303
var. *asterionelloides* 166, 173, 176, 177, 178, 179, 180, 185, 189, 190, 191, 283, 299, 303, 422
var. *fenestrata* 177, 299, 301, 303
flocculosa 166, 172, 180, 283
Tanypodinae 234, 235, 239
Tanytarsini 234, 235, 236, 238, 239, 427
Tanytarsus 235, 236, 238, 240, 242
lugens 281
miriforceps 242
niger 242
norvegicus 238, 242
sylvaticus 242
Tardigrada 213
Teesdalia nudicaulis 13
Testacea 213, 214–217, 243, 280, 500
Testudinella 218
Tetraedron 189
minimum 159, 161, 173, 176, 177, 189
Thecamoeba striata 213
Thelypteris palustris 139
Theodoxus fluviatilis 233
Thermocyclops oithonoides 221
Thienemannimyia 233
vitellina 235
Thyas pachystoma 280
Tilia 14
Tinca tinca 263
Tinodes waeneri 233
Tobrilus
allophysis 218
steineri 218
Tracheleuglypha dentata 216
Trachelius ovum 217
Trachelomonas 177
Trichocerca 218
capucina 255
Trichoptera 231, 233, 234, 235, 237, 238, 239
Tricladida 213
Trigonopyxis arcula 217
Trinema 216
enchylis 216
Tripyla glomerans 218
Tubificidae 240, 241
Turbellaria 213, 234, 235, 239, 240
Typha 214, 216, 217, 218, 227
angustifolia 11, 13, 133, 136, 138, 143, 144, 147, 149
latifolia 138
Typhetum angustifoliae 138

Ulmus 14
Unio pictorum 233
Uroglena 166, 170, 171
eustylis 280
Utricularia 235
intermedia 13, 137, 138–139
minor 13, 137, 138–139
vulgaris 133, 136, 137, 138–139, 143

Vahlkampfia limax 213
Valvata 236
cristata 236
piscinalis 236
Vaucheria 131, 238, 239
dichotoma 132, 143, 145, 148, 228
Vertebrata 280
Vejdovskyella comata 236
Viola riviniana 13
Volvocales 158, 179
Volvox 160
Vorticella
convallaria 217
monilata 217, 218
nebulifera
var. *similis* 218

Zannichellia palustris 135
Zygoptera 236, 241

Subject index

activity coefficient
  microphytobenthos 315, 316, 317, 318
  phytoplankton 300, 301, 302, 305, 306, 307
air temperature 43, 44
Alder-Birch-carr 12
Alder carr 11, 12, 140
algae
  number of taxa 277
  ultraplanktonic 248
algal taxa 280
Alleröd 14
Altglobsow (village) 17
amoebae protozoa 280
animal(s)
  number of taxa 277, 280
  nutrition 326
  size 323, 324, 328, 330–335, 338, 339
annelides 280
anthropogenic influences 472
assimilation of zooplankton 251, 328–333, 335, 338–342
  efficiency 340
  excess 323
  food 246
  mean maximum rate 328, 331
  of crustaceans 323, 325, 326
  rate 329
  selective 341
  size dependent 333
associations
  of Characeae 130–133
  of pondweeds 133–137
  of duckweeds 137
  of small bladderworts 137–138
  of reeds 138–140
Atlanticum stage 13, 14
atmospheric humidity 43–46
atmospheric phosphorus load and oligotrophy 97
aufwuchs (habitat of animals) 214–218
autumnal full (complete) circulation 56, 163

bacteria 197–211, 248, 251, 497
  conceptional approaches 197–199
  copiotrophic 198, 200, 207
  methods 199
  methylotrophic 368
  oligocarbophilic 198, 200
  oligotrophic 198, 199
  sulphate reducing 366
  taxonomical analysis 207, 208, 369
  zymogenic 198, 199
bacterial
  biomass 202, 207
  distribution 200–203
  number 199–203, 358, 359
  populations 207–208
  production 204–207
  respiration rate 352–354
  standing crop 203–204
Baltic Land Ridge 9
Baltic region 263
bathymetry 30, 32, 34, 35, 36, 37
Beech-forests 11, 13, 16
beetle 281
Berlin 18
biomanipulation 449–451
biomass 21
  animals 331
  bacterial 202, 207
  crustaceans 244–250, 252, 253
  macrophytes 145–148
  microphytobenthos 315, 316, 317, 318
  periphyton 149–157
  phytoplankton 166–172, 178–182, 187–188, 246, 252, 297, 299, 305, 306, 310, 313
  rotifers 244–248, 255

values 243
bioproduction 451, 461, 466
birds 269–275
breeding success 272
density of breeding 269
number of breeding 269, 270
of prey 272, 274
Bittern 272
Black-throated Diver 273
bladderworts 137, 138
Bleak 263
blue-green alga(e) 160, 166, 167, 168, 172, 174–175, 180, 181, 182, 183, 184, 185, 186, 187, 188, 189, 280
blue-greens 329
bog(s) 10
-pools 11
mosses 11
-Pine-wood 11
wooded raised 11, 12
Sphagnum 11
-dwellers, stenotopic 236, 243
boreal
distribution 280, 281, 283
species 13, 226
stage 14
boreo-alpine distribution 226, 280, 281, 283
Bream 263
buffering capacity 447
Burow (village) 17

caddis flies 236
calcite 376, 382
concentration 413, 414
content 382
crystals 79, 106, 418
deposit 382, 415
formation 416
re-dissolving 381
sedimentation 381, 416
self-flocculation of algae 419
solution 381
calcite precipitation 101, 102, 106, 314, 381, 382, 412–419, 440, 444, 447–449, 451
biogenic 101, 102
co-precipitation (PO_4) 413–414, 417
coupled reactions 413–415, 417
decrease of CO_2 414–415
ecological effects 314, 413–419
model-simulation 416–417, 444, 447–449
pelagic 101, 102, 382
rise in turbidity 414
role in the life of the lakes 412–419
Secchi depth 413
self-flocculation 414, 417, 444
self-purification 417–419
calcium 106
content of the epilimnion 106
in deep sediments 380
sediment 382
sedimentation 381
carbon 101
balance 341
budget 21, 101–110, 338
concentration 107
content 102, 107
dissolved inorganic (DIC) 102, 103
flow 199, 370
particulate inorganic (PIC) 102–106
-input 364
in sedimenting material 388
dissolved organic (DOC) 106–110
particulate organic (POC) 106–110, 334, 337, 341
sedimentation 392
turnover 23, 102, 335
carbonate(s)
calcium- 380
content of sediments 378, 381, 382
spatial distribution in lake sediments 380–382
trophic status 381
Carp 261, 263
catchment area, Lake Stechlin 29, 34, 35, 37, 41
catchment area evaporation 69
cell number of phytoplankton
Lake Stechlin 159–165, 172–178, 182–185
Lake Nehmitz-N 165–166, 185–187
Lake Nehmitz-S 166
cell size of phytoplankton 307, 310
chalk
sediments, free of 376
solution 106
Chara-beds 275
Chara-meadows 263, 265
Characeae carpets 237
Charophyceae, number of taxa 277
chlorophyll a, phytoplankton 82, 312, 313, 314
Chironomidae
number of species 231
Chironomini
medium abundance 234, 235, 239

ciliates 280
circulation (full, complete) 55–56
  autumnal 55, 56, 163, 412, 420
  spring 55, 56, 161, 172, 412, 420
cladocerans, reproductive cycle 220–221
climatic conditions 41–64
cloud cover 46
clouding 417
C:N ratio
  correlation with the trophic status 383–386
  dependent on sedimenting material 384
  from different depths 383
  in lake sediments 217, 364, 382–386
  in sedimenting material 387–389
  correlation with character of sediment 217, 229, 383–386
  result of autumnal leaf-fall 217, 383
C:N:P ratios 390, 394
Coleoptera
  medium abundance 234, 235
  number of species 231
community 243
  crustacean 246, 254
  rotifer 254
competitive exclusion 224
complexity
  of a model 433
  of an ecosystem 442
control
  mechanisms, ecological 442–447, 451
  variables 443–447, 449
cooling water (circulation, circuit) 19, 21, 50
  influence on:
  bacteria 424
  biocoenoses 421–423
  equilibrium, ecological 412
  circulation 420
  heat balance 61–63, 419–421
  ice-cover 63, 420
  lake metabolism 427–429
  macrophytic vegetation 140–145, 422
  macrozoobenthos 241–242, 423, 424
  microbial activity 357–361
  oxygen 122
  phosphorus 98–99, 421, 425
  phytobenthos 152–156, 316, 317, 424
  phytoplankton 177–178, 182, 185, 190–193, 422–424
  primary production 289, 290
  retention time 419
  stability 420
  stratification 56, 420
  temperature 50–51, 420, 425–426
  thermocline 56, 57, 420
  transparency 81, 82
  trophic status 427–429
  water balance 289, 290, 419–421, 426
coot(s) 269, 271, 272, 273, 274, 275
co-precipitation, of phosphorus 447
copepods 246, 248, 266, 268
  calanoid 244, 246, 247
  cyclopoid 244
Coregonus
  breeding of 266–268
  spawning of 264
  -lake 263
  larvae 266, 268
C:P ratio 394
Crustacea, planktonic 233
crustacean(s) 244, 246, 248, 249, 250, 251, 253, 254, 266, 325, 338–340, 342
  calanoid 247
  secondary production (assimilation) 328–335, 338–339
cryptodepression 30, 33
cyanophycean(s) 166

Dabbling ducks 272
Dagow (village) 17
dead ice
  area 10
  block(s) 10, 29, 31, 35
  lakes 29
decalcification
  biogenic 376, 459
  by assimilation 102
decision making 433
decomposition
  cages 401
  of fallen leaves 406–407
  of organic matter 385
  of organic substances 394
  of sedimenting seston 394, 395
deep water diversion 475
depth of the lakes studied 33
detritivores 222, 228
  cladocerans 223
  harpacticoids 222
  ostracods 222
  rotifers 340
detritus 239, 241, 248, 251, 252, 340, 341, 377, 440, 441, 447

content 340, 342
feeders 340
nutrition 340
nutritional value 340
particles 340
-rain 265, 362
sediment- 342
development of vegetation 14, 16
diatoms 150, 248, 376
biomass 160, 166, 167, 168, 169, 170, 171, 172, 174, 175, 178, 179, 180, 181, 182, 183, 184, 186, 187, 188, 192, 248, 306
cell number 160, 161, 164, 165, 166, 167, 168, 174, 175, 178, 181, 183, 184, 185, 186, 190
(eu)planktonic 277, 279
number of taxa 277
spring maximum 161, 169
vertical distribution 189
dimictic (lake) 55, 159
Diptera
medium abundance 234, 235, 239
dipteres 280
dishomogeneity, horizontal, of zooplankton 338
distribution, annual of zooplankton 244–248
distribution, horizontal of zooplankton 252–255
distribution of taxa
boreal 13, 226
boreo-alpine 226
continental 13
continental boreal 13
oceanic 13
distribution, vertical
phytoplankton 189–191
zooplankton 249–252, 253
Diving ducks 273
Dragon-fly 281
Dryas (stage) 14
duck farming 458, 474
Duckweeds 137
dy 217, 382, 383, 385, 401
dystrophic lake(s) 217, 243, 312, 382
dystrophy 217

earthquake(s) 6, 7
economy 17, 18
economic evaluation 470–472
ecosystem analysis 433
Eel 261, 263
European Cisco 261, 262, 263, 265, 266
energy
conversion factor 75
equivalent 297
sources 401
Ephemeroptera/Plecoptera
medium abundance 234, 235, 239
epibenthonic algae 218
eudynamic lake 364
eutrophic (condition, lake, state) 19, 91, 120, 130, 136, 192, 226, 241, 242, 312, 384, 385, 395, 397, 401, 415, 450, 460, 461, 465, 466
eutrophication 130, 133, 135, 192, 255, 269, 312, 314, 450, 456, 463, 471, 472, 479
control 447, 449–451
eutrophized lake 397, 422, 471
eutrophy (trophic state) 193, 217
evaporation 29, 64–67, 69–71
extinction coefficient 75, 77, 78, 82–85
extinction depth 439
exudation 197, 348

fallen leaves 401–409
amount 404–405, 407
break down 405–407
C and N input 407
decomposition 405–407
nutrient cycle 407
P-release 406
feedback control 451
feeding
experiments 324
behaviour 341
Feldberg (town) 21
feldspar 376
fen 35
Fenchelberg 29, 30, 31
feudal lords
Danish 15
German 15
Polish 15
filter-feeder(s) 232, 238, 246, 248, 252, 326
filtration 251
efficiency of net 243
filtrator 247
fish 261–268
breeding glasses 266
dominant species 261–263
predatory 445, 449–450
spawning sites (plakes) 265, 266
stock(s) 261, 445, 449–450
yield 261, 262, 263, 264
zooplankton-eating 445, 449

fishing 17, 261–263
bow net 261
catch site(s) 265, 266
drag net 261, 262
electro 261
gill net 261
sites 264
Fish-otter 280
flagellates 248, 299, 306
floating plants 136
floristic research 129
focusing of sediment 377
Fontane, Theodor 2, 3, 7, 11
food
concentration 323, 324, 327, 328, 330, 332, 337–340
conditions 248
distribution, niches 251
preferences 335
selection 252
supply 328, 337, 338
type 323, 327, 328, 331, 333, 335
forcing functions 438
Frederician colonization 17
full-circulation 55
autumnal 56
spring 56
fungi, number of taxa 277
funneling-effect 377, 391
Fürstenberg (town) 15, 18
Fürstenberger Staffel 9, 10, 29, 35

glacial deposits 9
glaciation, quaternary 9
Scandinavian
glass-work(s) 18
global radiation 57, 72, 73, 297
Goldeneye 269, 272, 274
Goosander(s) 269, 271
grazing 335, 336–338
selective 340
rotifers 323, 335
zooplankton 335, 341
maximum, mean potential 335
Great Crested Grebe 269, 272, 273, 274
Great Reed Waibler 269, 272
green algae 172, 184, 185, 188
microplanktonic 160, 167, 168, 174, 175, 183, 184, 186
nanoplanktonic 160, 167, 168, 174, 175, 183, 184, 186, 189
ultraplanktonic (μ-algae) 158, 160, 182, 184, 185, 186, 280, 299, 301, 303, 305, 307
Green Tench 263
gross primary production 297
Groß Zerlang (village) 17
ground-water 29, 31, 218, 229
growth
conditions of animals 246
rate 331
growth efficiency, of phytoplankton 300, 302
gyttja 132, 133, 236, 376, 383

habitats
lacustrine 224–226
semilacustrine 224–225
hard-water lake(s) 383, 413, 417–419, 447–449
calcite-precipitation 415, 417–419
C:N-ratio in sediments 383
self-protection 447–449
self-purification 419
heat (water)
balance 56–64
income 47, 48–50, 51, 59–60
latent 60–61
sensible 60–61
storage 60
heliozoan protozoa 280
Hemiptera, medium abundance 234, 235, 239
herbivores 222
heterotrophic activity 348
distribution pattern 348–352
and primary production 348, 350, 356
and thermal pollution 359
trophic status 356
heterotrophic CO_2-fixation 198, 206, 348, 351, 357
heterotrophic potential 198, 347, 356
holiday accomodation 479, 480
holomixis 31, 87
Horned Grebe 273
Hydracarina 231, 234, 236, 238
number of species 231
medium abundance 234, 235, 239, 240
humic standard 462
mesohumous 462
oligohumous 462
polyhumous 462
hydric predators 232
hydrography 457, 466
Hydrometeorological Station (Neuglobsow) 19, 42

Hydrozoa, medium abundance 234, 235
hypertrophic lake(s) 415, 460, 461
hypsometric curve(s)
  Lake Stechlin 30, 33
  Lake Nehmitz 34, 35
  Lake Große Fuchskuhle 33, 37
  Lake Dagow 34, 35

ice-calendar 63
I_k-value 302, 303, 307
incipient limiting level of food concentration 335, 337
indicator species 269
indicators 217, 218
ingestion 251, 326, 335, 336
  selective 341
input of organic matter
  allochthonous 31
input variables, of a model 435, 436, 438, 444
insects, number of taxa 280, 516
interstitial fluid 364
interstitial ground-water 229
invertebrates 232

kinetic
  phosphorus uptake 307
  velocity constant (k_1) 307, 308
K_s-value 307

lake classification 456–468
  hydrographic criteria 457–459
  pollutants 460–463
  territorial criteria 457–459
  trophic criteria 459–460
Lake Dagow
  bathymetric map 37
  classification 463–467
  figure 38
  hypsometric curve 36
  morphometry 35
  nitrogen budget 114
  oxygen budget 120–121
  phosphorus, annual course 91–92
  primary production 295–296
  sedimentation 388, 389, 390, 395, 397
  sediments 384
  silicon budget 118
lake management 455, 467–468
lake metabolism
  reaction of cooling water influence 423–427
  role of calcite precipitation 417–419
Lake Nehmitz
  bathymetric map 36
  Beckensee, kombinierter 10
  biomass, phytoplankton 171–172, 180–182, 188
  cell number phytoplankton 165–166, 185–187
  classification 463–467
  figures 34–35, 38
  macrophytic vegetation, change in 143–145
  morphometry 31–34
  nitrogen budget 113
  oxygen budget 118
  phosphorus, annual course 89–91
  phosphorus, inputs 96–100
  primary production 294–295
  sedimentation 386–397
  sediments 384
  silicon budget 118
lake(s) origin 10
Lake Stechlin
  bathymetric map 32
  biomass, periphyton 149–157
  biomass, phytoplankton 166–171, 178–180, 187–188
  birds, breeding site of 269–272
  birds, resting site of 272–275
  birds, wintering site of 272–275
  calcite precipitation 412–413
  cell number, phytoplankton 159–165, 172–178, 182–185
  classification 463–467
  ecological status 411–412
  figure 4–5
  hypsometric curve 33
  macrophytic vegetation, change in 110–143
  morphometry 29–31
  nature reserve 479–483
  nitrogen budget 111–113
  oxygen budget 119–120
  oxygen content 121–123
  phosphorus, annual course 89
  phosphorus content, horizontal distribution 94–95
  phosphorus inputs 96–100
  primary production 289–293, 314–319
  sedimentation 386–397
  sediments 375–377
  silicon budget 116–117
  vertical distribution of phytoplankton 189–191
  water volume 33
Lake Stechlin area
  aquatic macrophytes 129–149

economy 15, 17–18
fauna 277–280
flora 280–283
hydrometeorology 41–86
man-made modifications 472–475
map 8
morphogenesis 7–9
morphology 7–9
morphometry of lakes 29–40
natural potential 470–472
plant cover 11–15
recreation area 475–479
recreation property 476
settlement 15–17
watersheds 32
Langmuir-spirals 252
leaf-fall period 402
legend, of the Red Cock 3, 6, 7
life of the lake 411–431
ecological status 411–412
calcite precipitation 412–419
role of nuclear power plant 419, 427–429
light adaption of phytoplankton 298, 300, 302
light climate, underwater 75–79
light limitation of phytoplankton 300, 301
light penetration 72–86
light utilization efficiency, phytoplankton 297, 298, 299, 300, 301, 309, 310
Limnological Laboratory Stechlin 18, 19–24, 130, 157
Lindow (village) 17
litter fall 197, 364, 369
Little Bittern 272
littoral plankton 221
livestock units 459
load 457
characteristics 457, 458, 459, 466
nutrient, measures to curbe 472–475
orthophosphate 460
phosphorus 96–100, 451, 472
thermal 61
loading factor 468
nutrient 254
thermal 317

macrophytes
aquatic 19, 129–149, 277, 280
communities 129–140
habitat of animals 214–243
number of taxa 277
semi-aquatic 277
standing crop 145–149
macrozoobenthos 230–243
abundance of species 230–231
communities 231–242
indicator species 242–243
stock of species 230–231
Mallard 269, 272
manganese 380
mapping 465
Marsh Harrier 272
Mayfly 236
Mecklenburg Lake District 7, 106, 383
medieval agricultural expansion 15
meiozoobenthos 219–229
boreal species 226
bottom-dwellers 222–223, 228–229
cold-water-stenothermal species 221, 226, 229
communities 226–229
habitat analysis 221–226
iliobiontic species 222
iliophilic species 222
indicator species 225–226
lasiophilic species 229
microcrustacean communities 226–229
perennial species 220
phenology 220–221
phytophylic species 222–223, 225
psammobiontic species 222, 224
psammophilic species 222, 224
stygobiontic species 222
warm-water-stenothermal species 221
zoogeographical aspects 224
Menz (village) 9, 17
mesotrophic (condition, lake, state) 19, 226, 242, 312, 395, 415, 450, 460–466, 472
metalimnetic oxygen maximum 119, 120
methane 367
concentration 368
cycle 367
oxidation 368
production 369
mica 376
Michael-Menten kinetics 347
microbial activity
thermal pollution 356–362
phosphate mobilization 366–367
methane cycle 367–370
microcrustaceans
boreal and boreo-alpine species 280
lacustrine species 224–225, 227

semilacustrine species 225, 227
microfauna 214
microphytobenthos
activity coefficient 315, 316, 317, 318
biomass 149–152, 315–318
primary production 149–152, 314–317
vertical distribution 316, 317, 318
micro-turbellaries 280
microzoobenthos 213–218
Middle Age(s) 15, 16, 17
migration, phytoplankton 298
migration, zooplankton 251
daily 249
diurnal 249, 250, 253, 338
pattern 249
vertical 249
Minack, fisherman 6, 7
mite 280
model(s)
calcite precipitation 447–459
dynamic ecological 433–451
empirical 442, 447
general 434, 435
lake-specific 434
mathematical 338, 339
predictive 434
primary production 287, 288, 301, 302, 438–439
principles 433–435
role of artificial heating 443–445
SALMO 323, 435–438
structure 437
thermal stratification 443–446
uncertainty 434, 442
validation 434, 438–442
weak points of 450
Mollusc(a)
number of species 231
medium abundance 234, 235, 239
shells 376
monomixis 421
moraine(s) 29
ground 9
terminal 9
plateaus 15
morphometry of lakes 29, 30, 35, 412, 457, 466
mortality, of the zooplankton 445, 449
mussels 280, 281
Mute Swan 269

natural potential, area 470, 471
nature reserve 467, 479, 480
nauplii 266, 268, 341
Nematoda
medium abundance 234, 235, 239, 240
nematodes 280
neophytes 15
net cages 458, 474
net terrestrial radiation 59
Neubrandenburg (town) 21
Neuruppin, meteorological station 42
Neustrelitz, meteorological station 42
niche separation 248
nanoplankton 158, 304, 305, 306
nanoplanktonic algae 192
nanoplanktonic diatoms 160, 164–189, 248, 306
nanoplanktonic flagellates 306
nanoplanktonic phytoflagellates 160–180
nanoplanktonic primary production 304, 305, 306
net-planktonic primary production 304, 306
nitrate 113, 114
nitrogen
ammonium 113, 114
budget 21, 111–116
distribution
-seasonal 115
-spatial 112
-temporal 112
inorganic 441
in sedimenting material 388
limitation of bioproduction by 114–116
ratio to phosphorus 114–116
North Sea 10
nuclear power plant 19, 21, 22, 31, 34, 35, 140, 157, 159, 213, 252, 254, 255
changes in biocoenoses 421–427
macrophytes 140–145
macrozoobenthos 241–242
phytoplankton 164, 172–177, 185, 190–193
microphytobenthos 152–155, 316
changes in heat balance 49, 59–63, 419–421, 426
changes in matter balance 421
oxygen 122
phosphorus 94, 98, 101, 424
changes in water balance 48–51, 56, 419–421
effect on the life of the lake 427–429
lake metabolism 423–427
microbial activity 356–361
reaction of biocoenoses 423–427
yield of fish 265

nutrient balance 461, 466
nutrient limitation 114–116
nutrient load 472
nutrient release, sediment 451
nutritional
adaptation 326
biology of zooplankton 323
value 330, 340

Odonata
number of species 231
medium abundance 234, 235, 239
Oligochaeta
medium abundance 234, 235, 239
oligotrophication 397, 471
oligotrophic (condition, lake, state) 19, 217, 225, 242, 312, 319, 367, 370, 378, 384, 385, 397, 401, 412, 415, 421, 423, 429, 450, 451, 460, 461, 465, 466, 468,471, 475, 482
oligotrophy 225, 412
Orthocladiinae
medium abundance 235, 239, 274
orthophosphate 88–101, 394–397, 421, 425, 440–441, 460, 461, 466
concentration 425
consumption 395
increase 394, 395
input 421
load 460
Osprey 270, 272
ostracode shells 376
outwash 9
plain 29
over fishing 265, 266
oxygen
balance 461, 466
budget 21, 118–125
capacity, hypolimnetic 429
cooling water influence 426, 429
concentration, hypolimnetic 122, 352
consumption 119, 123, 412, 442, 451
deficit 123, 394
depletion, hypolimnetic 35
epilimnetic 123
horizontal distribution 121
hypolimnetic 122, 123, 395, 441, 448
regime 426

parameters, of a model 434, 435, 444–447, 450
particle catchers 238
Perch 261, 263, 266
periphyton, *see* microphytobenthos
phosphate
elimination 474
mobilization 366
phosphorus
accumulation, hypolimnetic 94
atmospheric input 96–97
balance, external 100–101
budget 87–101
cooling water influence 421
co-precipitation 413–419
depth relationship 380
export 100
horizontal distribution 94–95
import 96–100, 444, 475
input from sediment 396
in the sedimenting material 388
input through inflow 98–99, 421
limitation of bioproduction 87, 114–116, 309–313
load, external 96–100, 444, 446–447, 448, 449, 451, 472
orthophosphate *see* orthophosphate
output into sediment 394
^{32}P-phosphorus 307–311
pelagic, annual course 87–92
pelagic, content 92–94
release from sediment 395
spatial distribution in sediments 377
uptake, plankton fraction 310, 311
uptake, kinetic 307, 308
uptake, phytoplankton 307–311
uptake, seasonal distribution 307, 308, 310
uptake, vertical distribution 307, 309
photosynthetically active radiation 74, 76, 297, 300, 301
photosynthesis
light relation 301–303
phytoplankton 287–314
rate, specific 440, 445, 447, 450
physiography 29
phytoflagellates *see* nanoplanktonic phytoflagellates
phytoplankton 248, 251, 252
as food 248, 251, 252, 340–342
biomass 166–172, 178–182, 187–188, 297, 299, 305, 306, 438–440
cell number 159–166, 172–178, 182–187
chlorophyll *a* 82, 312, 313, 314
cooling water influence 190–193
fraction 304, 306, 307, 310, 311

groups, functional 435
losses 440
phosphorus uptake 307–311
primary production 287–314, 323, 338–342
seasonal periodicity 159–188, 192
species composition 159–188
standing crop 323, 338, 349
state variable 435, 438–440
vertical distribution 189–191, 438–440, 441, 450
Pike 261, 263
Pike-Perch 263
Pine-forests 11, 13
Pine-woods 11
pisids 238
planarians 238
plankter(s), algal, number of taxa
littoral- 277
tycho- 277
plankton rain 252
plankton feeder 263
plants, number of taxa 277
POC, size fractions 108–110
Pochard(s) 272, 274, 275
pollutants 462, 466
polymictic (lake) 458
polytrophic (condition, lake, state) 460, 461
Polzow
-brook 18
-canal 10, 18
Pondweeds 133–135
population equivalent 458
Post (Characeae) 18
Potsdam, meteorological series 42
precipitation 29, 68, 407
predation 450–451
predatory fish 445, 449–451
pre-seismograph 7
primary production of phytoplankton 21, 246, 248, 251, 252, 287–314, 439
annual rate 290, 291, 296, 297, 312
cooling water influence 289, 290, 427
daily rate 289–296
efficiency 297–301
gross 297
maximum rate (a_{opt}) 289, 294, 295, 301
nanoplanktonic 304, 305, 306
net-planktonic 304, 306
phosphorus uptake 307–311
population, and 304–307
seasonal distribution 289–297
trophic criteria 459–460, 461, 466
trophic status 312–314
vertical distribution 289–297, 300
protection of waters 24

quartz 376
quasi-dormant stock of species 177

radiation
balance 56–59
global 72, 73, 74, 297
photosynthetically active 72–74, 76, 297, 300, 301
surface incident radiation 72–75
rafting canal 18
rain: P content 96
Rainbow Trout 261
rate variables, of a model 435, 436
recreation
area 18, 475
day(s) 476
potential 471
property 476
use 458
Red Eye 263
Reed Bunting 269
reed(s) 134, 138–140, 142, 146, 147, 224–227, 230, 232
Reed Warbler 269
relicts, post-glacial 226
relief, preglacial 9
research
floristic 129–130
phytosociological 129–130
programme 19, 21, 23
project 19–24
work 21
residual lake 37
respiration
losses 251, 325, 327, 333, 340
rate 440, 445, 450
retrogression of settlements 15
Rheinsberg
Basin 9, 10
Gate 9
town 9, 18
rheophilous species 234
Roach 261, 263
Roofen (village) 17
rotifers 244, 245, 247, 248, 250, 251, 252, 253, 254, 255, 268, 323, 327, 335–338 (grazing),

340–342
runoff
surface 71
subsurface 71

salinity 462, 466
SALMO, model 435–438, 450
sandr 10
sanitation measures 472–475
Savi's Warbler 269
scenario
analysis 440, 442–447, 451
control 442–447
specification 442–447
seasonal clouding(s) 413
Schräder, Theodor 19, 21
Secchi depth 79–85
and calcite precipitation 413, 414
and biomanipulation 450
Secchi disc transparency 79–85
nuclear power plant 428
secondary production 21
of zooplankton 323–345
assimilation 328–338
carbon budget 338–339
nutritional situation 339–342
primary production 339–342
Sedge Warbler 269
sedimentation 386–397, 444
oscillations 391–392
phytoplankton 392
phosphorus balance 394–396
primary production 393
summer stagnation period 387–389
rate 363
vessels 380
winter 389–391
sediment(s) 214–218, 229, 243, 375–386, 401
balance 396
carbonates in 380–382
catcher 377
chemistry and lake depth 377–381
C:N ratio 229, 382–386
distribution of bacteria 365–366
impact in lakes 442, 450, 451
microbial processes in 362–370
nutrient release 442–451
submodel 442–450
sedimented masses 387–391
self-flocculation, of the algae 417, 418, 419, 447, 448, 449
self-governing power 412
self-purification 417–419
calcite precipitation 413
potential 468, 471
self-protection, of hard-water lakes 447, 448, 449
semilacustrine habitats 224–225
sensitivity analysis 438
seston 335, 342
composition 340
C:N ratio 384, 385
content 331
desegregation process 391
phytoplankton, source of 392
sedimenting 378, 385, 395
sinking 384
seston-filtrators 222
settlement
history 15
sewage
irrigation 474
treatment 472, 474
shell types, of Testacea 214
shore line 30, 33, 34, 35, 37, 405
silicon 21, 380
budget 21, 116–118
Silver Carp 261
simplicity, of a model 433
simulation of model 434, 438–440
snails 238
snow: P content 96
Sphagnum-belt(s) 39, 139, 217
Sphagnum-bog(s) 11, 12, 37
sport diving 476
spring circulation 55, 56, 161, 172, 412, 420
stagnation period (phase)
summer 55, 161, 162, 169, 173, 178, 189
winter 56, 159
Stammstechlin (village) 16
standard scenario 443–447
standing crop, of macrophyt 145–149
state variable(s), of a model 433, 435–438
Stechlin
novel 2, 3, 6
village 3
stratification, of the
aufwuchs 216
stratification, thermal 55, 56, 441, 445
structure, of a model 437
substrate-filtrators 222–223
sunshine duration 46

swamp-plants 13

Tanypodinae
  medium abundance 234, 235, 239
Tanytarsini
  medium abundance 234, 235, 239
Tanytarsus-lake 242
tardigrades 280
tar furnaces 18
testate amoeba 281
temperature, of air 43, 44
temperature, of water 47–56
  cooling water influence 48–51, 420, 423, 429, 443–447
  direct effects on 445, 446, 451
  increase in 444, 445, 446, 451
  indirect effects on 443, 446, 451
  mean annual course 48–50
  of water surface 48–51
  stratification 55–57, 421
  vertical distribution 51–55
  thermal pollution and microbial activity 356–361
  tolerance of with regard to biocoenoses 425–429
thermal pollution 356, 357
thermal stratification 55
thermocline 55–57, 189
thermophilic water plants 13, 129
tourist industry 18
transmission, of PAR 75–79
transparency (Secchi depth) 79–85
  decrease 427–428
Trichoptera
  number of species 231
  medium abundance 234, 235, 239
Trichtereffekt (funneling-effect) 377, 391
trophic criteria 459–463
trophic degree 460
trophic features 466
trophic state (status) 312
  heterotrophic activity 356
  heterotrophic potential 354
  primary production 312
Trout cultivation 458, 474
Tubificidae, medium abundance 290
Tufted Duck 272
Turbellaria
  medium abundance 234, 235, 239, 240
turbidity 447
trophogenic processes 457
tropholytic processes 457
turnover rate, carbon 329, 334, 335
turnover time
  glucose 347, 349, 350, 352, 354
  phosphorus 307–311

Uferflucht (avoidance of the shore) 252
ultraplankton(ic) (algae, μ-algae) 77, 79, 157, 158, 160, 182, 184, 185, 186, 187, 280, 299, 301, 303, 305, 307
uncertainty, of a model 434, 442
uptake rate
  glucose of bacteria 347
  phosphorus of phytoplankton 307–311

validation, of a model 434, 438–441
vegetation
  present real 11–13
  development 14, 16
  macrophytic, changes in 140–145
villages, deserted 15, 17

Waldhumussee 401
Waren, meteorological series 42
waste water 35
water balance
  components 67–72
  Lake Stechlin area 72
water level of lakes 29, 34, 35, 37
waterfowl
  food 274–275
  migrants 272–273
  phenology 273–274
  overwintering 272–275
water management 455, 466, 468
water mites 232, 236
water-plants 13
  thermophilic 13
water resource 468
water storage 71–72
water temperature 47–56
  measurements 47–48
  surface 48–50
  vertical distribution 51–55
water use 464
water utilization, strategy 468–470
'weak points', of a model 450
Weichselian, glacial period 9, 10, 29
Whitetailed Eagle 274
whiting(s) 77, 79, 381, 413
wind

direction 31
speed 46–47
winter stagnation 56, 159

Zechlinerhütte (village) 18
Zehdenick, meteorological station 42
zooplankton
age classes 435
assimilation 328–335, 338–339
biomass 439, 440–441
biomass, annual distribution 244–248
biomass, horizontal distribution 252–255
biomass, vertical distribution 249–252
biomass, crustaceans 245, 246, 248, 249, 253, 254
biomass, rotifers 245, 247, 254
body carbon 327, 331
carnivorous 449
community 243
eating fish 445, 449
egy development 435, 440
enrichment 266
functional groups 445, 449–451
grazing 335–339
level 254
migration of biomass 249, 250
mortality 445, 449–451
nutritional situation 339–342
population 243–259, 327
state variable 435, 440–441
Zugergläser 266, 267

GPSR Compliance

The European Union's (EU) General Product Safety Regulation (GPSR) is a set of rules that requires consumer products to be safe and our obligations to ensure this.

If you have any concerns about our products, you can contact us on ProductSafety@springernature.com

In case Publisher is established outside the EU, the EU authorized representative is:

Springer Nature Customer Service Center GmbH
Europaplatz 3
69115 Heidelberg, Germany

Batch number: 10823775

Printed by Printforce, the Netherlands